Perspectives in the Structure of Hadronic Systems

NATO ASI Series

Advanced Science Institutes Series

A series presenting the results of activities sponsored by the NATO Science Committee, which aims at the dissemination of advanced scientific and technological knowledge, with a view to strengthening links between scientific communities.

The series is published by an international board of publishers in conjunction with the NATO Scientific Affairs Division

| A | Life Sciences | Plenum Publishing Corporation |
| B | Physics | New York and London |

C	Mathematical and Physical Sciences	Kluwer Academic Publishers
D	Behavioral and Social Sciences	Dordrecht, Boston, and London
E	Applied Sciences	

F	Computer and Systems Sciences	Springer-Verlag
G	Ecological Sciences	Berlin, Heidelberg, New York, London,
H	Cell Biology	Paris, Tokyo, Hong Kong, and Barcelona
I	Global Environmental Change	

Recent Volumes in this Series

Volume 326 —Linking the Gaseous and Condensed Phases of Matter:
The Behavior of Slow Electrons
edited by Loucas G. Christophorou, Eugen Illenberger,
and Werner F. Schmidt

Volume 327 —Laser Interactions with Atoms, Solids, and Plasmas
edited by Richard M. More

Volume 328 —Quantum Field Theory and String Theory
edited by Laurent Baulieu, Vladimir Dotsenko, Vladimir Kazakov,
and Paul Windey

Volume 329 —Nonlinear Coherent Structures in Physics and Biology
edited by K. H. Spatschek and F. G. Mertens

Volume 330 —Coherent Optical Interactions in Semiconductors
edited by R. T. Phillips

Volume 331 —Hamiltonian Mechanics: Integrability and Chaotic Behavior
edited by John Seimenis

Volume 332 —Deterministic Chaos in General Relativity
edited by David Hobill, Adrian Burd, and Alan Coley

Volume 333 —Perspectives in the Structure of Hadronic Systems
edited by M. N. Harakeh, J. H. Koch, and O. Scholten

Series B: Physics

Perspectives in the Structure of Hadronic Systems

Edited by

M. N. Harakeh

Kernfysisch Versneller Instituut
Groningen, The Netherlands

J. H. Koch

NIKHEF–K
Amsterdam, The Netherlands

and

O. Scholten

Kernfysisch Versneller Instituut
Groningen, The Netherlands

Plenum Press
New York and London
Published in cooperation with NATO Scientific Affairs Division

Proceedings of a NATO Advanced Study Institute on
Perspectives in the Structure of Hadronic Systems,
held August 2–15, 1993,
in Dronten, The Netherlands

NATO-PCO-DATA BASE

The electronic index to the NATO ASI Series provides full bibliographical references (with keywords and/or abstracts) to more than 30,000 contributions from international scientists published in all sections of the NATO ASI Series. Access to the NATO-PCO-DATA BASE is possible in two ways:

—via online FILE 128 (NATO-PCO-DATA BASE) hosted by ESRIN, Via Galileo Galilei, I-00044 Frascati, Italy

—via CD-ROM "NATO Science and Technology Disk" with user-friendly retrieval software in English, French, and German (©WTV GmbH and DATAWARE Technologies, Inc. 1989). The CD-ROM also contains the AGARD Aerospace Database.

The CD-ROM can be ordered through any member of the Board of Publishers or through NATO-PCO, Overijse, Belgium.


```
           Library of Congress Cataloging-in-Publication Data


Perspectives in the structure of hadronic systems / edited by M.N.
  Harakeh, J.H. Koch, and O. Scholten.
       p.   cm. -- (NATO ASI series. Series B, Physics ; v. 333)
    "Published in cooperation with NATO Scientific Affairs Division".
    "Proceedings of a NATO Advanced Study Institute on Perspectives in
  the Structure of Hadronic Systems, held August 2-15, 1993, in
  Dronten, The Netherlands"--T.p. verso.
    Includes bibliographical references and index.
    ISBN 0-306-44739-8
    1. Hadron interactions--Congresses.  2. Hadron-nuclei
  interactions--Congresses.  3. Quarks--Congresses.  4. Quark-gluon
  plasma--Congresses.   I. Harakeh, M. N.  II. Koch, J. H.
  III. Scholten, O. (Olaf)  IV. North Atlantic Treaty Organization.
  Scientific Affairs Division.  V. NATO Advanced Study Institute on
  Perspectives in the Structure of Hadronic Systems (1993 : Dronten,
  Netherlands)  VI. Series.
  QC794.8.S8P48  1994
  539.7'216--dc20                                      94-34541
                                                          CIP
```


ISBN 0-306-44739-8

PREFACE

The last decade has been witness to many exciting and rapid developments in the fields of Nuclear Physics and Intermediate Energy Physics, the interface between Nuclear and Elementary Particle Physics. These developments involved to a large extent the subnucleonic degrees of freedom in nuclei. In deep inelastic lepton scattering from nuclei, for example, it was observed that the quark structure of the nucleon is influenced by the nuclear medium. Also, the spin-dependent structure function of the nucleon was found to differ from sum rules based on $SU(3)$ symmetry, a discrepancy referred to as the "spin crisis". In pion electroproduction at threshold and in the production of pions and other mesons in heavy ion collisions at intermediate energies interesting experimental results have been obtained, which triggered lively theoretical discussions. Furthermore, the search for the quark-gluon plasma phase of hadronic matter, a phase that is supposed to have existed in the first few seconds of the Big Bang, has been intensified.

Not only were these developments accompanied by technical developments, such as the building of new experimental facilities, but also extensive theoretical efforts have been directed towards understanding these phenomena. These concerted efforts will hopefully lead to an understanding of the transition from the non-perturbative QCD regime to the perturbative one, in which the quark structure of nucleons is better understood.

All of the aforementioned developments occur at a high pace, making it difficult to incorporate them into the courses offered to advanced students. The Dronten Summer School, a NATO Advanced Study Institute, therefore provided courses for young scientists (graduate students and postdocs) by experts in these research activities. Strong emphasis was placed not only on state of the art research, but also on the necessary physics background.

This volume contains the proceedings of the lectures given during the Summer School. The lectures focussed on several topics of current interest. The hadronic (sub) structure was addressed starting at low energies with meson production and nucleon resonance excitation in lepton and hadron scattering from nucleons and nuclei. The high energy muon scattering results of the NMC/SMC collaboration were discussed in connection with the quark and gluon content of the nucleon and its spin structure. Finally, the quest for the quark-gluon plasma in ultra-relativistic heavy-ion collisions was covered.

From the reaction of the participants it was clear that the lectures were clear and well geared to the level of the school. We wish to thank all speakers for their efforts and for spending extra time on problem and discussion sessions.

Furthermore, the organisers wish to thank Marijke Oskam-Tamboezer for her help in preparing and running the summer school.

Finally, we would like to acknowledge the financial support from the Science Committee of the North Atlantic Treaty Organization which made this school possible.

M. N. Harakeh
J. H. Koch
O. Scholten

CONTENTS

ELECTROMAGNETIC MESON PRODUCTION
AT LOW ENERGIES

Berthold H. Schoch

Physikalisches Institut
Universität Bonn
Germany

INTRODUCTION

Since the first pion-photoproduction experiments 40 years ago, the results extracted out of such experiments have played a key role in our understanding of the structure of the nucleon. Especially the large multipole amplitudes, like e.g. the M_{1+} amplitude responsible for the Δ production, have been determined with a precision of a few percent. However, due to the spin and isospin structure of the photon a wealth of other small production amplitudes are accessible, carrying information of current interest.

Very close to the production threshold, the angular momentum barriers, which inhibit low-energy pion emission for $l_\pi > 0$, favors s-wave production. This results in the dominance of E1 absorption, the only possibility that can give rise to s-wave pion emission. In the absence of any resonant $(j=1/2^-)$ process, the amplitudes for E1 absorption will be proportional to the electric dipole moments in the intermediate states[1, 2]. In addition, the wavelength λ of the incident photon is close to 9 fm, large compared to the diameter of the nucleon, so that the detailed structure of the nucleon is not expected to influence the threshold photoproduction amplitude. Therefore, like calculating the Thomson cross section in low-energy photon scattering, a classical approach can be used as a first-order approximation. Starting with the electric dipole moment $\mathbf{D}$:

$$\mathbf{D} = Q_\pi \mathbf{r}_\pi + Q_N \mathbf{r}_N \, , \tag{1}$$

the classical dipole moment can be calculated to have the form:

$$\mathbf{D} = (Q_\pi - \frac{m_\pi}{M} Q_N)\mathbf{r}_\pi \tag{2}$$

with $\mathbf{r}_\pi$ and $\mathbf{r}_N$ in the c.m. system of the (N-π) system. The charges can be expressed in terms of the third components of the isospin vectors:

$$Q_\pi = \tau_{3\pi} e \tag{3}$$

Perspectives in the Structure of Hadronic Systems
Edited by M.N. Harakeh *et al.*, Plenum Press, New York, 1994

$$Q_N = (\tau_{3N} + \tfrac{1}{2})e \tag{4}$$

Then the corresponding Hamiltonian has the form:

$$H_{E1} = C\{\tau_{3\pi} - \frac{m_\pi}{M}(\tau_{3N} + \tfrac{1}{2})\} \tag{5}$$

with C as a common factor to every production channel. Using this Ansatz, results for the ratios of the different channels are given in table 1.

Table 1. Dipole moments of the π-N systems

Reaction	Dipole moment	Relative Ratio
$\gamma + p \rightarrow \pi^+ + n$	$e \cdot r_\pi$	1
$\gamma + p \rightarrow \pi^0 + p$	$-\frac{m_\pi r_\pi e}{M}$	$-\frac{m_\pi}{\sqrt{2}M} = -0.1$
$\gamma + n \rightarrow \pi^- + p$	$-(1+\frac{m_\pi}{M})e \cdot r_\pi$	$-(1+\frac{m_\pi}{M}) = -1.15$
$\gamma + n \rightarrow \pi^0 + n$	0	0

These classical results find their expression in so-called low-energy theorems which provide powerful constraints for the photoproduction amplitudes. Starting from a kinematics – the production threshold – where these constraints should hold, the continuation of these amplitudes into other kinematic regimes is of great importance for testing various models of the nucleon.

The cross sections to be measured are small compared to other more prominent channels but well separated in energy. Fig. 1 shows the total photoabsorption cross section on the proton in the energy range from below pion threshold up to 1.5 GeV. Large resonance contributions determine the shape of the cross section. Fig. 2 shows

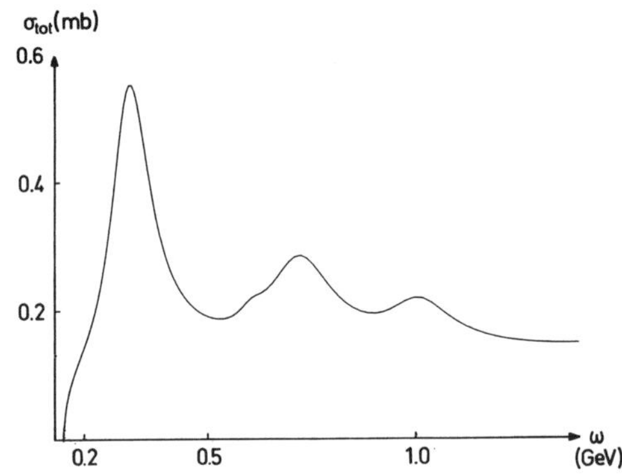

Figure 1. The total photoabsorption cross section on the proton

the cross section magnified close to the pion production threshold. This represents the region where the data described here have been measured and where by extrapolation to the threshold value a comparison with the low-energy-theorem (LET) predictions becomes possible.

The experimental investigations of charged pion production were performed, so far, by using untagged bremsstrahlung beams. In section 2 the data for positive pions and in section 3 the data and the method – as an example for the difficulties to acquire data for the neutron – for negative pions will be reported. Section 4 is devoted to the investigation of the production of neutral pions.

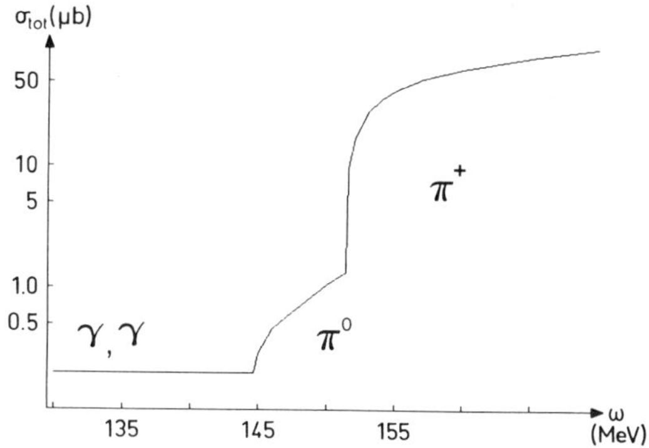

Figure 2. The total photoabsorption cross section on the proton in the threshold region.

THE PRODUCTION OF POSITIVELY CHARGED PIONS

The data available start with a photon energy of $E_\gamma = 153.4$ MeV. The threshold for this process is 151.3 MeV. The data[3] up to $E_\gamma = 160$ MeV are shown in fig. 3 for two angles in the laboratory. An emulsion technique has been applied to detect the

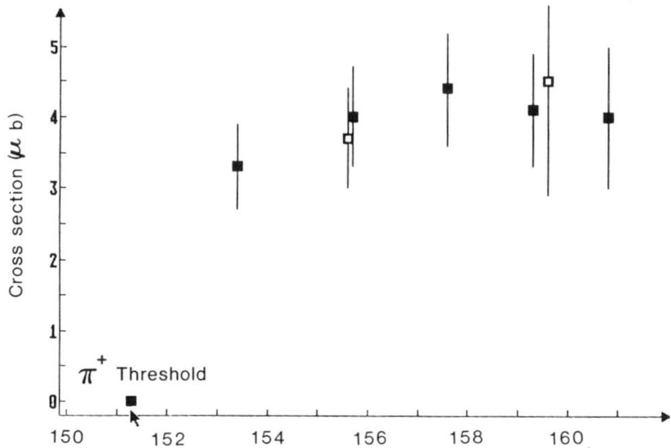

Figure 3. Cross sections for π^+ production at angles of 30^0 and 90^0(open squares) in the Lab. The x-axis gives the photon energy in MeV.

pions. Within the rather large error bars the data agree with an isotropic production. In order to extract with a good precision the electric dipole amplitude at threshold a fit including data up to $E_\gamma = 200$ MeV has been made. By extrapolating down to threshold the electric dipole amplitude $E_{0+} = 28.6 \pm 0.3 \cdot 10^{-3}(1/m_\pi)$ has been extracted.

THE PRODUCTION OF NEGATIVELY CHARGED PIONS

The experimental investigation of negatively charged pions needs special methods to carry out. In the absence of a target of free neutrons the bound neutrons in a nucleus can serve as a target. To keep the corrections small and – hopefully – managable the deuteron seems to be the best candidate to serve as a neutron target. Again the intention is to measure the cross section as close as possible to the production threshold.

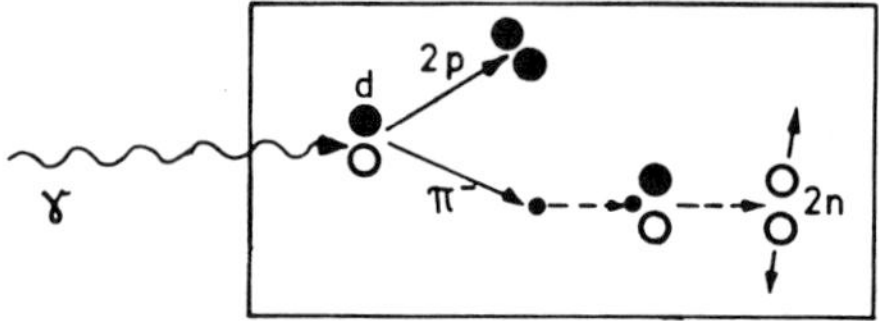

Figure 4. Schematic representation of the method of detection of negatively charged pions.

Fig. 4 shows schematically the method used in an experiment with the Mainz linear accelerator. A photon hits the neutron in the deuteron and produces a negatively charged pion. Because of the large extension of the target compared to the photon beam the low-energy pion gets stopped within the target. It then gets captured into a Bohr orbit by one of the deuterons in the target and, finally, gets absorbed by the deuteron nucleus. The absorption results in the emission of two monochromatic neutrons. In order to measure the total π^- cross section it is sufficient to determine the number of neutrons created by this process. However, there is a competing process, i.e. the photodisintegration of the deuteron. By selecting the "right" method this background contribution can be avoided. Conversely the photodisintegration of the deuteron can be used as a calibration reaction for the neutron detection.

Fig. 5 shows – schematically – the set-up used. An electron beam consisting of

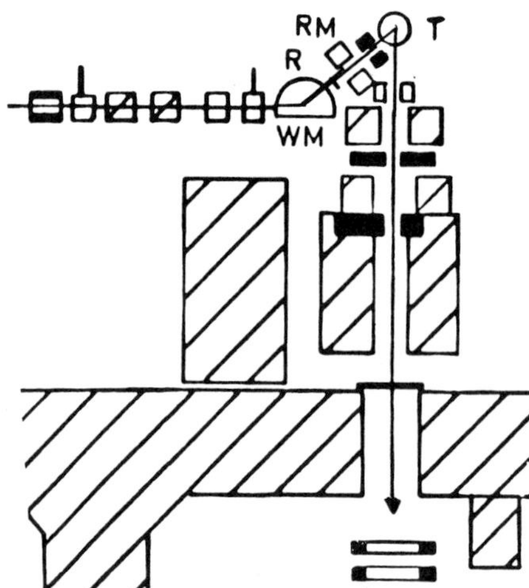

Figure 5. The overall set-up of the experiment for π^- photoproduction.

"short" (1ns) pulses prepared with the Mainz Linac gets deflected in a magnet (WM) and hits a radiator. The bremsstrahlung is collimated onto a deuteron target. The magnet RM sweeps away the electrons. The neutrons from the target are collimated and detected in a scintillator. The time of flight of the neutrons can be measured relative to the arrival times of the pulses of the electrons. The expected time of flight spectrum for infinitesimaly short pulses and perfect time resolution in the neutron detector would look like the spectrum shown in fig. 6. There is a line of monochromatic neutrons from the pion production and a neutron spectrum due to the photodisintegration of the deuteron which gets its shape mainly by the energy distributions of the photons within

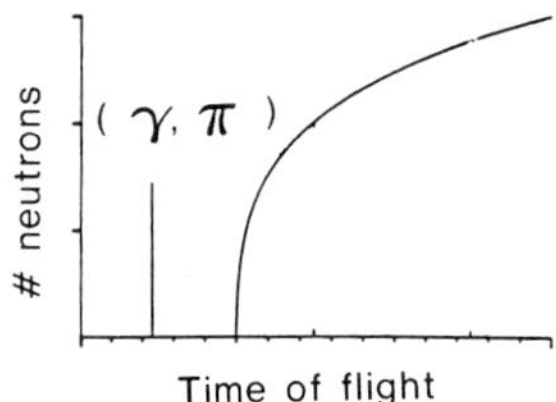

Figure 6. "Ideal" TOF spectrum: the clear separation of neutrons originating from the photodisintegration and pion production is demonstrated.

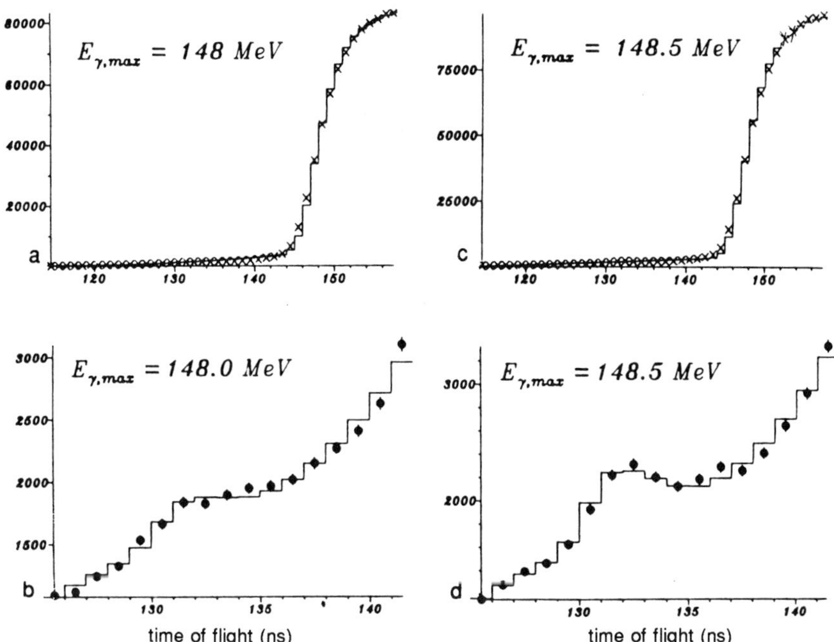

Figure 7. Measured TOF spectra close to threshold: b) and d) magnified.

the bremsstrahlung spectrum. Fig. 7 and 8 show neutron time-of-flight (TOF) spectra at different end-point energies as measured in the experiment[4]. A common fit connects the known cross section of the photodisintegration with the cross section of the pion production. Fig. 9 shows the yield of neutrons due to pion production as a function of the bremsstrahlung end-point energies. From this yield curve the total cross section for negative-pion production on the deuteron can be extracted.

Fig. 10 shows the total cross section as a function of the photon energy above threshold together with two points measured by[3] and the results obtained by the production of positive pions[5]. Qualitatively, the measurements support the results shown in table 1 based on classical arguments. In order to extract an accurate value of the E_{0+} amplitude for negatively charged pions a calculation based on the impulse approximation with corrections due to various effects e.g. final-state interactions must be performed. Such calculations do not exist up until now.

However, there exists another way to determine the E_{0+} amplitude. By measuring

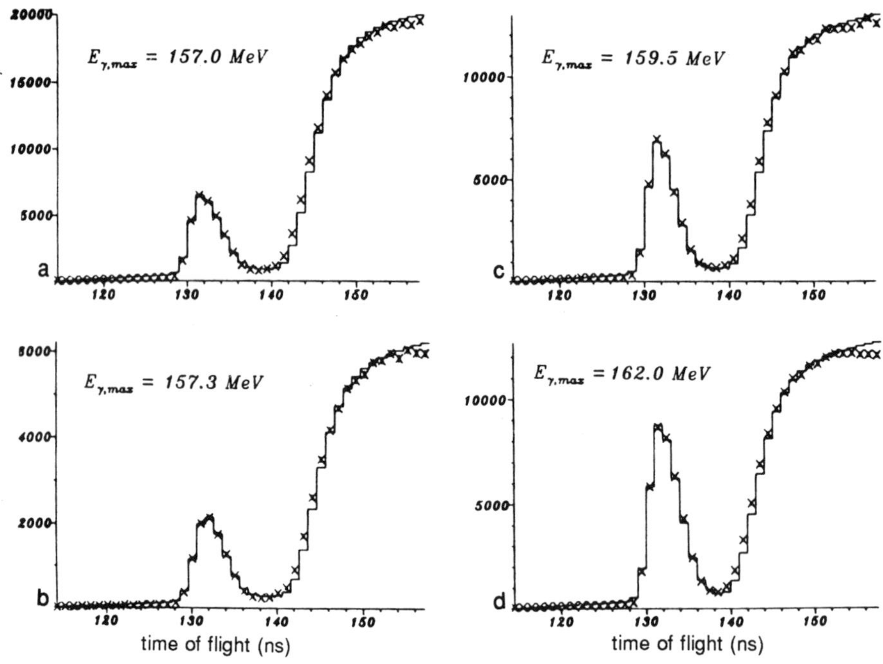

Figure 8. TOF spectra for higher photon end-point energies.

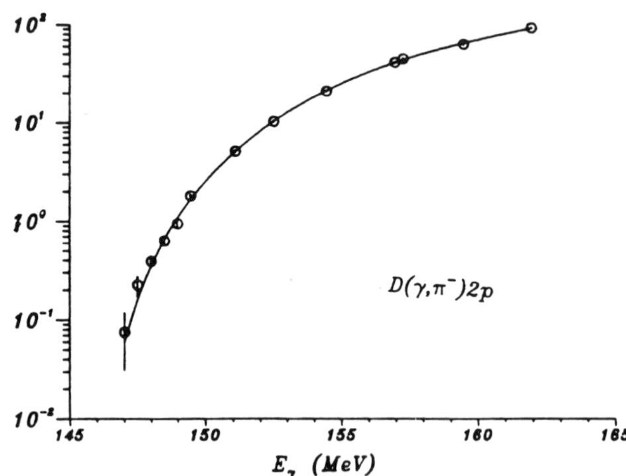

Figure 9. π^- Yields extracted from the peaks in the TOF spectra.

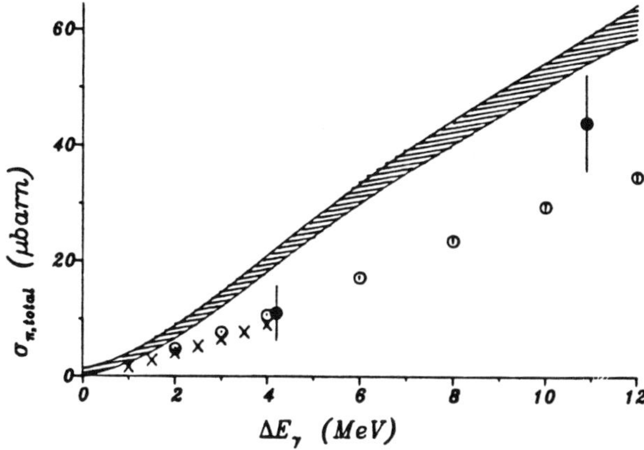

Figure 10. The cross sections for charged-pion photoproduction on the deuteron; negatively charged pions: our measurement (band), full points[3]; positively charged pions: crosses[6] and empty circles[5].

the Panofsky ratio

$$P = \frac{\pi^- + p \Rightarrow \pi^0 + n}{\pi^- + p \Rightarrow \gamma + n} \tag{6}$$

and the extraction of the s-wave part of the π-N exchange scattering cross section from pion-scattering experiments the amplitude can be determined. The result $E_{0+} = -31.5 \pm 1.1 \cdot 10^{-3}(1/m_\pi)$ compares quite well with the value for the positively charged pions when the accuracy is taken into account. Within the error bars the ratio agrees with the value given in table 1.

THE PRODUCTION OF NEUTRAL PIONS

Data of the differential and total cross sections continuously covering the first 10 MeV starting at threshold will be discussed in this section. The experiment has been performed at the Mainz Microtron MAMI-A[7], using a tagging facility[8] in combination with a π^0 spectrometer[9], see fig. 11. A c.w. electron beam of 183.5 MeV, provided by MAMI-A, was used to produce photons with well defined energy (FWHM=0.28 MeV) by means of bremsstrahlung tagging. In the energy range 131.4 – 157.2 MeV the tagged photon flux was 1×10^7/s.

Two blocks of 88 lead-glass detectors making up the π^0 spectrometer, have been placed, horizontally at both sides of the beam at a distance of 50 cm apart from a liquid hydrogen target. The angle between the blocks and the γ beam was chosen to be 80^0. In this configuration the complete π^0 angular distribution $0 \le \theta \le 180^0$ was covered for the given tagging range. The absolute detection efficiency as well as the angular acceptances have been determined using Monte-Carlo simulations. These have been checked by measurements of the ^{12}C(γ, π^0) reaction. The p-wave nature of this coherent process, within 4.4 MeV above the production threshold, provided for this purpose a characteristic angular distribution. The absolute energy calibration of the tagged photons has been checked by feeding the MAMI beam directly into the tagging spectrometer as well as by the determination of the threshold of the ^{12}C(γ,

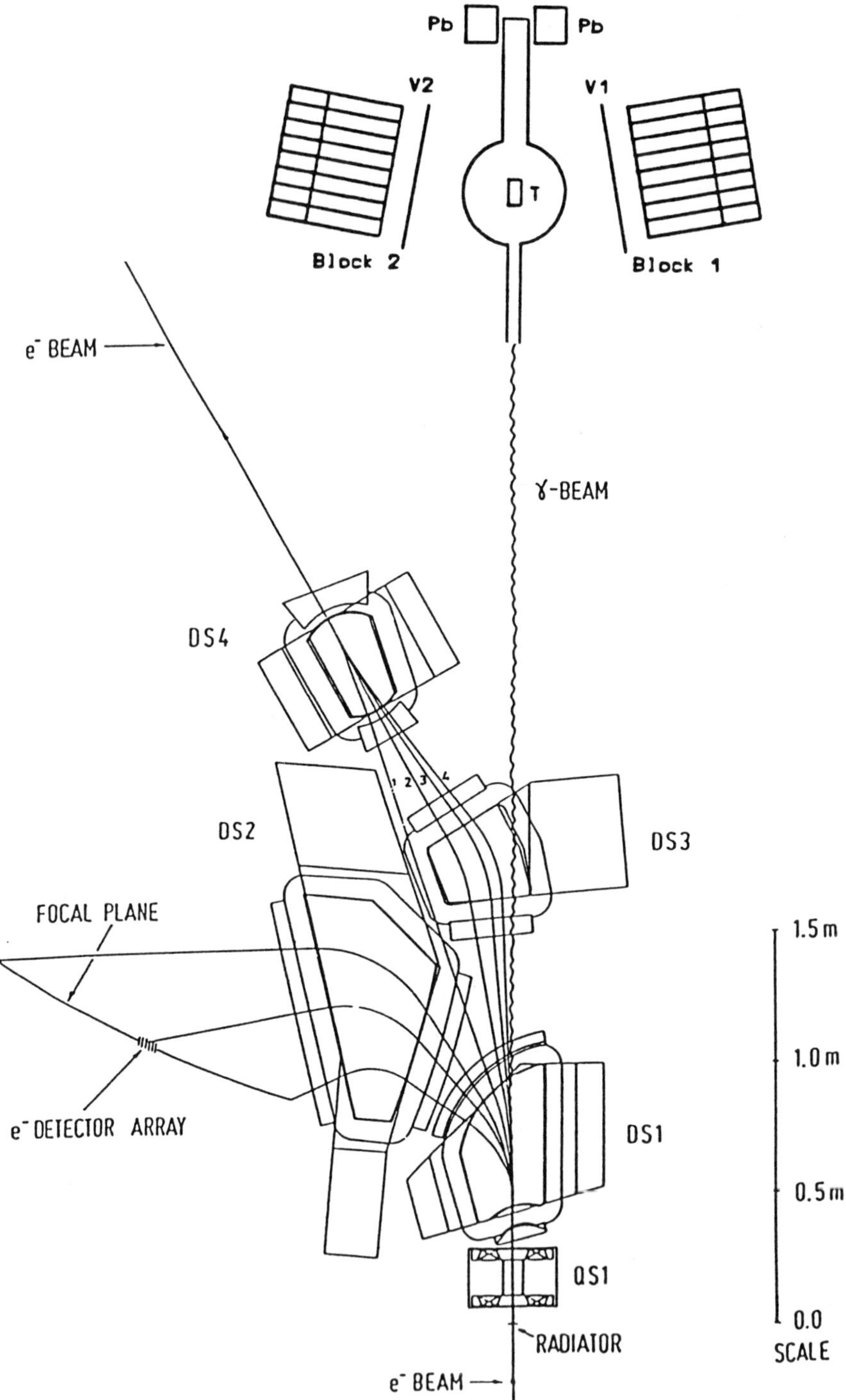

Figure 11. The experimental set-up: Tagging spectrometer and pion spectrometer are shown schematically. DS1-4: Dipole magnets.

$\pi^0)^{12}$C reaction. An agreement was found within 220 keV. The target thickness of 11 cm together with the tagged photon flux of 1×10^7/s provided a luminosity of 4.6 $\times10^{30}$ $s^{-1} \times cm^{-2}$.

Neutral pions were identified event by event by a cut in a two-dimensional plot of the tagged photon energy versus opening angle of the decay photons (see fig. 12). Random events were subtracted using appropriate cuts in the time spectra. Fig. 13

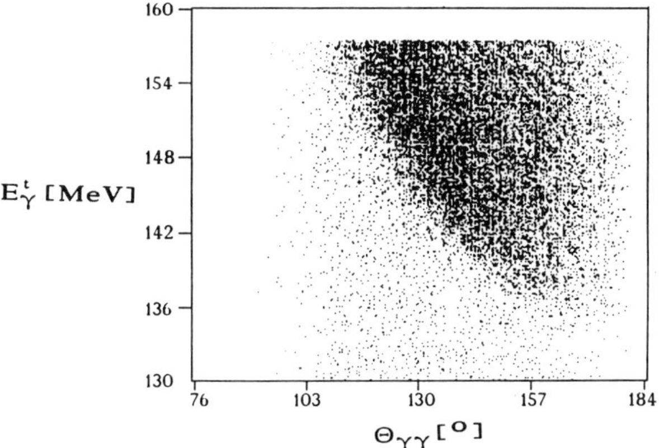

Figure 12. Scatter plot of the tagged photon-energy versus opening angle of the decay photons for ^{12}C$(\gamma, \pi^0)^{12}$C reaction.

shows the total cross section as a function of the photon energy together with the data obtained by Mazzucato et al.[10]. There is agreement within the experimental error bars.

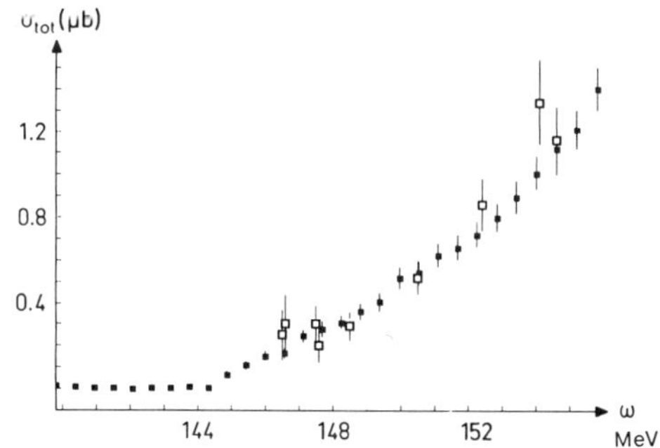

Figure 13. The total cross section as measured in Saclay (open squares) and Mainz.

The differential cross sections for five energy intervals around E_γ = 146.8, 149.1, 151.4, 153.7 and 156.1 MeV are shown in fig. 14. The absolute overall systematic error amounts to 7.5 percent, due to the uncertainty in the luminosity, the π^0-detection efficiency and dead-time corrections. The experiment has been performed in one setting of the spectrometer and incoming electron energy. Consequently, relative uncertainties in the energy dependence as well as the angular distribution were minimised. Based on this data base, a nearly model independent extraction of the E_{0+} amplitude as a

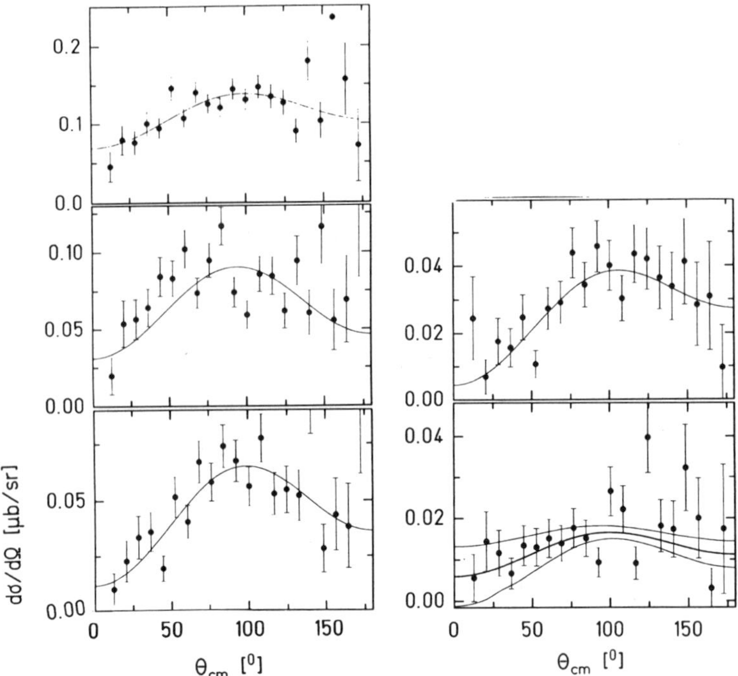

Figure 14. The differential cross sections for 5 energy intervals (see text).

function of energy has been accomplished. Since close to threshold only l=0,1 partial waves are important the differential cross section can be parametrised according to

$$\frac{d\sigma}{d\Omega} = \frac{p_\pi}{k_\gamma}(A + B \cdot cos\theta + C \cdot cos^2\theta) . \qquad (7)$$

Here p_π and k_γ are the momentum of the pion and the photon, respectively, and θ is the emission angle of the π^0 in the c.m. system relative to the photon momentum. The angular distributions, however, reveal in a much more transparent way the dominant feature of the cross section.

Figs. 15-17 show the results of a fit to the angular distributions. By choosing the scale of the x-axis $\sim (p_\pi \cdot k_\gamma)^2$ and $p_\pi \cdot k_\gamma$ for A as well as C, and B, respectively, the dependencies on p- and s-waves become more transparent:

$$A = |E_{0+}|^2 + |a \cdot p_\pi \cdot k_\gamma|^2 , \qquad (8)$$
$$B = 2ReE_{0+}(b \cdot p_\pi \cdot k_\gamma) , \qquad (9)$$
$$C = |c \cdot p_\pi \cdot k_\gamma|^2 . \qquad (10)$$

In a linear extrapolation to $p_\pi = 0$, $|E_{0+}|$ can be obtained out of A, provided E_{0+} does not depend on p_π.

Rescattering effects create p_π dependencies. In a linear extrapolation to p_π=0, C should be 0. B should yield a straight line too, provided there is no p_π dependence of E_{0+}. Three lines are shown together with the data in fig. 16:

1. $E_{0+} = 2.97 \cdot 10^{-3}$, the value obtained by the classical calculation.
2. $E_{0+} = 1.53 \cdot 10^{-3}$, extracted by a linear extrapolation out of A.
3. $E_{0+} = 0.5 \cdot 10^{-3}$, taken from[10].

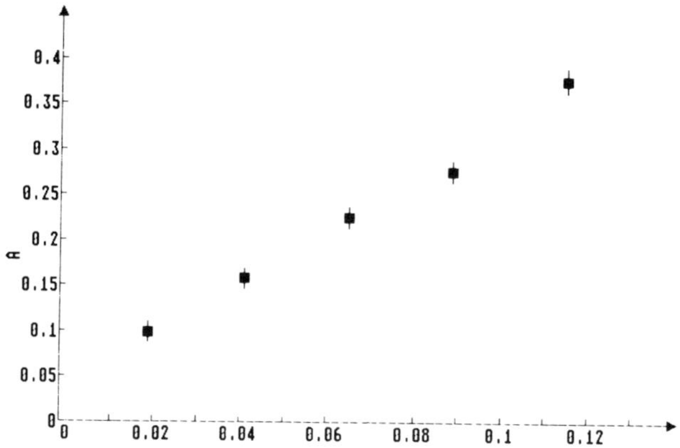

Figure 15. Coefficient A versus $(p_\pi \cdot k_\gamma)^2$.

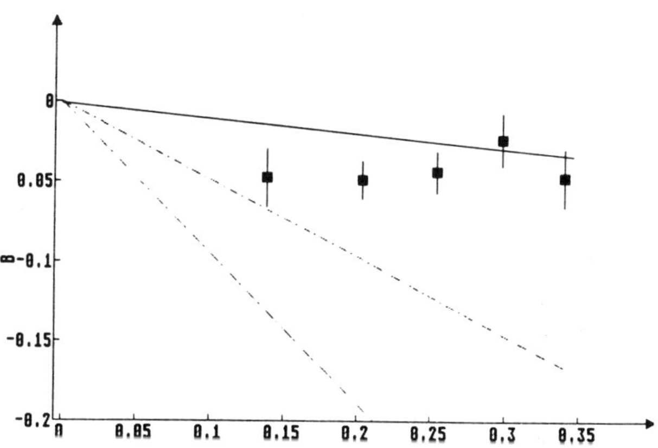

Figure 16. Coefficient B versus $p_\pi \cdot k_\gamma$.

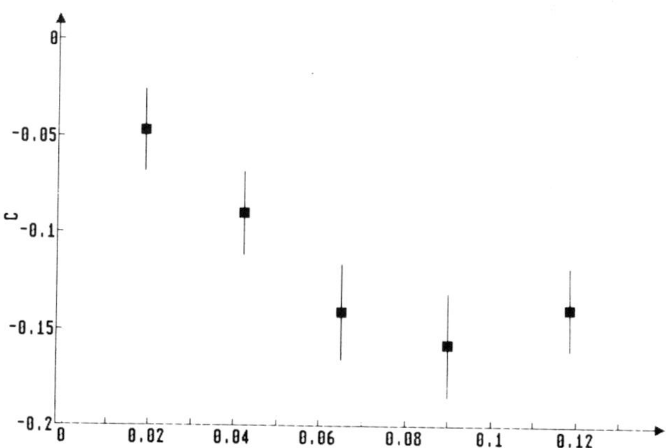

Figure 17. Coefficient C versus $(p_\pi \cdot k_\gamma)^2$.

For all three cases a p-wave amplitude of $8.0 p_\pi \cdot k_\gamma / m_\pi^2 \cdot 10^{-3} [1/m_\pi]$ has been used, assuming M_{1+} dominance for the p-wave. This dominance finds its expression in the ratio $\frac{C}{A}$, which would be -0.6 for for a M_{1+} contribution only. Fig. 18 shows the measured ratio.

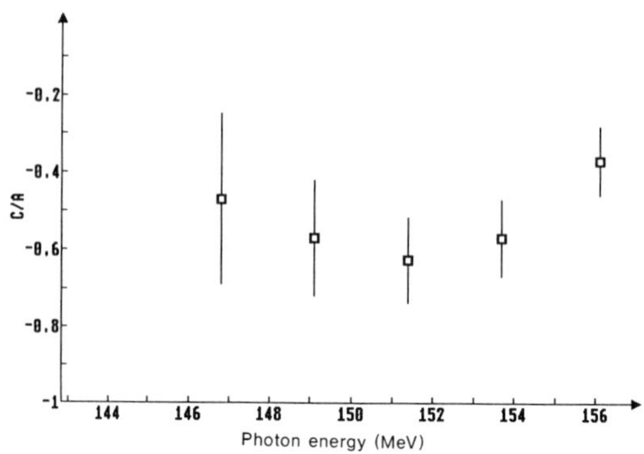

Figure 18. The C/A ratio as a measure for the M_{1+} contribution.

An important ratio is given by

$$\frac{2 \cdot B}{A + C} = \frac{2 \cdot E_{0+} \cdot M_1}{|E_{0+}|^2 + |M_1|^2} \tag{11}$$

where M_1 stands for the sum of the p-waves. Depending on the sign of the amplitudes a value of ± 1 is reached for $|E_{0+}| = |M_1|$ which becomes important for the solutions of equations (12) and (13). Fig. 19 shows this ratio which indicates that the values of these amplitudes become equal for photon energies close to $E_\gamma = 149.1$ MeV. A, B, and

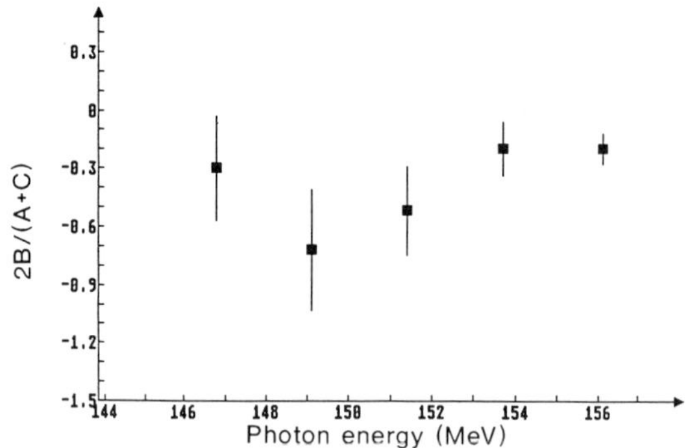

Figure 19. The ratio eq.(11) which shows at what energy s- and p-waves are equal.

C are functions of the complex multipoles E_{0+}, M_{1+}, M_{1-} and E_{1+}. The relations

$$A + C = |3E_{1+} + M_{1+} - M_{1-}|^2 + |E_{0+}|^2 \tag{12}$$

and

$$B = 2Re(E_{0+}(3E_{1+} + M_{1+} - M_{1-})^*) \tag{13}$$

show that E_{0+} can be determined as one solution of a quadratic equation assuming the imaginary part of the amplitudes can be neglected compared to the real one. However, it is not possible to extract all p-wave multipoles which are shown in the equation. Instead the sum (M_1) will be extracted without any further assumptions like e.g. the above mentioned momentum dependence of the p-waves. The numerical results of A, B and C are given in table 2.

Table 2. Numerical results for the coefficients A, B, C.

$E\gamma$ [MeV]	A [μb/sr]	B [μb/sr]	C [μb/sr]
146.8	.099±0.011	-0.0158±0.012	-0.047±0.021
149.1	0.158±0.011	-0.049±0.012	-0.09±0.022
151.4	0.225±0.012	-0.044±0.013	-0.141±0.025
153.7	0.277±0.012	-0.024±0.016	-0.158±0.027
156.1	0.376±0.014	-0.048±0.018	-0.139±0.033

After a discussion of the signs[11] and (11) the solution of equations (12) and (13) yields

$$E_{0+} = .5(\sqrt{A + B + C} - \sqrt{A - B + C}) \text{ for } E_\gamma \geq 151.4 \ MeV \qquad (14)$$

and

$$E_{0+} = .5(-\sqrt{A + B + C} - \sqrt{A - B + C}) \text{ for } E_\gamma < 151.4 \ MeV . \qquad (15)$$

In addition, from the total cross section only (see Table 3), two values very close to the threshold were extracted.

Fig. 20 shows the amplitudes E_{0+} and the total p-wave amplitude M_1. The E_{0+}

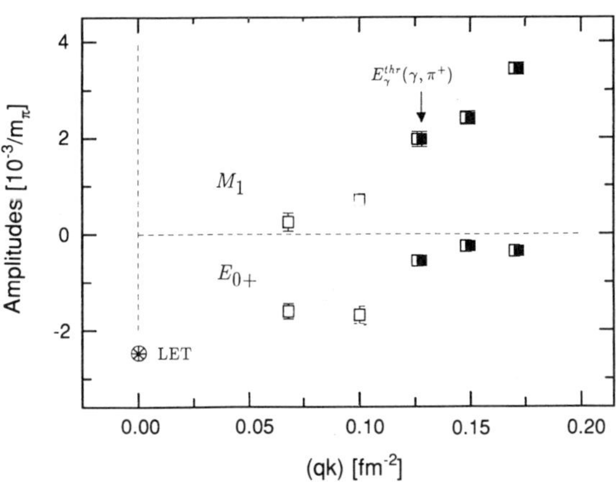

Figure 20. E_{0+} and M_1 amplitudes.

amplitude changes rapidly by crossing the π^+ threshold. There seems also to be an anomaly in the p-wave because the sum of the p-wave multipoles M_1 approaches the threshold not in a straight line.

CONCLUSIONS

The pion-photoproduction amplitudes at threshold are fixed points. All model calculations should produce results in accordance with these fixed amplitudes. Away from

Table 3. Measured E_{0+} amplitudes.

$E\gamma$ [MeV]	E_0^+ $[10^{-3}/m_\pi]$
144.87	-2.21±.5
145.43	-1.88±.3
146.8	-1.61±0.16
149.1	-1.69±0.18
151.4	-0.56±0.13
153.7	-0.26±0.17
156.1	-0.35±0.13

these points the sensitivity to the different models sets in. Therefore, high precision experiments at and close to threshold are very well suited to test models. With the new generation of accelerators more complete and more accurate experiments become possible. For the first time neutral pion production on the neutron should be feasible and, therefore, a test of the isospin decomposition of the multipoles becomes possible. Further theoretical discussions can be found in refs.[12, 13, 17, 14, 15, 16, 18] and the references therein.

References

[1] T. Ericson and W. Weise, Pions and Nuclei, Clarendon Press 1988, p. 276

[2] B.T.Feld, Models of Elementary Particles, Blaisdell Publishing Company 1969, p. 143

[3] M.I. Adamovich, Proceedings (Trudy) of the P.N. Lebedev, Physics Institute 54 (1976) 1

[4] F. Klein, PhD-Thesis, Mainz 1985, unpublished

[5] E.C. Booth et al., Phys. Rev. C20 (1979) 1217

[6] G. Audit et al., Phys. Rev. C16 (1977) 1517

[7] H. Herminghaus et al., Nucl. Instr. Meth. 138 (1976) 1

[8] J.D. Kellie et al., Nucl. Instr. Meth. Phys. Res. A241 (1985) 153

[9] H. Stroeher et al., Nucl. Instr. Meth. Phys. Res. A269 (1988) 568

[10] E. Mazzucato et al., Phys. Rev. Lett. 57 (1986) 3144

[11] R. Beck et al., Phys. Rev. Lett. 65 (1990) 1841

[12] D. Drechsel, M.M. Giannini, Rep. Prog. Phys. 52 (1989) 1083

[13] D. Drechsel, in "Structure of Hadrons and Hadronic Matter", World Scientific, Singapore, Editors: O. Scholten, J.H. Koch, 1990

[14] J.C. Bergstroem, Phys. Rev. C44 (1991) 1768

[15] A.M. Bernstein and B. Holstein, Comm. Nucl. Part. Phys. 20 (1991) 197

[16] H.W. Naus et al., Phys. Rev. C41 (1990) 2852

[17] T. Schaefer and W. Weise, Phys. Lett. B250 (1990) 6

[18] U.G. Meissner, Rep. Prog. Phys. 56 (1993) 903

PARITY VIOLATION IN ELECTRON SCATTERING

Robert D. McKeown

Kellogg Radiation Laboratory
California Institute of Technology
Pasadena, CA 91125, USA

INTRODUCTION

Recently there has been considerable speculation concerning the possibility of sizeable strange quark matrix elements in the nucleon. Although a constituent quark model of the nucleon contains only three constituent quarks and only up and down flavors, these constituent quarks have considerable sub-structure including gluons and quark-antiquark pairs. In this way, even the constituent quark model has the capability to incorporate the presence of strange quark-antiquark pairs in the quark "sea".[1] Much of this discussion is motivated by the existence of some experimental information indicating sizeable strange quark effects in various observable properties of the nucleon: e.g., mass and spin.

The strange quark contribution to the nucleon mass can be addressed by studying the "sigma term" in π-nucleon scattering.[2] One first obtains the value of the isospin even πN scattering amplitude (from experiment) extrapolated to the pole at $2m_\pi^2$ (the "Cheng-Dashen" point). Then one may combine this information with hyperon mass relations to arrive at a value of the contribution of $\bar{s}s$ pairs to the nucleon mass. Until recently, the result of such an analysis was that the $\bar{s}s$ contribution was of order 30%.[2] However, a more detailed analysis including higher order corrections now indicates that the correct value is more like 15%[3] with a rather large uncertainty.

The flavor structure of the nucleon spin can be addressed by studying spin-dependent deep-inelastic lepton scattering.[4] Much of this discussion has focussed on the recent measurement of the spin structure function of the proton via deep-inelastic muon scattering.[5] This result has been interpreted as an indication of a strange quark sea ($s\bar{s}$) strongly polarized opposite to the nucleon spin, leading to the conclusion that the total quark spin contributes very little to the total spin of the proton. A similar conclusion has been drawn from the results of low energy elastic neutrino-proton scattering.[1,6]

These considerations have led to a set of proposals for new experiments to measure the neutral weak form factors of the nucleon. In these lectures, I show

Perspectives in the Structure of Hadronic Systems
Edited by M.N. Harakeh *et al.*, Plenum Press, New York, 1994

how measurements of the neutral weak form factors are sensitive to strange quarks and how these form factors can be studied in parity violating electron scattering. Finally, I will discuss the experimental techniques developed for these measurements and the various proposals for future experiments in this subject.

ELECTROWEAK COUPLINGS AND NUCLEON FORM FACTORS

Throughout this discussion we will use electroweak couplings as prescribed within the Standard Model. The Standard Model accurately reproduces all known electroweak phenomena with impressive precision. Therefore the strategy is to assume the Standard Model to be sufficiently precise to extract the desired hadronic structure information. The precision of our knowledge of the electroweak couplings has recently been dramatically improved through the very detailed studies of the properties of the Z boson at CERN. For our purposes here, we use $\sin^2 \theta_W = 0.233$, as determined[7] from the recent measurements of the mass of the Z boson.[8] Nevertheless, one should keep in mind that it may still be possible to construct variations of the Standard Model with anomalous neutral weak quark couplings.

Table 1. Electroweak quark couplings.

Flavor	γ	Z vector	Z axial
u	$\frac{2}{3}$	$\frac{1}{4} - \frac{2}{3}\sin^2\theta_W$	$-\frac{1}{4}$
d	$-\frac{1}{3}$	$-\frac{1}{4} + \frac{1}{3}\sin^2\theta_W$	$\frac{1}{4}$
s	$-\frac{1}{3}$	$-\frac{1}{4} + \frac{1}{3}\sin^2\theta_W$	$\frac{1}{4}$

We begin by writing down the hadronic electromagnetic and neutral weak currents which couple to the photon and the Z boson, $\hat{V}^\mu_\gamma$, $\hat{V}^\mu_Z$ and $\hat{A}^\mu_Z$. Each term is constructed as a sum of the individual quark couplings. Contributions from quarks heavier than the strange quark are believed to be small and are ignored. The coefficients for each of the individual quark couplings are listed in Table 1. For example, the electromagnetic vector coupling is

$$\hat{V}^\mu_\gamma = \frac{2}{3}\overline{u}\gamma^\mu u - \frac{1}{3}\overline{d}\gamma^\mu d - \frac{1}{3}\overline{s}\gamma^\mu s \ . \tag{1}$$

It is convenient to express $\hat{V}^\mu_\gamma$ in terms of isoscalar and isovector pieces:

$$(\hat{V}^\mu_\gamma)_{T=1} = \frac{1}{2}\left(\overline{u}\gamma^\mu u - \overline{d}\gamma^\mu d\right) \tag{2a}$$

$$(\hat{V}^\mu_\gamma)_{T=0} = \frac{1}{6}\left(\overline{u}\gamma^\mu u + \overline{d}\gamma^\mu d\right) - \frac{1}{3}\overline{s}\gamma^\mu s \ . \tag{2b}$$

These are the only two linear combinations accessible in electromagnetic processes involving the neutron and proton. The elastic form factors are obtained by taking matrix elements of these current operators:

$$\langle N|\hat{V}^\mu_\gamma|N\rangle = \overline{u}_N\left(F_1\gamma^\mu + \frac{i}{2M_N}F_2\sigma^{\mu\nu}q_\nu\right)u_N \ . \tag{3}$$

Each of the form factors F_1 and F_2 contains the isospin structure indicated in Eq. (2).

Using the information in Table 1 we can then write the neutral weak vector current operator as follows:

$$\hat{V}_Z^\mu = (\frac{1}{4} - \frac{2}{3}\sin^2\theta_W)\bar{u}\gamma^\mu u + (-\frac{1}{4} + \frac{1}{3}\sin^2\theta_W)\cdot(\bar{d}\gamma^\mu d + \bar{s}\gamma^\mu s) . \qquad (4)$$

This can be re-expressed as a combination of the isoscalar and isovector electromagnetic currents in Eq. (2) plus an explicit contribution from strange quarks:

$$\hat{V}_Z^\mu = \left(\frac{1}{2} - \sin^2\theta_W\right)(\hat{V}_\gamma^\mu)_{T=1} - \sin^2\theta_W(\hat{V}_\gamma^\mu)_{T=0} - \frac{1}{4}\bar{s}\gamma^\mu s . \qquad (5)$$

We can similarly relate the neutral weak axial coupling to the charged weak axial current (relevant to neutron β decay) and a piece due to strange quarks:

$$\hat{A}_Z^\mu = -\frac{1}{2}(\hat{A}^\mu)_{T=1} + \frac{1}{4}\bar{s}\gamma^\mu\gamma^5 s . \qquad (6)$$

We can now take matrix elements of $\hat{V}_Z^\mu$ and obtain expressions for the neutral weak form factors analogous to the electromagnetic case in Eq. (3). These neutral weak form factors for the nucleon can then be expressed in terms of electromagnetic form factors and a strange contribution by using the flavor structure of Eqs. (2), (3), and (5).

$$F_{1,2}^Z = \left(\frac{1}{2} - \sin^2\theta_W\right)\left[\frac{F_{1,2}^p - F_{1,2}^n}{2}\right]\tau_3 - \sin^2\theta_W\left[\frac{F_{1,2}^p + F_{1,2}^n}{2}\right] - \frac{1}{4}F_{1,2}^s \qquad (7a)$$

$$G_1 = -\frac{1}{2}g_A\tau_3 + \frac{1}{4}G_1^s , \qquad (7b)$$

where $\tau_3 = +(-)1$ for protons (neutrons). The neutral pseudoscalar form factor G_2 also exists, but it contributes only to parity conserving processes with massive leptons. Thus it is inaccessible to experiment and for the purposes of this discussion we neglect it.

Static Form Factors

The electromagnetic form factors, $F_{1,2}^{p;n}$ are known at $Q^2 = 0$. They are related to the "Sachs" form factors $G_E = F_1 - \tau F_2$ and $G_M = F_1 + F_2$ where $\tau = Q^2/(2M_N)^2$. The value of the axial form factor at $Q^2 = 0$ is determined from neutron beta decay experiments,[9] $g_A(0) = 1.262$. Table 2 shows the $Q^2 = 0$ values of the neutral weak form factors for the proton and the neutron.

Table 2. Form factor values @ $Q^2 = 0$ ($\sin^2\theta_W = 0.233$)

	Proton	Neutron
F_1^Z	$0.017 - \frac{1}{4}F_1^s$	$-0.251 - \frac{1}{4}F_1^s$
F_2^Z	$0.508 - \frac{1}{4}F_2^s$	$-0.480 - \frac{1}{4}F_2^s$
G_1	$-0.631 + \frac{1}{4}G_1^s$	$+0.631 + \frac{1}{4}G_1^s$

Our interest here is in the strange form factors. $F_1^s(0)$ is fixed at 0 because $\bar{s}s$ pairs have no net charge. The vector form factor $F_2^s(0)$ has never been measured. The axial form factor $G_1^s(0)$ has been extracted from the data of Ref. 5 combined with the F/D ratio in Ref. 10, resulting in $G_1^s(0) = -0.38 \pm 0.11$. From the neutrino scattering data of Ref. 6 a value of $G_1^s(0) = -0.30 \pm 0.16$ is extracted which is consistent with the EMC result. However, in both of these determinations, several assumptions have been made which have been criticized in the literature. Hence, one should not conclude at this point that $G_1^s(0)$ is known with any precision.

Q^2 Dependence of Form Factors

For all but G_E^n, the Q^2 dependence of the electromagnetic form factors is well described by a dipole form,

$$G(Q^2) = \frac{G(0)}{\left(1 + \frac{Q^2}{M_V^2}\right)^2} \tag{8}$$

with $M_V = 0.843$ GeV/c. A more detailed and more theoretically based description of the form factors is given by Höhler et $al.$[11] In the range of Q^2 of interest to us $(0 < Q^2 < 1$ GeV$^2)$, the dipole description agrees with the formulation of Ref. 11 to within a few percent. The neutron electric form factor is only been precisely known at $Q^2 = 0$. We use the form of Galster et $al.$,[12]

$$G_E^n(Q^2) = -\frac{\mu_n \tau}{1 + 5.6\tau} G_E^p(Q^2) \, , \tag{9}$$

in our calculations. There are several experiments planned for the near future which should provide more precise measurements of $G_E^n(Q^2)$.

The Q^2 dependence of g_A is not as well known; it is also assumed to follow a dipole form, and the dipole mass parameter M_A has been measured in charge changing neutrino scattering.[13-15] Using all of the available neutrino data gives an average value of $M_A = 1.032 \pm 0.04$ GeV.

Before evaluating the sensitivity of various experimental approaches it is necessary to address the expected magnitudes of the strange form factors in more detail. For the strange vector form factors, we take the recent work of Jaffe[16] as a starting point. In his paper, he formulates the hypothesis that the Q^2 dependences of F_1^s and F_2^s are described by dispersion relations with poles corresponding to the ω meson, the ϕ meson and some third higher mass meson. His analysis uses the results of the Höhler et $al.$ (Ref. 11) description of the electromagnetic form factors. For the sake of discussion below, I will use fit "7.1" from Ref. 16, which results in specific values for the "strangeness radius", $r_s^2 \equiv -6 \left[dF_1^s/dQ^2\right]_{Q^2=0} = 0.16$ fm^2, and for the "strange anomalous magnetic moment", $\mu_s \equiv F_2^s(0) = -0.43$. It is important to point out, however, that predictions for $F_2^s(0)$ range from 0 to -0.88.[17] In Fig. 1 these form factors are plotted as a function of Q^2 to illustrate the general behavior. Note especially that $|F_1^s|$ remains small ($\lesssim 0.2$) throughout the Q^2 range. The Q^2 dependence of G_1^s is even more uncertain in that the neutral axial form factor is not well known. We use the EMC value of $G_1^s(0) = -0.38$ and make the assumption that $G_1^s(Q^2)$ has the same dipole form as $g_A(Q^2)$ with $M_A^s = 1.032$ GeV.

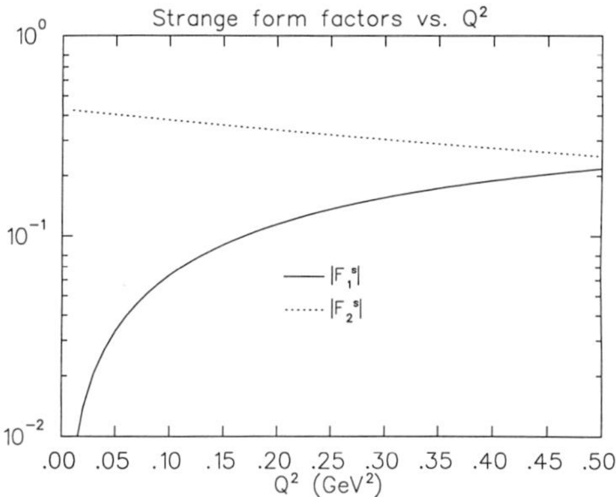

Figure 1. The Q^2 dependence of the vector strange form factors, calculated using fit "7.1" of Ref. 16.

ELASTIC NEUTRINO SCATTERING

Before discussing the role of parity violating electron scattering experiments, we will examine what has been and will be learned from elastic neutrino scattering. The elastic neutrino scattering cross section is given by the following expression:

$$\frac{d\sigma}{dQ^2} = \frac{G_F^2 M_p^2}{8\pi E_\nu^2} \left[A \pm BW + CW^2 \right] ,$$ (10)

where $W = (4E_\nu M_p - Q^2)/M_p^2$, the $+/-$ sign is for $\nu/\overline{\nu}$, and

$$A = 4\tau \left[G_1^2 (1 + \tau) - 4F_1^{Z\,2} (1 - \tau) + 4\tau F_2^{Z\,2} (1 - \tau) + 16\tau F_1^Z F_2^Z \right]$$ (11a)

$$B = -8\tau G_1 \left(F_1^Z + F_2^Z \right)$$ (11b)

$$C = \frac{1}{4}G_1^2 + F_1^{Z\,2} + \tau F_2^{Z\,2} .$$ (11c)

In Ref. 6 measurements of the $\nu_\mu p$ and $\overline{\nu}_\mu p$ cross sections were reported in the range $0.2 < Q^2 < 1.2\ \mathrm{GeV}^2$. I have chosen $Q^2 = 0.5\ \mathrm{GeV}^2$ for our analysis to illustrate the sensitivity of this experiment. Keeping only the linear terms in the strange form factors in Eqs. (10) and (11) and using the expressions for the neutral weak form factors from Eq. (7), the cross section can be broken into a nonstrange piece and the three strange contributions. For example, the νp cross section can be written as

$$\frac{d\sigma}{dQ^2} = \sigma_{\mathrm{nonstrange}} \left[1 - 0.72F_1^s(Q^2) - 0.78F_2^s(Q^2) - 1.23G_1^s(Q^2) \right] ,$$ (12a)

and the $\overline{\nu}p$ cross section can be written

$$\frac{d\sigma}{dQ^2} = \sigma_{\mathrm{nonstrange}} \left[1 + 0.83F_1^s(Q^2) + 0.62F_2^s(Q^2) - 2.11G_1^s(Q^2) \right] .$$ (12b)

21

In order to extract a value for $G_1^s(0)$ the authors of Ref. 6 make the assumption that $F_1^s = F_2^s = 0$ and $M_A^s = M_A = 1.032$. The result, $G_1^s = -0.30 \pm 0.16$, is in reasonable agreement with the EMC result. However, a more recent reanalysis of these data has been undertaken,[18] using fewer assumptions in extracting the strange form factors. As one can see in Eq. (12), these cross sections are quite sensitive to the combination $G_M^s = F_1^s + F_2^s$. The authors of Ref. 18 find that in fact this strange magnetic form factor is constrained to be quite small at the Q^2 of this experiment, and that G_1^s is rather poorly determined by these data. Clearly, higher quality neutrino scattering data would be quite helpful.

From Eq. (11) one can see that in the limit of $Q^2 \to 0$ the neutrino cross section becomes approximately proportional to $(G_1)^2$. Thus a measurement at lower energies would be extremely interesting as a probe of G_1^s. A measurement of $G_1^s(0)$ may come from the new LSND experiment at LAMPF.[19] Although the primary goal of the experiment is to study neutrino oscillations, a 20% measurement of the νp cross section at $Q^2 \simeq 0.05$ GeV^2 and $E_\nu \simeq 0.15$ GeV is also expected. Assuming the above parameterizations, the cross section has a 25% contribution from G_1^s and is relatively insensitive to F_1^s. This measurement would provide crucial new information on G_1^s, and since the measurement is at a much lower Q^2 an assumption about the value of M_A^s is not required.

A major problem with the LSND technique is that the target material is actually mineral oil (CH) so that most of the protons are in Carbon. At the low incident energies at LAMPF (~ 150 MeV) the quasielastic scattering from Carbon is a difficult nuclear physics problem and it is unclear whether the true elastic cross section can be extracted. One method is to use two different target materials in separate runs with different C:H ratios and subtract out the Carbon contribution. Unfortunately, this would compromise the statistical accuracy of the experiment. A better idea[20] is to study the ratio of (νp) to (νn) rates in quasielastic scattering from Carbon. It is easy to show that in the $Q^2 = 0$ limit the ratio for free nucleons is approximately

$$\frac{\sigma_p}{\sigma_n} \simeq 1 - 2\frac{G_1^s}{g_A} . \tag{13}$$

It has been shown that the nuclear corrections to this ratio are quite small.[13]

PARITY VIOLATION IN ELECTRON SCATTERING: THEORY

Another method of accessing the neutral weak form factors is through parity violation in electron scattering. The basic idea is to study the interference of the amplitudes for γ- and Z-exchange illustrated by the Feynman diagrams shown in Fig. 2. The cross section will consist of a helicity independent piece (dominated by the squared electromagnetic amplitude) and a term that depends on the electron helicity which violates parity (dominated by the interference of electromagnetic and neutral weak amplitudes). One usually quotes the ratio of helicity dependent to helicity independent cross sections, or the parity violating asymmetry:

$$A = \frac{d\sigma_R - d\sigma_L}{d\sigma_R + d\sigma_L} , \tag{14}$$

where σ_R and σ_L are the cross sections for right- and left-handed electrons, respectively.

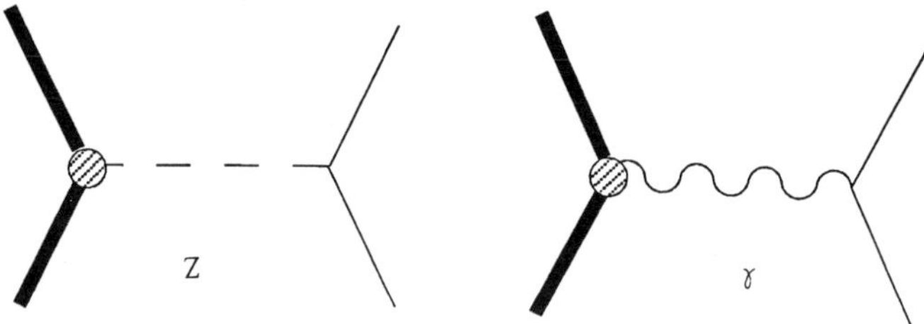

Figure 2. The amplitudes relevant to parity violating electron scattering. Generally one studies the interference of these two amplitudes.

Qualitative Overview

Before studying this problem in detail, it is useful to analyze the structure of the expected result in order to gain some qualitative understanding. Thus we may write the scattering amplitude as a sum of the amplitudes indicated in Fig. 2:

$$\mathcal{M} \sim \frac{\alpha}{q^2} V_e^\gamma V_h^\gamma + G_F (V_e^Z + A_e^Z)(V_h^Z + A_h^Z) \,, \tag{15}$$

where V (A) represents and vector (axial vector) current, e (h) corresponds to the electron (hadron), α is the fine structure constant and G_F is the Fermi weak coupling constant. Upon squaring this amplitude and keeping only the leading terms one obtains

$$\sigma \sim |\mathcal{M}|^2 \sim \sigma_\gamma \pm \frac{\alpha\, G_F}{q^2} V_e^\gamma V_h^\gamma (V_e^Z A_h^Z + A_e^Z V_h^Z) \,, \tag{16}$$

where the $\pm$ refers to positive or negative helicity electrons. Note that the parity odd piece has 2 terms and each contains one axial current matrix element that causes the parity violation. Note also that the first parity odd term contains the hadronic combination $V_h^\gamma A_h^Z$ whereas the second parity odd term contains $V_h^\gamma V_h^Z$. Thus we expect terms with electromagnetic form factors multiplied by hadronic axial form factors and terms with electromagnetic form factors multiplied by neutral weak vector form factors. Since the electromagnetic cross section has the form

$$\sigma_\gamma \sim \frac{\alpha^2}{q^4} (V_h^\gamma)^2 \,, \tag{17}$$

the parity violating asymmetry has the general structure:

$$A \sim \frac{G_F q^2}{\alpha} \frac{(...)V_h^\gamma A_h^Z + (...)V_h^\gamma V_h^Z}{(V_h^\gamma)^2} \,. \tag{18}$$

The parentheses (...) represent factors involving kinematic variables and coupling constants associated with the electron currents and we have kept the hadronic matrix element structure as well as the overall scale factor in front. Note that the structure of the hadronic matrix elements in the numerator follows the pattern discussed above. Also note that the factor in front governs the size of this parity violating effect, and it is proportional to the squared momentum transfer q^2. Thus we expect these asymmetries to be typically of order

$$A \sim (10^{-4} \text{ to } 10^{-3}) \times [Q^2 \text{ in } (\text{GeV}/c)^2] \,. \tag{19}$$

Therefore, one typically must measure parity violating asymmetries of order 10^{-5} or 10^{-6}, which represents a substantial experimental challenge. (It would seem that this problem could be solved by making measurements at higher Q^2, but we will see later that a correct statistical analysis leads to a different conclusion.)

Before moving on to treat this problem in more detail, I should mention that there are electroweak radiative corrections which must also be considered. Some of the important amplitudes are shown in Fig. 3. Although they generally appear to be of order α smaller than the Z-exchange ("tree level") amplitude in Fig. 2, sometimes the tree level amplitude is suppressed and these corrections are more significant.[14] In the following I will ignore these corrections and deal only with the tree level amplitudes.

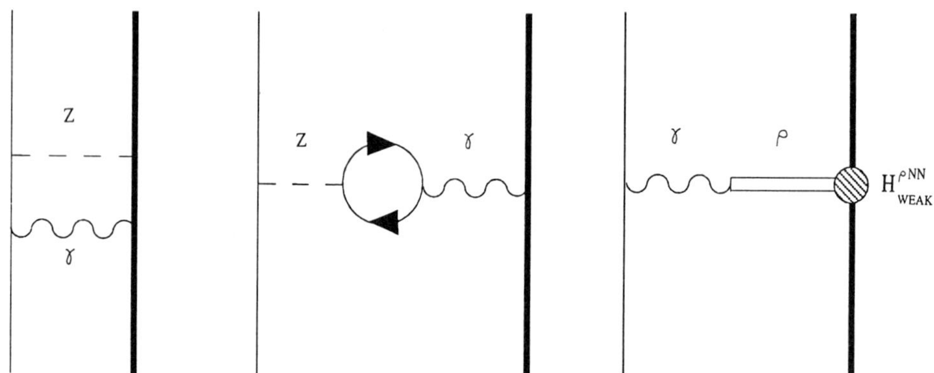

Figure 3. Examples of Feynman graphs illustrating electroweak radiative corrections.

Spinless Isoscalar Target

It is instructive to follow through a rather complete treatment of the simplest case: elastic scattering from a spin 0 and isospin 0 target. First consider the electromagnetic cross section, for which we need the photon exchange amplitude consisting of a product of the hadronic and electronic currents. The leptonic current is given by the usual expression

$$\mathcal{L}^\mu_\gamma = e\,\bar{u}(k')\gamma^\mu u(k) \,. \tag{20}$$

Since there is no spin in the hadronic system, there are only two available 4-vectors associated with the hadron to form the hadronic current:

$$\mathcal{J}^\mu_\gamma = \langle\Psi|\hat{V}^\mu_\gamma|\Psi\rangle = e[F^\gamma(q^2)P^\mu + G^\gamma(q^2)q^\mu] \,, \tag{21}$$

where the two vectors are $P = p+p'$ and $q = p-p'$, p and p' being the initial and final momenta of the hadron, respectively. Since the current must be conserved, we have $q_\mu \mathcal{J}^\mu_\gamma = 0$. Now $q \cdot P = 0$, so F^γ can be any function of q^2. However, since $q^2 \neq 0$ we must enforce $G^\gamma = 0$ to ensure current conservation. Generally, $F^\gamma(q^2 = 0) = Z$ the nuclear charge and $F^\gamma(q^2 < 0)$ represents the Fourier transform of the charge distribution. Since we now have $\mathcal{J}^\mu_\gamma = eF^\gamma P^\mu$, we can write the amplitude

$$\mathcal{M}_\gamma = \frac{\mathcal{J}^\mu_\gamma \mathcal{L}_{\mu\gamma}}{q^2} = \frac{e^2}{q^2}F^\gamma \bar{u}(k')\not{P}u(k) \,. \tag{22}$$

Finally, the squared amplitude can be expressed using the usual trace procedure

$$|\mathcal{M}_\gamma|^2 = \frac{e^4}{q^4}|F^\gamma|^2 \text{Tr}\{\not{k}\not{P}\not{k}\not{P}\} \ . \tag{23}$$

Evaluation of the trace leads to the usual Rosenbluth formula for elastic electron scattering from a spin-0 target. However, we will only need this form in this treatment.

For the neutral current (Z-exchange) amplitude, we have the following simple generalizations of the above.

$$\mathcal{J}_Z^\mu = \frac{g\sin^2\theta_W}{\cos\theta_W}F^Z P^\mu \ , \tag{24}$$

where the weak coupling constant is given by $g = M_W(8G_F/\sqrt{2})^{\frac{1}{2}}$ and F^Z is the corresponding weak form factor. The leptonic current is given by

$$\mathcal{L}_Z^\mu = e\,\bar{u}(k')[a\gamma^\mu + b\gamma^\mu\gamma^5]u(k) \ , \tag{25}$$

where the weak electronic couplings are given by $a = (1 - 2\sin^2\theta_W)$ and $b = 1$. The amplitude is then written

$$\mathcal{M}_Z = \frac{\mathcal{J}_Z^\mu \mathcal{L}_{\mu Z}}{M_Z^2} \ . \tag{26}$$

Now for longitudinally polarized electrons, we introduce the helicity projection operators into the leptonic currents

$$\mathcal{L}_\gamma^\mu(\pm) = e\,\bar{u}[\gamma^\mu\frac{(1\pm\gamma^5)}{2}]u \ , \tag{27a}$$

$$\mathcal{L}_Z^\mu(\pm) = e\,\bar{u}[(a\gamma^\mu + b\gamma^\mu\gamma^5)\frac{(1\pm\gamma^5)}{2}]u \ . \tag{27b}$$

Then the parity violating interference can be written

$$\begin{aligned}
2\delta\left[\text{Re}(\mathcal{M}_Z\mathcal{M}_\gamma^*)\right] &= 2\text{Re}[\mathcal{M}_Z(+)\mathcal{M}_\gamma^*(+) - \mathcal{M}_Z(-)\mathcal{M}_\gamma^*(-)]\\
&= 2\left\{\frac{\mathcal{L}_Z(+)\cdot\mathcal{J}_Z}{M_Z^2}\times\frac{\mathcal{L}_\gamma(+)\cdot\mathcal{J}_\gamma}{q^2} - \frac{\mathcal{L}_Z(-)\cdot\mathcal{J}_Z}{M_Z^2}\times\frac{\mathcal{L}_\gamma(-)\cdot\mathcal{J}_\gamma}{q^2}\right\}\\
&= \frac{ge}{4\cos\theta_W}\left\{[\bar{u}\gamma^\nu\gamma^5 u][\bar{u}(a\gamma^\mu + b\gamma^\mu\gamma^5)u] + [\bar{u}\gamma^\nu u][\bar{u}(a\gamma^\mu + b\gamma^\mu\gamma^5)\gamma^5 u]\right\}\\
&\quad\times\frac{\mathcal{J}_\mu^Z}{M_Z^2}\frac{\mathcal{J}_\nu^\gamma}{q^2}\\
&= \frac{g^2 e^2\sin^2\theta_W}{2\cos^2\theta_W M_Z^2 q^2}F^Z F^\gamma \text{Tr}\{\not{k}\not{P}\not{k}\not{P}\} \ . \tag{28}
\end{aligned}$$

Thus the parity violating asymmetry becomes

$$\begin{aligned}
A &= \frac{2\delta\left[\text{Re}(\mathcal{M}_Z\mathcal{M}_\gamma^*)\right]}{|\mathcal{M}_\gamma|^2}\\
&= \frac{g^2 e^2\sin^2\theta_W F^Z F^\gamma}{2\cos^2\theta_W M_Z^2 q^2}\times\frac{q^4}{e^4|F^\gamma|^2}\\
&= \frac{G_F q^2}{\sqrt{2}\pi\alpha}\frac{F^Z}{F^\gamma}\sin^2\theta_W \ . \tag{29}
\end{aligned}$$

This is the final result. Note the familiar factor (with an extra $\sqrt{2}\pi$ in the denominator to make the experiment a little more difficult!) in front. It is interesting to return briefly to Eq. (5) and think about the matrix element $\langle\Psi|\hat{V}_Z^\mu|\Psi\rangle$. Since in this case $|\Psi\rangle$ is an isoscalar system, only the isoscalar part of the operator can contribute. Then we see that

$$\langle\Psi|\hat{V}_Z^\mu|\Psi\rangle = -\sin^2\theta_W\langle\Psi|\hat{V}_\gamma^\mu|\Psi\rangle - \frac{1}{4}\langle\Psi|\bar{s}\gamma^\mu s|\Psi\rangle \, . \tag{30}$$

Thus if the strange quark matrix element vanishes, the weak matrix element is just $-\sin^2\theta_W$ times the electromagnetic matrix element. In terms of the form factors in Eq. (29) this implies $F^Z = F^\gamma$ (we have kept the $\sin^2\theta_W$ factored out separately in the normalization in Eq. (24)) and the asymmetry becomes independent of any hadronic form factors. Thus, a measured asymmetry that deviates from this prediction could imply the presence of the strange quark matrix element in Eq. (30).

Parity Violation and Nucleon Form Factors

A similar calculation (but more involved) can be carried out for a spin 1/2 hadron such as the nucleon. The resulting asymmetry in the cross section for elastic scattering of left- and right-helicity electrons from a nucleon is :

$$A = \frac{d\sigma_R - d\sigma_L}{d\sigma_R + d\sigma_L}$$
$$= -\frac{G_F Q^2}{\pi\alpha\sqrt{2}} \times \frac{\mathcal{N}}{\mathcal{D}} \, , \tag{31}$$

$$\mathcal{N} = F_1^\gamma F_1^Z + \tau F_2^\gamma F_2^Z + 2\tau\tan^2\frac{\theta}{2}\left(F_1^\gamma + F_2^\gamma\right)\left(F_1^Z + F_2^Z\right)$$
$$- \frac{E + E'}{2M_N}\tan^2\frac{\theta}{2}\left(1 - 4\sin^2\theta_W\right) G_1\left(F_1^\gamma + F_2^\gamma\right)$$
$$\mathcal{D} = F_1^{\gamma\,2} + \tau F_2^{\gamma\,2} + 2\tau\tan^2\frac{\theta}{2}\left(F_1^\gamma + F_2^\gamma\right)^2 \, ,$$

where E is the incident electron energy and E' and θ are the scattered electron energy and angle. (In this expression, I use the notation $Q^2 = -q^2 > 0$.) The denominator is the familiar Rosenbluth form for the electromagnetic scattering. One should note that the $V^\gamma V^Z$ terms on the numerator have the same structure as the $V^\gamma V^\gamma$ terms in the denominator. (Our result for the spinless object above is obtained by setting $F_2 = G_1 = 0$ in this expression, except of course for the $\sin^2\theta_W$ we factored out in Eq. (30).) Note also the term in the numerator containing G_1, which is suppressed by the factor $(1 - 4\sin^2\theta_W)$ (recall that $\sin^2\theta_W \sim \frac{1}{4}$). Therefore, while low Q^2 neutrino scattering is quite sensitive to G_1, parity violating electron scattering is rather insensitive to this form factor. Rather, it appears that parity violating electron scattering is more useful in determining the vector weak form factors. If we neglect the G_1 term in Eq. (31), then the numerator has the Rosenbluth form of the denominator and we see that the "magnetic" form factor $G_M = F_1 + F_2$ dominates at backward scattering angles ($\theta \to \pi$) whereas the "electric" form factor $G_E = F_1 - \tau F_2$ contributes at forward angles.

A similar analysis can be made for quasielastic electron scattering from a nucleus with Z protons and N neutrons, in which case the asymmetry is

$$A_{\text{nuc}} = -\frac{G_F Q^2}{\pi\alpha\sqrt{2}} \times \frac{N\mathcal{N}_n + Z\mathcal{N}_p}{N\mathcal{D}_n + Z\mathcal{D}_p} \, . \tag{32}$$

Finally, it is worth returning to the case of a $J = 0$ and $T = 0$ target for which we have derived an expression for the asymmetry (Eq. (29)). In fact, ^{4}He is such a target for which (as we will see later) parity violation measurements are quite feasible. If one employs the impulse approximation and assumes that the strange quark matrix element is associated with the strange quark content of the nucleon, then one obtains the following expression for the asymmetry:[15]

$$A = \frac{G_F q^2}{\sqrt{2}\pi\alpha} \left[\sin^2\theta_W + \frac{G_E^s}{2(G_E^p + G_E^n)} \right].$$

(33)

It turns out that this expression is valid even in the presence of the significant meson exchange current corrections, so its validity effectively goes beyond the impulse approximation.[21] Note that the only unknown quantity in this expression is G_E^s, so there is no contamination of the result with uncertainties associated with other form factors like G_M^s or G_1. The electroweak radiative corrections are small and under control in this case, and isospin violation effects are insignificant.[22] Thus it appears that there is a great advantage in studying G_E^s through the parity violation in elastic scattering of electrons from ^{4}He.

PARITY VIOLATION IN ELECTRON SCATTERING: EXPERIMENT

In this section, we examine various aspects of performing parity violating electron scattering experiments. Clearly, the experimental technique will differ considerably from the usual cross section measurement technique. On the other hand, many features of these parity violation experiments are relevant to other (parity conserving) electron helicity asymmetry experiments such as those involving polarized targets. The techniques associated with these parity violating electron scattering experiments have been developed in a series of pioneering experiments at three different laboratories.

Previous Experiments: A Brief Review

The first experiment was a measurement of parity violation in deep inelastic scattering at SLAC.[23] The result of this experiment was pivotal in establishing the violation of parity in the weak neutral current for the first time. The relatively large $Q^2 \sim 1(\text{GeV/c})^2$ yielded a rather substantial experimental asymmetry ($\sim 10^{-4}$), so that the precision achieved in the measurement $\sim 10^{-5}$ was in fact quite significant.

The second experiment was performed at the Mainz electron linac.[24] Using 300 MeV incident electrons, a measurement of the asymmetry in quasielastic scattering from ^{9}Be was performed; the precision achieved in this effort was about an order of magnitude better than the SLAC measurement, of order $\sim 10^{-6}$. The major motivation was a measurement of the electron axial-vector coupling a which is $(1 - 4\sin^2\theta_W)$, and in fact the asymmetry in this experiment is quite insensitive to the strange quark form factors of interest here. (This is easily demonstrated using the form factor dependence contained in Eq. (32).)

The third experiment was performed at the Bates/MIT linac,[25] and was a measurement of the asymmetry in elastic scattering from ^{12}C at very low Q^2. This experiment attained a precision of $\sim 10^{-7}$, and was limited by the statistical uncertainty. The experimenters claim that the systematic errors were reduced to a level about an order of magnitude smaller: $\sim 2 \times 10^{-8}$. At the very low Q^2 of this measurement there is essentially no sensitivity to strange quark effects.

Thus one can see very impressive progress in the development of the experimental technique associated with parity violating electron scattering over the last 15 years. The precision attained in these previous efforts seems to imply that new experiments to study strange quark form factors can be expected to yield very significant results. The next question to answer is: what new experiments are the most useful?

The Experimental Program for Neutral Weak Form Factors of the Nucleon

We have already established that parity violating electron scattering is better suited to studying the weak neutral vector form factors rather than the axial form factor. Thus we will focus our attention on the vector form factors and we will use Fig. 1 as a guide for the magnitudes and Q^2 dependences. First, we notice that F_1^s only becomes significant at $Q^2 \sim 0.5$ $(\text{GeV}/c)^2$. From Eq. (31) we know that the sensitivity to F_1^Z is in forward angle scattering from the nucleon. Recalling the expression for $Q^2 = 4EE' \sin^2(\theta/2)$, we see immediately that measurement of this quantity requires an incident energy of $E \sim 1$ GeV or greater.

The optimal situation for F_2^Z requires a bit more investigation. From Fig. 1, F_2^s is significant in the complete range of $0 \le Q^2 \le 0.5$ $(\text{GeV}/c)^2$. Although it is largest at $Q^2 = 0$, recall that the asymmetry will grow proportional to Q^2. However, one must also consider that the cross section falls rather strongly with increasing Q^2. It is easy to demonstrate that the optimal measurement of the asymmetry (i.e., the smallest fractional error in A) for a fixed amount of running time in a fixed geometry is achieved when the figure of merit $\mathcal{F}$ is maximized. $\mathcal{F}$ is defined as

$$\mathcal{F} \equiv A^2 \sigma \, . \tag{34}$$

Since Eq. (31) shows that the sensitivity to F_2^Z is maximized at backward angles, one can consider $\mathcal{F}$ at $\theta = \pi$ as a function of the incident electron energy. This is displayed in Fig. 4 where one can see that the optimal energy to study this form factor is at about 400 MeV.

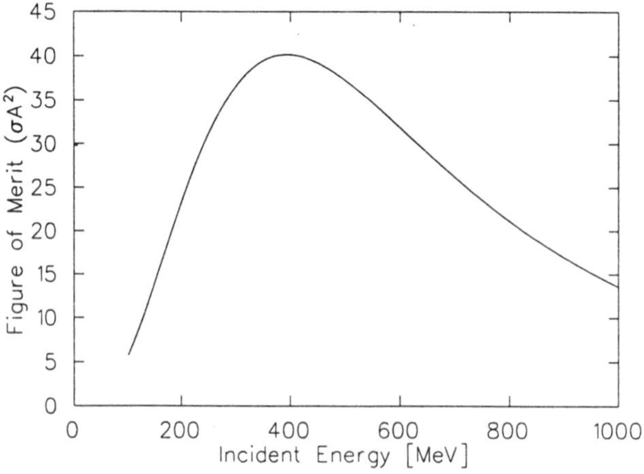

Figure 4. The figure of merit for e^--nucleon scattering at $\theta = \pi$ as a function of the incident electron energy.

In summary, it appears that a reasonable outline for a program to study the strange quark form factors of the nucleon would include:

- low Q^2 measurements of elastic neutrino-nucleon scattering which are primarily sensitive to the axial form factor G_1^s;
- backward angle measurements of parity violating electron scattering in the 200–500 MeV range to study the strange magnetic form factor G_M^s;
- forward angle measurements of parity violating electron scattering from the nucleon and ^{4}He at incident energies $E > 1$ GeV to gain sensitivity to the strange electric form factor G_E^s.

Experimental Techniques

The experimental technique associated with parity violating electron scattering requires both high luminosity (beam intensity $\times$ target thickness $\times$ detector acceptance) to achieve the high statistical precision necessary for observation of small asymmetries and excellent control of systematic effects that may generate false asymmetries in the experiment. Satisfying these requirements presents very challenging experimental problems.

I will begin with a discussion of the polarized electron beam technology. The goal is to generate a high intensity beam with high polarization and fairly rapid helicity reversal. The technique that is generally used is photoemission from GaAs. The GaAs surface is cleaned and coated with a thin layer of Cesium to decrease the work function at the surface and thus increase the quantum efficiency (typically one achieves a few percent quantum efficiency). The GaAs crystal is optically pumped with a circularly polarized laser beam and the photoemitted electrons have a longitudinal polarization that is correlated with the laser beam polarization. Thus the helicity is reversed by simply reversing the circular polarization of the laser beam. A piezoelectric $\frac{1}{4}$ wave plate ("Pockels cell") in the laser beam can reverse the polarization on a time scale of microseconds. The electron polarization is limited to $< 50\%$ due to the degeneracy of states in the valence band of the crystal. In practice the achievable polarization is of order $\sim 40\%$. New "strained" crystals are under development to generate higher polarization electron beams, but so far the quantum efficiency is rather poor for these crystals. Nevertheless, the "standard" polarized source can generate 1–100 μA of 40% polarized electrons and there are excellent prospects for improving the polarization in the next few years.

Usually the spin of the electrons precesses away from longitudinal in the acceleration/beam transport system due to the $g - 2$ precession. Various schemes, such as a Wien spin rotator, are used to prepare the spins in the polarized injector so that they rotate to the desired longitudinal orientation at the experiment. The polarization is usually measured near the experiment (typically just upstream) using Møller polarimetry. By scattering the electrons from the polarized electrons in a magnetized foil, and using the known (from QED) spin dependence of the $e^- - e^-$ scattering, one can infer the beam polarization. In this fashion, one can typically determine the beam polarization with about 3–5% precision and there are hopes that the uncertainty can be reduced further.

The basic technique is then to measure, for each helicity state of the beam, the ratio of detector response ("count rate") to beam intensity (monitored with various devices such as toroidal transformers). This ratio I will call S_+ (or S_-) for positive (or negative) helicity beam. One then forms the difference of the two ratios (for each helicity state) divided by the sum and the parity violating asymmetry is this quantity divided by the beam polarization (P_e):

$$A = \frac{1}{P_e} \frac{S_+ - S_-}{S_+ + S_-} \, .$$

(35)

Usually S_+ and S_- consist of many individual measurements closely spaced in time (associated with the frequency of helicity flip) to eliminate the effect of long term drifts in experimental parameters. In addition, the helicity is usually chosen randomly to avoid any resonant locking to particular frequencies in the laboratory (such as the power line frequency). One must be very careful to ensure that the measurement of S is not affected by the status of the electronics that controls the beam helicity. For example, ground-loops that change as a result of helicity reversal because of the Pockels cell voltage change must be avoided. One must develop tests to be sure that none of these effects creep into the experiment.

An extremely important class of systematic effects is associated with the fact that beam parameters (position, angle, or energy) can be correlated with the helicity of the beam. These effects can cause the detector response to be helicity correlated in a systematic way (for example the cross section is generally quite energy dependent) which would cause a false non-zero asymmetry. Therefore one must monitor these parameters of the beam (which I label $\{\alpha_i\}$) using various beam position monitors near the experiment and in a place where the beam is energy-dispersed (to monitor the beam energy). One is then in a position to measure the helicity correlations in these parameters:

$$\Delta\alpha_i^{RL} = \alpha_i^R - \alpha_i^L \, .$$

(36)

The first goal is then to tune the system so that these helicity correlations are minimized (for example by adjusting the polarized source laser optics). Secondly, one purposely varies the parameters ($\{\alpha_i\}$) by using steerers and other beam controls to measure the signal response derivatives:

$$\frac{\partial S}{\partial \alpha_i} \, .$$

(37)

This is, of course, done independent of helicity and generally the steps in a parameter α_i are large compared to the helicity correlations $\Delta\alpha_i^{RL}$. In fact, this can be done during the experiment as long as the parameters are not changed between adjacent pairs of beam helicity states. With all of this information one may then perform a final correction to the "raw" measured asymmetry:

$$A_{\mathrm{corr}} = A_{\mathrm{raw}} - \frac{1}{S} \sum \frac{\partial S}{\partial \alpha_i} \Delta\alpha_i^{RL} \, .$$

(38)

Of course it is desirable that $A_{\mathrm{corr}} - A_{\mathrm{raw}} \ll A_{\mathrm{raw}}$ so that these corrections are small. In addition to making the $\Delta\alpha_i^{RL}$ small as described above, one attempts to make the derivatives small (such as by designing a symmetric detector system that is insensitive to beam motion).

"SAMPLE" EXPERIMENT

The SAMPLE experiment[26] is being prepared at the Bates/MIT Linear Accelerator Center to measure the parity violation in elastic e^--proton scattering at backward angles. The goal is to measure the strange magnetic form factor

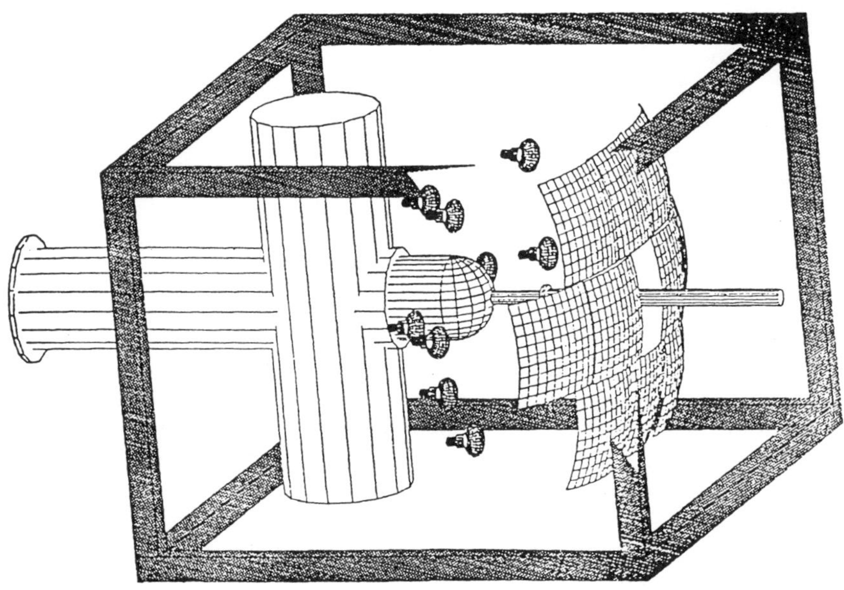

Figure 5 Layout of the SAMPLE experiment. The beam is incident from the right in the vacuum pipe. The LH$_2$ target is in the scattering chamber which has a hemispherical dome for the electrons scattering at large angles. The ten Cerenkov mirrors and phototubes are also shown.

$G_M^s = F_1^s + F_2^s$ at low $Q^2 \simeq 0.1(\mathrm{GeV}/c)^2$. The experiment will be performed using a 200 MeV polarized electron beam incident on a liquid hydrogen target. (We have chosen the lower energy because the detector background is expected to be lower at lower energy.) The scattered electrons are detected in a large solid angle (~ 2 sr) Cerenkov detector at backward angles $130° < \theta < 170°$. A schematic diagram of the experimental apparatus is shown in Fig. 5.

The expected asymmetry (assuming no strange quark effects) is 8×10^{-6}, and the goal of the experiment is a measurement of this quantity with $\sim 8\%$ relative uncertainty. Approximately half of the uncertainty is expected to come from systematic effects (measurement of the electron beam polarization should dominate) and about half should be due to statistics (assuming 24,000 μA-hours of production running).

The liquid hydrogen target has been constructed at Caltech. Due to its 40 cm length and the high-intensity (40–60 μA) of the electron beam used in the experiment, the target must withstand over 500 watts of heat deposited by the beam. The target is thus designed as a high flow-rate recirculating liquid hydrogen system with a heat exchanger to remove the heat. The primary coolant (He gas at 12°K) for the heat exchanger is supplied by a commercial refrigeration system. The refrigerator achieved over 800 watts of cooling during acceptance tests. Initial tests of this target system with up to 35 μA of beam incident were successful; the target maintained a sub-cooled condition (below boiling) even under these conditions.

The detector was constructed at the University of Illinois. It consists of 10 large ellipsoidal mirrors that reflect the Cerenkov light into 8 inch diameter photomultiplier tubes. Each photomultiplier is shielded from the target and room background by a cast lead shield. The apparatus has recently been installed on the beamline and is ready to begin tests with beam. We hope to perform preliminary measurements of the parity violating asymmetry during 1994.

In addition to studying the asymmetry for scattering from the proton, this apparatus could be used to measure the asymmetry in quasielastic scattering from deuterium. Through application of Eq. (32), one can show that this asymmetry is almost completely insensitive to strange quark effects. Thus, measurement of this asymmetry is an excellent test of the experimental technique. In fact, such a measurement takes less running time since both the cross section and the asymmetry are larger than for the proton.

PROPOSED CEBAF EXPERIMENTS

There are three parity violation experiments proposed for the CEBAF facility in Newport News, VA. These experiments are all aimed at studying the contribution of strangeness to the neutral weak form factors of the nucleon. They all have conditional approval at CEBAF which implies a commitment to pursue these experiments subject to some initial conditions (e.g., demonstration of the required beam properties after the accelerator begins operation) being satisfied. Clearly the higher beam energy and high expected beam quality (and intensity) at CEBAF make it a very attractive option for pursuing this program.

One proposal (91-010)[27] is to measure the forward angle asymmetry in elastic $e-p$ scattering. This proposal would utilize 2 high-resolution magnetic spectrometers which are under construction as part of the basic complement of experimental equipment in experimental Hall A at CEBAF. These spectrometers would both be employed at 12.5° (one on each side of the beamline) with a beam energy of 3.5 GeV, yielding a Q^2 of 0.5 (GeV/c)^2. The very high count rates for the elastic scattering require replacement of the standard detector package with a special Pb-glass counter system equipped with integrating electronics to give an analog signal proportional to the elastic count rate. This measurement measures a combination of G_E^s and G_M^s without separating them. If a large effect is seen in the experiment, a series of further measurements to separate the electric and magnetic effects are proposed.

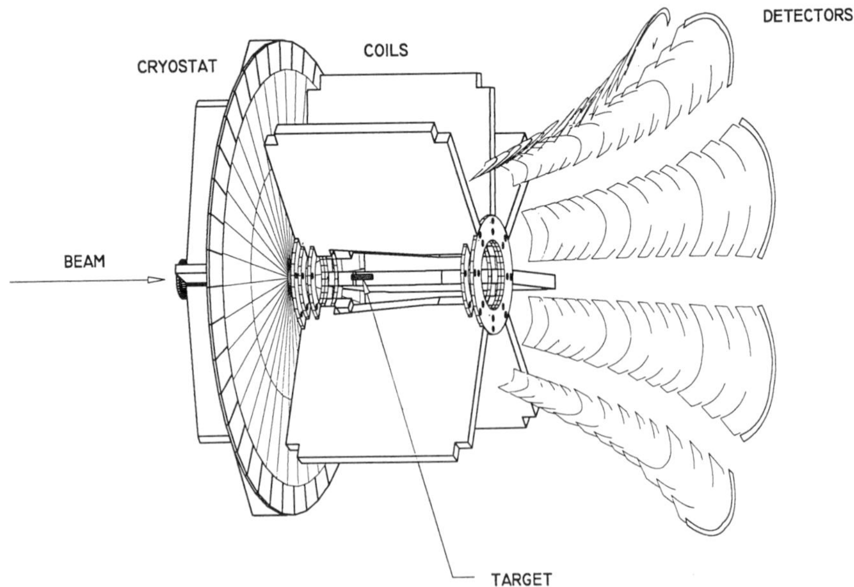

Figure 6. Toroidal spectrometer design for the proposed "G^0" experiment at CEBAF.

Another experiment (91-017, or "G^0")[28] has been proposed to study the elastic scattering from the proton. This experiment represents a more complete program to study the strange vector form factors of the proton by using a dedicated apparatus optimized to the task. The heart of the apparatus is a segmented iron-free toroidal magnetic spectrometer which has extremely large acceptance and very good azimuthal symmetry. A schematic diagram of this experiment is shown in Fig. 6. It will be used to detect backward scattered electrons with lower energy electron beams (300–600 MeV) for good sensitivity to G_M^s over a range of $Q^2 \sim 0.2 - 0.5(\mathrm{GeV})^2$. In addition, the toroid will be used to detect recoil protons in the angular range $60° < \theta_p < 77°$ with a higher energy incident electron beam (2.5 GeV). This is anticipated to be a particularly clean way of accessing the forward (electron) angle region. The anticipated precision for G_E^s and G_M^s are shown in Fig. 7.

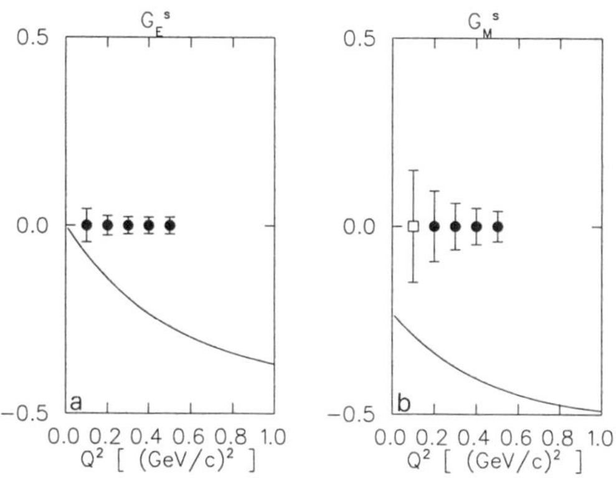

Figure 7. Anticipated uncertainties for (a) G_E^s and (b) G_M^s for the "G^0" proposal at CEBAF. The open square indicates the anticipated uncertainty from the SAMPLE experiment.

Finally, the third experiment (91-004)[29] proposes to measure the asymmetry in elastic scattering from ^{4}He at $Q^2 = 0.6(\mathrm{GeV}/c)^2$. Here the high-resolution spectrometers in Hall A are to be employed also, but with the standard detectors since the count rates are much lower. However, the expected asymmetry is quite large ($\sim 5 \times 10^{-5}$ for no strange quarks), and the sensitivity to G_E^s (see Eq. (33)) is huge. This is demonstrated in Fig. 8 where the effect of the Jaffe strange form factors[16] can be seen to even change the sign of the asymmetry at these kinematics!

CONCLUSIONS

Clearly there is a great deal of interest in studying the strangeness in the nucleon through the determination of the three strange elastic form factors G_E^s, G_M^s, and G_1^s. Parity violating electron scattering seems to have a significant (and perhaps unique) role in the study of the first two vector form factors at low Q^2. The theoretical formalism is now in place for the analysis of such experiments. The experiments are

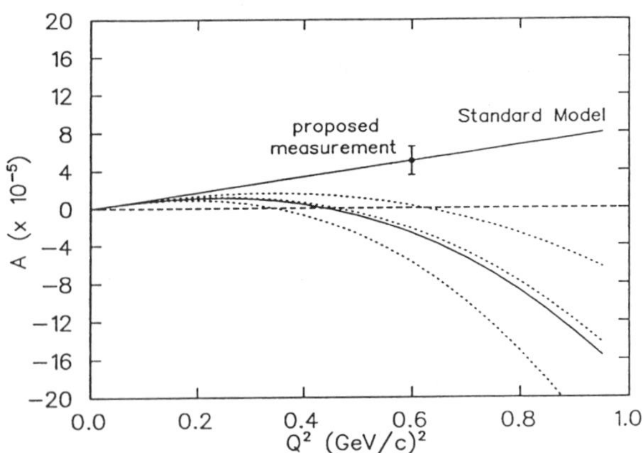

Figure 8. Sensitivity of the ^{4}He elastic asymmetry to the strange form factors from Ref. 16. The projected data point for CEBAF proposal (91-004) is shown. The curve labelled "Standard Model" is the prediction without strange quark effects.

quite challenging, but they do appear feasible. The SAMPLE experiment at Bates promises to give us our next glimpse at this subject through a measurement of G_M^s in the next year or two. A more ambitious program is planned for CEBAF in the future (there is also interest in starting a program at Mainz[30]). Hopefully a clearer understanding of the fundamental sub-structure of the nucleon will result from these efforts.

ACKNOWLEDGMENT

This work has been supported in part by the National Science Foundation, Grant No. PHY91-15574.

REFERENCES

1. D. Kaplan and A. Manohar, *Nucl. Phys. B* 310:527 (1988).
2. R. L. Jaffe and C. L. Korpa, *Comm. Nucl. Part. Phys.* 17:163 (1987).
3. J. Gasser, H. Leutwyler, and M. E. Sainio, *Phys. Lett. B* 253:163 (1991).
4. Tj. Ketel, these proceedings.
5. J. Ashman *et al.*, *Nucl. Phys. B* 328:1 (1989).
6. L. A. Ahrens *et al.*, *Phys. Rev. D* 35:785 (1987).
7. S. Fanchiotti and A. Sirlin, *Phys. Rev. D* 41:319 (1990).
8. F. Abe *et al.*, *Phys. Rev. Lett.* 63:720 (1990); G. S. Abrams *et al.*, *Phys. Rev. Lett.* 63:724 (1990); B. Adeva *et al.*, *Phys. Lett. B* 237:136 (1990).
9. S. Freedman, *Comm. Nucl. Part. Phys.* 19:209 (1990).
10. M. Bourquin *et al.*, *Z. Phys. C* 21:27 (1983).
11. G. Höhler *et al.*, *Nucl. Phys. B* 114:505 (1976).
12. S. Galster *et al.*, *Nucl. Phys. B* 32:221 (1971).
13. G. T. Garvey, E. Kolbe, K. Langanke, and S. Krewald, *Phys. Rev. C* (in press).
14. M. Musolf and B. Holstein, *Phys. Lett. B* 242:461 (1990).
15. D. H. Beck, *Phys. Rev. D* 39:3248 (1989).
16. R. L. Jaffe, *Phys. Lett. B* 229:275 (1989).

17. R. L. Jaffe and A. Manohar, *Nucl. Phys. B* 321:343 (1989).
18. G. T. Garvey, W. C. Louis, and D. H. White, *Phys. Rev. C* (in press).
19. LAMPF Proposal # 1173, W. C. Louis, contact person; and G. T. Garvey, private communication.
20. G. T. Garvey, S. Krewald, E. Kolbe, and K. Langanke, *Phys. Lett. B* 289:249, (1992).
21. M. Musolf and T. W. Donnelly, private communication.
22. T. W. Donnelly, J. Dubach, and I. Sick, *Nucl. Phys. A* 503:589 (1989).
23. C. Y. Prescott *et al.*, *Phys. Lett. B* 77:347 (1978).
24. W. Heil *et al.*, *Nucl. Phys. B* 327:1 (1989).
25. P. A. Souder *et al.*, *Phys. Rev. Lett.* 65:694 (1990).
26. Bates Proposal # 89-06, R. D. McKeown and D. H. Beck, contact people.
27. CEBAF Proposal # 91-007, P. Souder and J. M. Finn, contact people.
28. CEBAF Proposal # 91-017, D. H. Beck, contact person.
29. CEBAF Proposal # 91-004, E. J. Beise, contact person.
30. D. von Harrach, private communication.

POLARIZATION IN LEPTON-INDUCED RELATIONS

T.W. Donnelly

Center for Theoretical Physics, Laboratory for Nuclear Science and
Department of Physics
Massachusetts Institute of Technology
Cambridge, Massachusetts 02139

I. INTRODUCTION

These lecture notes are organized in the following way: in Sec. II the nature of the leptonic tensor is reviewed (see also Ref. [1]) and extensions to include both incident and scattered electron polarizations incorporated. The general problem was treated in Ref. [2] — here I have chosen to present some new results for the leptonic polarization transfer process. In Sec. III the nature of the hadronic tensor and the electromagnetic hadronic (nuclear) response functions for single–arm, inclusive electron scattering are discussed. Recent work on the inclusive process $A(\vec{e}, \vec{e}\,')$ is summarized after repeating the familiar arguments leading to the unpolarized inclusive electron scattering cross section. The subject of inclusive scattering from polarized targets, on the other hand, is not discussed in these lectures and the reader is directed to Ref. [2] for detailed treatment of those ideas. In Sec. IV the concepts involved in two–arm coincidence electron scattering are introduced from a general perspective. The relevant kinematics are treated in some detail and connections are made with inclusive electron scattering — the latter are especially important in circumstances where inclusive electron scattering in the quasielastic region is being studied. Following these general considerations, in Sec. V a specific simple model is discussed, namely the polarized plane–wave impulse approximation (PWIA), to provide a tool for treating the two–arm coincidence electron scattering of polarized electrons from polarized targets in the quasielastic region (the general formalism for such studies is presented in Ref. [3]). After introducing the model and briefly revisiting inclusive scattering, a few selected examples of polarized two–arm coincidence scattering are presented. The lectures conclude with a brief discussion of multi–particle coincidence electron scattering in Sec. VI and finally a summary of the material presented in these lectures is given in Sec. VII.

Before entering into the developments outlined above, let us recall the fundamental starting point for descriptions of electron scattering from nucleons or nuclei. The basic diagram required when studying parity–conserving electron scattering from hadronic systems in the one–proton–exchange or first Born approximation is shown in Fig. 1. Here is an electron with 4–momentum[†] $K = (\epsilon, \vec{k})$ and spin projection λ is scattered through an angle θ_e to 4–momentum $K' = (\epsilon', \vec{k}')$. The virtual photon exchanged in the process carries 4–momentum transfer $Q = (\omega, \vec{q})$ and, in interacting with the nucleus, causes it to proceed from state $|i\rangle$ with total 4–momentum P_i to state $|f\rangle$ with total 4–momentum P_f. Conservation of 4–momentum tells us that $Q = K - K' = P_f - P_i$. Furthermore, we have $Q^2 = \omega^2 - q^2 < 0$ for electron scattering corresponding to the exchange of spacelike virtual photons. The real–photon point involved in photoexcitation or gamma–decay corresponds to the limit $Q^2 \to 0$.

[†] Four–vectors are generally indicated with capital letters, K; 3–vectors are generally indicated with lower-case letters, $\vec{k}$, and their magnitudes with lower-case letters having no arrows, $k = |\vec{k}|$. The conventions used here are otherwise those of Bjorken and Drell[4] so that, for example, $K^2 = K_\mu K^\mu = \epsilon^2 - k^2 = m_e^2$.

Perspectives in the Structure of Hadronic Systems
Edited by M.N. Harakeh *et al.*, Plenum Press, New York, 1994

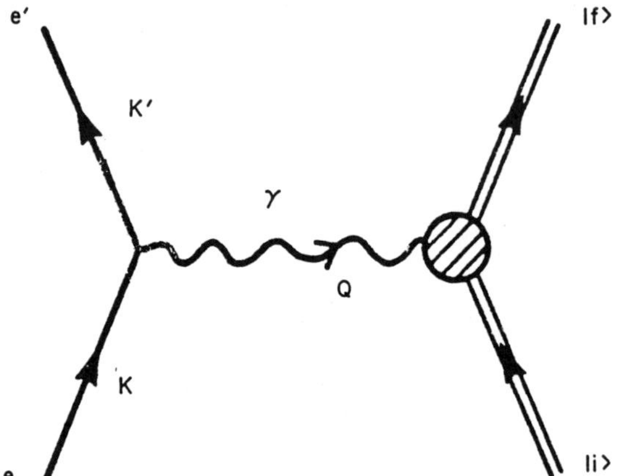
Fig. 1. Electron scattering in the one–photon–exchange approximation.

The cross section for the process shown in Fig. 1 follows by applying the Feynman rules (see Ref. [4]) and involves the square of the absolute value of the invariant matrix element, $\mathcal{M}_{fi}$, which is in turn made up as a product of three factors — the electron current j^μ, the photon propagator $g_{\mu\mu'}/Q^2$ and the nuclear current matrix element, $J_{fi}^{\mu'}$:

$$d\sigma \sim |\mathcal{M}_{fi}|^2 \sim \left| j_\mu \frac{1}{Q^2} J_{fi}^\mu \right|^2$$
$$= \frac{1}{Q^4} j_\mu^* j_\nu \cdot J_{fi}^{\mu*} J_{fi}^\nu \ . \tag{1}$$

Of course, we must perform the appropriate average–over–initial and sum–over–final states (indicated $\overline{\sum}$) in obtaining the cross section pertaining to the actual experimental conditions (*e.g.*, electron polarized or not, target polarized or not, inclusive or exclusive scattering, *etc.*). This yields the leptonic and hadronic tensors

$$\eta_{\mu\nu} \equiv \overline{\sum}_{\text{leptons}} j_\mu^* j_\nu \tag{2a}$$
$$W^{\mu\nu} \equiv \overline{\sum}_{\text{hadrons}} J_{fi}^{\mu*} J_{fi}^\nu \tag{2b}$$

whose contraction is involved in forming the cross section:

$$d\sigma \sim \frac{1}{Q^4} \eta_{\mu\nu} W^{\mu\nu} \ . \tag{3}$$

It proves useful to decompose both leptonic and hadronic tensors into pieces which are symmetric (s) or antisymmetric (a) under the interchange $\mu \leftrightarrow \nu$:

$$W^{\mu\nu} = W_s^{\mu\nu} + W_a^{\mu\nu} \tag{4a}$$
$$\eta_{\mu\nu} = \eta_{\mu\nu}^s + \eta_{\mu\nu}^a \ . \tag{4b}$$

Clearly, in contracting with the leptonic tensor no cross–terms are allowed:

$$\eta_{\mu\nu} W^{\mu\nu} = \eta_{\mu\nu}^s W_s^{\mu\nu} + \eta_{\mu\nu}^a W_a^{\mu\nu} \ . \tag{5}$$

Furthermore, we shall make use of the fact that both the leptonic and nuclear electromagnetic currents are conserved:

$$Q_\mu W_s^{\mu\nu} = Q_\mu W_a^{\mu\nu} = 0 \tag{6a}$$
$$Q^\mu \eta_{\mu\nu}^s = Q^\mu \eta_{\mu\nu}^a = 0 \ . \tag{6b}$$

Let us now turn in the next section to a detailed discussion of the leptonic tensor.

II.   LEPTONIC TENSOR

Our focus in the present section is on the leptonic tensor $\eta_{\mu\nu}$ and the results summarized here are taken from the detailed discussions presented in Ref. [2]. Substituting for the electron current we have

$$\eta_{\mu\nu} = \overline{\sum} [\bar{u}(K, \lambda)\gamma_\mu u(K', \lambda')][\bar{u}(K', \lambda')\gamma_\nu u(K, \lambda)] \ , \tag{7}$$

where properties[4] of the spinors u and gamma matrices γ have been used to obtain the complex conjugate of the electron current as the first group of three factors in Eq. (7). We can guarantee that electrons and not positrons occur by inserting projection operators $(\not{K}' + m_e)/2m_e$ and $(\not{K} + m_e)/2m_e$ in the appropriate places:

$$\eta_{\mu\nu} = \frac{1}{4m_e^2} \overline{\sum} \bar{u}(K, \lambda)\gamma_\mu(\not{K}' + m_e)u(K', \lambda')\bar{u}(K', \lambda')\gamma_\nu(\not{K} + m_e)u(K, \lambda) \ . \tag{8}$$

Unpolarized Electron Scattering

Let us first consider unpolarized electron scattering, in which case $\overline{\sum}$ corresponds to an unrestricted sum over all four components in the spinors divided by two for the initial–state spin average. This is in fact simply a trace which can be evaluated using standard techniques:[4]

$$\eta_{\mu\nu}^{\text{unpol}} = \frac{1}{8m_e^2} \, \text{Tr} \left\{ \gamma_\mu (\not{K}' + m_e) \gamma_\nu (\not{K} + m_e) \right\} \tag{9a}$$

$$= \frac{1}{2m_e^2} \left[K_\mu K'_\nu + K'_\mu K_\nu - g_{\mu\nu} \left(K \cdot K' - m_e^2 \right) \right] \quad . \tag{9b}$$

Note that this tensor is symmetric under the interchange $\mu \longleftrightarrow \nu$ and manifestly satisfies the current conservation condition, $Q^\mu \eta_{\mu\nu}^{\text{unpol}} = 0$. As we shall see in Secs. III–VI, various classes of response will arise when particular combinations of the Lorentz indices μ and ν are selected. For example, familiar unpolarized single–arm electron scattering (see also Sec. III) involves the combinations

$$\ell_L \equiv \eta_{00} \tag{10a}$$

$$\ell_T \equiv \frac{1}{2} [\eta_{11} + \eta_{22}] \quad , \tag{10b}$$

where here the labels L and T refer to projections of the current martix elements longitudinal and transverse to the virtual photon direction, respectively. In the extreme relativistic limit (ERL), where the electron's mass may be neglected with respect to its energy, one obtains the familiar expressions:

$$\ell_L \to v_0 v_L \tag{11a}$$

$$\ell_T \to v_0 v_T \quad , \tag{11b}$$

involving the longitudinal and transverse electron kinematic factors (see Ref. [2])

$$v_L = \rho^2 \tag{12a}$$

$$v_T = \frac{1}{2}\rho + \tan^2 \frac{\theta_e}{2} \quad , \tag{12b}$$

where $\rho \equiv -Q^2/q^2 = 1 - (\omega/q)^2 \to 0 \le \rho \le 1$ and $v_0 \equiv (\epsilon + \epsilon')^2 - q^2$.

Polarized Incident Electrons

With polarized electrons things are somewhat more complicated. Let us begin by assuming that the scattered electron's polarization is not measured, but that the incident electron beam is prepared with its spin pointing in some direction characterized by the 4–vector S^μ which must satisfy[4] $S^2 = -1$ and $S \cdot K = 0$. The more general situation with both incident and scattered electrons polarized is discussed below. We may then insert into Eq. (8) just after the factor γ_ν the spin projection operator $(1 + \gamma_5 \not{S})/2$ and once again now have a sum over all four spinor components and hence a trace:

$$\eta_{\mu\nu} = \frac{1}{16m_e^2} \, \text{Tr} \left\{ \gamma_\mu \left(\not{K}' + m_e \right) \gamma_\nu (1 + \gamma_5 \not{S}) (\not{K} + m_e) \right\} \tag{13a}$$

$$\equiv \frac{1}{2} \left[\eta_{\mu\nu}^{\text{unpol}} + \eta_{\mu\nu}^{\text{pol}} \right] \quad , \tag{13b}$$

where $\eta_{\mu\nu}^{\text{unpol}}$ is the leptonic tensor in Eq. (9b) and, upon evaluation of the trace,[2] the new piece which contains all reference to the electron spin is given by

$$\eta_{\mu\nu}^{\text{pol}} = \frac{1}{2m_e^2} \left(i\epsilon_{\mu\nu\alpha\beta} \, m_e S^\alpha Q^\beta \right) \quad . \tag{14}$$

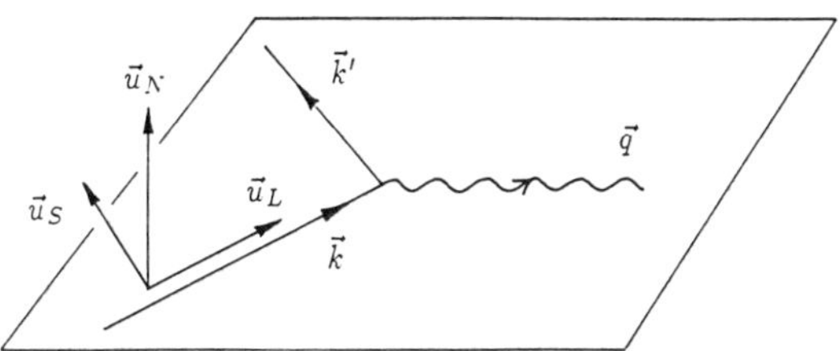

Fig. 2. Coordinate system for electron spin: $\vec{u}_L$ along $\vec{k}$, $\vec{u}_N$ normal to the electron scattering plane ($\vec{u}_N \sim \vec{k} \times \vec{k}'$) and $\vec{u}_S = \vec{u}_N \times \vec{u}_L$ in the "sideways" direction.

Note that this tensor is antisymmetric under the interchange $\mu \leftrightarrow \nu$ and separately satisfies the current conservation condition, $Q^\mu \eta^{\mathrm{pol}}_{\mu\nu} = 0$.

Now let us specify the spin direction of the incident electron by writing for the 3–vector $\vec{s}$ the following:

$$\vec{s} = hs\left(\cos\zeta\,\vec{u}_L + \sin\zeta\,\vec{u}_\perp\right) \quad , \tag{15a}$$

where $\vec{u}_L$ is a unit vector pointing along $\vec{k}$ and $\vec{u}_\perp$ is a unit vector in the plane perpendicular to $\vec{k}$:

$$\vec{u}_\perp = \cos\eta\,\vec{u}_S + \sin\eta\,\vec{u}_N \quad . \tag{15b}$$

The three orthogonal unit vectors ($L \leftrightarrow$ longitudinal, $S \leftrightarrow$ sideways, $N \leftrightarrow$ normal) are indicated in Fig. 2. The factor $h = \pm 1$ is added only for convenience in flipping the spin; alternatively the angles (ζ, η) can be chosen to have $\vec{s}$ point in two opposing directions. The conditions $S^2 = -1$ and $S \cdot K = 0$ imply that

$$s = |\vec{s}| = \frac{1}{\sqrt{1 - \beta^2\cos^2\zeta}} = \frac{\gamma}{\sqrt{\cos^2\zeta + \gamma^2\sin^2\zeta}} \tag{16a}$$

$$S^0 = h\beta s\cos\zeta \quad , \tag{16b}$$

where $\beta = k/\epsilon = \sqrt{1 - (m_e/\epsilon)^2}$ and $\gamma = 1/\sqrt{1 - \beta^2} = \epsilon/m_e$ are the familiar relativistic factors. It is convenient to define

$$\tan\mu \equiv \gamma\tan\zeta \tag{17}$$

and the spin 4–vector multiplied by m_e (as this is what occurs in Eq. (14)) is then

$$m_e S^\mu = h\epsilon\left(\beta\cos\mu,\ \cos\mu\,\vec{u}_L + \frac{1}{\gamma}\cdot\sin\mu\,\vec{u}_\perp\right) \quad . \tag{18}$$

If the angle μ is small, then the electron is longitudinally polarized; if near 90°, then it is transversely polarized. One must remember that $\gamma = \epsilon/m_e$ is very large in all practical studies in nuclear or particle physics and hence μ near 0° or 90° implies that ζ is extremely near 0° or 90°, respectively. For the latter the scale of how close to the limits "extremely near" really is depends on the actual value of $1/\gamma$. In the case of purely longitudinal polarized electrons we have

$$\text{longitudinal polarization:} \qquad m_e S^\mu = h\epsilon(\beta, \vec{u}_L) \quad , \tag{19a}$$

while in the case of purely transversely polarized electrons we have

$$\text{transverse polarization:} \qquad m_e S^\mu = \frac{1}{\gamma}\cdot h\epsilon(0, \vec{u}_\perp) \quad . \tag{19b}$$

Thus we see that transverse polarization effects are suppressed relative to longitudinal polarization effects by a factor $1/\gamma$. Henceforth in treating situations where only the incident electron beam is

41

polarized, we shall only consider purely longitudinally polarized electrons and shall take the ERL where $\beta \to 1$, $\gamma \to \infty$:

$$K^\mu \longrightarrow \epsilon(1, \vec{u}_L) \tag{20a}$$

$$(m_e S^\mu)_L \longrightarrow h\epsilon(1, \vec{u}_L) = hK^\mu \quad ; \tag{20b}$$

that is, $h = \pm 1$ becomes the electron helicity. The leptonic tensors under these conditions become

$$2m_e^2 \eta_{\mu\nu}^{\text{unpol}} \to K_\mu K_\nu' + K_\mu' K_\nu - g_{\mu\nu} K \cdot K' \tag{21a}$$

$$2m_e^2 \eta_{\mu\nu}^{\text{pol}} \to -ih\epsilon_{\mu\nu\alpha\beta} K^\alpha K'^\beta \ , \tag{21b}$$

and we see that they are in general comparable in magnitude, both being characterized by the product of the electron energies, $\epsilon\epsilon'$.

With these developments, we are in a position to proceed with discussions of electron scattering with or without a polarized beam. One important consequence of the arguments summarized in this section is that only *longitudinally polarized electrons* are relevant for most studies in nuclear or particle physics — although an exception to this statement is discussed below. In electron storage/stretcher rings this observation has important design implications if one wishes to use polarized electrons with internal targets, since then some provision must be made to counter the $g - 2$ precession of the electron's spin and restore it to a longitudinal direction at the target.

Polarized Incident and Scattered Electrons

Let us now consider the most general situation, namely, where the incident beam of electrons is polarized and where as well the polarization of the scattered electron is assumed to be measured (see Ref. [2] for details). We then have to add to Eq. (13a) another spin projection, $\left(1 + \gamma_5 \not{S}'\right)/2$, to yield:

$$\eta_{\mu\nu} = \frac{1}{16m_e^2} \text{Tr} \left\{ \gamma_\mu \left(1 + \gamma_5 \not{S}'\right) \left(\not{K}' + m_e\right) \gamma_\nu \left(1 + \gamma_5 \not{S}\right) \left(\not{K} + m_e\right) \right\} \quad . \tag{22}$$

The trace may be evaluated to give for the leptonic tensor

$$\eta_{\mu\nu} = \frac{1}{8m_e^2} \left[P_\mu P_\nu (1 - S \cdot S') + Q^2 g_{\mu\nu} (1 + S \cdot S' - 2\Sigma \cdot \Sigma') \right.$$

$$\left. + Q^2 (\Sigma_\mu \Sigma_\nu' + \Sigma_\mu' \Sigma_\nu) + (P_\mu U_\nu + U_\mu P_\nu) + 2i\epsilon_{\mu\nu\alpha\beta} m_e (S + S')^\alpha Q^\beta \right] \quad , \tag{23}$$

where, following the nomenclature in Ref. [2], the following quantities have been defined:

$$Q_\mu \equiv K_\mu - K_\mu' \tag{24a}$$

$$P_\mu \equiv K_\mu + K_\mu' \tag{24b}$$

$$\Sigma_\mu \equiv S_\mu - (Q \cdot S/Q^2)Q_\mu \tag{24c}$$

$$\Sigma_\mu' \equiv S_\mu' - (Q \cdot S'/Q^2)Q_\mu \tag{24d}$$

$$U_\mu \equiv (Q \cdot S')\Sigma_\mu - (Q \cdot S)\Sigma_\mu' \quad . \tag{24e}$$

By construction we have that $Q \cdot \Sigma = Q \cdot \Sigma' = 0$ and hence $Q \cdot U = 0$. Also note that $Q \cdot P = K^2 - K'^2 = 0$. Finally, note that any terms proportional to Q_μ or Q_ν in the leptonic tensor here or above have been dropped, since they would lead to vanishing contributions when the leptonic and hadronic tensors are contracted, that is, using $Q_\mu W^{\mu\nu} = Q_\nu W^{\mu\nu} = 0$ from current conservation.

If no final–state polarization is measured, then all terms containing S' should be set to zero; hence $\Sigma' \to 0$, $U \to 0$ and so the only surviving contributions are the following: symmetric terms

$$\eta_{\mu\nu}^s = \frac{1}{8m_e^2}[P_\mu P_\nu + Q^2 g_{\mu\nu}] = \frac{1}{2}\eta_{\mu\nu}^{\text{unpol}} \ , \tag{25a}$$

when terms proportional to $Q_\mu Q_\nu$ are dropped (see above), and antisymmetric terms

$$\eta_{\mu\nu}^a = \frac{1}{4m_e^2}[i\epsilon_{\mu\nu\alpha\beta} \, m_e S^\alpha Q^\beta] = \frac{1}{2}\eta_{\mu\nu}^{\text{pol}} \ . \tag{25b}$$

Hence we recover the above results. If the scattered electron's polarization is measured, but the incident electron beam is unpolarized, then these same answers occur except now with S'^α replacing S^α in Eq. (25b).

What is considerably more complicated is the general situation in Eq. (23) where *both* incident and scattered electron polarizations are specified. The antisymmetric terms involve incident or scattered electron polarizations, but not both. The symmetric terms, on the other hand, have contributions where no polarizations are specified (the unpolarized results discussed above) and contributions that correspond to leptonic polarization transfer reactions, $(\vec{e}, \vec{e}\,')$. Let us examine this last class of reactions.

In the general situation the incident electron beam is polarized in some general direction specified by the angles (ζ, η) (see Eqs. (15) and Fig. 2). The scattered electron's polarization may similarly be characterized using the spin 3–vector

$$\vec{s}' = h's'\left(\cos\zeta'\vec{u}_{L'} + \sin\zeta'\vec{u}_{\perp'}\right) \ , \tag{26a}$$

where, in analogy with the treatment of the incident electron polarization, $\vec{u}_{L'}$ is a unit vector pointing along $\vec{k}'$ and $\vec{u}_{\perp'}$ is a unit vector in the plane perpendicular to $\vec{k}'$:

$$\vec{u}_{\perp'} = \cos\eta'\,\vec{u}_{S'} + \sin\eta'\,\vec{u}_{N'} \ . \tag{26b}$$

In this case, the three orthogonal unit vectors ($L' \leftrightarrow$ longitudinal, $S' \leftrightarrow$ sideways, $N' \leftrightarrow$ normal) are defined with respect to the scattered electron's momentum $\vec{k}'$ with angles (ζ', η') in analogy with the picture in Fig. 2 for the incident electron. While $\vec{u}_{N'}$ has been chosen to be equal to $\vec{u}_N$, namely, normal to the electron scattering plane, the other unit vectors $\vec{u}_{L'}$ and $\vec{u}_{S'}$ are rotated by θ_e from $\vec{u}_L$ and $\vec{u}_S$, respectively. As before, the factor $h' = \pm1$ is added only for convenience in flipping the spin. The previous results are easily extended to yield:

$$s' = |\vec{s}'| = \frac{1}{\sqrt{1 - \beta'^2 \cos^2\zeta'}} = \frac{\gamma'}{\sqrt{\cos^2\zeta' + \gamma'^2 \sin^2\zeta'}} \tag{27a}$$

$$S'^0 = h'\beta's'\cos\zeta' \ , \tag{27b}$$

where $\beta' = k'/\epsilon' = \sqrt{1 - (m_e/\epsilon')^2}$ and $\gamma' = 1/\sqrt{1 - \beta'^2} = \epsilon'/m_e$. As for polarized incident electrons, it is convenient to define

$$\tan\mu' \equiv \gamma'\tan\zeta' \tag{28}$$

and the spin 4–vector multiplied by m_e is then

$$m_e S'^\mu = h'\epsilon'\left(\beta'\cos\mu', \ \cos\mu'\,\vec{u}_{L'} + \frac{1}{\gamma'}\cdot\sin\mu'\,\vec{u}_{\perp'}\right) \ . \tag{29}$$

Substituting Eqs. (18) and (29) into (23) yields the general leptonic tensor. Rather than writing out the complete result let us specialize to the following situation: (1) retain only two particular combinations of Lorentz indices (see below) and (2) again take the ERL.[†] For the cases defined in Eqs. (10) we obtain

$$\ell_L = \frac{1}{2}v_0 v_L\left[1 + hh'\left\{\cos\mu\,\cos\mu' + \sin\mu\,\sin\mu'\,\cos(\eta - \eta')\right\}\right] \tag{30a}$$

$$\ell_T = \frac{1}{2}v_0\left[v_T\left(1 + hh'\left\{\cos\mu\,\cos\mu' + \sin\mu\,\sin\mu'\,\cos(\eta - \eta')\right\}\right)\right.$$
$$\left. - hh'\tan^2\theta_e/2\,\sin\mu\,\sin\mu'\,\cos(\eta - \eta')\right] \ . \tag{30b}$$

We shall return to discuss some implications of these results in the next section. Here let us only note the following: (1) both transverse and longitudinal polarizations now play a role, in contrast to the single–polarized situations where the former are suppressed; (2) only specific polarization transfers

---

[†] I would like to thank M. A. Titko for re–deriving some of the formalism in this section.

occur, namely, $[L \rightarrow L']$ involving terms with cosine factors and $[S \rightarrow S'] + [N \rightarrow N']$ involving terms with sine factors; (3) as above, μ or μ' small corresponds to ζ or ζ' extremely small and μ or μ' near 90° corresponds to ζ or ζ' extremely close to 90° because of the large factors γ and γ'. As a consequence of (3) the usual situation for polarization transfer is *purely transverse* — only for both ζ and ζ' extremely close to zero (typically a small fraction of a milliradian) do the LL' terms contribute significantly. For purely transversely polarized electrons we have

$$\ell_L = \frac{1}{2} v_0 v_L \left[1 + hh' \cos \Delta \eta \right] \tag{31a}$$

$$\ell_T = \frac{1}{2} v_0 \left[v_T + \frac{1}{2} |Q^2/q^2| hh' \cos \Delta \eta \right] \ . \tag{31b}$$

As we shall see in the next section, in principle it is possible to use the dependence on $\Delta \eta \equiv \eta - \eta'$ above (which is different in ℓ_L and ℓ_T) to separate responses without changing the electron scattering angle.

III. INCLUSIVE ELECTRON SCATTERING

Unpolarized Single–Arm Electron Scattering

Let us begin our discussions of hadronic response functions by repeating the familiar arguments (see Refs. [2,5–9]) which pertain in the case of single–arm parity–conserving electron scattering from unpolarized targets, $A(e, e')$ and $A(\vec{e}, e')$. The nuclear vertex involved is shown in Fig. 3. Here the virtual photon from the electron scattering (Fig. 1) brings in 4–momentum Q and causes the nucleus to go from state $|i\rangle$ to state $|f\rangle$. These hadronic states have 4–momenta P_i and P_f, respectively, as shown in the figure. Thus, we must build the hadronic tensors from Q, P_i and P_f. In fact, we can use momentum conservation to eliminate one, say $P_f = P_i + Q$ leaving two independent 4–momenta: $\{Q, P_i\}$. The possible Lorentz scalars in the problem are Q^2, P_i^2 and $Q \cdot P_i$. Since we presumably know what the target is and since $P_i^2 = M_i^2$, we are left with two independent scalars to vary: $\{Q^2, Q \cdot P_i\}$. Moreover, since $Q^2 = \omega^2 - q^2$ and $Q \cdot P_i = \omega M_i$ in the laboratory system, we can regard our hadronic tensors to be functions of $\{Q^2, Q \cdot P_i\}$ or $\{q, \omega\}$.

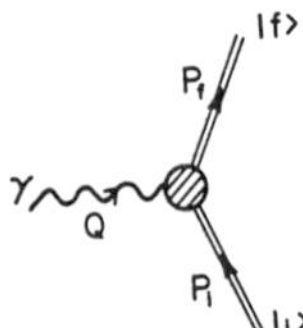

Fig. 3.   Hadronic vertex for inclusive scattering.

Now we wish to write $W_s^{\mu\nu}$ and $W_a^{\mu\nu}$ in terms of the two independent 4–vectors Q^μ and P_i^μ. Alternatively, instead of P_i it turns out to be useful to employ the 4–vector

$$V_i^\mu \equiv \frac{1}{M_i} \left\{ P_i^\mu - \left(\frac{Q \cdot P_i}{Q^2} \right) Q^\mu \right\} \ , \tag{32}$$

which is especially convenient because $Q \cdot V_i = 0$, by construction. To begin with, $W^{\mu\nu}$ must be a second–rank Lorentz tensor and so we can write the following general expansions:

$$W_s^{\mu\nu} = X_1 g^{\mu\nu} + X_2 Q^\mu Q^\nu + X_3 V_i^\mu V_i^\nu + X_4 \left(Q^\mu V_i^\nu + V_i^\mu Q^\nu \right) \tag{33a}$$

$$W_a^{\mu\nu} = i Y_1 \left(Q^\mu V_i^\nu - V_i^\mu Q^\nu \right) + i Y_2 \epsilon^{\mu\nu\alpha\beta} Q_\alpha V_{i\beta} \ , \tag{33b}$$

where the scalar response functions (the X's and Y's) depend on the scalars discussed above:

$$X_\ell = X_\ell \left(Q^2, Q \cdot P_i \right) \ , \qquad \ell = 1, \ldots, 4$$

$$Y_\ell = Y_\ell \left(Q^2, Q \cdot P_i \right) \ , \qquad \ell = 1, 2 \ .$$

Next we note that, in the absence of parity–violating effects from the weak interaction, the hadronic electromagnetic current matrix elements are polar vectors and so the tensors here must have specific properties under spatial inversion. In particular, the ϵ–terms in Eq. (33b) have the wrong behavior and so Y_2 must vanish. Finally, we must make use of the current conservation conditions, Eqs. (6a). These lead us to the following expressions (recall that $Q \cdot V_i = 0$)

$$Q_\mu W_s^{\mu\nu} = 0 = (X_1 + X_2 Q^2)Q^\nu + (X_4 Q^2)V_i^\nu \tag{34a}$$

$$Q_\mu W_a^{\mu\nu} = 0 = (Y_1 Q^2)V_i^\nu \tag{34b}$$

and so $X_1 + X_2 Q^2 = 0$, $X_4 = 0$ and $Y_1 = 0$, using the linear independence of Q^ν and V_i^ν. Defining $W_1 \equiv -X_1$ and $W_2 \equiv X_3$, to use more common nomenclature, we have then rederived the familiar results

$$W_s^{\mu\nu} = -W_1 \left(g^{\mu\nu} - \frac{Q^\mu Q^\nu}{Q^2} \right) + W_2 V_i^\mu V_i^\nu \tag{35a}$$

$$W_a^{\mu\nu} = 0 \ . \tag{35b}$$

Contracting these hadronic tensors with the leptonic tensors obtained above (see Sec. II, Eqs. (21)), we obtain for unpolarized electron scattering

$$\eta_{\mu\nu}^{\mathrm{unpol}} W_s^{\mu\nu} \sim W_2 + 2W_1 \tan^2 \frac{\theta_e}{2} \ , \tag{36a}$$

where of course only the symmetric responses enter. This familiar form for inclusive unpolarized electron scattering (see Ref. [9]) contains only two independent response functions W_1 and W_2 which may be separated at fixed q and ω (or, equivalently, fixed Q^2 and $Q \cdot P_i$) by varying the electron scattering angle to make a Rosenbluth decomposition. On the other hand, for polarized electron scattering

$$\eta_{\mu\nu}^{\mathrm{pol}} W_a^{\mu\nu} = 0 \ , \tag{36b}$$

where now only the antisymmetric responses enter. Thus, if only parity–conserving interactions are considered and if the target is unpolarized, no differences will be seen when the electron's helicity is flipped from $+1$ to -1.

In passing, let us go a little further with the unpolarized scattering problem. When we contract the unpolarized leptonic tensor developed in the preceding section with the general hadronic tensor for inclusive electron scattering from unpolarized nuclei we obtain a sum involving projections of the current matrix elements. It is convenient to choose these to be transverse (T) or longitudinal(L) with respect to the direction $\vec{q}$. Note that "L" used here is not the same as "L" used in Sec. II; the correct choice is usually clear from the context. Thus we obtain a structure of the form $\ell_L W^L + \ell_T W^T$ (see Eqs. (10,11)) and can write the inclusive cross section in the laboratory system as

$$\left(\frac{d\sigma}{d\Omega_e d\omega} \right)^{\mathrm{unpol}} = \frac{1}{M_i} \sigma_{\mathrm{Mott}} \left\{ v_L W^L + v_T W^T \right\} \tag{37a}$$

$$= \frac{\rho}{2M_i \mathcal{E}} \sigma_{\mathrm{Mott}} \left\{ 2\rho \mathcal{E} W^L + W^T \right\} \ , \tag{37b}$$

using the ERL and hence Eqs. (12). Here the Mott cross section is given by

$$\sigma_{\mathrm{Mott}} = \left\{ \frac{\alpha \cos \frac{\theta_e}{2}}{2\epsilon \sin^2 \frac{\theta_e}{2}} \right\}^2 \tag{38}$$

and we have introduced the quantity

$$\mathcal{E} \equiv \left[1 + \frac{2}{\rho} \tan^2 \frac{\theta_e}{2} \right]^{-1} \ , \tag{39}$$

usually referred to as the photon's degree of longitudinal polarization to represent the "virtualness" of the exchanged photon. W^L and W^T are related to W_1 and W_2 by

$$W^L = \left(W_2 - \rho W_1\right)/\rho^2 \tag{40a}$$

$$W^T = 2W_1 \ . \tag{40b}$$

If discrete states are involved, then a single–differential cross section can be obtained (see Refs. [2,5]):

$$\left(\frac{d\sigma}{d\Omega_e}\right)^{\text{unpol}} = 4\pi\sigma_{\text{Mott}}f_{\text{rec}}^{-1}\left\{v_L F_L^2(q) + v_T F_T^2(q)\right\} \tag{41a}$$

$$= 4\pi\sigma_{\text{Mott}}f_{\text{rec}}^{-1}F^2(q,\theta_e) \ , \tag{41b}$$

where the recoil factor is given by $f_{\text{rec}} = 1 + \frac{2\epsilon}{M_i}\sin^2\frac{\theta_e}{2}$ and where the form factors F_L^2 and F_T^2 result from integrating the responses W^L and W^T (the form factors are only functions of q, where ω is fixed by the excitation energy and the momentum transfer). For given angular momentum and parity quantum numbers only a finite set of multipole form factors occur:

$$F_L^2(q) = \sum_{J \geq 0} F_{CJ}^2(q) \tag{42a}$$

$$F_T^2(q) = \sum_{J \geq 1} \left\{F_{EJ}^2(q) + F_{MJ}^2(q)\right\} \ , \tag{42b}$$

where the sums are restricted to $|J_f - J_i| \leq J \leq J_f + J_i$ (in addition to $J \geq 0$ for Coulomb multipoles and $J \geq 1$ for transverse electric and magnetic multipoles), where natural parity multipoles CJ/EJ are present only when $(-1)^J = \pi \equiv \pi_f \pi_i$ (non–natural parity multipoles MJ only when $(-1)^J = -\pi$), and where no electric multipoles are present at all for elastic scattering ($f = i$). For more detailed discussions see Refs. [2,5,6]. By fixing q (and hence ω) and varying θ_e, it is possible to make a Rosenbluth decomposition of Eq. (41a) and separately determine $F_L^2(q)$ and $F_T^2(q)$. However, in general more terms than one occur in the sums in Eqs. (42) and so the individual multipole form factors cannot be extracted.

As discussed in Ref. [2], when polarized targets are involved or when final–state nuclear polarizations are determined, a much richer variety of polarization observables becomes accessible. In general these contain interferences between the various form factors and consequently a complete decomposition into the underlying electromagnetic matrix elements can in principle be achieved (up to a simple phase ambiguity in the arbitrary spin case — see Ref. [2]).

The subject of inclusive electron scattering from polarized targets and the related subject of scattering with final–state polarizations determined were discussed in previous Summer School proceedings (see for example Ref. [8]) and so will not be repeated here. Instead, let us turn to a rather specialized aspect of single–arm electron scattering from unpolarized targets.

The Reaction $A(\vec{e},\vec{e}\,')$: Leptonic Polarization Transfer

Let us briefly consider the form taken by the inclusive cross section when *both* incident and scattered electrons have their polarizations determined; specifically, let us assume that the incident beam is transversely polarized and that the scattered electron's polarization is measured (the general case discussed at the end of Sec. II). We again have a structure similar to Eq. (37), namely, $\sim \ell_L W^L + \ell_T W^T$ involving the same longitudinal and transverse response functions as in the unpolarized case. Here the difference is that the leptonic factors are those given in Eqs. (31) (or in general in Eqs. (30)) rather than the unpolarized quantities given in Eqs. (11). The polarization transfer double–differential cross section then becomes (Cf. Eq. (37)):

$$\left(\frac{d\sigma}{d\Omega_e d\omega}\right)^{\text{pol}} = \frac{\sigma_{\text{Mott}}}{2M_i}\left\{v_L[1 + hh'\cos\Delta\eta]W^L + [v_T + hh'\frac{1}{2}\rho\cos\Delta\eta]W^T\right\} \ , \tag{43}$$

which may be written using Eqs. (40) in the form

$$\left(\frac{d\sigma}{d\Omega_e d\omega}\right)^{\text{pol}} = \frac{\sigma_{\text{Mott}}}{2M_i}\left\{[W_2 + 2W_1\tan^2\frac{\theta_e}{2}] + hh'\cos\Delta\eta W_2\right\} \ . \tag{44}$$

Let us fix $h' = +1$ and define the asymmetry obtained by flipping the incident electron's helicity:

$$\begin{aligned}
\mathcal{A}(\Delta\eta) &\equiv \frac{\sigma(h = +1, \Delta\eta) - \sigma(h = -1, \Delta\eta)}{\sigma(h = +1, \Delta\eta) + \sigma(h = -1, \Delta\eta)} \\
&= \frac{\cos\Delta\eta\ W_2}{W_2 + 2W_1\tan^2\frac{\theta_e}{2}} \\
&= \cos\Delta\eta\left(\frac{\mathcal{E}(1+x)}{1+\mathcal{E}x}\right) \ ,
\end{aligned} \tag{45}$$

where $x \equiv 2\rho\ W^L/W^T$. As $\mathcal{E} \to 0$ $(\theta_e \to 180°)$ the asymmetry goes to zero; as $\mathcal{E} \to 1$ $(\theta_e \to 0°)$ the asymmetry goes to $\cos\Delta\eta$. Consequently, to obtain information on the longitudinal–to–transverse ratio x using transversely polarized electrons, it is necessary to work at some scattering angle intermediate between 0° and 180°. In principle, by measuring the asymmetry defined here it is possible to determine x and hence the ratio of W^L and W^T at a fixed angle and so avoid making a Rosenbluth decomposition (with its attendant systematic errors).

IV. TWO–ARM COINCIDENCE ELECTRON SCATTERING

Exclusive–1 Electron Scattering, No Hadronic Polarizations

Next let us repeat the arguments that lead to the general form for the reactions $A(e, e'x)$ and $A(\vec{e}, e'x)$ — see also Refs. [3,7]. Here a particle with 4–momentum P_x is detected in coincidence with the scattered electron. Let us again assume that the total final nuclear state has 4–momentum P_f, so that all but the particle in coincidence (*i.e.*, the unobserved particles) have 4–momentum P_0 (see Fig. 4 — for generalization to multi–particle coincidence in Sec. VI, P_x is labelled P_1 in Figs. 4 and 5). Momentum conservation allows us to eliminate these two ($P_f = P_i + Q$ and $P_0 = P_f - P_x$), so that we are left with three independent 4–momenta from which to build the hadronic tensor: $\{Q, P_i, P_x\}$.

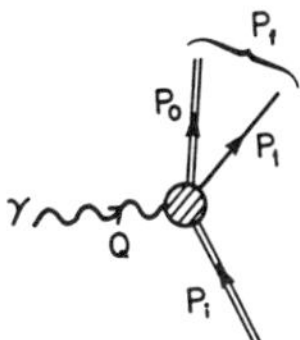

Fig. 4. Hadronic vertex for exclusive–1 electron scattering, $(e, e'x)$.

The possible Lorentz scalars in the problem are Q^2, P_i^2, P_x^2, $Q\cdot P_i$, $P_x\cdot P_i$ and $Q\cdot P_x$. Again, we know the target mass, M_i and the mass of the detected particle, M_x, and so, since $P_i^2 = M_i^2$ and $P_x^2 = M_x^2$ are fixed, we are left with four independent scalars to vary: $\{Q^2, Q\cdot P_i, P_x\cdot P_i, Q\cdot P_x\}$. It is useful to express the kinematical dependence in the response functions in terms of laboratory system variables. We shall use the coordinate system in Fig. 5 to describe exclusive–1 electron scattering in which the detected particle has 3–momentum $p_x = |\vec{p}_x|$ and is detected at angles (θ_x, ϕ_x) with respect to the chosen basis. The four scalar variables may be re–expressed in this laboratory system:

$$\begin{aligned}
Q^2 &= \omega^2 - q^2 \\
Q\cdot P_i &= \omega M_i \\
P_x\cdot P_i &= E_x M_i \\
Q\cdot P_x &= \omega E_x - q p_x\cos\theta_x \ ,
\end{aligned} \tag{46}$$

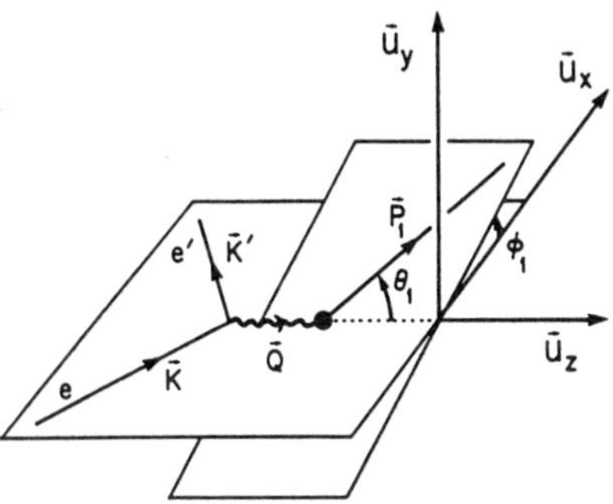

Fig. 5. Two–arm coincidence reactions, $A(e, e'x)$ and $A(\vec{e}, e'x)$. The direction in which particle x is detected is specified by the angles (θ_x, ϕ_x).

where $E_x = \sqrt{p_x^2 + M_x^2}$ is the total energy of the detected particle. Note that the azimuthal angle ϕ_x, does not occur here; that is, the internal functional dependence involves θ_x, but not ϕ_x. Thus, the hadronic tensors for double coincidence reactions can be regarded to be functions of the scalar variables $\{Q^2,\, Q \cdot P_i,\, P_x \cdot P_i,\, Q \cdot P_x\}$ or of the laboratory quantities $\{q,\, \omega,\, E_x,\, \theta_x\}$.

Now, as in the case of inclusive scattering, we wish to write $W_s^{\mu\nu}$ and $W_a^{\mu\nu}$ in terms of the independent 4–momenta in the problem. Instead of $\{Q^\mu,\, P_i^\mu,\, P_x^\mu\}$ we use the equivalent set $\{Q^\mu,\, V_i^\mu,\, V_x^\mu\}$, where (see Refs. [7,8])

$$V_i^\mu \equiv \frac{1}{M_i}\left\{ P_i^\mu - \left(\frac{Q \cdot P_i}{Q^2}\right) Q^\mu \right\} \tag{47a}$$

$$V_x^\mu \equiv P_x^\mu - \left(\frac{Q \cdot P_x}{Q^2}\right) Q^\mu \ , \tag{47b}$$

and these have the convenient properties, $Q \cdot V_i = Q \cdot V_x = 0$. Since $W^{\mu\nu}$ must be a second–rank Lorentz tensor, we have the following expansions:

$$\begin{aligned}
W_s^{\mu\nu} &= X_1 g^{\mu\nu} + X_2 Q^\mu Q^\nu + X_3 V_i^\mu V_i^\nu + X_4 V_x^\mu V_x^\nu \\
&\quad + X_5 (Q^\mu V_i^\nu + V_i^\mu Q^\nu) + X_6 (Q^\mu V_x^\nu + V_x^\mu Q^\nu) \\
&\quad + X_7 (V_i^\mu V_1^\nu + V_1^\mu V_i^\nu) \\
W_a^{\mu\nu} &= iY_1 (Q^\mu V_i^\nu - V_i^\mu Q^\nu) + iY_2 (Q^\mu V_x^\nu - V_x^\mu Q^\nu) \\
&\quad + iY_3 (V_i^\mu V_x^\nu - V_x^\mu V_i^\nu) \ ,
\end{aligned} \tag{48a, 48b}$$

where we have not written any ϵ–terms (Cf. Eq. (33b)) using the parity properties of the electromagnetic tensor ($\sim VV$). The scalar response functions depend on the scalars discussed above:

$$\begin{aligned}
X_\ell &= X_\ell(Q^2,\, Q \cdot P_i,\, P_x \cdot P_i,\, Q \cdot P_x) \ , \quad \ell = 1,\ldots,7 \\
Y_\ell &= Y_\ell(Q^2,\, Q \cdot P_i,\, P_x \cdot P_i,\, Q \cdot P_x) \ , \quad \ell = 1,\ldots,3 \ .
\end{aligned}$$

Finally, we use the current conservation conditions (Eqs. (6a)) to obtain the constraints

$$Q_\mu W_s^{\mu\nu} = 0 = (X_1 + X_2 Q^2) Q^\nu + (X_5 Q^2) V_i^\nu + (X_6 Q^2) V_x^\nu \tag{49a}$$

$$Q_\mu W_a^{\mu\nu} = 0 = (Q^2 Y_1) V_i^\nu + (Q^2 Y_2) V_x^\nu \ , \tag{49b}$$

48

and so $X_1 + X_2 Q^2 = 0$, $X_5 = 0$, $X_6 = 0$, and $Y_1 = 0$, $Y_2 = 0$ using the linear independence of Q^ν, P_i^ν and P_x^ν. For the symmetric hadronic tensor we obtain

$$
\begin{aligned}
W_s^{\mu\nu} = X_1 \left(g^{\mu\nu} - \frac{Q^\mu Q^\nu}{Q^2} \right) + X_3 V_i^\mu V_i^\nu \\
+ X_4 V_x^\mu V_x^\nu + X_7 (V_i^\mu V_x^\nu + V_x^\mu V_i^\nu) \; ,
\end{aligned}
\tag{50a}
$$

in which there are four independent response functions, to be compared with inclusive scattering (Eq. (35a)) in which only two appeared. For the antisymmetric tensor we obtain

$$
W_a^{\mu\nu} = i Y_3 (V_i^\mu V_x^\nu - V_x^\mu V_i^\nu) \; ,
\tag{50b}
$$

in which a fifth response function appears, in contrast to inclusive scattering (Eq. (35b)) where the antisymmetric tensor vanishes.

When these hadronic tensors are contracted with the previous results given in Eqs. (21) for the electron scattering leptonic tensor (we assume the ERL here for simplicity), we obtain

$$
\eta_{\mu\nu} W^{\mu\nu} = \eta_{\mu\nu}^s W_s^{\mu\nu} + \eta_{\mu\nu}^a W_a^{\mu\nu} \; ,
\tag{51}
$$

where for the symmetric contributions to exclusive–1 scattering we have

$$
\eta_{\mu\nu}^s W_s^{\mu\nu} \sim v_L \, W_{(x)}^L + v_T \, W_{(x)}^T + v_{TL} \, W_{(x)}^{TL} \cos\phi_x + v_{TT} \, W_{(x)}^{TT} \cos 2\phi_x,
\tag{52a}
$$

where the notation "(x)" is used to indicate that *one* particle $x = p, n, \pi, \gamma, K, \ldots$ is detected in coincidence with the scattered electron. Now, in addition to longitudinal (L) and transverse (T) pieces (Cf. Eq. (37)), we also have a transverse–transverse interference (TT) and a transverse–longitudinal interference (TL). Of course, the four response functions here are linear combinations of X_1, X_3, X_4 and X_7, and depend on the four scalar quantities discussed above. For the antisymmetric contribution we have

$$
\eta_{\mu\nu}^a W_a^{\mu\nu} \sim h v_{TL'} \, W_{(x)}^{TL'} \sin\phi_x \; ,
\tag{52b}
$$

which is directly proportional to Y_3. The general set of electron kinematical factors (the v's) needed are the following (Cf. Eqs. (12) and see Ref. [2]):

$$
\begin{aligned}
v_L &= \rho^2 \\
v_T &= \frac{1}{2}\rho + \tan^2 \frac{\theta_e}{2} \\
v_{TL} &= -\frac{1}{\sqrt{2}}\rho \sqrt{\rho + \tan^2 \frac{\theta_e}{2}} \\
v_{TT} &= -\frac{1}{2}\rho \\
v_{T'} &= \sqrt{\rho + \tan^2 \frac{\theta_e}{2}} \, \tan \frac{\theta_e}{2} \\
v_{TL'} &= -\frac{1}{\sqrt{2}}\rho \tan \frac{\theta_e}{2} \; .
\end{aligned}
\tag{53}
$$

To obtain the inclusive cross section from these results we must integrate over the angle dependence (θ_x, ϕ_x) and sum over all open channels. Indeed, as is apparent from Eqs. (52), integrating over the explicit ϕ_x–dependence immediately causes the TT, TL, and TL' contributions to vanish, leaving only the L and T pieces as expected (see Eq. (37)).

Kinematics of Exclusive–1 Electron Scattering; Inclusive Scattering

The various kinematic quantities involved in 2–arm coincidence electron scattering reactions are indicated in Fig. 6. Here we shall discuss the general kinematics of A(e,e′N), with N = p or n — below in Sec. V we return to outline the ideas behind the plane–wave impulse approximation[10–12] which provides a simple, tractable model for the more general reaction process. The hadronic variables are the following: $P_A^\mu \equiv (M_A, \vec{0})$, $P_B^\mu \equiv (E_B, \vec{p}_B)$, $P_N^\mu \equiv (E_N, \vec{p}_N)$ are the 4–momenta of the target, residual nucleus and emitted nucleon, respectively. We have the relationships $E_B = \sqrt{p_B^2 + M_B^2}$ and $E_N = \sqrt{p_N^2 + M_N^2}$, where M_N is the nucleon mass, M_B is the rest–mass of the residual nucleus (and includes any internal excitation energy in that system), $p_B \equiv |\vec{p}_B|$ and $p_N \equiv |\vec{p}_N|$. The general cross section for exclusive electron scattering in the laboratory system (see the preceding section) can be written as

$$\frac{d\sigma}{d\Omega_e\, d\omega\, d\Omega_N\, dE_N} = \frac{2\alpha^2}{Q^4}\left(\frac{\epsilon'}{\epsilon}\right)\frac{p_N M_N M_B}{(2\pi)^3 E_B}\,\eta_{\mu\nu} W^{\mu\nu}\,\delta(E_N + E_B - M_A - \omega) \ . \tag{54}$$

Here $\eta_{\mu\nu}$ is the familiar leptonic tensor in the ERL discussed in Sec. II, whereas $W^{\mu\nu}$ is the hadronic tensor containing all of the nuclear structure and dynamics information, treated above. The latter is given in terms of the nuclear electromagnetic transition currents in momentum space by (see Eq. (2b))

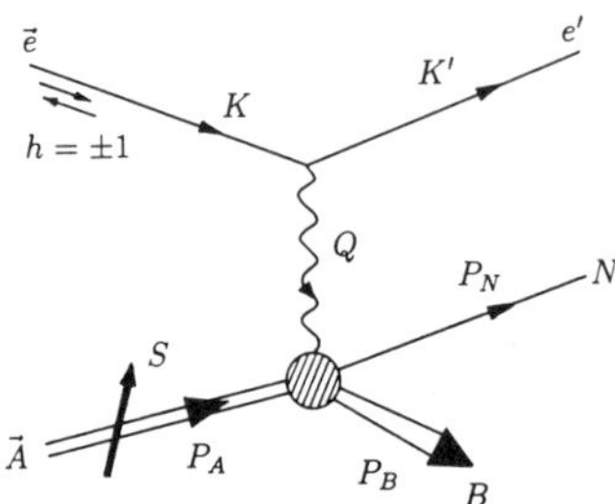

Fig. 6.   Feynman diagram for the exclusive electromagnetic $\vec{A}(\vec{e}, e'N)B$ electron scattering process in the one–photon–exchange approximation.

$$W^{\mu\nu} = \overline{\sum_{i,f}}\langle f|\hat{J}^\mu|i\rangle^*\langle f|\hat{J}^\nu|i\rangle \ , \tag{55}$$

where $|i\rangle$ represents the nuclear target state and $|f\rangle$ the final state characterized by the emitted nucleon and the residual nucleus. Conservation of the nuclear electromagnetic current implies that only three of the components of the 4–current J^μ are independent quantities:

$$Q^0 J^0(\vec{q})_{fi} - \vec{q}\cdot\vec{J}(\vec{q})_{fi} = \omega\rho(\vec{q})_{fi} - q J^z(\vec{q})_{fi} = 0 \ , \tag{56}$$

where the z–axis has been chosen to be along $\vec{q}$. Integrating the delta function of energy over E_N, we obtain

$$\frac{d\sigma}{d\Omega_e\, d\omega\, d\Omega_N} = \frac{2\alpha^2}{Q^4}\left(\frac{\epsilon'}{\epsilon}\right)\frac{p_N M_N M_B}{(2\pi)^3 M_A}\,f_{\rm rec}^{-1}\eta_{\mu\nu} W^{\mu\nu} \ , \tag{57}$$

where now p_N is specified by solving the energy balance equation

$$\sqrt{p_N^2 + M_N^2} + \sqrt{p_N^2 - 2p_N q\cos\theta_N + q^2 + M_B^2} = M_A + \omega \tag{58}$$

with θ_N being the angle between $\vec{p}_N$ and $\vec{q}$ (see Fig. 7). The general expression for the contraction of the electron and hadronic tensors can be written

$$2\eta_{\mu\nu} W^{\mu\nu} = v_0\overline{\sum_{i,f}}[R_{fi} + h R'_{fi}] \ . \tag{59}$$

Using the general properties of the leptonic tensor, as discussed above, it may be shown that these total responses have the following decompositions:

$$R_{fi} = v_L W_{fi}^L + v_T W_{fi}^T + v_{TL} W_{fi}^{TL} + v_{TT} W_{fi}^{TT} \tag{60a}$$

$$R'_{fi} = v_{T'} W_{fi}^{T'} + v_{TL'} W_{fi}^{TL'} \quad , \tag{60b}$$

where the W's are discussed in the last section (Cf. also Eqs. (52)). The differential cross section in Eq. (57) may now be written

$$\frac{d\sigma^h}{d\omega d\Omega_e d\Omega_N} = \frac{p_N M_N M_B}{(2\pi)^3 M_A} \sigma_{\text{Mott}} f_{\text{rec}}^{-1} \left\{ R_{fi} + h R'_{fi} \right\} \tag{61}$$

$$\equiv \Sigma_{fi} + h \Delta_{fi} \quad ,$$

containing the helicity–sum (electron unpolarized) cross section

$$\Sigma_{fi} = \frac{1}{2} \left[\frac{d\sigma^+}{d\omega d\Omega_e d\Omega_N} + \frac{d\sigma^-}{d\omega d\Omega_e d\Omega_N} \right] \tag{62a}$$

and the helicity–difference (electron polarized) cross section

$$\Delta_{fi} = \frac{1}{2} \left[\frac{d\sigma^+}{d\omega d\Omega_e d\Omega_N} - \frac{d\sigma^-}{d\omega d\Omega_e d\Omega_N} \right] \quad . \tag{62b}$$

In particular, later we shall be interested in the situation where only the initial electron and target nucleus are possibly polarized, with no polarization in the final state (indicated by fi). Furthermore, the only final–state information (aside from knowing that q and ω have been transferred to the nucleus and hence that only certain final states are accessible) is contained in the fact that a nucleon is presumed to be detected at specific angles (θ_N, ϕ_N) and has energy E_N (see Fig. 7). Under these conditions it is possible to show (see Ref. [3]) that the dependences on the azimuthal angles ϕ_N and ϕ^* (the azimuthal angle of the target polarization axis of quantization) are the following:

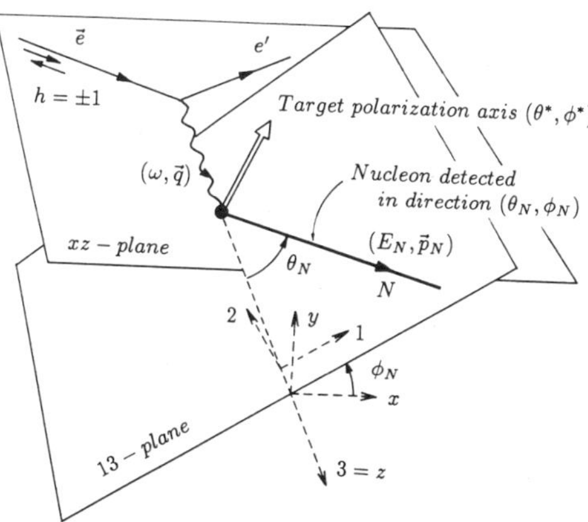

Fig. 7. Kinematics for two–arm coincidence reactions with polarized targets. $\vec{u}_z$ is along $\vec{q}$, $\vec{u}_y$ is normal to the electron scattering plane and $\vec{u}_x = \vec{u}_y \times \vec{u}_z$ lies in the scattering plane. The target polarization direction is specified by the angles (θ^*, ϕ^*) in this coordinate system. The unit vectors $\vec{u}_1$, $\vec{u}_2$, $\vec{u}_3$ are constructed by rotating the previous ones $(\vec{u}_x, \vec{u}_y, \vec{u}_z)$ by an angle ϕ_N: $\vec{u}_3 \equiv \vec{u}_z$, $\vec{u}_1$ lies in the nucleonic plane and $\vec{u}_2$ is perpendicular to it.

$$\Sigma_{fi} \sim v_L W^L_{fi}(\Delta\phi) + v_T W^T_{fi}(\Delta\phi)$$
$$+ v_{TL}\left(\cos\phi_N W^{TL}_{fi}(\Delta\phi) + \sin\phi_N \tilde{W}^{TL}_{fi}(\Delta\phi)\right) \tag{63a}$$
$$+ v_{TT}\left(\cos 2\phi_N W^{TT}_{fi}(\Delta\phi) + \sin 2\phi_N \tilde{W}^{TT}_{fi}(\Delta\phi)\right)$$

and

$$\Delta_{fi} \sim v_{T'} \tilde{W}^{T'}_{fi}(\Delta\phi)$$
$$+ v_{TL'}\left(\sin\phi_N W^{TL'}_{fi}(\Delta\phi) + \cos\phi_N \tilde{W}^{TL'}_{fi}(\Delta\phi)\right) \ , \tag{63b}$$

where each response depends on q, ω, p_N (or E_N) and θ_N as well as on the target polarization angles θ^* and $\Delta\phi$. It should be noted that these results have been expressed in terms of $\Delta\phi \equiv \phi^* - \phi_N$, where the direction of the target polarization is referred to the nucleon plane rather than to the electron plane.

Having set up the basic form for the exclusive–1 cross section, let us next consider the kinematics in more detail. In this section we shall simply use energy–momentum conservation to inter–relate the various energies and momenta; no assumption is made about the reaction mechanism. In the next section we shall return to discuss the PWIA as a simple realization of these more general considerations. We have seen above that the cross section depends on a set of kinematic variables. Explicit dependences on the electron scattering angle θ_e (through the generalized Rosenbluth factors v_K in Eq. (53)) and an azimuthal angle ϕ_N (see Eqs. (52); in the present discussions we shall suppress the target polarization dependences, although they are straightforward to incorporate — see Eqs. (63)) have been isolated and will no longer be discussed. More complicated are the dependences on the set of "dynamical" (in contrast to "geometrical") variables $\{q, \omega, E_N, \theta_N\}$ or equivalently $\{Q^2, Q\cdot P_A, P_N\cdot P_A, Q\cdot P_N\}$ (compare Eqs. (46)), since these dependences involve detailed aspects of the nuclear dynamics. However, while these sets of dynamical variables are, of course, completely usable and indeed natural from an experimental point of view, as we shall see in the following it turns out that alternative sets of four quantities are more convenient when studying the specifics of the cross section. Let us begin with the set $\{q, \omega, E_N, \theta_N\}$ and make a sequence of changes of variable. First, let us define $\vec{p} \equiv -\vec{p}_B = \vec{p}_N - \vec{q}$ so that the daughter energy becomes $E_B = \sqrt{M_B^2 + p^2}$. Instead of E_N we will now use p. This is completely general and has nothing to do with the PWIA, although in the next section we shall see that this is a convenient variable to use. Clearly this momentum characterizes the split in momentum flow between the detected nucleon and the unobserved daughter nucleus. We need some corresponding energy variable to characterize the split in energy; several possibilities come to mind. One choice is to introduce the energy

$$\mathcal{E}^* \equiv M_B - M_B^0 \geq 0 \ , \tag{64}$$

to represent the excitation of the residual nucleus; here M_B includes the internal excitation energy of the B system while M_B^0 is its rest mass when it is in its ground state. By construction $\mathcal{E}^*$ is greater than or equal to zero — and equal to zero when the daughter nucleus is left in its ground state. As we shall see below, we could now use $\mathcal{E}^*$ and p in place of E_N and θ_N, although it may be shown that still another choice for the energy is preferable for certain purposes than $\mathcal{E}^*$, namely

$$\mathcal{E} \equiv E_B - E_B^0 \geq 0$$
$$= \mathcal{E}^* \times \left[\frac{M_B + M_B^0}{\sqrt{M_B^2 + p^2} + \sqrt{M_B^0{}^2 + p^2}}\right] \tag{65}$$
$$\rightarrow \mathcal{E}^* \quad \text{for } p^2 \ll M_B^0{}^2 \ ,$$

where as before $E_B = \sqrt{M_B^2 + p^2}$ and now also $E_B^0 = \sqrt{M_B^0{}^2 + p^2}$. As the last line in Eq. (65) demonstrates, usually $\mathcal{E}$ and $\mathcal{E}^*$ do not differ very much. Let us call $\mathcal{E}^*$ the "daughter excitation energy" and $\mathcal{E}$ the "daughter energy difference".

Another energy in the problem is the separation energy (or "Q–value"), given by $E_s \equiv M_N + M_B^0 - M_A$, the minimum energy needed to separate the nucleus A into a nucleon and the residual nucleus B in its ground state. If the recoil kinetic energy corresponding to the (in general excited) residual nucleus is denoted $T_{\text{rec}} = E_B - M_B$ and likewise $T_{\text{rec}}^0 = E_B^0 - M_B^0$ for the residual nucleus in its ground state, then what is sometimes called the missing energy (note that there are still other definitions for this quantity) is given by

$$
\begin{aligned}
E_m &= M_N + M_B - M_A \\
&= E_s + \mathcal{E}^* \\
&= E_s + \mathcal{E} + (T_{\text{rec}}^0 - T_{\text{rec}}) \\
&\cong E_s + [1 - \frac{p^2}{2M_B^0 M_B}]^{-1}\mathcal{E} \ ,
\end{aligned}
\tag{66}
$$

where the last approximation pertains when $p^2 \ll 2M_B^0 M_B$ (which is usually the case). Now overall energy conservation yields a value for E_N for given ω:

$$
E_N = M_A + \omega - E_B \ .
\tag{67}
$$

Since $E_B = E_B^0 + \mathcal{E}$, we have an equation for $\mathcal{E}$ in terms of q, ω, p and the angle θ (between $\vec{p}$ and $\vec{q}$):

$$
\mathcal{E} = M_A + \omega - \sqrt{M_N^2 + p^2 + q^2 + 2pq\cos\theta} - \sqrt{{M_B^0}^2 + p^2} \ ,
\tag{68}
$$

which just a re-writing of Eq. (58). Thus there are clear relationships between the sets $\{E_N, \theta_N\}$ and $\{p, \theta\}$ and hence $\{\mathcal{E}, p\}$. Instead of the first set, we shall now use the last set as a pair of dynamical variables.

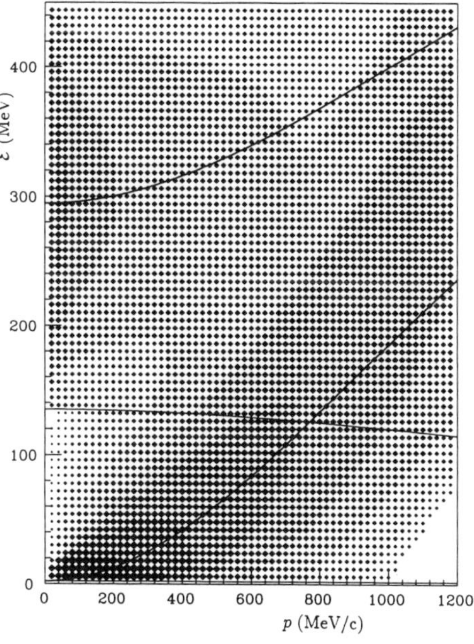

Fig. 8.   Representation of the coincidence cross section for the reaction ^{3}He(e,e$'$p) at fixed q and ω as a function of $\mathcal{E}$ and p. The symbol size represents the logarithm of the cross section. The various regions and lines are discussed in the text.

With these preliminaries in hand let us look at a rough representation of the coincidence cross section as a function of $\mathcal{E}$ and p for fixed q and ω (and of course fixed θ_e and ϕ_N). In Fig. 8 such a density plot is shown for the reaction ^{3}He(e,e$'$p) under typical conditions. Two–body breakup leaving the daughter nucleus in its ground state, namely ^{3}He $+$ e $\rightarrow$ e$'$ $+$ p $+$ ^{2}H corresponds to a δ–function line at $\mathcal{E} = 0$; three–body breakup into the continuum, ^{3}He $+$ e $\rightarrow$ e$'$ $+$ p $+$ p $+$ n, begins at the line shown at $\mathcal{E}$ slightly above zero; pion production in the daughter channel, ^{3}He $+$ e $\rightarrow$ e$'$ $+$ p $+$ π $+$ X, begins at a threshold shown in the figure as a line somewhat below $\mathcal{E} = m_\pi$. The two upward–curving lines in the figure correspond to simple assumptions[13] for quasifreely knocking two nucleons out of the nucleus, one of which is detected, (lower line) or quasifreely knocking a nucleon and a delta out of the nucleus, with the nucleon being detected (upper line). Clearly the modeling of the cross section appears roughly to have behaviors of this sort. While this is meant to be the cross section, it is in fact a representation of the spectral function (see the PWIA discussions in the next section) that has been used to arrive at the figure.

Now let us see what kinematic flexibility we have in probing this cross section, as indeed we have not yet required that the kinematic relationships discussed above should be satisfied. When we do so, we find that only selected regions in the generic plot shown in Fig. 8 are accessible. Noting that Eq. (68) yields a curve of $\mathcal{E}$ versus p in the $(\mathcal{E}, p)$–plane for each choice of θ, let us see what constraint the requirement that $-1 \leq \cos\theta \leq +1$ imposes on the kinematics. First, consider "ω small" (to be specified completely below) and plot the trajectory when $\cos\theta = -1$. A curve rising from negative $\mathcal{E}$ to intersect $\mathcal{E} = 0$ at $p = p_{\min} > 0$ which peaks at some value of p and then falls to intersect $\mathcal{E} = 0$ again, this time at $p = p_{\max} > p_{\min}$, is generally obtained. All physically allowable values of $\mathcal{E}$ and p must lie below this curve and, of course, above $\mathcal{E} = 0$. To obtain the other extreme, $\cos\theta = +1$, one can simply replace p by $-p$ in Eq. (68); the physically allowable values of $\mathcal{E}$ and p must lie above this curve. For "ω small", no physically allowable values at all occur near the latter curve and the physical region is completely defined by the $\cos\theta = -1$ curve and $\mathcal{E} = 0$. Following past work[13] we shall call the minimum value of momentum $p_{\min} \equiv -y$ and the maximum value $p_{\max} \equiv +Y$. The formal definition of "ω small" then becomes "$y < 0$". We can set $\mathcal{E} = 0$ in Eq. (68) and solve for y and Y, yielding

$$y(q,\omega) = \frac{1}{W^2}\left\{ (M_A + \omega)\sqrt{\Lambda^2 - M_B^{0\,2} W^2} - q\Lambda \right\} \tag{69a}$$

$$Y(q,\omega) = \frac{1}{W^2}\left\{ (M_A + \omega)\sqrt{\Lambda^2 - M_B^{0\,2} W^2} + q\Lambda \right\} \tag{69b}$$

with

$$W = \sqrt{(M_A + \omega)^2 - q^2} \tag{70a}$$

$$\Lambda = \frac{1}{2}\left(W^2 + M_B^{0\,2} - M_N^2 \right) \quad . \tag{70b}$$

A useful relationship is the following:

$$\omega = \sqrt{(q + y)^2 + M_N^2} + \sqrt{y^2 + M_B^{0\,2}} - M_A \quad . \tag{71}$$

In particular, assuming that the quasielastic peak occurs when $y = 0$, one has

$$\omega_{\mathrm{QE}} \equiv \left\{ \sqrt{q^2 + M_N^2} - M_N \right\} + E_s = |Q^2|/2M_N + E_s \quad . \tag{72}$$

Furthermore, the equation for the upper boundary of the allowed region (*i.e.*, corresponding to $\cos\theta = -1$) is given by

$$\mathcal{E}_M = \sqrt{M_N^2 + (q + y)^2} - \sqrt{M_N^2 + (q - p)^2} + \sqrt{M_B^{0\,2} + y^2} - \sqrt{M_B^{0\,2} + p^2} \quad . \tag{73}$$

When the momentum transfer become very large this goes to the *finite* asymptotic limit

$$\mathcal{E}_M \to \mathcal{E}_M^\infty = y + p - \left[\sqrt{M_B^{0\,2} + p^2} - \sqrt{M_B^{0\,2} + y^2} \right] \quad . \tag{74}$$

Henceforth, instead of the sets $\{q, \omega, E_N, \theta_N\}$ or $\{Q^2, Q \cdot P_A, P_N \cdot P_A, Q \cdot P_N\}$ we shall use the set $\{q, y, \mathcal{E}, p\}$ to characterize the general two–arm coincidence cross section.

In Fig. 9 are shown families of curves of $\mathcal{E}_M$ versus p for specific values of q and $y < 0$. The physical regions lie below these curves for chosen kinematics. Clearly, by comparing these kinematic constraints with the generic cross section shown in Fig. 8 it is possible to see what features of the

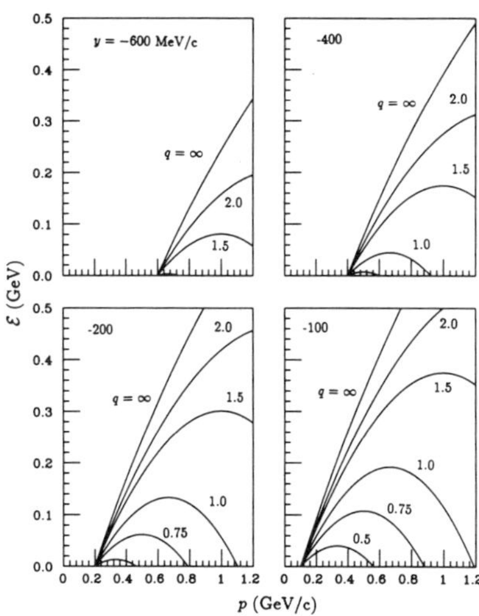

Fig. 9. Plots of $\mathcal{E}_M$ (see Eq. (73)) for various values of q (numbers — given in GeV/c) and $y < 0$ for the reaction ^{3}He(e,e$'$p).

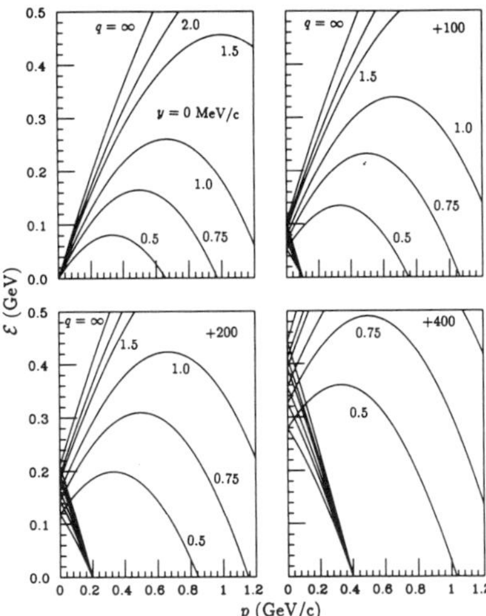

Fig. 10. As for Fig. 9, but now for $y \geq 0$.

dynamics are or are not accessible in the $y < 0$ region. Note that even when $q \to \infty$ only a limited part of the dynamics is accessible. Also note that inclusive scattering corresponds to integrating over the entire accessible region for q and y (or equivalently ω) fixed, and summing over all allowed particle species (protons, neutrons, ...). In the next section we shall return briefly to discuss inclusive scattering in the $y < 0$ region ("y-scaling" — see Ref. [13]).

These developments can be extended rather easily to the "ω large" region, which becomes equivalent to $y > 0$. Again the curves of $\mathcal{E}$ versus p when $\cos\theta = \pm 1$ define boundaries. The $\cos\theta = -1$ curve (namely, $\mathcal{E} = \mathcal{E}_M$ above) is much as before, except that now $p_{\min}$ is negative and so $y \equiv -p_{\min}$ is positive. Reflecting $p \to -p$ to obtain the $\cos\theta = +1$ curve from the $\cos\theta = -1$ curve as before now yields a nontrivial result: the physically allowable region must lie below the $\cos\theta = -1$ curve and above the $\cos\theta = +1$ curve, and since the latter lies in the quadrant where $\mathcal{E} \geq 0$ and $p \geq 0$, this provides a new boundary. In Fig. 10 results similar to those in Fig. 9 are shown, except now for $y \geq 0$.

The physically accessible region in each case lies *above* the lines extending from $p = y$ to the $\mathcal{E}$-axis and *below* the curves extending from the $\mathcal{E}$-axis to peak at some value of p and fall again, eventually intersecting the $\mathcal{E} = 0$ line at $p_{\max} = Y$. Again we see that specific parts of the generic cross section in Fig. 8 are accessible for these kinematics.

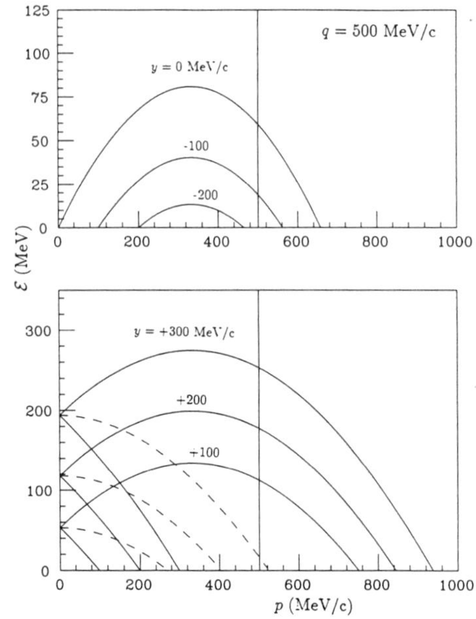

Fig. 11. Plots of the trajectories for parallel (solid: left of the vertical line), anti–parallel (solid: right of the vertical line) and perpendicular (dashed) kinematics (see text), all for $q = 500$ MeV/c and various values of y.

Note finally that parallel kinematics ($\theta_N = 0° \Rightarrow \theta = 0°$ or $180°$) corresponds to a line in the $(\mathcal{E}, p)$–plane; likewise anti–parallel kinematics ($\theta_N = 180°$) corresponds to a continuation of that line (see Fig. 11). In particular, for $y > 0$ the line goes from $\mathcal{E} = 0$ at $p = y$ upwards to the left to the $\mathcal{E}$-axis and then continues upwards and to the right to the vertical line shown in the figure. At that point the kinematics go from parallel to anti–parallel and the line continues back down to the p-axis. Experiments are frequently performed in parallel kinematics (for instance, to eliminate the TL and TT responses; see above) and yet the "natural" description of the underlying dynamics may not correspond to such trajectories in the $(\mathcal{E}, p)$–plane, but rather to slices at constant p or constant $\mathcal{E}$. Also shown in Fig. 11 are lines corresponding to perpendicular kinematics ($\theta = 0°$), which occur only for $y > 0$.

V.   POLARIZED PLANE–WAVE IMPULSE APPROXIMATION

In the present section the focus is on a specific tractable model for 2–arm coincidence electron scattering, $\vec{A}(\vec{e}, e'N)$, where $N = p$ or n. In particular, in recent work done together with Juan Caballero and Greg Poulis (see Ref. [10] for more detail) initial aspects of the polarization degrees of freedom in the problem have been explored. In this section we shall limit our attention to the hadronic structure probed in the quasi–free region where, for a given value of q one has $\omega \approx \sqrt{q^2 + M_N^2} - M_N = |Q^2|/2M_N$. Such kinematics roughly correspond to interacting with a nucleon in the nucleus that is at rest and ejecting it with momentum $\vec{q}$. Accordingly, one hopes to be probing essentially the single–nucleon content of the nucleus. At sufficiently high momentum transfers the process is presumed to become "quasi–free" and hence relatively mildly influenced by final–state interactions and various exchange effects which are usually neglected or, at best, only treated approximately (see Refs. [11,12]).

Specification of the Model

We begin with the basic approximation which underlies most treatments of electromagnetic coincidence reactions, *viz.*, the impulse approximation (IA) in which the electromagnetic current is taken to be a one–body operator:

$$\hat{J}_\mu = \sum_{mm'} \sum_{\tau\tau'} \int \int d\vec{p}\, d\vec{p}\,' \, \langle \vec{p}\,', m'(S'), \tau' | \Gamma_\mu | \vec{p}, m(S), \tau \rangle a^+_{\vec{p}\,'m'\tau'} a_{\vec{p}m\tau} \ , \tag{75}$$

where $|\vec{p}, m(S), \tau\rangle$ is an on–shell spinor state characterized by 3–momentum $\vec{p}$ (with corresponding energy $\sqrt{p^2 + M_N^2}$) and 4–spin S^μ. The label $m(S)$ denotes the spin projection ($\pm 1/2$) referred to the axis of quantization characterized by S^μ. Since one has occasion to refer spin projections to different axes of quantization in different Lorentz frames (see Ref. [10]), it is important to keep track not only of the m–values but also the directions of these axes. The label $\tau = \pm 1/2$ carries the isospin of the nucleon (proton or neutron, respectively). The normalization convention for the creation–annihilation operators is taken to be $\left\{ a^+_{\vec{p}\,'m'\tau'}, a_{\vec{p}m\tau} \right\} = \delta(\vec{p} - \vec{p}\,')\delta_{mm'}\delta_{\tau\tau'}$. In specializing to Eq. (75) we do not allow for two–body meson–exchange currents; our main focus here is on the "quasi–free" region where we expect such MEC effects to be small. The quantity Γ_μ is presumed to embody the physics at the γNN vertex and will be discussed in detail below.

Let us now turn to the plane–wave impulse approximation where, in addition to restricting the currents to one–body operators, one makes several more stringent assumptions to simplify the problem. Specifically, first one takes the emitted nucleon to be a plane wave, *i.e.*, the nucleon is ejected from the nucleus without any further interaction with the residual nuclear system. Secondly, one assumes that the nucleon detected in the coincidence reaction is the one to which the virtual photon is attached (see Fig. 12).

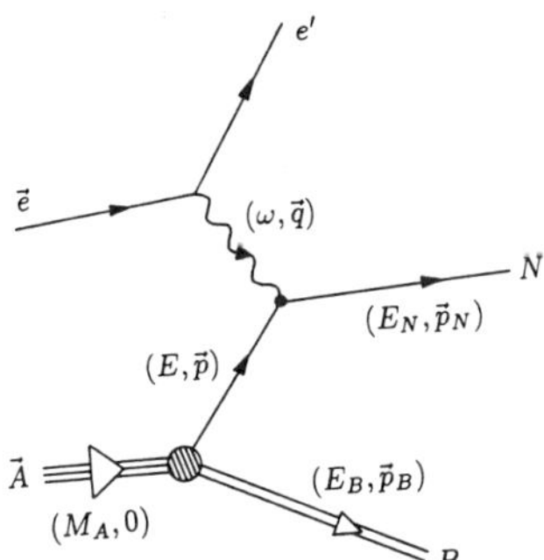

Fig. 12.   One–photon–exchange diagram for the reaction $\vec{A}(\vec{e}, e'N)B$ in the plane–wave impulse approximation (PWIA).

In so–doing various exchange effects are neglected: for instance, the photon can eject a nucleon, leaving a residual nucleus which is sufficiently excited that a second nucleon is ejected — that second nucleon can be the one that is detected in the coincidence reaction. In the plane–wave impulse approximation, PWIA (and also in the factorized distorted–wave impulse approximation, DWIA) the cross section for coincidence processes of the type $(e, e'N)$ can be factorized into two basic terms, the electron–nucleon cross section and the spectral function, as discussed in Refs. [10–12]. The latter factor provides the probability that a nucleon is to be found in the nucleus with given energy and momentum, while the former deals directly with the interaction between the incident electrons and the bound nucleons inside the nucleus which takes place via exchange of a virtual photon. The interactions are assumed to occur only with the individual nucleons (hence, "impulse" approximation). In the PWIA the final state $|f\rangle$ is then characterized simply by the product involving the state of the residual nucleus $|B\rangle$ and the on–shell knocked–out nucleon spinor state $|\vec{p}_N, m_N(S_N), \tau_N\rangle$:

$$\langle f|a^+_{\vec{p}'m'\tau'} = \delta(\vec{p}' - \vec{p}_N)\delta_{m'm_N}\delta_{\tau'\tau_N}\langle B| \quad . \tag{76}$$

The momentum of the struck nucleon (and hence the momentum of the residual nucleus) is fixed by the momentum transfer and the momentum of the emitted nucleon, $\vec{p} = \vec{p}_N - \vec{q} = -\vec{p}_B$. Neglecting charge–exchange effects (thus only one flavor of nucleon is involved — this is governed by which type of reaction we are studying, $(e, e'p)$ or $(e, e'n)$) and removing the dependence on the isospin variables, we then have for the many–body current matrix elements

$$\langle f|\hat{J}_\mu|i\rangle = \sum_m \int d\vec{p} \; \langle \vec{p}_N, m_N(S_N)|\Gamma_\mu|\vec{p}, m(S)\rangle\langle B|a_{\vec{p}m}|A\rangle \quad . \tag{77}$$

Inserting this result in the expression for the hadronic tensor we can write

$$W^{\mu\nu} = (2\pi)^3 \sum_{mm'} \mathcal{W}^{\mu\nu}_{mm'}(\vec{p}; \vec{q})n_{mm'}(\vec{p}, \Omega^*) \quad , \tag{78}$$

where we have introduced two new quantities: the single–nucleon tensor

$$\mathcal{W}^{\mu\nu}_{mm'}(\vec{p}; \vec{q}) = \sum_{m_N}\langle \vec{p}_N, m_N(S_N)|\Gamma^\mu|\vec{p}, m'(S')\rangle^*\langle \vec{p}_N, m_N(S_N)|\Gamma^\nu|\vec{p}, m(S)\rangle \quad , \tag{79}$$

which represents the part of the total hadronic tensor that depends directly on the γNN vertex, and the polarized momentum distribution

$$n_{mm'}(\vec{p}, \Omega^*) = \frac{1}{(2\pi)^3}\sum_A p(A)\sum_B \langle B|a_{\vec{p}m'}|A\rangle^*\langle B|a_{\vec{p}m}|A\rangle \quad , \tag{80}$$

which characterizes the probability that the polarized target nucleus contains a nucleon with momentum $\vec{p}$. Since we assume that no final–state polarizations are measured, here there is a sum over B involving all possible nuclear states. On the other hand, the initial state is assumed to be polarized and this is represented in Eq. (80) by the sum over A with the weighting factor $p(A)$, *viz.*, the probability that specific projections of the ground–state angular momentum occur. For example, if the nucleus is unpolarized, then $p(A) = 1/(2J_A + 1)$ and accordingly one has an average over initial states in Eq. (80); if $p(A) = 1$ when $M_{J_A} = J_A$ and zero otherwise, then the target is 100% polarized. Note that in the general case where the target is polarized it is necessary to specify an axis of quantization for the target spin and consequently the polarized momentum distribution in Eq. (80) depends on the angular variables $\Omega^* = (\theta^*, \phi^*)$ which specify this direction (see Fig. 7).

In terms of the kinematic variables defined above the spin–dependent spectral function is given by

$$S_{mm'}(\vec{p}, \mathcal{E}, \Omega^*) = n_{mm'}(\vec{p}, \Omega^*)\delta(E + E_B^0 + \mathcal{E} - M_A) \quad , \tag{81}$$

and represents the probability of finding a nucleon of momentum $\vec{p}$, energy $E = E_N - \omega = M_A - E_B^0 - \mathcal{E}$ and spin projection mm' in the initial nucleus. This quantity is a density matrix in spin space and in general involves off–diagonal matrix elements. Note that the representation of the exclusive–1 cross section shown in Fig. 8 now becomes a plot of the spectral function when the PWIA is invoked. In fact, for the polarized PWIA there should be four such plots corresponding to the different matrix elements of the (matrix) spectral function. As remarked above, it is not necessary to assume the PWIA to display the cross sections as functions of $\{q, y, \mathcal{E}, p\}$, since those kinematic quantities arise simply from overall energy–momentum conservation. When the PWIA is assumed, then the factorization below occurs and the choice of kinematic variables is especially advantageous. While not considered further here, it is noted that the distorted–wave impulse approximation (DWIA) also has this factorization — only the interpretation of the (distorted) spectral function is different.

Introducing the above results in the general expression for the cross section we obtain

$$\frac{d\sigma}{d\Omega_e\,d\omega\,d\Omega_N\,dE_N} = \frac{2\alpha^2}{Q^4}\left(\frac{\epsilon'}{\epsilon}\right)\frac{p_N M_N M_B}{E_B}\sum_{mm'}\eta_{\mu\nu}\mathcal{W}^{\mu\nu}_{mm'}(\vec{p};\vec{q})S_{mm'}(\vec{p},\mathcal{E},\Omega^*)$$
$$= \frac{p_N M_N M_B}{E_B}\sum_{mm'}\sigma^{eN}_{mm'}S_{mm'}(\vec{p},\mathcal{E},\Omega^*) \tag{82}$$

or after integrating over E_N:

$$\frac{d\sigma}{d\Omega_e\,d\omega\,d\Omega_N} = \frac{p_N M_N M_B}{M_A}f^{-1}_{rec}\sum_{mm'}\sigma^{eN}_{mm'}n_{mm'}(\vec{p},\Omega^*) \quad , \tag{83}$$

where by analogy with the work of de Forest (see Ref. [14]) we have introduced an "off–shell polarized electron–nucleon cross section":

$$\sigma^{eN}_{mm'} = \frac{2\alpha^2}{Q^4}\left(\frac{\epsilon'}{\epsilon}\right)\eta_{\mu\nu}\mathcal{W}^{\mu\nu}_{mm'}(\vec{p};\vec{q}) \quad , \tag{84}$$

where the cross section depends on the electron variables $(\theta_e,\epsilon,\epsilon')$ or equivalently (θ_e,q,ω) and the outgoing nucleon variables $(p_N$ or $E_N,\theta_N,\phi_N)$. The kinematics of the struck nucleon are then specified, as discussed above.

Imposing current conservation for the single–nucleon current and following the same procedures as in the nuclear case discussed in Secs. III and IV, the off–shell polarized electron–nucleon cross section can be written

$$\sigma^{eN}_{mm'} = \sigma_{\text{Mott}}\left\{\sum_K v_K\mathcal{R}^K_{mm'} + h\sum_{K'} v_{K'}\mathcal{R}^{K'}_{mm'}\right\} \quad , \tag{85a}$$

where the single–nucleon response functions $\mathcal{R}^{K,K'}_{mm'}$ are labelled as usual by $K = L,T,TL,TT$ and $K' = T',TL'$. Note that, since the nucleonic tensor considered here is spin–dependent, it can be shown (see Ref. [10]) that both real and imaginary parts contribute for the responses when off–diagonal spin matrix elements are considered. The nuclear response functions can now be written in terms of the quantities defined above:

$$W^K_{fi} = (2\pi)^3\sum_{mm'}\mathcal{R}^K_{mm'}n_{mm'}(\vec{p},\Omega^*) \quad . \tag{85b}$$

Half–Off–Shell Unpolarized Single–Nucleon Responses

Within the context of the impulse approximation one has to deal with the half–off–shell γNN vertex. By imposing Lorentz invariance together with correct parity and time reversal transformation properties and applying the Ward–Takahashi identities to ensure gauge–invariance for such a vertex, it is possible to see that it generally involves four form factors (to be contrasted with the on–shell case where only two are needed). More discussion on this point may be found in Ref. [15]. In this section the procedures involved in obtaining the single–nucleon response functions to be used in the PWIA and DWIA are discussed. In so–doing it will be necessary to introduce various prescriptions for the half–off–shell γNN vertex. Relying on simple *ad hoc* prescriptions becomes inevitable, since no tractable approach to treating the off–shell dependences rigorously presently exists. In practice it has become common practice to use specific off–shell extrapolations of the on–shell vertex, especially those extrapolations given by de Forest, although there is no *a priori* reason favoring one over another. Since two of the unpolarized prescriptions of de Forest (called CC1 and CC2) are in common use, in Ref. [10] we chose to generalize those particular prescriptions for the polarized case, at the same time cautioning the reader about their universality. Following de Forest, let us introduce "on–shell" quantities $\bar{P}^\mu \equiv (\bar{E},\vec{p})$ and $\bar{Q}^\mu \equiv P^\mu_N - \bar{P}^\mu = (\bar{\omega},\vec{q})$, where P^μ_N is the 4–momentum of the detected nucleon. The quantity $\bar{P}^\mu$ is the 4–momentum for on–shell kinematics having the same 3–momentum as the struck nucleon $(\vec{p})$, but on–shell energy $\bar{E} \equiv \sqrt{M_N^2 + p^2}$. $\bar{Q}^\mu$ is the corresponding 4–momentum transfer, with $\bar{Q}^2 = \bar{\omega}^2 - q^2$ and $\bar{\omega} \equiv E_N - \bar{E}$. Then the de Forest prescriptions (see Ref. [14]) involve three steps:

a) Treat the spinors as free (on–shell) ones.

b) Employ the following two forms for the vertex operator corresponding to the CC1 and CC2 prescriptions:

$$1: \quad \Gamma^\mu_{CC1} = (F_1 + F_2)\gamma^\mu - \frac{F_2}{2M_N}(\bar{P} + P_N)^\mu \tag{86a}$$

$$2: \quad \Gamma^\mu_{CC2} = F_1\gamma^\mu + \frac{iF_2}{2M_N}\sigma^{\mu\nu}Q_\nu \ , \tag{86b}$$

with F_1 and F_2 the on–shell Pauli and Dirac form factors, respectively.

c) Finally there is the issue of current conservation. The total many–body current is conserved; however the impulse approximation involves only one–body contributions. Consequently, there is no *a priori* reason to expect the IA to lead to a conserved current. For these recipes one finds that matrix elements of operators 1 and 2 violate conservation of the one–body current according to the amount of off–shellness, that is, $Q_\mu J^\mu$ is of order $\omega - \bar{\omega}$. The cross sections resulting from these recipes are denoted NCC1 and NCC2, respectively, (for not–current–conserving) in the present work. The so called CC1/CC2 (current–conserving) recipes enforce current conservation by eliminating the 3– or 0–components of the current matrix elements by means of the relation $qJ^{(3)} = \omega J^{(0)}$ (with $\vec{q}$ along the $\hat{z}$–direction, as before). In de Forest's work the 3–components are eliminated by employing Siegert's theorem to that purpose. In Ref. [10] both options were considered: depending on whether the 0– or 3–components are eliminated the corresponding nomenclature CC1$^{(0)}$ or CC1$^{(3)}$ (or analogously, CC2$^{(0)}$ or CC2$^{(3)}$) is used.

It is now straightforward to employ the specific forms for Γ^μ given in Eqs. (86) in obtaining specific answers for the single–nucleon CC1$^{(0)}$ and CC2$^{(0)}$ responses. In particular the polarization–independent responses which result are the following:

$$\mathcal{R}^L = \frac{\kappa^2}{\bar{\tau}}\left\{\chi^2\left[F_1^2 + \bar{\tau}F_2^2\right] + \left[F_1 - \bar{\tau}F_2\right]^2\right\} \tag{87a}$$

$$\mathcal{R}^T = \chi^2\left[F_1^2 + \rho_2\bar{\tau}F_2^2\right] + 2\bar{\tau}\left[F_1 + F_2\sqrt{(\rho_1 + \rho_2)/2}\right]^2 \tag{87b}$$

$$\mathcal{R}^{TL} = 2\sqrt{2}\xi\chi\left[F_1^2 + \bar{\tau}F_2^2\sqrt{(\rho_1 + \rho_2)/2}\right]\cos\phi \tag{87c}$$

$$\mathcal{R}^{TT} = -\chi^2\left[F_1^2 + \rho_1\bar{\tau}F_2^2\right]\cos 2\phi \ , \tag{87d}$$

where the struck and detected nucleon kinematical variables are related through

$$\phi = \phi_N \tag{88a}$$

$$\chi \equiv \frac{p}{M_N}\sin\theta = \frac{p_N}{M_N}\sin\theta_N \tag{88b}$$

$$\chi' \equiv \frac{p}{M_N}\cos\theta = \frac{p_N}{M_N}\cos\theta_N - 2\kappa \ , \tag{88c}$$

where following Ref. [10] we employ the following dimensionless momentum transfers: $\tau \equiv |Q^2|/4M_N^2 = \kappa^2 - \lambda^2$, $\bar{\tau} \equiv |\bar{Q}^2|/4M_N^2 = \kappa^2 - \bar{\lambda}^2$ and $\kappa \equiv q/2M_N$. Here dimensionless energies have been introduced:

$$\lambda \equiv \frac{\omega}{2M_N} \tag{89a}$$

$$\bar{\lambda} \equiv \frac{\bar{\omega}}{2M_N} \tag{89b}$$

$$\xi \equiv \frac{E_N + \bar{E}}{2M_N} = \sqrt{\frac{\kappa^2}{\bar{\tau}}(1 + \bar{\tau} + \chi^2)} \ . \tag{89c}$$

The quantities χ and χ' are related by

$$\chi' = \bar{\lambda}\sqrt{[1 + \bar{\tau} + \chi^2]/\bar{\tau}} - \kappa \ . \tag{89d}$$

For compactness we have also introduced the two parameters ρ_1 and ρ_2 in Eqs. (87); for the CC1$^{(0)}$ prescription one has $\rho_1 = \rho_2 = 1$ and for the CC2$^{(0)}$ prescription one has $\rho_1 = \tau/\bar{\tau}$ and $\rho_2 = 2[1 + \bar{\lambda}(\bar{\lambda} - \lambda)/\bar{\tau}]^2 - \rho_1$.

Inclusive Scattering in the PWIA: y–Scaling

Let us make a brief excursion into the subject of y–scaling as described within the context of the PWIA (see Ref. [13]). Here we need the inclusive cross section which involves a sum (integral) over all allowed final states of Eq. (54). Specifically in the PWIA for unpolarized scattering Eq. (82) yields

$$\frac{d\sigma}{d\Omega_e d\omega d\Omega_N dE_N} = \frac{p_N M_N M_B}{E_B}\, \sigma_0^{eN} S^0(\vec{p}, \mathcal{E}) \tag{90a}$$

$$\equiv p_N E_N\, \tilde{\sigma}_0^{eN}\, \tilde{S}^0(\vec{p}, \mathcal{E}) \ , \tag{90b}$$

where, following de Forest's treatment in Ref. [14], we define

$$\tilde{\sigma}_0^{eN} \equiv \left(\frac{M_N^2}{\bar{E}E_N}\right)\sigma_0^{eN} \ , \tag{91a}$$

involving the unpolarized off–shell single–nucleon cross section discussed above, and also

$$\tilde{S}^0(\vec{p}, \mathcal{E}) \equiv \left(\frac{M_B}{E_B}\right)\left(\frac{\bar{E}}{M_N}\right) S^0(\vec{p}, \mathcal{E}) \approx S^0(\vec{p}, \mathcal{E}) \ . \tag{91b}$$

The inclusive unpolarized cross section then becomes

$$\frac{d\sigma}{d\Omega_e d\omega} = \sum_{i=1}^{A} \int d\vec{p}_N \int dE_N\, \delta\big(E_N - \sqrt{p_N^2 + M_N^2}\,\big)\, \tilde{\sigma}_{0i}^{eN}\, \tilde{S}_i^0(\vec{p}, \mathcal{E}) \ , \tag{92}$$

where we have summed over all nucleons (protons and neutrons — labeled i) and integrated over final–state energy and angles. The integral over $\vec{p}_N$ may be converted to an integral over $\vec{p}$ using $\vec{p}_N = \vec{p} + \vec{q}$ and the integral over E_N may be converted to an integral over $\mathcal{E}$ in Eq. (65). We may define

$$< \tilde{\sigma}_{0i}^{eN} > \equiv \frac{1}{2\pi}\int_0^{2\pi} d\phi\, \tilde{\sigma}_{0i}^{eN} \tag{93}$$

and write the δ–function in the form

$$\delta\big(E_N - \sqrt{p_N^2 + M_N^2}\,\big) = \delta(\omega - \Omega) \tag{94a}$$

with

$$\Omega \equiv \left[\sqrt{M_N^2 + (\vec{p} + \vec{q})^2} - M_N\right] + \left[\sqrt{M_B^{0\,2} + p^2} - M_B^0\right] + E_s + \mathcal{E} \ , \tag{94b}$$

using Eq. (68). Converting the integral over $\cos\theta$, where θ is the angle between $\vec{p}$ and $\vec{q}$, to an integral over $\mathcal{E}$ we obtain

$$\frac{d\sigma}{d\Omega_e d\omega} = \sum_{i=1}^{A} \int_{-y}^{Y} p^2\, dp \int_0^{\mathcal{E}_M} d\mathcal{E} < \tilde{\sigma}_{0i}^{eN} > 2\pi \tilde{S}_i^0(p, \mathcal{E})\, |\partial\Omega/\partial\cos\theta|^{-1} \ . \tag{95}$$

Here the specific limits of integration have already been introduced above (see Eqs. (69,73)) and $|\partial\Omega/\partial\cos\theta| = pq/E_N$.

Finally, let us make two other approximations: (1) assume that the spectral function is isospin independent and (2) evaluate the off–shell single–nucleon cross section at the smallest value of p that is accessible when $y < 0$, namely, for kinematics where $p = -y$ (and where $\mathcal{E} = 0$) — the latter is expected to be the kinematic point where the integrand in Eq. (95) is largest. We then have

$$\frac{d\sigma}{d\Omega_e d\omega} \simeq \tilde{\Sigma}_0^{eN} \times F(q, y) \tag{96}$$

where

$$\tilde{\Sigma}_0^{eN} \equiv \left[\frac{E_N}{q} \sum_{i=1}^{A} < \tilde{\sigma}_{0i}^{eN} >\right]_{p=-y,\,\mathcal{E}=0} \tag{97a}$$

and

$$F(q,y) \equiv 2\pi \int_{-y}^{Y} p \, dp \, \tilde{n}(q,y;p) \tag{97b}$$

$$\tilde{n}(q,y;p) \equiv \int_{0}^{\mathcal{E}_M} d\mathcal{E} \, \tilde{S}^0(p,\mathcal{E}) \ . \tag{97c}$$

As discussed in Ref. [13], when $q \to \infty$ (the scaling limit) it is possible to take $Y \to \infty$ as well, since the spectral function is expected to fall off when p becomes very large; furthermore, $\mathcal{E}_M \to \mathcal{E}_M^\infty$ (see Eq. (74)) and then

$$\tilde{n}(q,y;p) \to \tilde{n}(\infty,y;p) = \int_{0}^{\mathcal{E}_M^\infty} d\mathcal{E} \, \tilde{S}^0(p,\mathcal{E}) \tag{98a}$$

$$F(q,y) \to F(\infty,y) \equiv F(y) = 2\pi \int_{-y}^{\infty} p \, dp \, \tilde{n}(\infty,y;p) \ , \tag{98b}$$

where the last is the y–scaling function (independent of q). Experimentally, one takes the measured inclusive cross section $d\sigma/d\Omega_e d\omega$ and divides by $\tilde{\Sigma}_0^{eN}$ in Eq. (97a): the resulting function $F(q,y)$ is plotted versus y and, if the results approach a universal result $F(y)$ when q becomes large enough one says that y–scaling has been attained. Typical results are presented in Ref. [13] and indeed they are seen to scale in the region where $y < 0$.

Half–Off–Shell Polarized Single–Nucleon Responses

Let us now turn briefly to consider the polarized analogs of Eqs. (87). As discussed in Ref. [10], in the PWIA we have only time–reversal–even responses which means that only the double–polarized cross sections are nonzero (*i.e.*, in the polarized situation we shall be discussing the reactions $\vec{A}(\vec{e},e'N)$). Furthermore, it is convenient to label the three T' and TL' cases by "ℓ" (longitudinal), "n" (normal) and "s" (sideways), that is, by spin projections along $\vec{q}$, along $\vec{q} \times \vec{p}_N$ and along $(\vec{q} \times \vec{p}_N) \times \vec{q}$, respectively (see Fig. 7):

$$\mathcal{R}_\ell^{T'} = \Omega_2 + \Omega_1 \chi' \tag{99a}$$

$$\mathcal{R}_s^{T'} = \Omega_1 \chi \cos\phi \tag{99b}$$

$$\mathcal{R}_n^{T'} = \Omega_1 \chi \sin\phi \tag{99c}$$

$$\mathcal{R}_\ell^{TL'} = \sqrt{2}[\Omega_4 + \Omega_3 \chi'] \cos\phi \tag{99d}$$

$$\mathcal{R}_s^{TL'} = \sqrt{2}[\Omega_5 + \Omega_3 \chi \cos^2\phi] \tag{99e}$$

$$\mathcal{R}_n^{TL'} = \sqrt{2}\Omega_3 \chi \sin\phi \cos\phi \ . \tag{99f}$$

It is possible to obtain the quantities Ω_i in Eqs. (99) corresponding to the $CC1^{(0)}$ and $CC2^{(0)}$ prescriptions, just as one does for the unpolarized situation discussed above. Specifically we express these basic results in the following relatively compact forms (Cf. Eqs. (87–89) above):

$$\begin{aligned}
\Omega_1 = \frac{2}{1 + \bar{E}/M_N} \Big\{ &\kappa[F_1 + F_2][F_1 - \bar{\tau}F_2\sqrt{(\rho_2 + \rho_1)/2}\,] \\
&+ \frac{\bar{\tau}}{\kappa}\xi[F_1 + \bar{\lambda}F_2][F_1 + F_2\sqrt{(\rho_2 + \rho_1)/2}\,] \\
&+ \frac{\bar{\tau}}{\kappa^2}F_2\sqrt{(\rho_2 - \rho_1)/2}\Big[(\kappa^2 + \bar{\tau}\xi - \bar{\lambda}\xi^2)F_1 \\
&+ \bar{\tau}F_2\sqrt{(\rho_2 + \rho_1)/2}\,(-\kappa^2 + \xi + \xi^2)\Big]\Big\}
\end{aligned} \tag{100a}$$

$$\begin{aligned}
\Omega_2 = -2\Big\{ &\bar{\lambda}[F_1 + F_2][F_1 - \bar{\tau}F_2\sqrt{(\rho_2 + \rho_1)/2}\,] \\
&+ \bar{\tau}F_2[F_1 + F_2\sqrt{(\rho_2 + \rho_1)/2}\,]\,\xi \\
&+ \frac{\bar{\tau}}{\kappa}F_2\sqrt{(\rho_2 - \rho_1)/2}\Big([F_1 - \bar{\tau}F_2\sqrt{(\rho_2 + \rho_1)/2}\,] - F_1\xi^2\Big)\Big\}
\end{aligned} \tag{100b}$$

$$\Omega_3 = \frac{2}{1 + \bar{E}/M_N}\bigg\{[F_1 + F_2][F_1 + \bar{\lambda}F_2]$$
$$+ \frac{\bar{\tau}}{\kappa}\sqrt{(\rho_2 - \rho_1)/2}\,[F_1 + (1 + \xi)F_2]F_2\bigg\}\kappa\chi \tag{100c}$$

$$\Omega_4 = -2\bigg\{\kappa^2[F_1 + F_2] - \frac{\bar{\tau}}{\kappa}F_1\xi\sqrt{(\rho_2 - \rho_1)/2}\bigg\}F_2\chi \tag{100d}$$

$$\Omega_5 = 2\bigg\{\kappa[F_1 + F_2][F_1 - \bar{\tau}F_2\sqrt{(\rho_2 + \rho_1)/2}\,] - \frac{\bar{\tau}}{\kappa}F_1F_2\xi^2[1 - \sqrt{(\rho_2 + \rho_1)/2}\,]\bigg\}. \tag{100e}$$

In Ref. [10] various prescriptions for the single–nucleon cross sections were explored in detail. Given a description of the polarized spectral function defined in Eq. (81) together with the unpolarized and polarized off–shell single–nucleon responses in Eqs. (87) and (99,100), we are in a position to

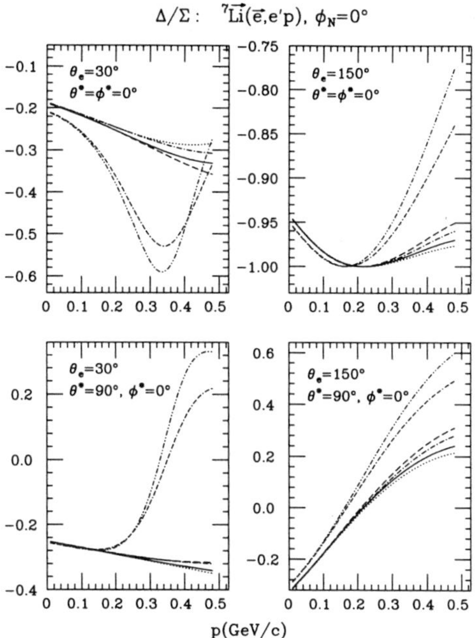

Fig. 13. Polarization ratio Δ/Σ (see Eqs. (62)) versus the inbound nucleon momentum for the reaction $^7\vec{Li}(e,e'p)^6He(g.s.)$. The results are taken from Ref. [10]: here $q = 500$ MeV/c and $y = 0$. Various off–shell single–nucleon prescriptions are employed, labelled CC1$^{(0)}$ (solid), CC2$^{(0)}$ (dashed), NCC1 (dotted), NCC2 (dot-dashed), CC1$^{(3)}$ (dot-double dashed), and CC2$^{(3)}$ (dash-double dotted).

predict the nuclear $\vec{A}(\vec{e}, e'N)$ cross sections. Here, for brevity, only one simple representative case is presented (see Fig. 13); the reader is directed to Ref. [10] for more discussion using simple models for the nuclear spectral function and to Ref. [16] for recent work on the interesting case of polarized electron scattering from a polarized, *deformed* nucleus, ^{21}Ne.

VI.   MULTI–ARM COINCIDENCE ELECTRON SCATTERING

Exclusive–n Electron Scattering, No Hadronic Polarizations

As discussed in Refs. [7,8], the above arguments can be generalized to include descriptions of the structure of the reactions $A(e, e'x_1 \cdots x_n)$ and $A(\vec{e}, e'x_1 \cdots x_n)$. The treatments given in the two previous subsections are then special cases of this more global formulation of the problem: $n = 0 \leftrightarrow$ inclusive, single–arm scattering; $n = 1 \leftrightarrow$ exclusive–1, 2–arm coincidence scattering.

In the general case of exclusive–n electron scattering there are $3n + 1 + \delta_{n0}$ independent scalar quantities; equivalently the following set of laboratory variables may be used:

$$\{q, \omega\} \ , \quad n \geq 0$$
$$\{E_k, \theta_k, \ k = 1, \ldots, n\} \ , \quad n \geq 1$$
$$\{\Delta\phi_{12}, \Delta\phi_{23}, \ldots, \Delta\phi_{n-1 n}\} \ , \quad n \geq 2 \ ,$$

where $\Delta\phi_{kk'} \equiv \phi_k - \phi_{k'}$. Only the azimuthal angle differences occur as dependences contained within the response functions. This leaves the average azimuthal angle

$$\Phi \equiv \frac{1}{n}(\phi_1 + \ldots + \phi_n) \ , \quad n \geq 1 \tag{101}$$

as the one kinematical aspect of the detected particles' momenta which does not occur as an argument in the response functions, but appears explicitly in the cross section. When we build the tensor $W^{\mu\nu}$ from the momenta $\{Q^\mu, P_i^\mu, P_1^\mu \ldots P_n^\mu\}$, we immediately observe that a certain saturation has occurred. At the level of exclusive–2 scattering we have four independent 4–momenta; however, for $n \geq 3$ we would be trying to use five or more independent vectors in a 4–dimensional space. We cannot do so, since the space is spanned by only four. In the case of $n = 3$, for example, we can write $P_3^\mu = aQ^\mu + bP_i^\mu + cP_1^\mu + dP_2^\mu$, where a, b, c and d are scalar quantities. Thus, we are back to having only the four momenta that were used for the $n = 2$ case. Indeed, this is true for all $n \geq 3$. The general result for the scattering of electrons (polarized or not) where no nuclear polarizations are specified is then

$$\eta_{\mu\nu}^s W_s^{\mu\nu} \sim v_L W_{(x_1 \cdots x_n)}^L + v_T W_{(x_1 \cdots x_n)}^T$$
$$+ v_{TL} \left(W_{(x_1 \cdots x_n)}^{TL} \cos\Phi + \tilde{W}_{(x_1 \cdots x_n)}^{TL} \sin\Phi \right)$$
$$+ v_{TT} \left(W_{(x_1 \cdots x_n)}^{TT} \cos 2\Phi + \tilde{W}_{(x_1 \cdots x_n)}^{TT} \sin 2\Phi \right) \tag{102a}$$
$$\eta_{\mu\nu}^a W_a^{\mu\nu} \sim v_{T'} \tilde{W}_{(x_1 \cdots x_n)}^{T'}$$
$$+ v_{TL'} \left(W_{(x_1 \cdots x_n)}^{TL'} \sin\Phi + \tilde{W}_{(x_1 \cdots x_n)}^{TL'} \cos\Phi \right) \ . \tag{102b}$$

Here the notation $(x_1 \cdots x_n)$ is used to indicate that the n particles $x_1, x_2 \ldots x_n$ are detected in coincidence with the scattered electron. For inclusive scattering $(n = 0)$ only the L and T terms contribute (see Eqs. (37)). For exclusive–1 scattering, $\Phi = \phi_x$ and the response functions $\tilde{W}_{(x)}^{T'}$, $\tilde{W}_{(x)}^{TT}$, $\tilde{W}_{(x)}^{TL}$ and $\tilde{W}_{(x)}^{TL'}$, are all absent (see Eqs. (52)). As before, in considering exclusive–n scattering, if we integrate over the angle dependence (θ_n, ϕ_n) and sum over open channels insofar as particle n is conserved, then we shall recover the exclusive–$(n-1)$ results.

In the general case we may imagine using the helicity dependence to separate the T' and TL' terms from the L, T, TT, and TL terms. After this, the Φ–dependence may be used to separate $v_L W_{(x_1 \cdots x_n)}^L + v_T W_{(x_1 \cdots x_n)}^T$, $W_{(x_1 \cdots x_n)}^{TT}$, $\tilde{W}_{(x_1 \cdots x_n)}^{TT}$, $W_{(x_1 \cdots x_n)}^{TL}$ and $\tilde{W}_{(x_1 \cdots x_n)}^{TL}$ and to separate $\tilde{W}_{(x_1 \cdots x_n)}^{T'}$, $W_{(x_1 \cdots x_n)}^{TL'}$ and $\tilde{W}_{(x_1 \cdots x_n)}^{TL'}$. Finally, the θ_e–dependence in v_T may be used in making a Rosenbluth decomposition of $W_{(x_1 \cdots x_n)}^L$ and $W_{(x_1 \cdots x_n)}^T$. Thus, all nine response functions are, in principle, experimentally accessible. Note that the original tensor $W^{\mu\nu}$ was constructed from bilinear products of the electromagnetic current matrix elements (Eq. (2b)). In turn, the currents are 4–vectors which satisfy the continuity equation, $Q_\mu J_{fi}^\mu = 0$; this implies* that $\omega J_{fi}^0 = q J_{fi}^3$, so that only three components of J_{fi}^μ are independent (say J_{fi}^1, J_{fi}^2 and J_{fi}^3, with $J_{fi}^0 = (q/\omega)J_{fi}^3$). Thus, there should be $3 \times 3 = 9$ independent terms in the cross section, and that agrees with the structure seen in Eqs. (102). A similar analysis of the weak interaction hadronic tensor leads to a general structure with $4 \times 4 = 16$ terms, since in that case we have axial–vector as well as vector currents and the former are not conserved.

---

* In fact, the nomenclature "longitudinal" used above is really not accurate: both longitudinal ($\mu = 3$) and "time" (charge, $\mu = 0$) components enter and are related by this current conservation identity.

General Coincidence Electron Scattering with Hadronic Polarizations

The underlying general formalism for treating $\vec{A}(\vec{e}, e'x)$ and $A(\vec{e}, e'\vec{x})$ reactions is given in Ref. [3]. In the present lectures we shall not repeat that material; rather we shall conclude with some comments on electromagnetic processes in general, with or without hadronic polarizations. When one considers the most general parity–conserving electron scattering reaction in the ERL, the 6 classes of response discussed above continue to represent the nature of the problem. That is, one has 4 classes labeled L, T, TL and TT for unpolarized electrons and an additional 2 classes labeled T' and TL' for polarized electrons. Let us go into these L/T decompositions a little further. Before performing $\overline{\sum}_{if}$ we have for specific initial and final nuclear states three independent current matrix elements, $\rho_{fi}(\vec{q})$, $J^x_{fi}(\vec{q})$ and $J^y_{fi}(\vec{q})$, with $J^z_{fi}(\vec{q}) = (\omega/q)\rho_{fi}(\vec{q})$. Equivalently, we can choose to deal with the three independent quantities

$$J^0_{fi}(\vec{q}) \equiv J^z_{fi}(\vec{q}) \tag{103a}$$

$$J^{\pm 1}_{fi}(\vec{q}) \equiv \mp \frac{1}{\sqrt{2}}\left(J^x_{fi}(\vec{q}) \pm iJ^y_{fi}(\vec{q})\right) \;, \tag{103b}$$

which transform as a rank–1 spherical tensor under rotations. With our conventions for the electron kinematical factors in Eqs. (53) we have for the hadronic tensors, for specific states i and f, the following

$$\begin{aligned}
W^L_{fi} &= |\rho_{fi}(\vec{q})|^2 = (q/\omega)^2|J^0_{fi}(\vec{q})|^2 \\
W^T_{fi} &= |J^{+1}_{fi}(\vec{q})|^2 + |J^{-1}_{fi}(\vec{q})|^2 \\
W^{TL}_{fi} &= -2\,\mathrm{Re}\left\{\rho_{fi}(\vec{q})^*\left(J^{+1}_{fi}(\vec{q}) - J^{-1}_{fi}(\vec{q})\right)\right\} \\
&= -2(q/\omega)\,\mathrm{Re}\left\{J^0_{fi}(\vec{q})^*\left(J^{+1}_{fi}(\vec{q}) - J^{-1}_{fi}(\vec{q})\right)\right\} \\
W^{TT}_{fi} &= 2\,\mathrm{Re}\left\{J^{+1}_{fi}(\vec{q})^*J^{-1}_{fi}(\vec{q})\right\} \\
W^{T'}_{fi} &= |J^{+1}_{fi}(\vec{q})|^2 - |J^{-1}_{fi}(\vec{q})|^2 \\
W^{TL'}_{fi} &= -2\,\mathrm{Re}\left\{\rho_{fi}(\vec{q})^*\left(J^{+1}_{fi}(\vec{q}) + J^{-1}_{fi}(\vec{q})\right)\right\} \\
&= -2(q/\omega)\,\mathrm{Re}\left\{J^0_{fi}(\vec{q})^*\left(J^{+1}_{fi}(\vec{q}) + J^{-1}_{fi}(\vec{q})\right)\right\} \;.
\end{aligned} \tag{104}$$

We can also define three more contributions by changing "Re" to "Im" in these equations, although they do not enter in the ERL. By rewriting these expressions in cartesian components we can verify that

$$L, T, TT, TL \quad \leftrightarrow \quad \text{symmetric under } \mu \leftrightarrow \nu$$

$$T', TL' \quad \leftrightarrow \quad \text{antisymmetric under } \mu \leftrightarrow \nu \;.$$

Furthermore, we can explore the properties of Eqs. (104) under rotations about the z–direction (the direction of $\vec{q}$). Rotating the x– and y–axis (see Fig. 5, for instance) is equivalent to varying the average azimuthal angle Φ (Eq. (101)), where now Φ includes not only dependence from the azimuthal angles for each particle detected in coincidence with the electron $\{\phi_1 \ldots \phi_n\}$, but also all azimuthal angles for initial– and final–state polarizations. It is straightforward to show that W^L_{fi}, W^T_{fi} and $W^{T'}_{fi}$ have no Φ–dependence, to show that the TL–interferences have the following structures,

$$\begin{aligned}
W^{TL}_{fi}(\Phi) &= \mathrm{Re}\,A^{TL}_{fi} \cdot \cos\Phi + \mathrm{Re}\,\tilde{A}^{TL}_{fi} \cdot \sin\Phi \\
W^{TL'}_{fi}(\Phi) &= -\mathrm{Im}\,A^{TL}_{fi} \cdot \sin\Phi + \mathrm{Im}\,\tilde{A}^{TL}_{fi} \cdot \cos\Phi
\end{aligned} \tag{105a}$$

and to show that the TT–interference has the following structure,

$$W^{TT}_{fi}(\Phi) = \mathrm{Re}\,A^{TT}_{fi} \cdot \cos 2\Phi + \mathrm{Im}\,A^{TT}_{fi} \cdot \sin 2\Phi \;. \tag{105b}$$

Here A^{TL}, $\tilde{A}^{TL}$ and A^{TT} are (generally complex) bilinear combinations of the electromagnetic current matrix elements. Now, in addition to the 6 classes of response defined in Eqs. (104) we can also define

3 more (see Refs. [2,8]):

$$W_{fi}^{TL} = -2\,\mathrm{Im}\left\{\rho_{fi}(\vec{q})^*\left(J_{fi}^{+1}(\vec{q}) + J_{fi}^{-1}(\vec{q})\right)\right\}$$
$$= -2(q/\omega)\,\mathrm{Im}\left\{J_{fi}^{0}(\vec{q})^*\left(J_{fi}^{+1}(\vec{q}) + J_{fi}^{-1}(\vec{q})\right)\right\}$$
$$W_{fi}^{TL'} = -2\,\mathrm{Im}\left\{\rho_{fi}(\vec{q})^*\left(J_{fi}^{+1}(\vec{q}) - J_{fi}^{-1}(\vec{q})\right)\right\} \tag{106}$$
$$= -2(q/\omega)\,\mathrm{Im}\left\{J_{fi}^{0}(\vec{q})^*\left(J_{fi}^{+1}(\vec{q}) - J_{fi}^{-1}(\vec{q})\right)\right\}$$
$$W_{fi}^{TT} = 2\,\mathrm{Im}\left\{J_{fi}^{+1}(\vec{q})^* J_{fi}^{-1}(\vec{q})\right\}\;.$$

Symmetry under rotations about the z–axis allows us to write the analogs of Eqs. (105):

$$W_{fi}^{TL}(\Phi) = \mathrm{Re}\,A_{fi}^{TL}\cdot\sin\Phi - \mathrm{Re}\,\tilde{A}_{fi}^{TL}\cdot\cos\Phi$$
$$W_{fi}^{TL'}(\Phi) = \mathrm{Im}\,A_{fi}^{TL}\cdot\cos\Phi + \mathrm{Im}\,\tilde{A}_{fi}^{TL}\cdot\sin\Phi$$
$$W_{fi}^{TT}(\Phi) = -\mathrm{Re}\,A_{fi}^{TT}\cdot\sin2\Phi + \mathrm{Im}\,A_{fi}^{TT}\cdot\cos2\Phi\;. \tag{107}$$

which involve the same quantities A^{TL}, $\tilde{A}^{TL}$ and A^{TT}. Hence, if measurements of W^{TL}, $W^{TL'}$ and W^{TT} are performed at different angles Φ, it is possible to determine $W^{\underline{TL}}$, $W^{\underline{TL}'}$ and $W^{\underline{TT}}$ completely. In fact,

$$W^{\underline{TL}}(\Phi) = -W^{TL}(\Phi + \pi/2)$$
$$W^{\underline{TL}'}(\Phi) = -W^{TL'}(\Phi + \pi/2)$$
$$W^{\underline{TT}}(\Phi) = W^{TT}(\Phi + \pi/4)\;, \tag{108}$$

and consequently the underlined responses ($\underline{TL}, \underline{TL}', \underline{TT}$) are redundant.

The basic responses in Eqs. (104,106) can also be written in terms of Cartesian projections of the currents, $J_{fi}^{x} = -(J_{fi}^{+1} - J_{fi}^{-1})/\sqrt{2}$, $J_{fi}^{y} = i(J_{fi}^{+1} + J_{fi}^{-1})/\sqrt{2}$ and $J_{fi}^{z} = J_{fi}^{0}$:

$$W_{fi}^{L} = (q/\omega)^2|J_{fi}^{z}(\vec{q})|^2$$
$$W_{fi}^{T} = |J_{fi}^{x}(\vec{q})|^2 + |J_{fi}^{y}(\vec{q})|^2$$
$$W_{fi}^{TT} = -|J_{fi}^{x}(\vec{q})|^2 + |J_{fi}^{y}(\vec{q})|^2$$
$$W_{fi}^{T'} = -2\mathrm{Im}\left\{J_{fi}^{x}(\vec{q})^* J_{fi}^{y}(\vec{q})\right\}$$
$$W_{fi}^{\underline{TT}} = 2\mathrm{Re}\left\{J_{fi}^{x}(\vec{q})^* J_{fi}^{y}(\vec{q})\right\}$$
$$W_{fi}^{TL} = 2\sqrt{2}(q/\omega)\mathrm{Re}\left\{J_{fi}^{z}(\vec{q})^* J_{fi}^{x}(\vec{q})\right\}$$
$$W_{fi}^{TL'} = -2\sqrt{2}(q/\omega)\mathrm{Im}\left\{J_{fi}^{z}(\vec{q})^* J_{fi}^{y}(\vec{q})\right\}$$
$$W_{fi}^{\underline{TL}} = 2\sqrt{2}(q/\omega)\mathrm{Re}\left\{J_{fi}^{z}(\vec{q})^* J_{fi}^{y}(\vec{q})\right\}$$
$$W_{fi}^{\underline{TL}'} = 2\sqrt{2}(q/\omega)\mathrm{Im}\left\{J_{fi}^{z}(\vec{q})^* J_{fi}^{x}(\vec{q})\right\}\;. \tag{109}$$

It is now clear that the T–response involves transverse projections of the current in a form corresponding to unpolarized photon exchange. The TT–response enters when the photon is linearly polarized; the T'–response enters when it is circularly polarized (see Eqs. (104)).

For completeness, let us rewrite the results in Eqs. (102) in an alternative form. To condense things somewhat, let us absorb the explicit Φ–dependence and define response functions σ^K (in the present discussion, for clarity we suppress the other labels on the responses):

$$\sigma^{L} \equiv W^{L},\quad \sigma^{T} \equiv W^{T},\quad \sigma^{T'} \equiv \tilde{W}^{T'}$$
$$\sigma^{TL} \equiv W^{TL}\cos\Phi + \tilde{W}^{TL}\sin\Phi$$
$$\sigma^{TL'} \equiv W^{TL'}\sin\Phi + \tilde{W}^{TL'}\cos\Phi \tag{110}$$
$$\sigma^{TT} \equiv W^{TT}\cos2\Phi + \tilde{W}^{TT}\sin2\Phi\;.$$

Then, the generalized forms for Eqs. (102) become

$$\Sigma \sim v_L \; \sigma^L + v_T \; \sigma^T + v_{TL} \; \sigma^{TL} + v_{TT} \; \sigma^{TT}$$
$$\Delta \sim v_{T'} \; \sigma^{T'} + v_{TL'} \; \sigma^{TL'} \; . \tag{111}$$

Now finally let us remove a factor v_T from every term to obtain

$$\frac{d\sigma}{d\omega \, d\Omega_e \, d\Omega_1 \ldots} \sim v_T \left\{ \left[\mathcal{E}(2\rho \; \sigma^L) + (\sigma^T) \right. \right.$$
$$\left. - \mathcal{E}(\sigma^{TT}) - \frac{1}{\sqrt{2}} \sqrt{\mathcal{E}(1+\mathcal{E})} (\sqrt{2\rho} \; \sigma^{TL}) \right]$$
$$\left. + h \left[\sqrt{1-\mathcal{E}^2}(\sigma^{T'}) - \frac{1}{\sqrt{2}} \sqrt{\mathcal{E}(1-\mathcal{E})} (\sqrt{2\rho} \; \sigma^{TL'}) \right] \right\} \; , \tag{112}$$

where $\mathcal{E}$ is defined in Eq. (39). If all kinematical variables except the electron scattering angle θ_e are fixed then a plot versus $\mathcal{E}$ will permit a decomposition of the cross section in Eq. (112) into individual responses. Note, however, that the L and TT responses are both multiplied by $\mathcal{E}$ and so cannot be separated in this manner; in fact, the Φ–dependence in the TT responses (see Eqs. (110)) must be used in this case. Note also that this convenient rewriting of the general formula is accomplished by adjoining a factor $\sqrt{2\rho}$ to each response in which there is an occurrence of the longitudinal projection of the current (and so a factor of 2ρ for the L term where this projection occurs bilinearly). In some work these factors are absorbed in new definitions (for example, $2\rho\sigma^L \to \sigma^L$) and so some caution must be exercised in interpreting just which type of response is meant.

VII. SUMMARY

In summary, in these lecture notes I have attempted to cover several aspects of the formalism required when studying polarization degrees of freedom in lepton–induced reactions relatively completely, while only briefly touching on other aspects of the problem. The treatment of the leptonic tensor is basic and consequently that subject has been dealt with in some detail. Naturally most of the leptonic developments have straightforward extensions to include the full electroweak interaction (see, for example, Refs. [1,7,17] and, in particular, Ref. [18] which contains a review of the subject of parity–violating electron scattering), although space and time did not permit their being discussed here. Where inclusive scattering is concerned, I chose to focus on a new topic, leptonic polarization transfer, as it helps in understanding the nature of polarized, ultra–relativistic lepton scattering. The subject of inclusive electron from polarized targets was omitted from these lectures (again for lack of space) and the reader is directed to Ref. [2] for detailed discussion in that area. Likewise, while the very general nature of coincidence electron scattering was developed here and, specifically, the polarized plane–wave impulse approximation was discussed, the formal treatment of coincidence and polarization to be found in Ref. [3] again goes beyond the context of this lecture series. Where the polarized PWIA is concerned only some of the basic ideas were presented (and no extensions beyond PWIA, such as the DWIA, were treated) — once again the reader is pointed to recent literature[10,16] for more thorough exposition of the requisite ideas.

ACKNOWLEDGMENT

This work is supported in part by funds provided by the U. S. Department of Energy (D.O.E.) under contract #DE-AC02-76ER03069.

REFERENCES

1. T. W. Donnelly, "Electromagnetic and Weak Currents in Nuclei," in *Symmetries in Nuclear Structure*, ed. K. Abrahams, K. Allaart and A. E. L. Dieperink (Plenum, 1983), p. 1.

2. T. W. Donnelly and A. S. Raskin, *Ann. Phys.* **169** (1986) 247.

3. A. S. Raskin and T. W. Donnelly, *Ann. Phys.* **191** (1989) 78.

4. J. D. Bjorken and S. D. Drell, *Relativistic Quantum Mechanics* (McGraw-Hill, N.Y., 1964).

5. T. deForest, Jr. and J. D. Walecka, *Adv. in Phys.* **15** (1966) 1.

6. T. W. Donnelly and J. D. Walecka, *Ann. Rev. Nucl. Sci.* **25** (1975) 329.

7. T. W. Donnelly, *Prog. in Particle and Nuclear Physics* **13** (1985) 183.

8. T. W. Donnelly, in *New Vistas in Electro-Nuclear Physics*, E. L. Tomusiak, H. S. Caplan, and E. T. Dressler, eds. (Plenum, 1986), p. 151.

9. R. von Gehlen, *Phys. Rev.* **118** (1960) 1445; M. Gourdin, *Nuovo Cim.* **21** (1961) 1094; J. D. Bjorken, (unpublished, 1960).

10. J. A. Caballero, T. W. Donnelly and G. I. Poulis, *Nucl. Phys.* **A555** (1993) 709.

11. A. L. Dieperink and T. de Forest, *Ann. Rev. Nucl. Sci.* **25** (1975) 1.

12. S. Frullani and J. Mougey, *Adv. in Nucl. Phys.* **14** (1984).

13. D. B. Day, J. S. McCarthy, T. W. Donnelly and I. Sick, *Ann. Rev. Nucl. and Part. Sci.* **40** (1990) 357.

14. T. de Forest, *Nucl. Phys.* **A392** (1983) 232.

15. H. W. L. Naus, S. J. Pollock, J. H. Koch and U. Oelfke, *Nucl. Phys.* **A509** (1990) 717 and references therein.

16. J. A. Caballero, T. W. Donnelly and G. I. Poulis, MIT preprint CTP#2098.

17. T. W. Donnelly and R. D. Peccei, *Phys. Rep.* **50** (1979) 1.

18. M. J. Musolf, T. W. Donnelly, J. Dubach, S. J. Pollock, S. Kowalski and E. J. Beise, CEBAF preprint #TH-93-11 (submitted to *Physics Reports*).

QUARK STRUCTURE OF THE NUCLEON
AND NUCLEON RESONANCES

Bernard Metsch

Institut für Theoretische Kernphysik
Nußallee 14-16
D-5300 Bonn
Federal Republic of Germany

INTRODUCTION

Strong interaction physics deals with the structure and dynamics of hadrons, which can suitably be classified according to the baryon number B. The simplest hadronic systems accordingly are the mesons ($B = 0$) and the baryons ($B = 1$). The internal structure of these objects is reflected by their excitation spectra and their decay properties. Ultimately it is expected, that these can be calculated from the fundamental theory of strong interactions: QuantumChromoDynamics. So far, however, it has not been possible to rigorously apply this theory even to the simplest hadrons and to achieve a realistic description of their excitation patterns. Accordingly, the explanation of the hadronic spectra is the realm of a wide variety of QCD-inspired, but nevertheless phenomenological models, among which the most prominent are:

- constituent quark models:

  - non-relativistic potential models;

  - bag-models.

- topological models based on non-linear meson dynamics, like e.g. the Skyrme-model.

In the present context, we will not try to compare these different approaches, but will try to use rather consistently the framework of one of the most successful models, the (non-relativistic) constituent quark model (CQM)[1-27], where hadrons are understood as bound states of quarks, resulting from a Schrödinger equation, to arrive at a description of the properties of all hadronic systems, mesons and baryons alike.

After a brief survey and classification of hadronic systems we will introduce the constituent quark model by a discussion of the charmonium and bottomonium spectra, where the justification of the various approximations is not too questionable. Here

Perspectives in the Structure of Hadronic Systems
Edited by M.N. Harakeh *et al.*, Plenum Press, New York, 1994

we will emphasize the two basic ingredients in any constituent quark model: *i.e.* the modeling of confinement (here by a linearly rising potential, reflecting a string between quarks), and the modeling of the quark-(anti)quark residual interaction.

We will demonstrate, that, although for the heavy quarkonia in analogy to the positronium spectrum, the latter can be suitably be taken as the non-relativistic approximation to the One-Gluon-Exchange (*i.e.* the Fermi-Breit Potential), the extrapolation with this interaction to light mesons does not lead to a consistent description. Fortunately, there exists an alternative QCD-based interaction, stemming from instanton effects, that will be exploited as an effective quark interaction for the light hadronic systems, and that indeed is found to lead to a very efficient description of the light meson spectra. Here, we will also treat the calculation of additional observables in the present framework, such as *e.g.* the electromagnetic properties of hadrons.

The very same modeling of the confinement and the residual quark-interaction will then be applied in a description of the spectra and other properties of baryons, the main topic of this lecture. Here we will in particular emphasize an important additional aspect, namely that virtually all hadrons are not bound states, but resonances that exhibit rather large widths for hadronic decay. In a simple model for this hadronic decay we will account for this by the construction of an effective, energy dependent Hamilton operator, that describes the bound states, such as the nucleon and the resonances on equal footing. In this formulation it is also possible to compute the scattering amplitudes for two-particle reactions, and we will discuss the results for a quark-model based description of elastic pion-nucleon scattering and photo-production of mesons at intermediate energies.

We will try to give a critical evaluation of the CQM, and some implications for future work, as well as for the extension to the description in the relevant energy domain of more complex hadronic systems, like nuclei.

HEAVY QUARKONIA

In this section we will briefly review the structure of the so called heavy quarkonia, i.e. the bottomonium and charmonium system, within the framework of the constituent quark model. The main purpose here is to introduce some notation, some technicalities, and to check the validity of the various approximations by a comparison to some experimental numbers.

The similarity of the mass spectra of charmonium and bottomonium with the positronium spectrum makes it plausible to try to explain in these systems the mass splittings by a quark interaction that is very similar to the non relativistic approximation of the one photon exchange, i.e. the Breit-Fermi interaction. We will show, that together with a parameterization of linear confinement potential , that is based on a scalar interaction, and that will be taken in the same order of (p/m), one can obtain a satisfactory description of most properties of the heavy quarkonia.

The Hamiltonian

If we restrict the description of the heavy mesonic states to $q\bar{q}$-configurations the Hamiltonian of the non-relativistic quark model is given by

$$H_{q\bar{q}} = m_q + m_{\bar{q}} + T + V_c + W_{q\bar{q}} \tag{1}$$

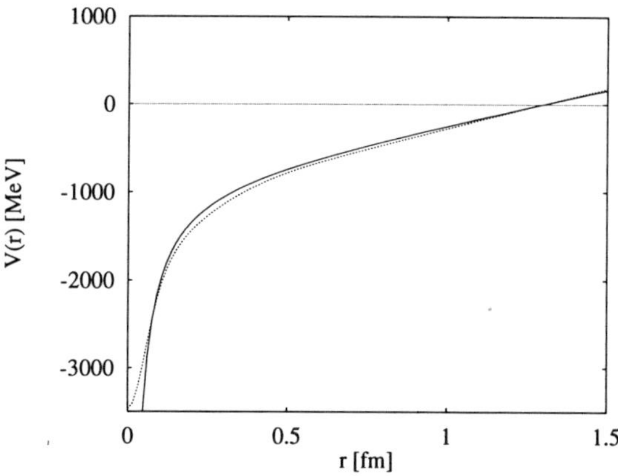

Figure 1. The original (solid curve) and the regularized (dashed curve) central potential for the heavy quarkonia.

with m the so called constituent quark mass

$$T = \frac{\vec{p}^2}{2\mu} \, with \, \mu = \frac{m_q m_{\bar{q}}}{m_q + m_{\bar{q}}} \tag{2}$$

the kinetic energy of the relative motion, with $\vec{p}$ the relative momentum, and where

$$V_c = V_c^C + V_c^{LS} + V_c^D \tag{3}$$

with

$$V_c^C(r) = a + br, \qquad r = |x|, \tag{4}$$

is the linear confinement potential with string tension b; x is the relative coordinate. The other terms one obtains in the expansion up to order $(p/m)^2$ and are called spin-orbit term and Darwin term,respectively.

For the quark-(anti)quark residual interaction $W_{q\bar{q}}$ one takes the Breit-Fermi potential, by expanding the One-Gluon-Exchange up to order $(p/m)^2$:

$$W_{BF} = W_{BF}^C + W_{BF}^{SS} + W_{BF}^T + W_{BF}^{LS} + W_{BF}^{LL} + W_{BF}^D \tag{5}$$

with the Coulomb potential

$$W_{BF}^C(r) = -\frac{4}{3}\alpha_s \frac{1}{r} \tag{6}$$

where α_s is the effective strong coupling constant and $-4/3$ is the expectation value of the color dependence $(\lambda^a \cdot \lambda^a)$ for a quark-antiquark color singlet state. The velocity dependent terms are the spin-spin interaction, the tensor interaction, the spin-orbit interaction, the Darwin-term of the Breit-Fermi interaction and the orbit-orbit interaction, respectively.

As a consequence of the non-relativistic expansion this interaction is too singular at $r = 0$ and in this form not suited for a non-perturbative treatment. We therefore

Table 1. Interaction parameters for bottomonium and charmonium

m_c	1907	MeV
m_b	5306	MeV
a	-864	MeV
b	740	MeV/fm
α_s	0.474	
r_0	0.14	fm

regularize this interaction by expanding the Coulomb potential in a interval $r_0 < r < 4r_0$ by a sum of gaussians:

$$\bar{V}^C_{BF}(r) = -\frac{4}{3}\alpha_s \frac{1}{2r_0}\sum_k c_k \exp\left(-\frac{\gamma_k r}{2r_0}\right)^2 \tag{7}$$

This makes the central potential finite at $r = 0$ and lets the velocity dependent terms vanish at the origin. The same is achieved for the velocity dependent terms of the confinement potential by setting

$$\bar{V}^C_c(r) = a + br\left(1 - \exp-\left(\frac{r}{2r_0}\right)^2\right). \tag{8}$$

The resulting central potential is depicted in Fig. 1.

The matrix elements of the Hamiltonian 1 are calculated in a finite harmonic oscillator basis

$$\langle \vec{x}\,|n(ls)jm\rangle = \sum_{m_s,m_l} \langle s\,m_s\,l\,m_l\,|j\,m\rangle\, \xi^s_{m_s} Y^l_{m_l}(\Omega_x) R^{(b)}_{nl}(|x|), \tag{9}$$

encompassing up to 20 oscillator excitations. Spectra are obtained by applying the Ritz variational principle: The Hamiltonian is diagonalized in the oscillator basis and the energy expectation value is minimized with respect to the oscillator length parameter

$$b^2 = \frac{\hbar}{mc}. \tag{10}$$

The Hamiltonian contains 6 parameter

$$H_{q\bar{q}} = M[m] + T[m] + V_c[a,b,r_0] + W_{q\bar{q}}[m,\alpha,r_0] \tag{11}$$

with $m = m_c, m_b$, to be adjusted to the experimental spectrum, the values obtained are given in Table 1.

Bottomonium and Charmonium Spectra

A comparison of the experimental and calculated spectrum of charmonium can be found in Fig. 2. The description of bottomonium is of the same quality. With the 6 parameter we get a reasonable description of the experimental situation: the mean deviation is about 18 MeV; the largest deviation of about 30 MeV is found for the η_c- and the Υ-resonance.

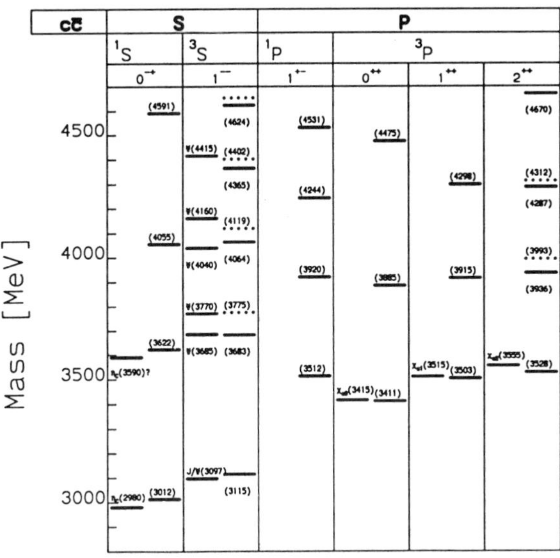

Figure 2. Charmonium spectrum. For fixed J^{PC} in each column the experimental states (left) are compared to the calculated states (right). The latter are classified according to the dominant orbital angular momentum, the states with $L = J + 1$ are denoted by dots.

Electromagnetic Properties

As an example for the procedure to calculate the electromagnetic decay widths we consider the leptonic decay: $Q\bar{Q} \to e^+ e^-$. We start from the following representation of the mesonic state:

$$\langle nLSJ; P \,|M_\mu\rangle - \int d^3p\, R_{\mu l}(p) \left[Y_l(\Omega_p) \times \left[a^\dagger(\tfrac{1}{2}P + p) \times b^\dagger(\tfrac{1}{2}P - p) \right]^S \right]^J |\rangle \qquad (12)$$

where $R(p)$ is the radial momentum space wave function and $a^\dagger(p)$ and $b^\dagger(p)$ are the creation operators for a quark and antiquark with momentum p, respectively. Using the full free Dirac spinors to calculate the lowest order Feynman diagram for this process we find for the decay width of the vector mesons:

$$\Gamma_{ll} = \frac{16\pi\alpha^2 e_q^2}{M^2} 4\pi \int p^2 dp\, R_{\mu l}(p) \left| \frac{m + 2\sqrt{m^2 + p^2}}{3\sqrt{m^2 + p^2}}\delta_{l,0} + \frac{\sqrt{m^2 + p^2} - m}{2\sqrt{m^2 + p^2}}\delta_{l,2} \right|^2 \qquad (13)$$

where M is the meson mass, $\alpha = \frac{1}{137}$ the electromagnetic fine structure constant and e_q, m_q the quark charge and mass, respectively. If one makes the usual non relativistic reduction and keeps only the momentum independent terms, the d-wave contribution vanishes and one finds:

$$\Gamma_{ll} = \frac{16\pi\alpha^2 e_q^2}{M^2} |\Psi(0)|^2 \qquad (14)$$

where $\Psi(0)$ is the value of the coordinate space wave function for the relative motion at the origin. This approximation is known as the van Royen-Weißkopf formula. Both prescriptions for the calculation of leptonic decay widths are compared to experimental data in Table 2. In a similar manner one can also calculate the electric and magnetic dipole transitions.

Table 2. Electromagnetic decay widths for bottomonium and charmonium

initial-state	final-state	Exp.[a]	model rel.[b]	nonrel.[c]
			Γ [keV]	
$\Upsilon(1s)$	e^+e^-	1.34(4)	1.24	1.32
$\Upsilon(2s)$	e^+e^-	0.59(3)	0.51	0.56
	$\chi_{b0}\gamma$	1.9(6)	1.1	1.4
	$\chi_{b1}\gamma$	2.9(7)	1.9	2.3
	$\chi_{b2}\gamma$	2.9(7)	1.8	2.2
$\Upsilon(3s)$	e^+e^-	0.44(3)	0.35	0.38
	$\chi_{b0}\gamma$	1.2(4)	1.2	1.4
	$\chi_{b1}\gamma$	2.9(7)	2.2	2.6
	$\chi_{b2}\gamma$	3.1(8)	2.3	2.7
$\Upsilon(4s)$	e^+e^-	0.24(5)	0.28	0.31
$J/\Psi(1s)$	e^+e^-	4.7(4)	5.3	5.7
	$\eta_c\gamma$	0.9(3)	1.5	1.8
$\Psi(2s)$	e^+e^-	2.2(2)	2.3	2.6
	$\eta_c\gamma$	0.7(2)	0.9	5.9
	$\chi_{c0}\gamma$	23(5)	24	34
	$\chi_{c1}\gamma$	21(4)	31	44
	$\chi_{c2}\gamma$	19(4)	26	37
$\Psi(2d)$	e^+e^-	0.26(4)	0.01	0.002
$\Psi(3s)$	e^+e^-	0.75(2)	1.59	1.85
$\Psi(3d)$	e^+e^-	0.77(2)	0.02	0.003
$\Psi(4s)$	e^+e^-	0.47(1)	1.14	1.35
χ_{c0}	$J/\Psi(1s)\gamma$	92(40)	189	317
χ_{c1}	$J/\Psi(1s)\gamma$	230(50)	264	315
χ_{c2}	$J/\Psi(1s)\gamma$	243(40)	271	321

[a]see[28]
[b]see Eq. 13
[c]see Eq. 14

It is found, that in spite of the large constituent quark masses in these systems, the corrections to the non-relativistic expression are appreciable, especially for the magnetic dipole transitions. This is related to a node in the wave function, that makes the result strongly depend on details of calculation. Although the corrections seem to be too large for bottomonium we thus obtain a overall satisfactory description of the electromagnetic properties.

Summary and Evaluation

In this section we described the dynamics of the heavy quark-antiquark systems by a non-relativistic approximation of all terms of the order of $(p/m)^2$ to the one gluon exchange, viz. the full Breit-Fermi potential, and a linear confinement potential assumed to be of scalar origin. It was found that the inclusion of the spin-independent, momentum dependent terms are important for a realistic description of electromagnetic decays.

In the description of electromagnetic properties it was found, that relativistic effects, especially taking into account the lower components of the Dirac spinors, are quantitatively important. It should be stressed, however, that the prescription presented here is of course not a covariant treatment. The magnitude of the relativistic corrections indicates, that a covariant description of the heavy quarkonia, e.g. within the framework of the Bethe-Salpeter equation, should be necessary for an improved description of these systems.

These remarks apply of course *a forteriori* for the light mesons and baryons, which are built of (u, d, s)-quarks. Even if one assumes, that these systems are to be described with so called constituent quark masses of about a few 100 MeV and not with the current quark masses (10 MeV for (u, d) and about 150 MeV for s), the mass splittings relative to the total mass are so large, that a non-relativistic treatment seems not very appropriate. Nevertheless, in view of fact, that a covariant treatment of many body bound states is still not established, it is useful to apply the basically non-relativistic treatment presented here also to the light hadronic systems.

LIGHT MESONS

A Constituent Quarkmodel for Light Mesons

As for the heavy quarkonia the Hamiltonian for mesons as $q\bar{q}$-configurations is of the form

$$H_{q\bar{q}} = m_q + m_{\bar{q}} + T + V_c + W_{q\bar{q}} \tag{15}$$

However, it was found that the Breit-Fermi potential, even after regularization that guarantees, that the Hamiltonian is a bounded operator, is not a suited residual interaction for the light systems: Especially the spin-orbit interaction is much too large to account for the experimental spectrum. Moreover, on the basis of a flavor independent interaction, it is not possible to explain the observed mass splittings of the low-lying pseudoscalar mesons π-η-η'.

Fortunately, there is an other QCD-based candidate for a residual quark interaction, that is based on instanton effects. This interaction was proposed by 't Hooft[29] to solve the so-called $U_A(1)$-problem; in this context this affects the mass splitting of the pseudoscalar mesons. He showed, that an expansion of the (euclidian) action around the one-instanton solutions of the gauge fields and assuming dominance of the zero modes of the fermion fields leads to an effective three-fermion interaction, that after normal ordering gives a contribution to the quark-antiquark interaction, that can be written as

$$
\begin{aligned}
V(2) = {} & -\frac{1}{4} g_{\mathrm{eff}}(j) \sum_{\alpha\gamma>0,\beta\delta<0} \varepsilon_{j,i_\alpha i_\beta}\varepsilon_{ji_\gamma i_\delta} q_\alpha^\dagger q_\beta^\dagger q_\gamma q_\delta \\
& \left\{ \left\langle \alpha\beta \left| (\gamma_0\cdot\gamma_0 + \gamma_0\gamma_5\cdot\gamma_0\gamma_5)(2 - \frac{3}{8}\lambda^a\cdot\lambda^a) \right| \gamma\delta \right\rangle \right. \\
& \left. - \left\langle \alpha\beta \left| (\gamma_0\cdot\gamma_0 + \gamma_0\gamma_5\cdot\gamma_0\gamma_5)(2 - \frac{3}{8}\lambda^a\cdot\lambda^a) \right| \delta\gamma \right\rangle \right\}
\end{aligned}
\tag{16}
$$

The interaction strength g_{eff} stems from an integration over the instanton density in one or two-loop approximation[30]. To apply this interaction in the constituent quark model we expand the quark fields in (p/m). In lowest order, keeping only the momentum independent terms, one finds for the matrix elements for mesonic quark-antiquark color singlet states:

$$\left\langle S'L'(FT)'1_c \left| V_{q\bar{q}}^{(2)} \right| SL(FT)1_c \right\rangle = \langle (FT)' | O_F(g,g') | (FT) \rangle\, \mathcal{M}\, \delta_{S'S}\delta_{S0}\delta_{L'L}\delta_{L0} \tag{17}$$

with $\mathcal{M}$ the radial matrix element and the flavour dependence given by

$$\langle (FT)' \mid O_F(g,g') \mid (FT) \rangle = \delta_{F'F}(-8g\delta_{T1} - 8g'\delta_{T\frac{1}{2}}) + \frac{8}{3}\delta_{T0}\langle F' \mid \mathcal{F}(g,g') \mid F \rangle \quad (18)$$

with the flavor matrix

$$\mathcal{F}(g,g') = \begin{pmatrix} g - 4g' & \sqrt{2}(g - g') \\ \sqrt{2}(g - g') & 2(g + 2g') \end{pmatrix} \quad (19)$$

where $F = \{8, 1\}$ is the dimension of the $SU(3)_F$-representation and where

$$g = \frac{3}{8}g^{eff}(s), g' = \frac{3}{8}g^{eff}(n). \quad (20)$$

This force acts in the present approximation only on pseudoscalar mesons, where it is attractive for isovector mesons and repulsive for the isoscalar states and also mixes the flavour singlet and flavour octet states. This property is unique for the 't Hooft interaction: It preserves within a relativistic context the chiral symmetry while at the same time it maximally breaks $U_A(1)$-symmetry.

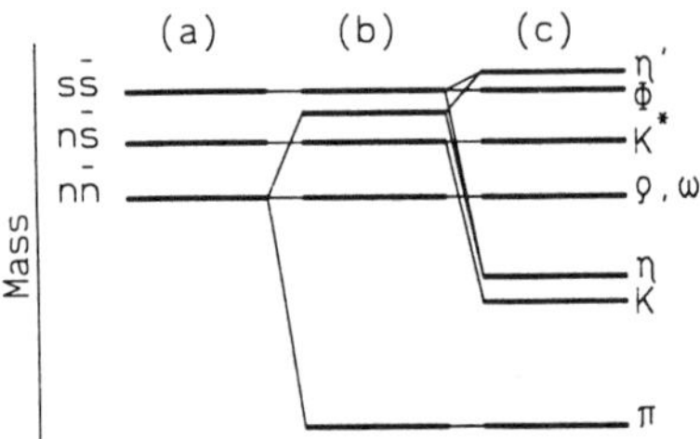

Figure 3. Schematic representation of the instanton induced mass splittings for light mesons. Column (a) shows the spectrum for $g = g' = 0$, column (b) for $g' = 0$ and column (c) for the full interaction.

In a matrix notation the flavour dependence can be given explicitly for the pseudoscalar mesons by

$$\begin{pmatrix} \pi^0 & \eta_n^0 & \eta_s^0 \end{pmatrix} \begin{pmatrix} -8g & 0 & 0 \\ 0 & 8g & 8g'\sqrt{2} \\ 0 & 8g'\sqrt{2} & 0 \end{pmatrix} \begin{pmatrix} \pi^0 \\ \eta_n^0 \\ \eta_s^0 \end{pmatrix} \quad (21)$$

where we denoted the pseudoscalar, isoscalar states with non-strange and strange quarks by $|\eta_n^0\rangle$ and $|\eta_s^0\rangle$, respectively. For the strange pseudoscalar mesons one finds the following non-vanishing matrix elements:

$$\langle q\bar{q}; S, L, T \mid W \mid q\bar{q}; S, L, T \rangle_{|S^*=1|} = -8g'\delta_{S,0}\delta_{L,0}W \quad (22)$$

In lowest order the instanton induced interaction yields the splitting of the low-lying vector- and pseudoscalar mesons as sketched in Fig. 3: The π-ρ-splitting is due to the interaction between non-strange quarks with strength g, the splitting of the ω-ϕ and

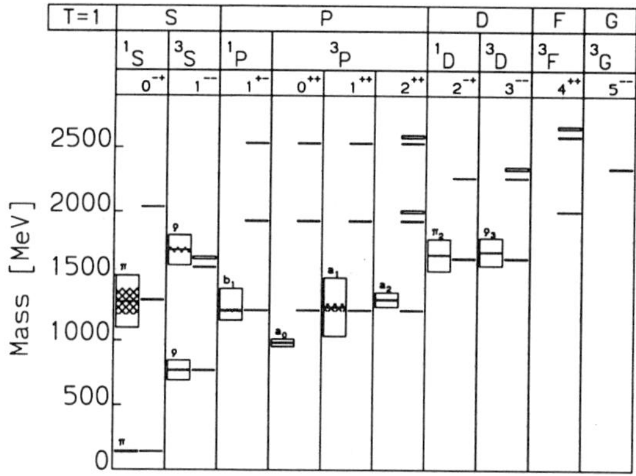

Figure 4. Isovector meson spectrum. The right hand side of each column shows the experimental resonance position[28]; the left hand side is the result from a constituent quark model with a linear confinement potential and an instanton induced force, that acts only in the pseudoscalar sector.

η-η' is due to the interaction between a non-strange and a strange quark with strength g', that also induces the $K - K^*$-splitting.

As it stands, the 't Hooft interaction is a pure contact force, as a consequence of the semi-classical approximation used. As such it would lead to a unbounded Hamiltonian. It is therefore necessary to regularize this force, e.g. by replacing the δ-distribution by a Gaussian:

$$\delta^3(r) \rightarrow \frac{1}{\lambda^3 \pi^{3/2}} \exp - \left(\frac{x}{\lambda}\right)^2 \tag{23}$$

where we interpret λ as the range of the force. It is to be expected, that an improved treatment of the instanton effects indeed would lead to an interaction with a finite range.

The spectra of the light mesons are obtained with the calculational procedure described in the previous section.

Spectra of Light Mesons

The resulting spectra for the light isoscalar, isovector and strange mesons are compared to the experimental known resonances in Figs. 4, 5 and 6, respectively.

With the instanton induced interaction the explanation of the mass splittings is in fact rather simple: Due to the selection rules this interaction acts exclusively in the pseudoscalar sector. Therefore the spectra in the other sectors are determined by the kinetic energy and the confinement potential alone. From the Figs. 4–6 it can be seen, that in this manner indeed an acceptable description is obtained. An exception is the position of the scalar mesons ($^3P_0$-states). It should be stressed that also the confinement potential in the present approximation only has a central component. We will come back to this problem below. The instanton induced interaction is attrac-

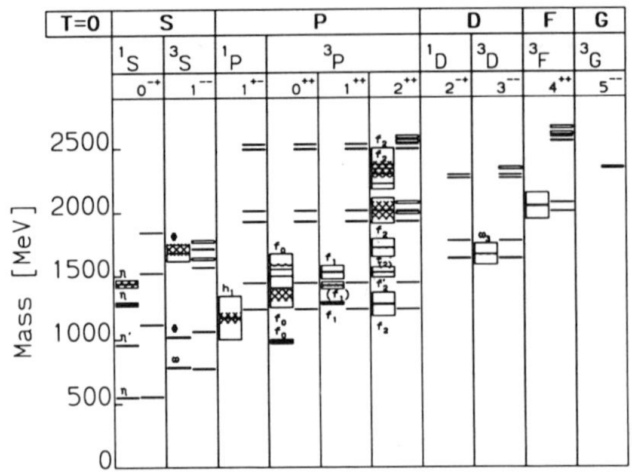

Figure 5. Isoscalar meson spectrum. See also caption to Fig. 4.

Table 3. Electromagnetic decay widths for light vector mesons

initial-state	final-state	Exp.[a]	rel.[b]	nonrel.[c]
			model	
ρ	e^+e^-	6.77(32)	3.26	4.65
ω	e^+e^-	0.60(2)	0.35	0.50
ϕ	e^+e^-	1.37(5)	1.06	1.32

[a] see[28]
[b] see Eq. 13
[c] see Eq. 14

tive for isovector states and repulsive for isoscalar states. Furthermore it mixes the $s\bar{s}$-configurations and the $n\bar{n}$-configurations, through the nonstrange-strange quark interaction, that also determines the splitting of the pseudoscalar and vector K-mesons. In this way a unified description of the $\rho - \pi$-, the $\pi - \eta - \eta'$- and the $K - K^*$-splittings determined by the parameters g and g' is obtained. Taken together with the other sectors and in view of the simplicity we thus obtain a good description of the gross proprties of the light meson spectrum.

Although the majority of the known meson resonances can be explained in terms of $q\bar{q}$-configurations, there are some exceptions, the most prominent being the low position of the a_0 and f_0 mesons. Indeed there is some evidence, that these mesons have a more complicated structure: in a coupled channel approach of the $\pi\pi$-scattering in a meson exchange framework the Jülich group[31] showed, that the strong increase of the scalar, isoscalar S_{11}-phase shift at $E_{cm} = 980$ MeV can be explained as due to the coupling of the $\pi\pi$-channel to the $K\bar{K}$-channel, the interaction in both channels through vector meson exchange being crucial. Furthermore it was found that the increase of the phase shift beyond this energy can be explained if one assumes a scalar s-channel resonance

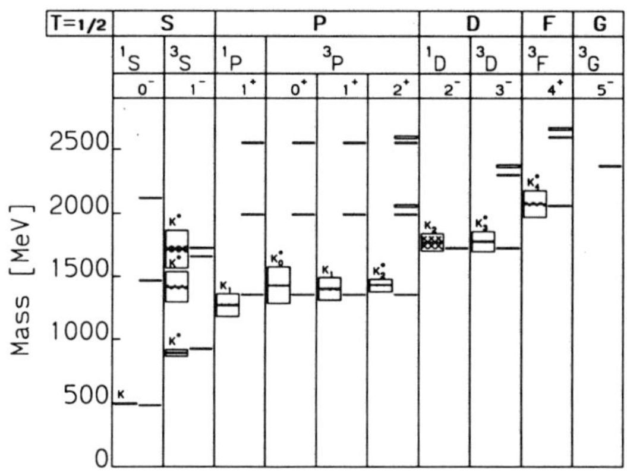

Figure 6. strange meson spectrum. See also caption to Fig. 4.

Table 4. Weak decay constants for light pseudoscalar mesons

Meson	Exp.[a]	model	
		rel.[b]	nonrel.[c]
π	131.7(2)	465	2248
K	161(2)	347	1171

[a] see[28]
[b] as in Eq. 13
[c] as in Eq. 14

at 1400 MeV. This would imply, that the scalar quark-antiquark state would, as the other P-wave mesons, be positioned at higher excitation energies. This would support the results discussed above. Similar conclusions were also found by Weinstein and Isgur[32, 33] in a variational calculation with subsequent projection on a meson-meson coupled channel problem.

Such calculations, taken together with the observation, that the majority of the mesons are in fact resonances, that couple strongly to multi-meson states show, that our understanding of the meson spectrum is still rather incomplete. Indeed, the large decay widths indicate, that the mixing between $q\bar{q}$, $q^2\bar{q}^2$ and $q^3\bar{q}^3$ configurations could be appreciable. We will touch upon this problem again when we treat the baryonic spectra.

Electromagnetic Properties of Light Mesons

As for the heavy quarkonia we will test the quality of the wave functions obtained by a calculation of some electroweak decays, see Table 3.

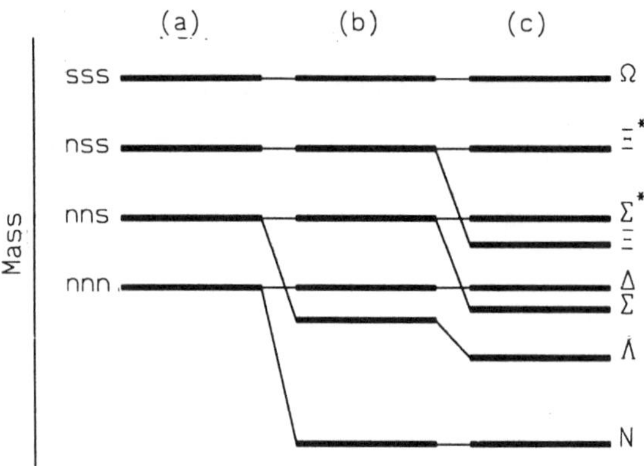

Figure 7. Schematic representation of the instanton induced mass splittings of the baryons in the low-lying flavour octet and decuplet states. Column (a) shows the spectrum for $g = g' = 0$, column (b) for $g' = 0$ and column (c) for the full interaction.

Although the relativistic correction in view of the small quark masses are rather moderate, we find that the absolute decay widths are in general too small. For the weak decays $\pi^+ \to \mu^+ + \nu_\mu$ and $K^+ \to \mu^+ + \nu_\mu$ defined by

$$\langle 0 \, | \, \psi_\pi(0)\gamma_\mu\gamma_5\psi_M(0) \, | \, \pi; p \rangle = f_M \cdot p_\mu \tag{24}$$

one finds the results given in Table 4.

The corrections found for these observables are huge and can be traced back to destructive contributions of the upper and lower components of the Dirac spinors, very similar to results found in so-called bag models. At this point we obviously have reached the limits of the (non-relativistic) constituent quark model.

BARYONS

Within the same model that was used for the description of the light mesons we will now calculate the baryon spectrum, where we will emphasize a consistent description of both systems.

We start again with the constituent quark model Hamiltonian. For baryons as q^3-configurations this can be written as:

$$H_b = M + T_{\rm rel} + V + W \tag{25}$$

where

$$M = \sum_{i=1}^{3} m_i \tag{26}$$

is the sum of the constituent quark masses and

$$T_{\rm rel} = \sum_{i=1}^{3} \frac{p_i^2}{2m_i} - \frac{P^2}{2M} \tag{27}$$

is the kinetic energy of the relative motions. Confinement will be parameterized by a three body force, with a potential that is proportional to the length of the shortest

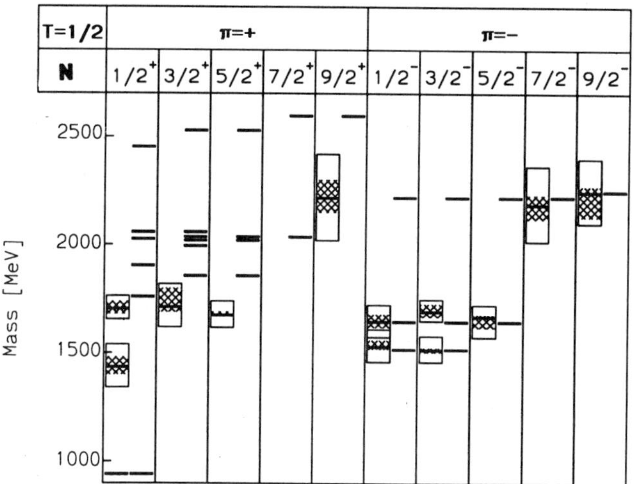

Figure 8. Nucleon spectrum. In each column for fixed J^{PC} the experimentally known states (left)[28] are compared to the calculated levels (right). The experimental resonance position is indicated by a line, the hatched area indicating the experimental uncertainty. The total decay width is represented by a rectangle.

string between three quarks[34, 35, 36]:

$$V(x_1, x_2, x_3) = a_{q^3} + b\min(x_0)\sum_{i=1}^{3}|x_i - x_0| \tag{28}$$

As residual interaction we will again employ the instanton induced force, which for this case is a sum of two body interactions:

$$W = \sum_{i<j=1}^{3} W(i,j) \tag{29}$$

where

$$W(i,j) = -2g\mathcal{P}_{\bar{3}}^{f}\left(2\mathcal{P}_{\bar{3}}^{c} + \mathcal{P}_{6}^{c}\right)\delta^{(3)}(x_i - x_j) \tag{30}$$

with $\mathcal{P}_{\bar{3}}^{c}(\mathcal{P}_{6}^{c})$ the projection operator on the color antitriplet (sextet) state and $\mathcal{P}_{\bar{3}}^{f}$ the projection operator on the flavour-antisymmetric two quark states. The term with $\mathcal{P}_{6}^{c}$ does not contribute to baryons, since the three quark states are in a color singlet state and thus two quarks are always in a antitriplet state. As for the mesons this force is regularized in order to guarantee that the resulting Hamiltonian is a bound operator.

In lowest order of the non-relativistic approximation we find the following non-vanishing matrix elements:

$$\left\langle q^2; \bar{3}_c, S, L, F, T \mid W \mid q^2; \bar{3}_c, S, L, F, T \right\rangle =$$
$$\delta_{S,0}\delta_{L,0}\left(-4g\delta_{F,\bar{3}}\delta_{T,0} - 4g'\delta_{F,\bar{3}}\delta_{T,\frac{1}{2}}\right) \tag{31}$$

where we have labeled the two-quark states by the dimension of the color representation, their spin, their angular momentum and the dimension of the flavour representation

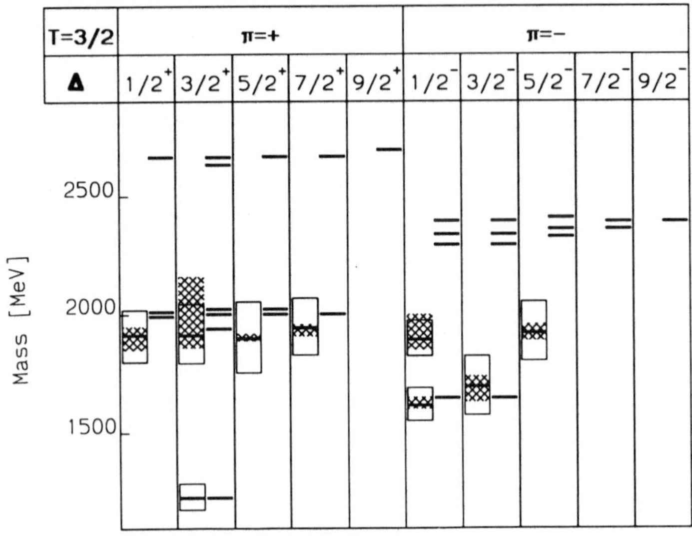

Figure 9. Δ-resonance spectrum. See also caption to Fig. 8.

and isospin, respectively. In this approximation this interaction is pairing force, that correlates two-quark states with trivial spin and orbital angular momentum, with a scale, which is determined by the range λ of the force, if we regularize the interaction as for the mesons. Consequently only those baryons will be affected by this force, that contain such diquark-components. This implies, that the flavour decuplet states will not be influenced by this interaction. For the octet states the shift of the nucleonic states is largest, since the diquark correlation for non-strange quarks is larger than that for diquarks with strangeness. This also explains the position of the Λ-resonance below the the Σ-resonance. The splittings are sketched in Fig. 7.

The matrix elements of the Hamiltonian of Eq. 25 are calculated in a SU(6)-spin-flavor, O(3)-oscillatorbasis of completely antisymmetric three-quark states, which can be written as a linear combination of product states:

$$\left\langle \vec{\rho}, \vec{\lambda} \,|(n_1 l_1)(n_2 l_2)L, S, J, F\right\rangle =$$
$$R^{b_\rho}_{n_1 l_1}(\rho) R^{b_\lambda}_{n_2 l_2}(\lambda) \left[\left[Y^{l_1}(\Omega_\rho) \times Y^{l_2}(\Omega_\lambda)\right]^L \times \chi^S\right]^J \Theta^F \Xi^c_0 \tag{32}$$

where the basis is truncated by the condition that the number of oscillator excitations is limited by

$$N = 2n_1 + l_1 + 2n_2 + l_2 \le 5 \tag{33}$$

Here the so-called Jacobi-coordinates are defined by

$$\vec{\rho} = \vec{x}_1 - \vec{x}_2$$
$$\vec{\lambda} = \frac{m_1 \vec{x}_1 - m_2 \vec{x}_2}{m_1 + m_2} - \vec{x}_3$$

In this finite basis the Hamiltonian is diagonalized and the expectation value of the energy of the lowest states is minimized with respect to the oscillator length.

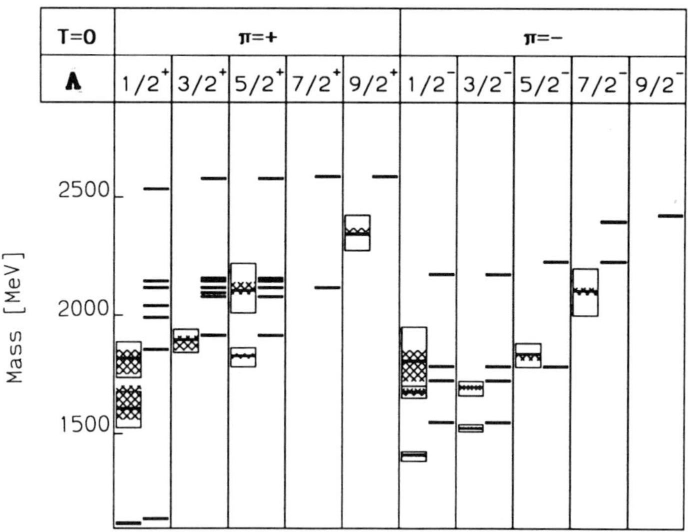

Figure 10. Λ-resonance spectrum. See also caption to Fig. 8.

Baryon Spectra

The calculated spectra for nucleons, Δ-, Λ-, Σ- and Ξ-, Ω-resonances are compared to the experimental data in Figs. 8-12.

As discussed in the preceding section the instanton induced force does not act on the flavor decuplet states and consequently the Δ- and Ω-resonances are determined by the kinetic energy and the confinement potential alone. As can be seen from Figs. 9 and 12, the positions of the positive parity resonances can be reproduced satisfactory in these sectors. The same applies to the lowest $\Delta(\frac{1}{2}^-)$ and $\Delta(\frac{3}{2}^-)$-resonances, but the calculated positions of the excited states with these quantum numbers is too high, a deficiency also found in other calculations in a similar framework.

For the other flavors the instanton induced interaction yields mass splittings according to the number of flavour antisymmetric quark pairs with trivial spin (diquarks). With the same parameters (see Table 6) as for the mesons the $N - \Delta$, $\Lambda - \Sigma$ and $\Xi - \Xi^*$ splittings for the ground state multiplets can be reproduced quantitatively, whereas the $\Sigma - \Sigma^*$-splitting is slightly too small. The position of the first excited $1/2^+$-resonance in the N-, Λ- and Σ-spectra is calculated too high by a few 100 MeV, although a selective lowering due to the instanton induced force can be observed. We will come back to this problem in the next section.

As for the mesons we will test the quality of the description by calculating some selected electroweak properties of the baryons.

Electromagnetic Form Factors

To calculate the baryonic current operators we essentially use the prescription for estimating some relativistic effects already given for the currents of the mesons. Here we will merely discuss some results.

The axial coupling constant g_A is calculated in any nonrelativistic treatment as

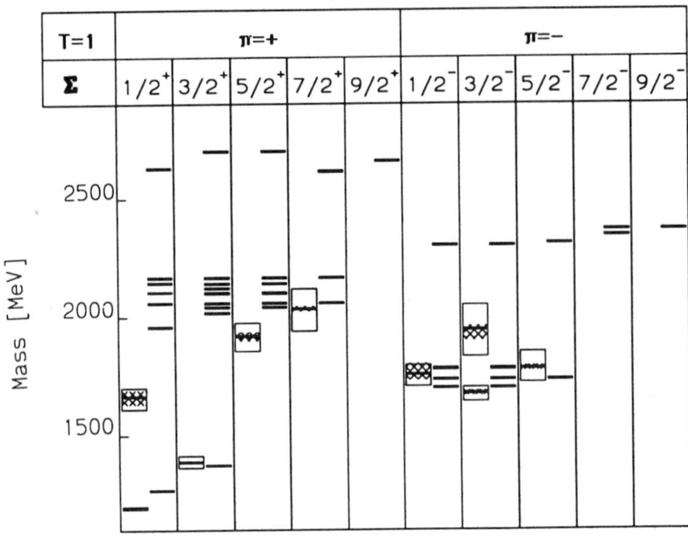

Figure 11. Σ-resonance spectrum. See also caption to Fig. 8.

Table 5. static properties of nucleons

	non rel.	rel.	exp.
$\mu_p[\mu_N]$	3.59	2.19	2.792847386(63)
$\mu_n[\mu_N]$	-2.37	-1.47	-1.91304275(45)
$\langle r_A^2\rangle^{\frac{1}{2}}[fm]$	0.61	0.63	0.66
$\langle r_{pE}^2\rangle^{\frac{1}{2}}[fm]$	0.61	0.76	0.862(12)
$\langle r_{pM}^2\rangle^{\frac{1}{2}}[fm]$	0.58	0.72	0.858(56)
$\langle r_{nE}^2\rangle^{[}fm^2]$	-0.05	-0.06	-0.1134(24)
			-0.1194(18)
$\langle r_{nM}^2\rangle^{\frac{1}{2}}[fm]$	0.60	0.74	0.858(56)

$g_A = \frac{5}{3}$, where effects from some d-wave admixtures only give some corrections of the order of a few percent. Using the full free Dirac-current we find $g_A = 1.29$, which is remarkably close to the experimental value $g_A = 1.261 \pm 0.007$[28].

Appreciable corrections are also found for the magnetic moments and the various radii, see also Table 5.

At this point it should be noted that taking into account the coupling to meson-baryon channels, as will be discussed in the next section, which will lead automatically to a meson-baryon component in the wave function, might quantitatively improve the description of the experimental data. In a non-relativistic estimate it was found that the charge radius of the proton is increased by about 15% and the quadratic charge radius of the neutron is even dominated by a $p - \pi^-$-component. To the absolute value of the magnetic moments the meson-baryon components contribute about $0.5\mu_N$.

Two examples for the calculated form factors are given in Figs. 13,14. It is found that the relativistic corrections also improve the momentum dependence in the range up to $q^2 = 0.5$ GeV^2 substantially. At higher momenta the deviations are due to the incorrect high momentum behavior of the harmonic oscillator basis functions used here.

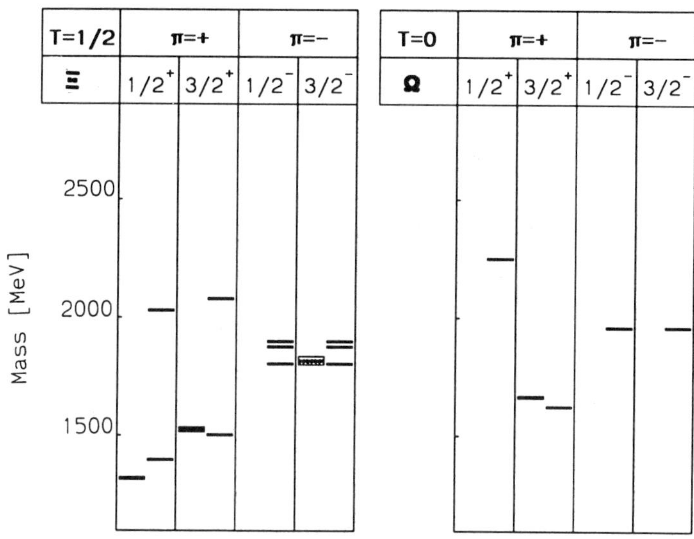

Figure 12. Ξ- and Ω-resonance spectrum. See also caption to Fig. 8.

Table 6. Comparison of empirical and calculated quark masses and coupling constants

	emp.		calc.		QCD-Par.		
$m_u = m_d$	300	MeV	270	MeV		9	MeV
m_s	540	MeV	330	MeV		150	MeV
b	850	MeV/fm			$\langle \bar{n}n \rangle^{(1/3)}$	-225	MeV
$a_{q\bar{q}}$	-892	MeV			$\langle \bar{s}s \rangle$	$0.8\,\langle \bar{n}n \rangle$	
a_{q^3}	-1534	MeV			Λ_{QCD}	200	MeV
g	$0.16\,10^{-4}$	MeV^{-2}	$0.17\,10^{-4}$	MeV^{-2}	ρ_c	0.45	fm
g'	$0.11\,10^{-4}$	MeV^{-2}	$0.09\,10^{-4}$	MeV^{-2}			
λ	0.37	fm					

Relation to QCD-Parameter

After presenting the spectra of light mesons and baryons it seems interesting to explore whether the parameter that we have found empirically by a fit to the experimental spectra can be related to QCD parameter. After normal ordering the contribution of instantons to the Lagrangian contains a single particle term that leads to constituent quark masses:

$$\Delta m_n = \int_0^{\rho_c} d\rho \frac{d_0(\rho)}{\rho^5} \frac{4}{3}\pi^2\rho^3 \left(m_n\rho - \frac{2}{3}\pi^2\rho^3\langle \bar{q}_n q_n \rangle \right) \left(m_s\rho - \frac{2}{3}\pi^2\rho^3\langle \bar{q}_s q_s \rangle \right) \tag{34}$$

and

$$\Delta m_s = \int_0^{\rho_c} d\rho \frac{d_0(\rho)}{\rho^5} \frac{4}{3}\pi^2\rho^3 \left(m_n\rho - \frac{2}{3}\pi^2\rho^3\langle \bar{q}_n q_n \rangle \right)^2 \tag{35}$$

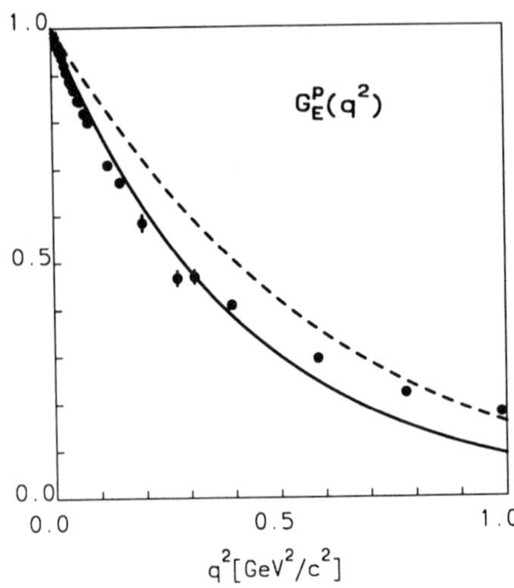

Figure 13. Charge form factor of the proton. The dashed curve is the result with a current operator in the nonrelativistic limit. The solid curve was calculated with full free Dirac spinors. The experimental data are from [37, 38, 39] and the reference given therein.

and a two-body interaction with an effective coupling constant given by

$$g_{\text{eff}}(i) = \int\limits_0^{\rho_c} d\rho \frac{d_0(\rho)}{\rho^5} \left(\frac{4}{3}\pi^2\rho^3 \right)^2 \left(m_i\rho - \frac{2}{3}\pi^2\rho^3 \langle \bar{q}_i q_i \rangle \right) \tag{36}$$

Here,

$$d_0(\rho) = (3.63 \, 10^{-3}) \left(\frac{8\pi^2}{g^2(\rho)} \right)^6 \exp\left(-\frac{8\pi^2}{g^2(\rho)} \right) \tag{37}$$

is the instanton density and in two-loop approximation one finds

$$\left(\frac{8\pi^2}{g^2(\rho)} \right) = 9\ln\left(\frac{1}{\Lambda\rho} \right) + \frac{32}{9}\ln\ln\left(\frac{1}{\Lambda\rho} \right) \tag{38}$$

with Λ the QCD scale parameter.

In Table 6 the values calculated with the formulas given above are compared to the fitted values.

For the critical instanton size used here the differences between the calculated values in one- and two-loop approximation are about 20%. In view of the approximation made, in particular the non-relativistic approximation and the parameterization of the confinement, the results are very satisfactory. Also the non-strange constituent mass can be accounted for by and large, whereas the strange quark mass is definitely too low. It should be stressed, however, that the instanton induced force not necessarily is the only source of constituent quark masses. It is therefore hard to judge, whether the differences found constitute a serious discrepancy.

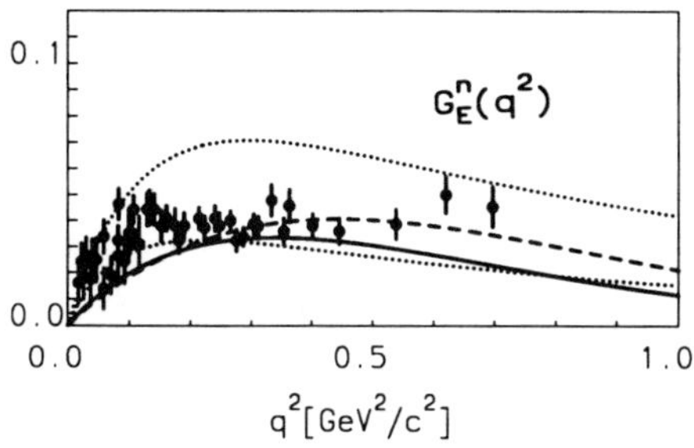

Figure 14. Electric form factor of the neutron. See also caption to 13. Experimental data are from [40, 41] and the references given therein. The upper and lower dotted line indicate the model dependence of the analysis with a Reid soft core and a Nijmegen NN-potential, respectively, see also[40, 41].

BARYONS AS RESONANCES

In the foregoing sections we emphasized the internal quark dynamics of the hadrons, by describing within the the constituent quark model the meson and baryon resonances as bound $q\bar{q}$- and q^3-configurations respectively. In such a framework the fact that the excited states are resonances, that decay via the strong interaction with a characteristic width of the order of 100 MeV, is completely ignored. Because of these large decay widths a perturbative treatment of the strong decay seems hardly justified and it is to be expected, that the observed mass splittings are not only determined by the internal quark dynamics, but also by the external hadron dynamics in the form of hadronic vertices. The latter reflect of course the internal quark dynamics again.

To take this feature into account, we will use a rather hybrid model for the structure of baryons: Within a formalism of two particle scattering, taking the internal structure of the particles into account, we will construct an explicitly energy dependent baryonic Hamiltonian with eigenmodes, that will be interpreted as either bound states or resonances. In this model the internal structure is given by the internal quark dynamics, containing a modeling of confinement and a residual quark interaction as discussed in the preceding section. This is extended by taking into account mesonic self energy corrections, that reflect the coupling to meson-baryon channels, where we will treat in a first, simple approximation the mesons as elementary fields. The very same formalism will then also allow for a description of resonant meson-baryon scattering and photoproduction of mesons.

An Effective Hamiltonoperator for Baryons

The Hamiltonian for a coupled system of baryons and mesons, where in principle both particles can have an intrinsic structure can generically be written as:

$$H = H_b + H_m + V_{bM} + V_{Mb} \tag{39}$$

where H_b (H_m) describes the internal (quark-)dynamics of the baryons (mesons) as bound states $|\nu\rangle$, $(|\mu\rangle)$ where we set

$$H_b|\nu, P\rangle = \sqrt{M_\nu^2 + P^2}\,|\nu, P\rangle \qquad H_m|\mu, k\rangle = \sqrt{m_\mu^2 + k^2}\,|\mu, k\rangle \tag{40}$$

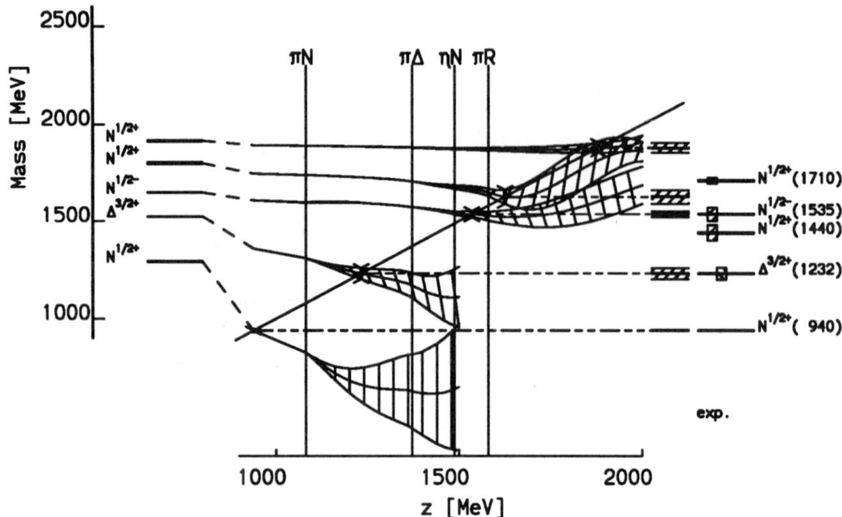

Figure 15. Energy dependence of the eigenvalues of the effective Hamiltonian of Eq. 43. The solid curve represents the real part, the hatched area is $\Re\lambda \pm \frac{1}{2}\Im\lambda$. The vertical lines indicate the thresholds of the channels taken into account.

and where V_{bM} and V_{Mb} describe the creation and annihilation of mesons. P and k denote the total momentum of the particles. The indices b, m and M refer to the projection on the one-baryon-space $\mathcal{H}_b$, the one-meson space $\mathcal{H}_m$, and the one-meson-baryon space $\mathcal{H}_M = \mathcal{H}_b \oplus \mathcal{H}_m$, respectively.

Limiting the description to $\mathcal{H}_M$, the Lippmann-Schwinger equation for meson-baryon scattering can be written as

$$T_{MM} = V_{Mb}\, G_b^{\text{eff}}(z)\, V_{bM}, \tag{41}$$

with the effective resolvent

$$G_b^{\text{eff}}(z) = \left[z - H_b^{\text{eff}}(z)\right]^{-1} \tag{42}$$

and the effective Hamilton operator

$$H_b^{\text{eff}}(z) = H_b + \delta H_b = H + V_{bM}(z - H)^{-1}V_{Mb} \tag{43}$$

Here H_b describes the internal structure of the baryons and δH_b the coupling to the meson baryon continua. In this framework the baryons are interpreted as bound states embedded in a continuum. The effective Hamiltonian $H_b^{\text{eff}}(z)$ is manifestly energy dependent and in general not hermitian, since the propagator exhibits poles above the first meson-production threshold. In the restframe of the baryons the energy dependent, in general complex, eigenvalues and the right and left eigenstates are written as

$$H_b^{\text{eff}}(z)\,|\nu; z\rangle \;=\; \lambda(z)\,|\nu; z\rangle \tag{44}$$

$$\langle\bar\nu; z|\,H_b^{\text{eff}}(z) \;=\; \lambda(z)\,\langle\bar\nu; z| \tag{45}$$

The interpretation of the properties of this effective Hamiltonian follow from the spectral representation of the resolvent:

$$\langle\mu', \nu', k'\,|\,T_{MM}\,|\,\mu, \nu, k\rangle =$$
$$\sum_{\hat\nu}\left\langle\mu', \nu', k'\,|\,V_{Mb}\,|\,\bar{\hat\nu}(z)\right\rangle (z - \lambda_{\hat\nu}(z))^{-1}\left\langle\hat\nu(z)\,|\,V_{bM}\,|\,\mu, \nu, k\right\rangle \tag{46}$$

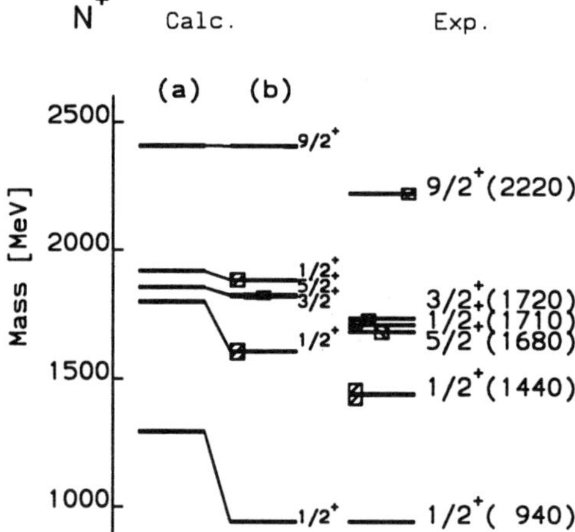

Figure 16. Comparison of calculated and experimental positive parity nucleon resonances: (a) Spectrum of H_b, (b) Spectrum of H_{eff}. The hatched rectangle represents the decay width in the channels taken into account.

For $z = E \in I\!\!R$ with $E = \Re\left[\lambda(E)\right]$ the T-matrix is imaginary, provided the resonances are well isolated. This is interpreted as a resonance with position

$$M = \Re\left[\lambda(E)\right] \tag{47}$$

and width

$$\Gamma(E) = -2\Im\left[\lambda(E)\right] \tag{48}$$

If $\Gamma(E) = 0$ we obviously have a bound state. In this manner we achieve a unified treatment of bound states and resonances: the effective Hamiltonian $H_b^{\mathrm{eff}}(z)$ describes implicitly a coupled meson-baryon system. Below the first meson production threshold it leads to a description of stable baryons as superpositions of q^3- and q^3-meson configurations. Above this threshold the Hamiltonian possesses complex eigenvalues to be interpreted as resonances, see also Fig. 15

Coupling to Meson-Baryon Channels

In the calculation of δH_b we make the following simplifications: First of all we limit the two-particle intermediate states to the $\pi N(940)$-, $\eta N(940)$- and – as a parameterization of the $\pi\pi N(940)$-channel – the $\pi\Delta(1232)$- and the $\pi N(1440)$-channel. Furthermore we will neglect the internal structure of the mesons and treat them as elementary fields, coupling locally to the quarks in the baryon via a non-relativistic version of the usual pseudoscalar coupling:

$$V_{bM} + V_{Mb} = H_c^\dagger + H_c \tag{49}$$

with

$$H_c = \sum_{i=1}^{3} H_c(i) \tag{50}$$

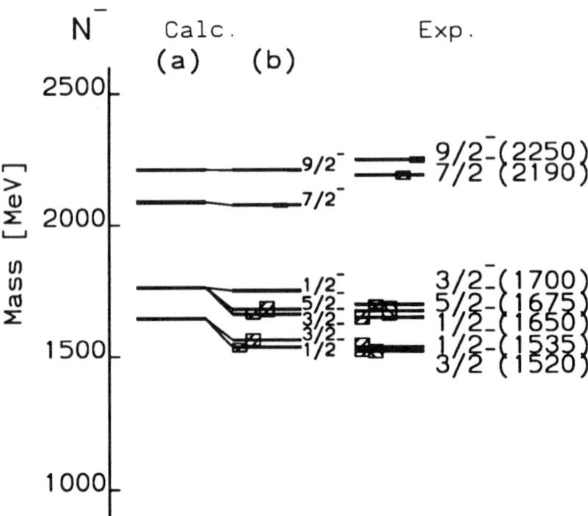

Figure 17. Comparison of calculated and experimental negative parity nucleon resonances, see also caption to Fig. 16.

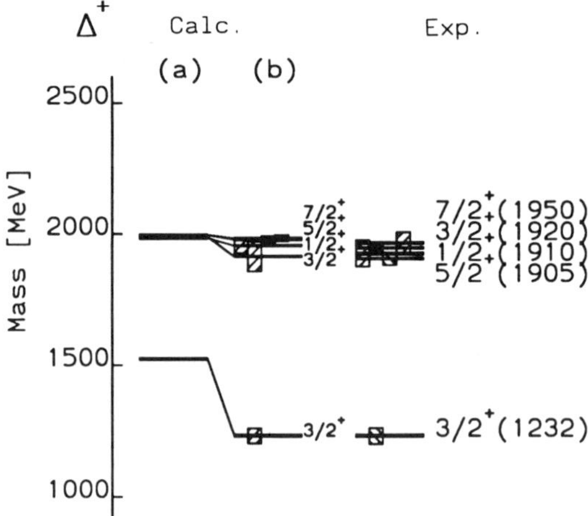

Figure 18. Comparison of calculated and experimental negative parity nucleon resonances, see also caption to Fig. 16.

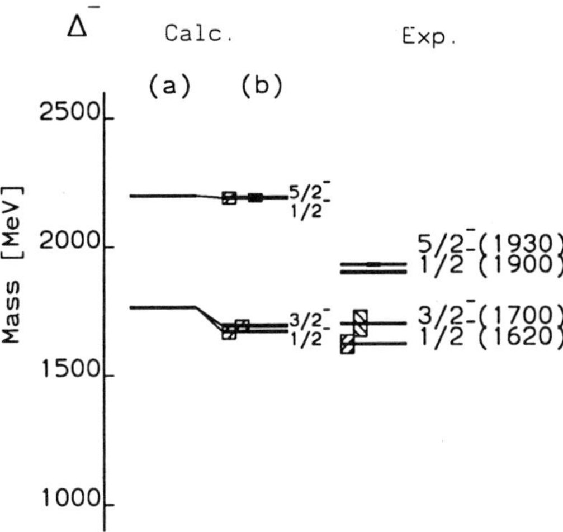

Figure 19. Comparison of calculated and experimental negative parity nucleon resonances, see also caption to Fig. 16.

where

$$H_c(i) = \sum_{\mu=\pi,\eta} -i\,\frac{g_{\mu qq}}{2m_i}\,\frac{1}{\sqrt{(2\pi)^3 2\omega_\mu(k_\mu)}}\left(\vec{\sigma}_i \cdot \vec{K}_\mu\right) \mathcal{F}_\mu(i)\delta^{(3)}(\vec{p}_i - \vec{p}_i{}' - \vec{K}_\mu) \qquad (51)$$

with $\vec{p}$ and $\vec{k}$ the quark-, and meson momentum, respectively. Here we put $K = k_\mu - (m_i/M)p'_i$, partly to parameterize recoil effects[42, 43]. Furthermore

$$\mathcal{F}_\mu(i) = \underline{\tau} \cdot \underline{\phi}^\dagger_\pi, \qquad \mathcal{F}_\eta(i) = \underline{\phi}^\dagger_\eta, \qquad (52)$$

are the flavor-operators for π- and η-meson creation, τ are the Pauli-isospin matrices and $\omega(k)$ is the (relativistic) energy of the meson. We use the following prescription for the matrix elements of the resolvent G_{eff}

$$\left\langle \mu,\nu,k \,\middle|\, G_{\text{eff}}(z) \,\middle|\, \mu,\nu,k \right\rangle = \left[z - \sqrt{M_\nu^2 + k^2} - \omega_k\right]^{-1} \qquad (53)$$

where M_ν is the experimental baryon mass in the decay channel. Here we assumed that the term $H - H_b$ can suitably be approximated by inserting the physical baryon mass. This guarantees that the effective Hamiltonian will have the correct energy dependence, in particular in the vicinity of the various thresholds. Finally, since the vertex operator of Eq. 51 does not lead to a bound Hamiltonian, it is necessary to introduce a quark-meson form factor:

$$F(k_\mu) = \exp -(r_0 k_\mu)^2 \qquad (54)$$

which can be interpreted as reflecting a finite meson size or an effective size for the constituent quarks.

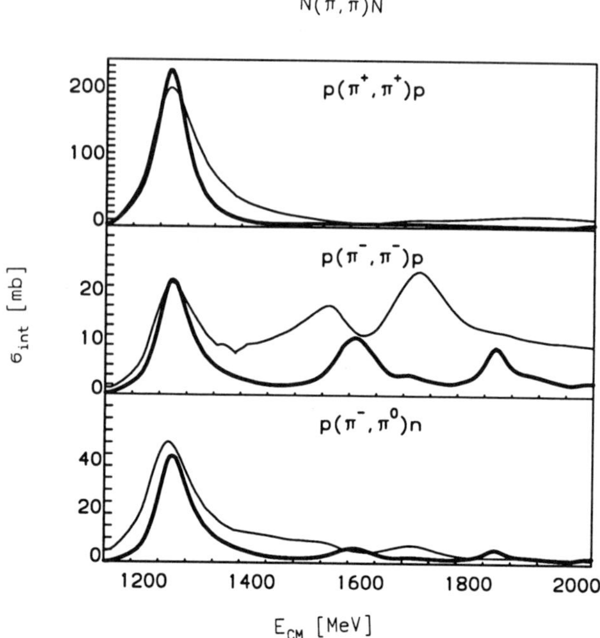

Figure 20. Angular integrated cross section for pion-nucleon scattering. The bold line is the calculated result, the solid line is the phase shift analysis of[44].

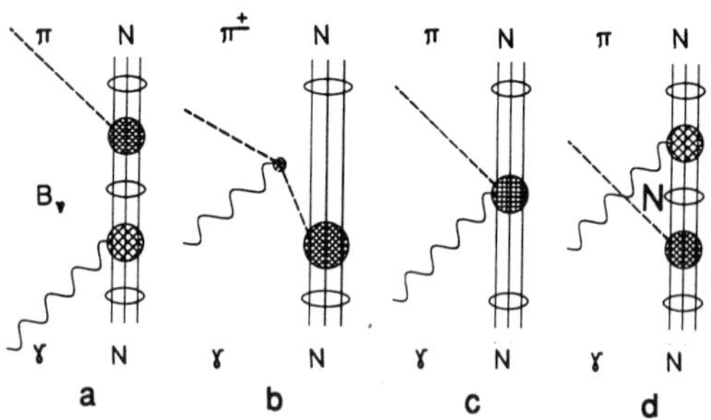

Figure 21. Contributions to photo-induced pion production on the nucleon. (a) Resonant contribution, (b) pion-pole contribution, (c) contact term, (d) crossed contribution.

Baryon Resonance Spectra

A comparison of the calculated baryon resonances with the experimental results as quoted by the *Particle Data Group*[28] is given in Figs. 16-19. Concerning the positive parity nucleon resonances, we find that, although the absolute values of the excited states are about 200 MeV too large, the splittings, due to the combined effect of the instanton induced interaction and the coupling to decay channels, are now correct, see e.g. the splitting of the excited $1/2^+$ states. It is remarkable that the position of the positive parity excited Δ-states can be accounted for. It is not clear whether this asymmetry is due to the limited model space, that can not efficiently account for the correlations of the instanton force. In contrast to this the negative parity states are calculated too high. In general we find a satisfactory albeit not excellent description of both the mass splittings and the decay widths, although the latter become too small with increasing excitation energy. This is probably due to the extreme simplifications we have made for the calculation of the meson baryon vertex: especially for highly excited states and the consequently large momenta associated with the decay the approximation by a single quark operator is highly questionable. Moreover it is to be expected that relativistic effects are important and finally we took into account only single meson-baryon channels and the multi meson decay channels are only parameterized by including $\pi - N(1440)$ and $\pi - \Delta$-channels. We thus conclude, that a more microscopic treatment that especially better reflects the multiparticle nature of meson-absorption and creation, and, moreover takes into account the internal structure of the mesons is necessary. Such a treatment could be given on the basis of the instanton induced force.

PHOTON AND PION SCATTERING

Elastic Pion Nucleon Scattering

As discussed above the scattering amplitude for resonant meson-baryon scattering can be written as

$$\langle \mu', \nu', k' \, | \, T_{MM} \, | \, \mu, \nu, k \rangle = \\ \sum_{\hat{\nu}} \langle \mu', \nu', k' \, | \, V_{Mb} \, | \, \overline{\hat{\nu}}(z) \rangle \, (z - \lambda_{\hat{\nu}}(z))^{-1} \, \langle \hat{\nu}(z) \, | \, V_{bM} \, | \, \mu', \nu', k' \rangle \qquad (55)$$

It should be noted, that here the vertex functions, apart from the usual momentum dependence of the form factors in addition are manifestly energy dependent, as are the resonant positions and widths of the intermediate resonances. In this manner the unitarity of the scattering matrix, also above particle production threshold is guaranteed. This resonant contribution to pion-nucleon scattering was calculated by taking into account 40-80 intermediate resonances. The resulting total elastic cross sections are shown in Fig. 20.

Although the description in the region of the Δ-excitation is quite satisfactory the cross section at higher energies are too low, in consonance with the observation made in the preceding section for the widths of the higher lying resonances. In addition some resonances are calculated too high and thus their contribution is suppressed for kinematical reasons. Similar observations[28] are also found for the differential cross sections, where especially the structure at backward angles cannot be reproduced.

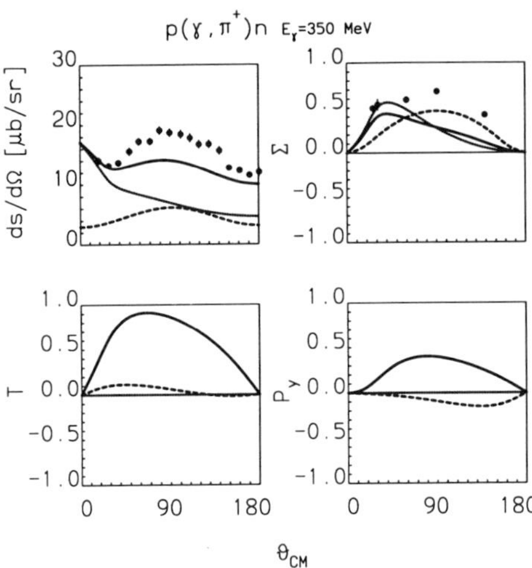

Figure 22. Contributions to the angular distributions of charged pion production: the dotted line gives the contributions of graphs (a) and (d), the dashed line the contribution from (b) and (d), see also 21, the solid curve is the total contribution.

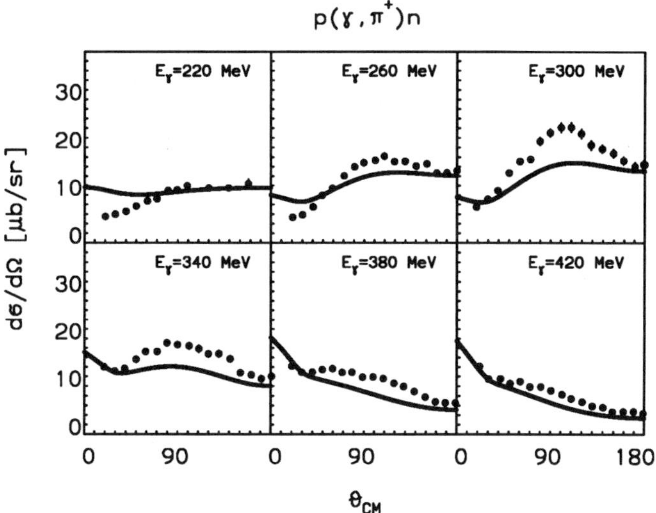

Figure 23. Differential cross section of photo induced positive charged pions.

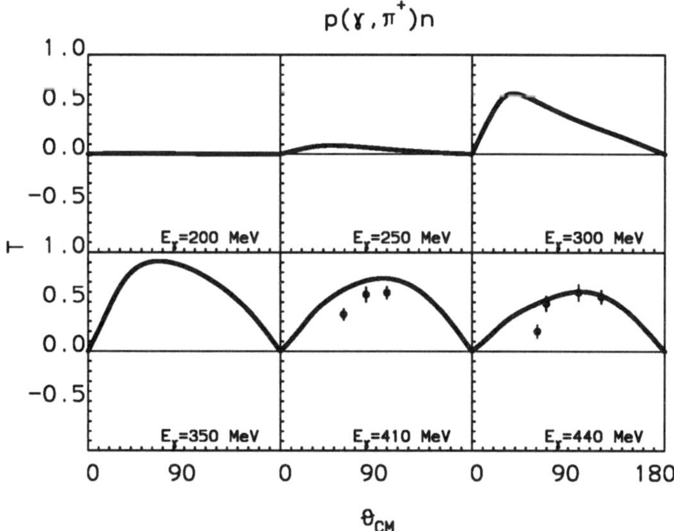

Figure 24. Target asymmetry of photo-induced positive charged pions.

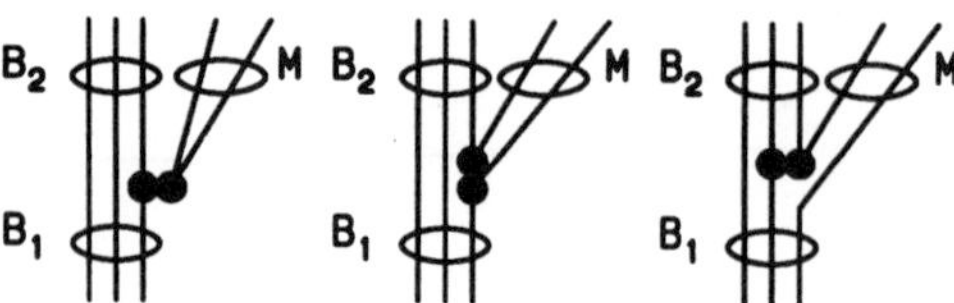

Figure 25. Generic contribution of an instanton induced interaction to the decay of a baryon in a meson-baryon channel.

Photoproduction of Mesons

Within the same framework it is also possible to calculate the resonant contribution for the photoproduction of pions. However, in particular for charged particle production the dynamics must be supplemented by some nonresonant contributions, which can be calculated without any additional parameters. Starting from the Lagrangian for coupled Dirac (Ψ), pseudoscalar (ϕ) and electromagnetic fields (A or (E, B)) one finds via a standard Foldy-Wouthuysen transformation for the single-particle terms of the coupling in linear approximation in g and $1/m_q$:

$$
\begin{aligned}
H_c^{(1)} \;=\; & \int d^3x\, \Psi^\dagger(x)\Big[\frac{g}{2m_q}\Big(\vec{\sigma}\cdot(\vec{\nabla}\phi)(x)\underline{\tau} - (\vec{\sigma}\cdot\vec{A}(x))(\underline{\tau}\times\underline{\phi}(x))_0\Big) \\
& +\frac{e_q}{m_q}\Big(\frac{1}{2}\vec{\sigma}\cdot\vec{B}(x) - i\vec{A}(x)\cdot\vec{\nabla}\Big) + \frac{e_q^2\vec{A}(x)^2}{2m_q}\Big]\Psi(x)
\end{aligned}
\tag{56}
$$

where e_q is the charge quark charge operator and g the pseudoscalar coupling constant. The first (third) term is the nonrelativistic form of the pion-quark (photon-quark) coupling as discussed above. The second term is a coupling to the pion current and contributes to the transition amplitude of charged pions, see also Fig. 21. The last term is a contribution to compton scattering.

For the neutral meson production only the resonant (and crossed) term contribute to the transition amplitude. However for charged particle production the the angular distribution is determined also by the non-resonant terms, see Fig. 22 where the contributions are shown for the $p(\gamma,\pi^0)p$-reaction. In particular at forward angles the rise stems from the interference of the pion contact term and the pion-pole contribution. In Fig. 23 the differential cross section and, as an example of polarization parameters in Fig. 24 the target asymmetry

$$
T = \frac{\sigma_\uparrow - \sigma_\downarrow}{\sigma_\uparrow + \sigma_\downarrow},
\tag{57}
$$

compared to experimental data[45, 46, 47, 48] for various energies in the range of the Δ-resonance excitation. In general the resonant contribution turns out to be too small, which is probably due to the simplified photon interaction used here. At this point we like to stress again, that the main objective is to demonstrate how observables can be related to the underlying quark structure of the baryons. Also the non resonant contributions constitute in this sense a test of the wave functions used. It is noted that a slight modification of the photon quark vertex, by an estimation of some vector dominance effects in fact improves the description considerably.

The main purpose of the present section was to explore how in a constituent quark model with an effective Hamiltonian that was extended by taking into account the strong coupling to decay channels a uniform description of spectral and reaction observables can be obtained. Indeed up to energies of about 1400 MeV such a semi-realistic picture is indeed feasible. At higher energies the description fails, mainly because of an inaccurate parameterization of the hadronic coupling vertices. It is interesting that the instanton induced force, when taking into account the terms linear in the momentum yields a genuine two-body operator, that could be used to describe these vertices.

The coupling of a baryon to a meson-baryon channel is induced by three terms as given in Fig. 25.

The dominant contribution is proportional to the pointlike coupling of a meson to a quark, as used in this section. The two other contributions are also significant in as much that the ratio of the pion-coupling for Δ-decay and the pion-nucleon coupling with this interaction is

$$\left(\frac{f_{\pi\Delta N}}{f_{\pi NN}}\right)^2_H = 3.4 \tag{58}$$

which better agrees with the experimental value

$$\left(\frac{f_{\pi\Delta N}}{f_{\pi NN}}\right)^2_{exp} = 4.3 \tag{59}$$

than the value that follows in the $SU(6)$-limit, which would result from the elementary coupling operator alone:

$$\left(\frac{f_{\pi\Delta N}}{f_{\pi NN}}\right)^2_{SU(6)} = \frac{72}{25} = 2.9 \tag{60}$$

CONCLUSION

In the present treatment we explored the (non-relativistic) constituent quark model for simple hadrons like mesons and baryons with respect to the relation of assumptions about the underlying quark dynamics – concerning e.g. the parameterization of confinement, the residual quark interactions and the hadronic creation and annihilation vertices – to spectra, static properties and scattering observables in exclusive two-body reactions.

In the first part we showed that for light hadrons, built of u, d and s quarks, a residual quark interaction based on instanton effects in QCD provides a suitable alternative (to the traditional Breit-Fermi interaction from One-Gluon-Exchange) for explaining the observed mass splittings. With the instanton induced interaction it is possible to arrive at consistent description of the gross properties of both the meson- and the baryon excitation spectrum. By an examination of electromagnetic properties we investigated the range of applicability of the non-relativistic treatment.

In the second part we emphasized another important aspect in hadron spectroscopy: the fact that most hadrons are short-lived resonances that couple strongly to decay channels. It was found that this coupling induces shifts for the resonance positions that are comparable to the splittings due to the internal quark interactions and in addition provides a semi-realistic description of the strong decay widths for the lower baryon resonances. The analysis of meson-baryon scattering at higher energies indicates that it is necessary to improve the treatment presented here (it was assumed that mesons couple pointlike to the quarks in the baryon) by a more microscopic treatment of the hadronic vertices, based on quark dynamics, by taking into account two-meson channels and considering relativistic effects.

It should be emphasized, that the issues discussed here will be relevant for a consistent extension to more complex hadronic systems, like the excitation spectrum of the deuteron and the intermediate energy excitations of nuclei.

References

[1] A. DeRújula, H. Georgi, S. Glashow, *Phys. Rev.* D12:147 (1975).

[2] W. Celmaster, *Phys. Rev.* D15:1391 (1977).

[3] D. Gromes, I.O. Stamatescu, *Nucl. Phys.* B112:213 (1976).

[4] D. Gromes, *Nucl. Phys.* B130:18 (1977).

[5] D. Gromes, *in*: "The Quark Structure of Matter", Proceedings of the YUKON Advanced Study Institute. N. Isgur, G. Karl, P.J. O'Donnell, eds., World Scientific, Singapore, pp.1-56 (1985).

[6] U. Ellwanger, *Nucl. Phys.* B139:422 (1978).

[7] L.J. Reinders, *J. Phys.* G4:1241 (1978).

[8] C.P. Forsyth, R.E. Cutkosky, *Phys. Rev. Lett.* 46:576 (1981).

[9] C.S. Kalman, R.L. Hall, *Phys. Rev.* D25:217 (1982).

[10] C.S. Kalman, *Phys. Rev.* D26:2326 (1982).

[11] Y. Nogami, N. Ohtsuka, *Phys. Rev.* D25:2394 (1982).

[12] C.P. Forsyth, R.E. Cutkosky, *Z. Phys.* C 18:219 (1983).

[13] R.K. Bhaduri, B.K. Jennings, J.C. Waddington, *Phys. Rev.* D29:2051 (1984).

[14] M.V.N. Murthy, M. Dey, J. Dey, R.K. Bhaduri, *Phys. Rev.* D30:152 (1984).

[15] M.V.N. Murthy, R.K. Bhaduri, R.K., *Phys. Rev. Lett.* 54:745 (1985).

[16] M.V.N. Murthy, M. Brack, R.K. Bhaduri, B.K. Jennings, *Z. Phys.* C 29:385 (1985).

[17] K.G. Horacsek, Y. Iwamura, Y. Nogami, *Phys. Rev.* D32:3001 (1985).

[18] B. Silvestre-Brac, C. Gignoux, *Phys. Rev.* D32:743 (1985).

[19] M.V.N. Murthy, R.K. Bhaduri, E. Tabarah, *Z. Phys.* C 31:81 (1986).

[20] N. Isgur, G. Karl, *Phys. Lett.* 72B:109 (1977).

[21] N. Isgur, N., G. Karl, *Phys. Rev.* D18:4187 (1978).

[22] N. Isgur, G. Karl, *Phys. Lett.* 74B:353 (1978).

[23] N. Isgur, G. Karl, R. Koniuk, *Phys. Rev. Lett.* 41:1269 (1978).

[24] N. Isgur, G, Karl, *Phys. Rev.* D19:2653 (1979).

[25] N. Isgur, G. Karl, *Phys. Rev.* D20:1191 (1979).

[26] N. Isgur, R. Koniuk, *Phys. Rev. Lett.* 44:845 (1980).

[27] N. Isgur, R. Koniuk, *Phys. Rev.* D21:1868 (1980).

[28] Particle Data Group, *Phys. Lett.* 204B:1 (1988).

[29] G. 't Hooft, *Phys. Rev.* D14:173 (1976).

[30] E.V. Shuryak, *Nucl. Phys.* B203:93 (1982).

[31] Büttgen, R., Holinde, K., Lohse, D., Müller-Groeling, A., Speth, J., Wyborny, P., *Z. Phys. C* 46:167 (1990).

[32] J. Weinstein, N. Isgur, *Phys. Rev.* D27:588 (1983).

[33] J. Weinstein, N. Isgur, *Phys. Rev.* D41:2236 (1990).

[34] J. Carlson, J. Kogut, V.R. Pandharipande, *Phys. Rev.* D27:233 (1983).

[35] J. Carlson, J. Kogut, V.R. Pandharipande, *Phys. Rev.* D28:2807 (1983).

[36] S. Capstick, N. Isgur, *Phys. Rev.* D34:2809 (1986), *and references quoted there.*

[37] G.G. Simon, PhD Thesis, Mainz KPH12/78 (1978).

[38] G.G. Simon, Ch. Schmitt, V.H. Walther, *Nucl. Phys.* A333:381 (1980).

[39] G.G. Simon, Ch. Schmitt, V.H. Walther, *Nucl. Phys.* A364:285 (1981).

[40] S. Platchkov *et al.* in: Few Body Systems, Suppl. 2 (1987).

[41] S. Platchkov *et al. Nucl. Phys.* A510:740 (1990).

[42] A.N. Mitra, R. Ross, *Phys. Rev.* 158:1630 (1967).

[43] W.H. Blask, S. Furui, R. Kaiser, B.C. Metsch, M.G. Huber, *Z. Physik A* 337:451 (1990).

[44] R. Koch, E. Pietarinen, *Nucl. Phys.* A336:331 (1980).

[45] D. Menze, W. Pfeil, R. Wicke, *ZAED Compilation of pion photoproduction data*, Universtät Bonn (1977)

[46] F.A. Berends, A. Donnachie, *Nucl. Phys.* B84:342 (1975).

[47] W. Pfeil, D. Schwela, *Nucl. Phys.* B45:379 (1972).

[48] A. Bagheri, K.A. Aniel, E. Entezami, M.D. Hasinoff, D.F. Measday, J.M. Poutisson, M. Salomon, F.C. Robertson, *Phys. Rev.* C38:875 (1988).

LEPTONIC PRODUCTION OF BARYON RESONANCES

Volker D. Burkert

Continuous Electron Beam Accelerator Facility
12000 Jefferson Avenue, Newport News
Virginia 23606, USA

INTRODUCTION

The quest for understanding the structure and interaction of hadrons has been the main motivation of strong interaction physics for decades. The advent of Quantum Chromodynamics (QCD) has led to a theoretical description of the strong interaction in terms of the fundamental constituents, quarks and gluons. However, we are still a long way from having the general theoretical framework in place to understand the strong force as it is manifest in the structure of complex hadronic systems such as baryons. At very high energies, perturbative methods have proven very effective. At low energies, simulations on the QCD lattice have generated remarkable results about the static properties of hadrons. Even calculations of electromagnetic form-factors have been possible at low momentum transfers. However, for light quark baryons such as nucleons, the small quark masses are difficult to implement, and crude approximations have been necessary.

Chiral perturbation theory is an attempt to apply QCD at low energies. In a well defined kinematical regime, for example close to the pion production threshold, it is possible to perform model independent calculations. However, even there resonance contributions give sizeable corrections, which at the present time cannot be calculated reliably.

In the forseeable future, models of the hadron structure will likely continue to play an important role and provide theoretical guidance to experimenters. The QCD inspired non-relativistic quark model[1] has been extended to include relativistic corrections in a consistent way[2] and applied successfully to the calculation of photoproduction amplitudes[3]. Other, conceptually completely different approaches are being pursued such as algebraic models[4]. The Skyrme model applied to photoproduction[5] can be used to calculate the complete partial wave amplitudes of the process $\gamma N \to \pi N$. However, in this description the electromagnetic vertex cannot be isolated from the hadronic vertex and the power of the electromagnetic interaction as a probe of the hadronic structure is lost, at least to some degree.

Perspectives in the Structure of Hadronic Systems
Edited by M.N. Harakeh *et al.*, Plenum Press, New York, 1994

Today's challenge in strong interaction physics at intermediate energies is to obtain a *detailed* understanding of the hadronic structure of matter. In other words, given QCD, how are hadrons made up of fundamental constituents, quarks and gluons?

Nucleons are probably the most common form of hadronic matter in the universe. Understanding their internal structure will give us insight into how the real world works. Nucleons can be probed in a number of ways which are sensitive to different aspects of the nucleon substructure:

- electromagnetic and weak elastic formfactors describe the electromagnetic and weak charge and current distributions of the ground state proton and neutron, providing information about the ground state wavefunction.

- the nucleon excitation spectrum is another important source of information about its internal structure. Existence or non-existence of certain states gives information about the underlying symmetry properties of the nucleon. The excitation spectrum may be probed with pion beams for those states that have an appreciable coupling to the πN channel. Electromagnetic excitation of resonances yields both overlapping and complementary information to pion scattering as some states may couple only weakly to pions or to photons.

- electroproduction of resonances is a unique means of probing the quark wavefunction of the excited states, provided the wavefunction of the ground state is known. In addition, photons (real or virtual) allow probing the spin structure of the resonance transition. Electro-excitation allows us to map out the spatial and spin structure at varying distances by choosing the four momentum transfer to the resonance.

In these lectures, I will focus on the electromagnetic transition between nonstrange baryon states. This sector received much attention in the early 1970's after the development of the first dynamical quark models. However, experimental progress was slow, partly because of the low rates associated with electromagnetic interactions, and partly because of the lack of guidance by theoretical models that went beyond the simplest quark models. It was also difficult for experiments to achieve the precision needed for a detailed analysis of the entire resonance region in terms of the fundamental photocoupling amplitudes over a large range in momentum transfer.

More realistic models were developed after the major electron accelerators used in these studies had been shut down in the wake of the J/ψ discovery in 1974. With the construction of continuous wave (CW) electron accelerators in the GeV and multi-GeV region, this situation is changing in a significant way. For example, use of large acceptance detectors at high luminosities of up to $10^{34} cm^{-2} sec^{-1}$ appears feasible, allowing measurement of several reaction channels simultaneously[6], over a large kinematic range, and with statistical accuracy comparable to that achieved with hadronic probes. Moreover, with 100% duty cycle, high statistics coincidence measurements can be conducted for exclusive channels with small cross sections. *To a large degree the statistics will not be limited by the luminosity achievable in these measurements but rather by the speed of the data acquisition system and the data analysis process.* It is interesting to note that hadronic reactions will no longer enjoy their traditional rate advantage over electromagnetic processes. This will bring to bear the full power of the electromagnetic interaction as a probe of the internal structure of hadrons and the strong interaction.

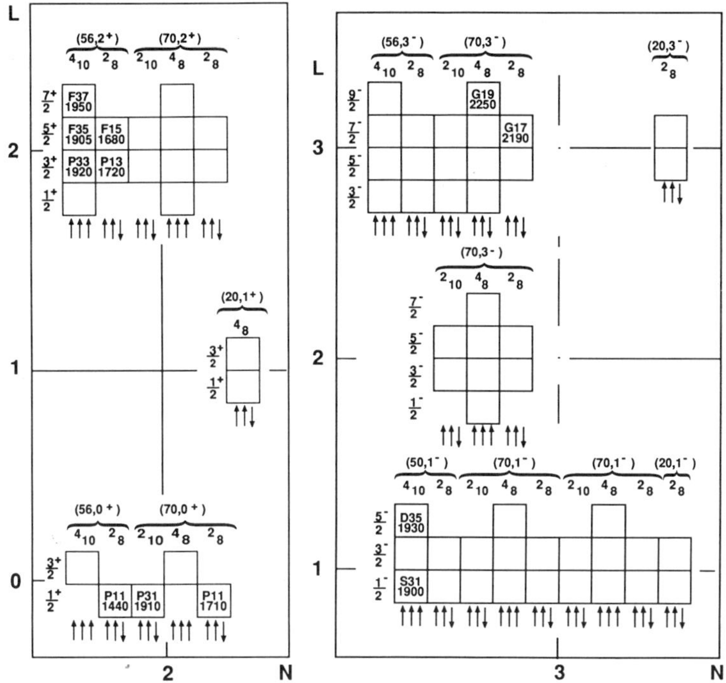

Figure 1. *Experimental status of the $N = 2$, $L_{3q} = 1$, and $N = 3$, $L_{3q} = 2$, quark model states in the $SU(6) \otimes O(3)$ basis.*

The lectures are organized as follows. Before reviewing the experimental status of electromagnetic transitions of baryons resonances, I will briefly discuss some aspects of baryon spectroscopy and what we have learned about the baryon structure from inclusive electron scattering in the nucleon resonance region. Then I will introduce some basic formalism which is necessary for understanding the language and analysis techniques. In the main part of this lecture I will discuss what we have learned in the past from exclusive electroproduction of low mass nucleon resonances. Finally I will discuss topics in nucleon resonance physics which are of current interest and which need to be addressed in future experiments, and outline an experimental strategy for how some of these problems may be tackled.

TOPICS IN LIGHT QUARK BARYON SPECTROSCOPY

In the course of the past decade, very little has happened in experimental light quark baryon spectroscopy, especially when considering its importance for the understanding of the structure of baryons. The field appears to be still in its infancy. The 1992 edition of the Review of Particle Properties[7] lists 19 established N^* or Δ states, and 26 candidate states with insufficient experimental evidence. This is only a small fraction of the states predicted by the most accepted QCD inspired non-relativistic constituent quark models (NRCQM). Much of the information we have on baryons states is the result of partial wave analyses of elastic pion-nucleon scattering measurements $\pi N \to \pi N$. The NRCQM allows the association of all established states with a level in the $SU(6) \otimes O(3)$ representation, where $SU(6)$ describes the symmetry properties of the spin-flavor wavefunction, and $O(3)$ is the symmetry group for the orbital angular momentum of the q^3 system. According to their symmetry properties, principal energy levels and orbital angular momentum, the 3-quark states can be associated with supermultiplets. The ground states and all states associated with the $[70, 1^-]_1$ super multiplet have been observed experimentally . However, several of the

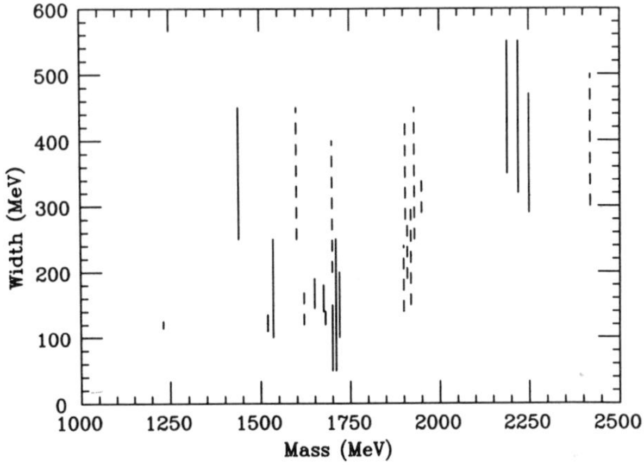

Figure 2. *Widths (including experimental uncertainties) of non-strange baryon states. Solid and dashed lines indicate isospin $\frac{1}{2}$ and $\frac{3}{2}$ states, respectively.*

N=2 states, and most of the N=3 and N=4 states have not been seen in $\pi N \to \pi N$ reactions. Figure 1 summarizes the experimental situation for the N=2 and N=3 super multiplets.

Experimental uncertainties in the widths of many states are large (Figure 2). Even for well established states such as the $P_{11}(1440)$, the uncertainty is nearly a factor of 2. The significantly different values found for the widths of many of these states in different analyses has impact on the accuracy of the helicity amplitudes extracted in electroproduction experiments.

Missing Baryon States

It has been suggested[8] that the problem of the "missing" states in the q^3 quark model may be of experimental nature related to the lack of data in the inelastic channels. Many of the missing states are predicted to couple weakly to the πN channel due to QCD mixing effects. The $\pi N \to \pi N$ process becomes rather ineffective in searching for these states. However, many of the N=2 and N=3 states are predicted to couple strongly to $\pi\Delta$, ρN, and ω, and if the πN channel does not totally decouple from the resonance, the process $\pi N \to \pi\pi N$ may offer a better chance for detecting these states. If they decouple completely from the πN final state, the electromagnetic transitions may be the only way to search for these states.

Table 1. Decay modes of some Baryon States [%]

State	πN	ηN	$N\pi\pi$
$P_{11}(1710)$	10 - 20	20 - 40	20 - 50
$P_{13}(1720)$	10 - 20	2 - 6	≥ 35
$G_{17}(2190)$	10 - 20	1 - 3	20 - 40
$H_{19}(2220)$	10 - 20	0.5 - 1	?
$G_{19}(2250)$	5 - 15	1 - 3	?
$I_{11,1}(2600)$	5 - 10	?	?
$P_{31}(1910)$	15 - 30	-	70 - 85
$P_{33}(1920)$	5 - 20	-	?
$D_{35}(1950)$	10 - 20	-	not seen
$F_{37}(1950)$	35 - 40	-	15 - 40
$H_{3,11}(2420)$	5 - 15	-	?

Obviously, our picture of baryon structure could change dramatically if these states do not exist. An extensive search for at least some of these states is therefore important and urgent. The quark cluster model[9] and the algebraic model[4] can accommodate the known baryon spectrum, but predict a fewer number of additional states. In these models the quarks are not in a spherically symmetric configuration but rather in a diquark-quark configuration with reduced number of degrees of freedom (Figure 3). Electroproduction of nucleon resonances should be a very sensitive tool to distinguish between these alternative configurations.

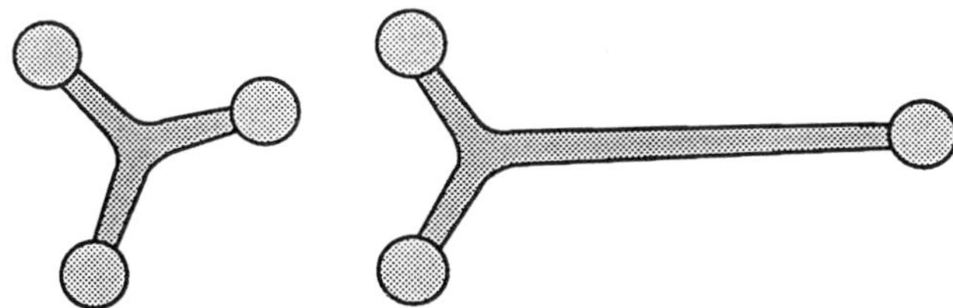

Figure 3. *Spherically symmetric q^3, and non-spherical $q^2 - q$ baryon configuration.*

Inelastic channels are not well determined experimentally, due to the lack of sufficiently detailed data in the $\pi N \rightarrow \pi\pi N$, $\pi N \rightarrow \pi\pi\pi N$, and $\pi N \rightarrow \eta N$ channels (table 1). The lack of knowledge of fundamental resonance properties has serious consequences regarding systematic uncertainties in the extraction of photocoupling amplitudes, where properties of the hadronic vertex are used as input.

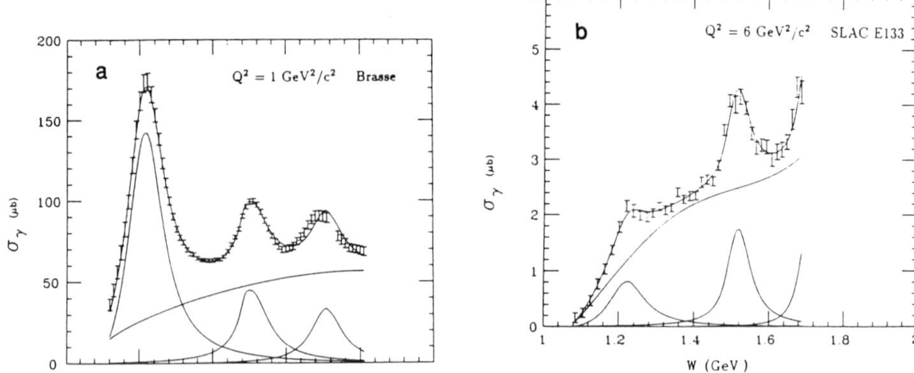

Figure 4. *The total photoabsorption cross section at various values of Q^2. Data compiled by P. Stoler[13].*

Another category of baryons about which we have no experimental information is the gluonic excitations or hybrids q^3G. These states have been discussed within the framework of the bag model[10],[11]. Low lying states such as $P_{11}(1440)$, or $P_{33}(1600)$ are possible candidates for gluonic excitations. The problem is how to distinguish gluonic excitations from regular q^3 excitations. Some of the hybrid states, e.g. the P_{13}^G, or the P_{31}^G are not allowed in the q^3 model at these low masses, therefore their mere existence would already signal that the standard q^3 model cannot be complete. Experimental indications for these states are weak and have been found in inelastic channels only[12]. As we will discuss in lecture 3, measurements of the electromagnetic transition form factors are a powerful tool in determining the nature of these states.

INCLUSIVE ELECTRON SCATTERING

Unpolarized Electron Scattering

The inclusive electron scattering cross section $eN \rightarrow eX$ reveals a few broad bumps, clearly indicating the excitation of resonances in the mass region below 2 GeV (Figure 4). However, their broad widths and close spacing make it impossible to separate them in inclusive production reactions. Nonetheless, one can obtain some global information about the Q^2 dependence of the dominant states by subtracting the non-resonant background contribution. This technique has been used by various groups [14], most recently to study the photocoupling amplitudes of the prominent states at very high momentum transfers[13],[15]. Even without such a subtraction procedure Fig. 4 reveals that the various resonances may have very different Q^2 dependences: the first excited state, the $P_{33}(1232)$ (or $\Delta(1232)$) falls off much more rapidly with increasing Q^2 than the dominant higher mass states. A closer look shows that the position of the second and third peak are not fixed at the same mass value with varying Q^2, indicating that they are composed of several resonant states with different Q^2 dependent formfactors.

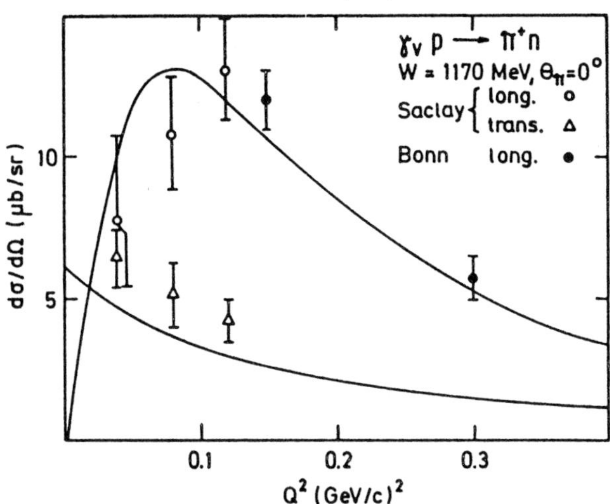

Figure 5. σ_L and σ_T for the π^+ production off the proton below the $P_{33}(1232)$ resonance peak.

The inclusive unpolarized electron scattering cross section can be written as:

$$\frac{d\sigma}{d\Omega_e\, dE'_e} = \Gamma_T \cdot \sigma_{TOT} \tag{1}$$

$$\sigma_{TOT} = (\sigma_T + \epsilon\sigma_L) \ , \tag{2}$$

where Γ_T is the virtual photon flux

$$\Gamma_T = \frac{\alpha}{4\pi^2 Q^2}\frac{(W^2 - M^2)E'}{ME}\frac{1}{1-\epsilon}, \tag{3}$$

and ϵ describes the photon polarization. σ_T and σ_L are the transverse and longitudinal total absorption cross sections, respectively. The transverse total absorption cross section σ_T can be expressed in terms of the total absorption for helicity $\frac{1}{2}$ and helicity $\frac{3}{2}$ in the initial state photon-nucleon system:

$$\sigma_T = \frac{1}{2}(\sigma^T_{1/2} + \sigma^T_{3/2}) \ . \tag{4}$$

σ_L contains only helicity $\frac{1}{2}$ components, and is typically small. For example, in the deep inelastic region $\sigma_L/\sigma_T \simeq 0.15$ due to the primordial transverse momentum of the quarks inside the nucleon.

In the resonance region, a significant σ_L could be evidence for non-quark degrees of freedom, such as photon absorption on spin-0 objects like pions or diquark configurations inside the nucleon. A strong longitudinal coupling to the proton is indeed observed in a kinematical regime where contributions from the pion cloud are expected to dominate: near threshold and at small values of Q^2 (Figure 5).

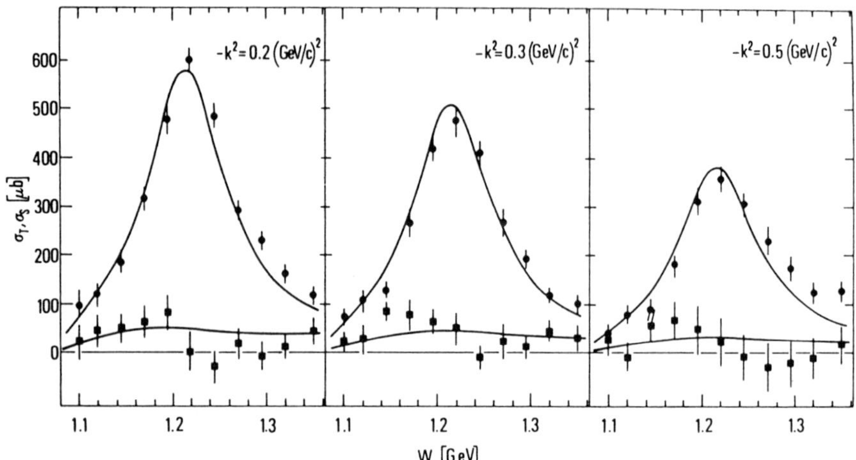

Figure 6. *Transverse and longitudinal cross section in the $P_{33}(1232)$ region ($-k^2 = Q^2$). Data from Bonn[16].*

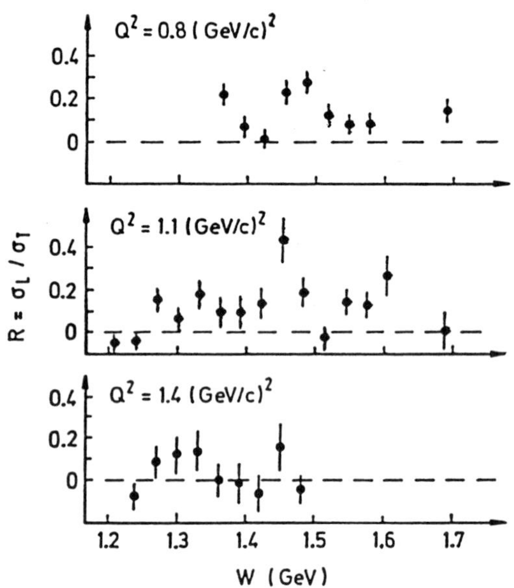

Figure 7. *Ratio σ_L/σ_T in the resonance region[17].*

Quark models predict small values of σ_L/σ_T for the resonance couplings, with possible exceptions for radial excitations, such as the $P_{11}(1440)$. Accurate separate measurements of σ_L and σ_T can therefore give important global information about the internal structure of baryons.

The longitudinal part can be separated from the transverse part by measuring the ϵ dependence of the total absorption cross section. This has been done for a few kinematical points only. The most accurate separation of σ_L and σ_T has been done at small Q^2 in the $P_{33}(1232)$ region (Figure 6), with the result that no resonant longitudinal coupling was found. Of course, the total absorption cross section is not very sensitive to relatively small amplitudes since they enter in quadrature. Therefore, small resonant longitudinal contributions are not excluded, and measurements which are more sensitive to these terms should be carried out. Results of measurements at higher masses yield similarly small values for σ_L/σ_T (Figure 7).

In a simple analysis of the inclusive cross section one may assume an incoherent superposition of resonances and non-resonant background. At fixed Q^2

$$\sigma_{TOT}(W) = \sigma_R(W) + \sigma_{NR}(W) , \tag{5}$$

with a Breit-Wigner parameterization for the resonant part:

$$\sigma_R(W) = A_R \cdot \frac{\vec{Q}_R^{*2}}{\vec{Q}^{*2}} \cdot \frac{W_R^2 \cdot \Gamma \cdot \Gamma_\gamma}{(W_R^2 - W^2)^2 + \Gamma^2 W_R^2} , \tag{6}$$

and a polynomial for the non-resonant contribution:

$$\sigma(W) = \sqrt{W - W_{thr}} \cdot \sum_{i=0}^{m} a_i \cdot (W - W_{thr})^i . \tag{7}$$

Γ and Γ_γ are the total width and the radiative width, respectively. $\vec{Q}^*$ is the photon momentum in the hadronic rest frame. This expression provides an s-wave energy behaviour near threshold. The least-square fit to the data yields A_R. For the higher mass resonances A_R is a superposition of different states with different widths, and different masses. The results are therefore ambiguous, and can give only indications of the global behaviour. The $P_{33}(1232)$, however, is an isolated resonance, and the fit should yield information about the $P_{33}(1232)$ resonance only. Since σ_L is small in the $P_{33}(1232)$ region, the resonant cross section can be expressed in terms of the transverse transition form factors $G_M^{*\Delta}$ and G_E^Δ only[28]:

$$\sigma_R \simeq \sigma_R^T = \frac{4\pi\alpha}{\Gamma} \cdot \frac{|\vec{Q}_L|^2}{W(W^2 - M^2)} \cdot (|G_M^\Delta|^2 + 3|G_E^\Delta|^2) \tag{8}$$

where $\vec{Q}_L$ is the photon momentum vector in the lab system. If G_E^Δ is small, which, as we shall see later is the case for moderate values of Q^2, one can determine $|G_M^\Delta|$. A compilation of the results from various groups is shown in Figure 8. The comparison with quark models demonstrates a longstanding problem: at the photon point, quark models underestimate the magnetic multipole transition amplitude by 20 to 30%.

Polarized Structure Functions in Inclusive Electron Scattering

If the target is polarized in the electron scattering plane the double polarized inclusive electron scattering cross section may be written as:

$$\frac{d\sigma}{d\Omega dE'} = \Gamma_T\{\sigma_T + \epsilon\sigma_L \pm \sqrt{1 - \epsilon^2} \cos\psi \sigma_T A_1 \pm \sqrt{2\epsilon(1 - \epsilon)} \sin\psi \sigma_T A_2\} , \tag{9}$$

where ψ is the polar angle of the target polarization vector relative to the photon 3-momentum vector. σ_T and σ_L are the transverse and longitudinal total photon

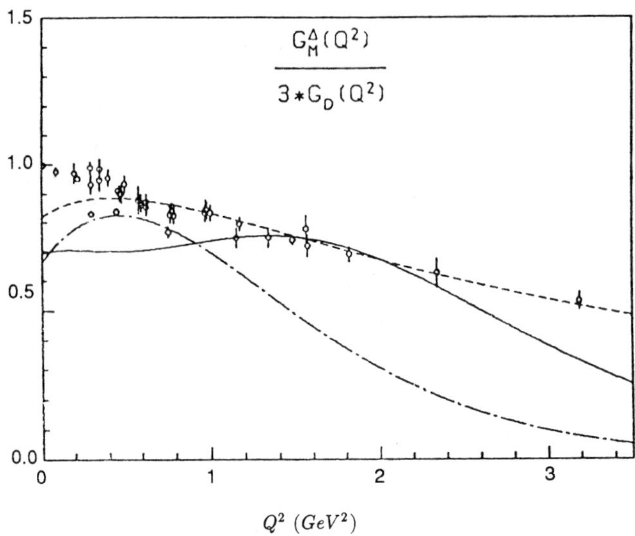

Figure 8. *The magnetic transition formfactor G_M^Δ normalized to the dipole form-factor. The curves are results of quark model calculations*[18],[19].

absorption cross section, and the sign $\pm$ is related to the sign of the product of beam and target polarization (assumed to be unity). A_1 and A_2 are the polarized asymmetries:

$$A_1 = \frac{\sigma_{1/2}^T - \sigma_{3/2}^T}{\sigma_{1/2}^T + \sigma_{3/2}^T} \tag{10}$$

$$A_2 = \frac{\sigma^{TL}}{\sigma_{1/2}^T + \sigma_{3/2}^T} \ , \tag{11}$$

where $\sigma_{1/2}^T(Q^2,\nu)$ and $\sigma_{3/2}^T(Q^2,\nu)$ are the transverse total absorption cross sections for total helicity $\lambda_{\gamma N} = \frac{1}{2}$ and $\lambda_{\gamma N} = \frac{3}{2}$, respectively. A_1 is limited to:

$$-1 \leq A_1 \leq +1 \ .$$

A_2 is a transverse-longitudinal interference term with an upper bound of:

$$A_2 \leq \sqrt{\frac{\sigma_L}{\sigma_T}} \ . \tag{12}$$

Since σ_L/σ_T is small throughout the resonance region, A_2 will remain relatively small too. The two polarization structure functions can be separated by polarizing the target in the scattering plane, and by varying the polarization angle ψ for fixed electron kinematics.

$A_1(W,Q^2)$ contains global information about the helicity structure of the nucleon resonances, and their Q^2 dependences. Knowledge of A_1 is also needed as an ingredient for determining the Q^2 evolution of the Gerasimov-Drell-Hearn sum rule.

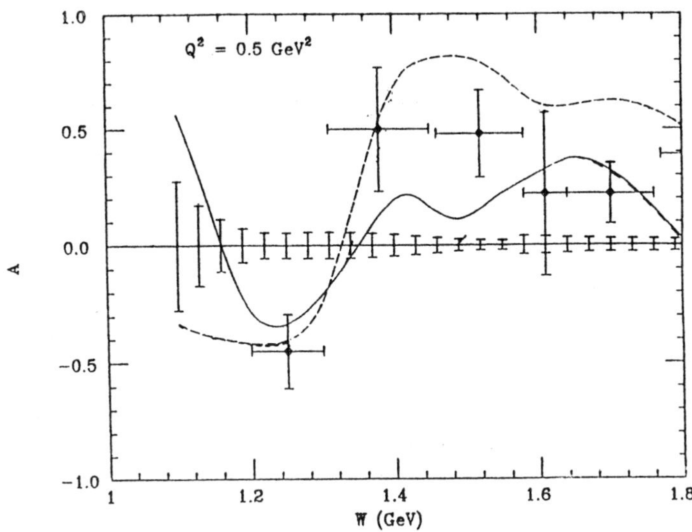

Figure 9. *Polarized inclusive structure function $A_1 + \eta A_2$ as measured at $SLAC$[30]. The bars around the horizontal axis are projected errors of an approved experiment at CEBAF.*

An experiment at SLAC measured a combination of A_1 and A_2 (Figure 9). Although the errors are large, an interesting Q^2 dependence was revealed; the helicity asymmetry in the region of $W = 1.5 - 1.8$ GeV, showed a positive value corresponding to helicity $\frac{1}{2}$ dominance at $Q^2 = 0.5$ GeV^2, whereas at the photon point the dominant resonant states $D_{13}(1520)$ and $F_{15}(1680)$ are dominantly excited by helicity $\frac{3}{2}$. This "helicity switch" is discussed in more detail in lecture 2.

The asymmetry in the mass region of the $P_{33}(1232)$ remains unchanged at $\simeq$ -0.5. For a magnetic dipole transition $\sigma_{1/2}/\sigma_{3/2} = \frac{1}{3}$, and therefore $A_1 = -\frac{1}{2}$, in good agreement with the data.

SINGLE PION ELECTROPRODUCTION

Inclusive measurements do not allow separation and identification of the various excited states. The three or four enhancements in the inclusive cross section correspond to about 20 excited states that may contribute for invariant masses below 2 GeV/c^2. The situation is illustrated in Figure 10. Three states may contribute to the second enhancement, and at least six states may contribute to the third bump, while even more are hidden under the rather smooth and uninteresting looking mass region around 2 GeV/c^2. As we will see later, several of the so-called "missing resonances" may be located in this mass region.

An unambiguous identification can only be accomplished by explicit measurement of the decay products such as πN, ηN, ρN, $\pi \Delta$. The $\gamma_v NN^*$ vertex for the transition into a specific state is described by two (for $J = \frac{1}{2}$ states) or three (for $J \geq \frac{3}{2}$ states) amplitudes, $A_{1/2}(Q^2)$, $A_{3/2}(Q^2)$, and $S_{1/2}(Q^2)$, where A and S refer to the transverse and scalar coupling, respectively, and the subscripts refer to the total helicity of the $\gamma_v N$ system (Figure 11). Spin and isospin can be extracted by

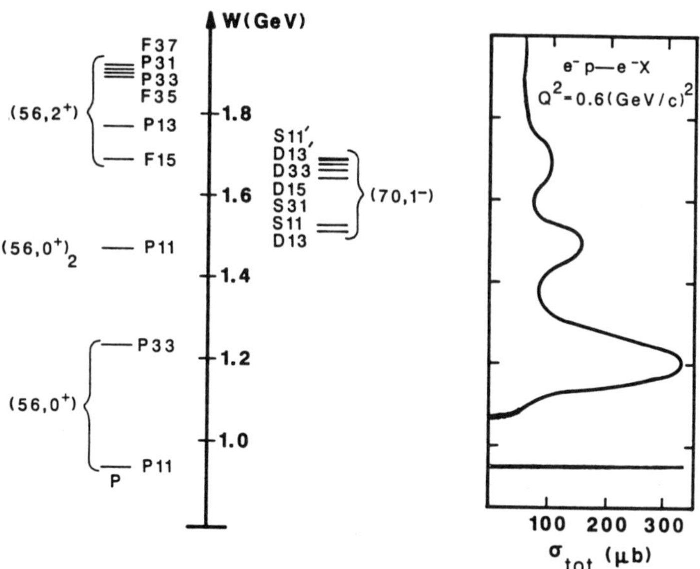

Figure 10. *The nucleon resonances with masses below 2 GeV (left) and the shape of the inclusive cross section at $Q^2 = 0.6$ GeV^2 (right).*

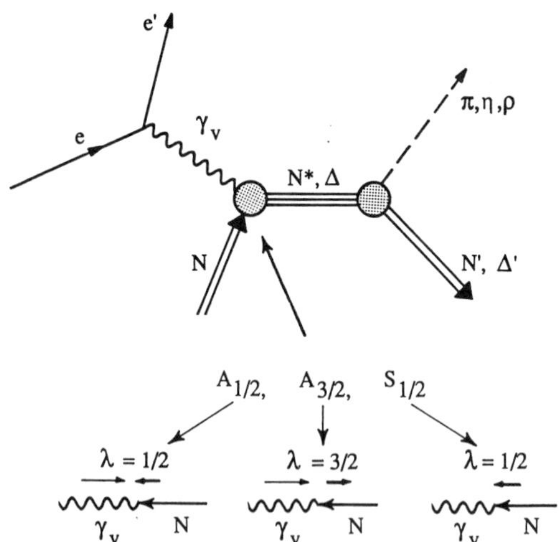

Figure 11. *Electroproduction of hadronic final states via s - channel resonance decays. The $\gamma_v NN^*$ vertex is described by the photocoupling helicity amplitudes $A_{1/2}$, $A_{3/2}$, and $S_{1/2}$, which are functions of Q^2 only.*

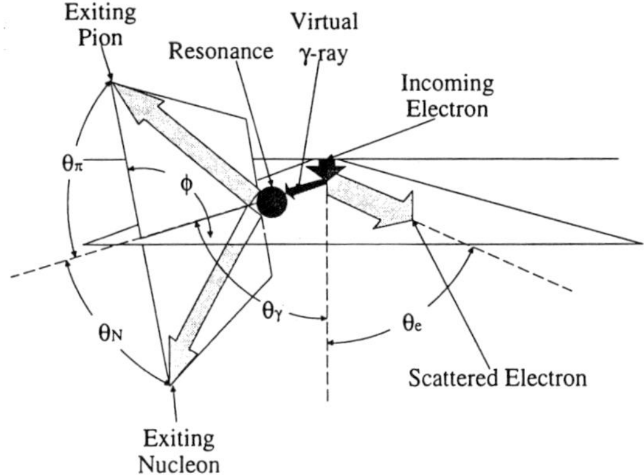

Figure 12. *Kinematics of single pion electroproduction off a nucleon.*

measuring the angular distribution in different isospin channels. Many of the low lying resonances decay primarily into the πN or ηN channels. Experiments have therefore concentrated on single π and η production. In the following section our current knowledge of the resonance transition amplitudes resulting from these reactions is reviewed. Very little is known about multiple pion production processes. Some global information about the reactions $p(e, e'p\rho)$, $p(e, e'\pi\Delta)$ has been obtained from a DESY streamer chamber experiment[22]. However, the data are not sufficiently detailed to allow the extraction of helicity amplitudes for specific resonances.

Multipoles and Partial Wave Helicity Element

Single pion production kinematics is shown in Figure 12. The differential cross section in single meson production contains four response functions:

$$\frac{d\sigma}{d\Omega} = \sigma_T + \epsilon\sigma_L + \epsilon\sigma_{TT}\cos 2\phi + \sqrt{2\epsilon(1+\epsilon)}\sigma_{TL}\cos\psi \ , \tag{13}$$

where ϕ is the azimuthal angle of the hadronic decay plane with respect to the electron scattering plane. The observables of the process $\gamma_v N \rightarrow \pi N$, where γ_v denotes the virtual photon, can be expressed in term of six parity conserving helicity amplitudes[26]:

$$H_i := \ <\lambda_\pi; \lambda_N |T|\lambda_{\gamma_v}; \lambda_p> \ = \ <0; \pm\frac{1}{2}|T| \pm 1, 0; \pm\frac{1}{2}> \ , \tag{14}$$

where λ denotes the helicity of the respective particle. The H_i are complex functions of Q^2, W, θ_π^*. The response functions in (13) are given by:

$$\sigma_T = \frac{|\vec{p}_\pi|W}{2KM} \cdot (|H_1|^2 + |H_4|^2 + |H_4|^2 + |H_4|^2)$$

$$\sigma_L = \frac{|\vec{p}_\pi|W}{KM} \cdot (|H_5|^2 + |H_6|^2)$$

$$\sigma_{TT} = \frac{|\vec{p}_\pi|W}{KM} \cdot Re(H_2 H_3^* - H_1 H_4^*) \tag{15}$$

$$\sigma_{TL} = \frac{|\vec{p}_\pi|W}{\sqrt{2}KM} \cdot Re(H_5^*(H_1 - H_4) + H_6^*(H_2 + H_3)) \ ,$$

where $\vec{p}_\pi$ is the pion momentum in the hadronic cms system, and K is the equivalent real photon lab energy for producing a state with mass W:

$$K = \frac{W^2 - M^2}{2M} \ .$$

The H_i can be expanded in terms of Legendre polynomials:

$$H_1 = \frac{1}{2}\sqrt{2}\sin\theta\cos\frac{\theta}{2}\sum_{l=1}^{\infty}(B_{l+} - B_{(l+1)-})(P_l'' - P_{l+1}'')$$

$$H_2 = \sqrt{2}\cos\frac{\theta}{2}\sum_{l=0}^{\infty}(A_{l+} - A_{(l+1)-})(P_l' - P_{l+1}')$$

$$H_3 = \frac{1}{2}\sqrt{2}\sin\theta\sin\frac{\theta}{2}\sum_{l=1}^{\infty}(B_{l+} + B_{(l+1)-})(P_l'' + P_{l+1}'')$$

$$H_4 = \sqrt{2}\sin\frac{\theta}{2}\sum_{l=0}^{\infty}(A_{l+} + A_{(l+1)-})(P_l' + P_{l+1}')$$

$$H_5 = \sqrt{2}\cos\frac{\theta}{2}\sum_{l=0}^{\infty}(C_{l+} - C_{(l+1)-})(P_l' - P_{l+1}')$$

$$H_6 = \sqrt{2}\sin\frac{\theta}{2}\sum_{l=0}^{\infty}(C_{l+} + C_{(l+1)-})(P_l' + P_{l+1}') \ ,$$

$$(16)$$

$A_{l\pm}$ and $B_{l\pm}$ are the transverse partial wave helicity elements for $\lambda_{\gamma N} = \frac{1}{2}$ and $\lambda_{\gamma N} = \frac{3}{2}$, respectively. $C_{l\pm}$ are the longitudinal partial wave helicity elements. In the subscript $l+$, and $l+1,-$ are the π (η) orbital angular momenta, and $\pm$ is related to the total angular momentum, $J = l_\pi \pm \frac{1}{2}$. The partial wave helicity elements are linear combinations of the electromagnetic multipoles:

$$M_{l+} = \frac{1}{2(l+1)}(2A_{l+} - (l+2)B_{l+})$$

$$E_{l+} = \frac{1}{2(l+1)}(2A_{l+} + lB_{l+})$$

$$M_{l+1,-} = \frac{1}{2(l+1)}(2A_{l+1,-} + lB_{l+1,-})$$

$$E_{l+1,-} = \frac{1}{2(l+1)}(-2A_{l+1,-} + (l+2)B_{l+1,-})$$

$$S_{l+} = \frac{1}{l+1}\sqrt{\frac{\vec{Q}^{*2}}{Q^2}}C_{l+}$$

$$S_{l+1,-} = -\frac{1}{l+1}\sqrt{\frac{\vec{Q}^{*2}}{Q^2}}C_{l+1,-} \ ,$$

$$(17)$$

$\vec{Q}^*$ is the photon 3-momentum in the hadronic rest frame. The partial wave helicity elements contain both non-resonant and resonant contributions. An analysis must be performed to separate the resonant parts $\hat{A}_{l\pm}$, $\hat{B}_{l\pm}$, and $\hat{C}_{l\pm}$ of the amplitudes. In a final step the known hadronic properties of a given resonance can be used to determine the photocoupling helicity amplitudes which characterize the electromagnetic vertex:

$$\hat{A}_{l\pm} = \mp F \cdot C^I_{\pi N} \cdot A_{1/2}$$

$$\hat{B}_{l\pm} = \pm F \sqrt{\frac{16}{(2j-1)(2j+3)}} C^I_{\pi N} A_{3/2} \, , \tag{18}$$

$$F = \sqrt{\frac{1}{(2j+1)\pi} \frac{K}{p_\pi} \frac{M}{W_R} \frac{\Gamma_\pi}{\Gamma^2}}$$

where the $C^I_{\pi N}$ are isospin coefficients. The total absorption cross section for the transition into a specific resonance is given by:

$$\sigma_T = \frac{2M}{W_R \Gamma}(A^2_{1/2} + A^2_{3/2}) \, . \tag{19}$$

The $A_{1/2}$ and $A_{3/2}$ are the quantities often used to connect theoretical calculations to the experimental analysis. However, some models such as the Skyrme model, make direct predictions for the partial wave helicity elements[5].

Isospin Decomposition

Nucleon resonances are eigenstates of the isospin, with quantum numbers $I = \frac{1}{2}, \frac{3}{2}$. The final states in electromagnetic pion production are not eigenstates of the isospin. In order to identify the isospin of the intermediate resonant state, measurement of various channels with different isopin in the initial or final state is usually necessary. The photon transfers $\Delta I = 0, 1$ resulting in three isospin amplitudes for single pion production: T^s, T^v_1, T^v_3, where T_s is the isoscalar, T^v_1 the isovector amplitude with $I_{\pi N} = \frac{1}{2}$, and T^v_3 the isovector amplitude with $I_{\pi N} = \frac{3}{2}$. Assuming isospin symmetry, the decompositions for the various production channels are:

$$< \pi^+ n |T| \gamma_v p > = \sqrt{\tfrac{1}{3}}T^v_3 - \sqrt{\tfrac{2}{3}}(T^v_1 - T^s)$$

$$< \pi^0 p |T| \gamma_v p > = \sqrt{\tfrac{2}{3}}T^v_3 + \sqrt{\tfrac{1}{3}}(T^v_1 - T^s)$$

$$< \pi^- p |T| \gamma_v n > = \sqrt{\tfrac{1}{3}}T^v_3 - \sqrt{\tfrac{2}{3}}(T^v_1 + T^s)$$

$$< \pi^0 n |T| \gamma_v n > = \sqrt{\tfrac{2}{3}}T^v_3 + \sqrt{\tfrac{1}{3}}(T^v_1 + T^s)$$

Note that because of the differences in the masses of charged and neutral pions and nucleons, isospin symmetry is broken near pion threshold. Production of η and ω mesons selects directly $I = \frac{1}{2}$ contributions and allows isolation of the N^* resonances.

The Transition γ_v p $\rightarrow$ P$_{33}$(1232)

In $SU(6)$ symmetric quark models, this transition is described by a simple quark spin-flip in the $L_{3Q} = 0$ ground state, corresponding to a magnetic dipole transition M_{1+}. The electric and scalar quadrupole transitions are predicted to be:

$$E_{1+} = S_{1+} \equiv 0 \, . \tag{20}$$

In more elaborate QCD based models which include color magnetic interactions arising from the one-gluon exchange at small distances, the $P_{33}(1232)$ acquires an

$L_{3Q} = 2$ component. This leads to small electric and scalar contributions. Dynamical quark model calculations predict E_{1+}/M_{1+} to remain small (≤ 0.1) over a large Q^2 range, and to have a weak dependence on Q^2. At very high Q^2, helicity conservation requires[15] $E_{1+}/M_{1+} \to 1$, and $S_{1+}/M_{1+} \to 0$. Precise measurements of these contributions from $Q^2 = 0$ to very large Q^2 are obviously important for the development of realistic models of the nucleon.

In the region of the $P_{33}(1232)$, one may expect that only s- and p- waves with $J \leq \frac{3}{2}$ contribute. In this approximation, the partial wave expansion in the differential cross section leads to:

$$\frac{d\sigma}{d\Omega} = A_0 + \epsilon B_0 + \cos\theta(A_1 + \epsilon B_1) + \cos^2\theta(A_2 + \epsilon B_2)$$
$$+ \epsilon C_2 \sin^2\theta \cos 2\phi + \sqrt{\epsilon(1+\epsilon)}\sin\theta\cos\theta(D_0 + D_1\cos\theta), \tag{21}$$

which contains nine measurable quantities $A_0, B_0, A_1, ...$ The unknown multipoles in this approximation are: E_{0+}, M_{1-}, E_{1+}, M_{1+}, S_{0+}, S_{1-}, S_{1+}. Since the multipoles are complex quantities there are 14 unknown numbers, which cannot be determined unambiguously from the partial wave expansion coefficient without further assumptions. In the region of the $P_{33}(1232)$ the process $\gamma_v p \to \pi^0 p$ is dominated by a M_{1+} magnetic dipole transition. Therefore retaining only terms containing the M_{1+} may be a reasonable approximation near the resonance peak. The cross section then reduces to:

$$\frac{d\sigma}{d\Omega} \simeq \frac{|\vec{p}_\pi W|}{KM} \cdot \left[\frac{5}{2}|M_{1+}|^2 - 3Re(M_{1+}E_{1+}^*) + Re(M_{1+}M_{1-}^*)\right.$$
$$+ 2\cos\theta Re(E_{0+}M_{1+}^*)$$
$$+ \cos^2\theta\left(-\frac{3}{2}|M_{1+}|^2 + 9Re(M_{1+}E_{1+}^*) - 3Re(M_{1-}M_{1+}^*)\right) \tag{22}$$
$$+ \epsilon\sin^2\theta\cos 2\phi\left(-\frac{3}{2}|M_{1+}|^2 - 3Re(M_{1+}E_{1+}^*)\right)$$
$$\left. - \sqrt{2\epsilon_L(\epsilon+1)}\sin\theta\cos\phi\left(Re(S_{0+}M_{1+}^*) + 6\cos\theta Re(S_{1+}M_{1+}^*)\right)\right] \ ,$$

where θ is the pion cms polar angle, and $\epsilon_L = Q^2/\vec{Q}^{*2}\epsilon$. The sensitivity to the electric and scalar quadrupole transitions rests with the interference terms of E_{1+} and S_{1+} with M_{1+}.

In this approximation the resonant terms

$$|M_{1+}|, \ Re(M_{1+}E_{1+}^*), \ Re(M_{1+}S_{1+}^*)$$

and the non-resonant terms

$$Re(E_{0+}M_{1+}^*), \ Re(S_{0+}M_{1+}^*), \ \text{and} \ Re(M_{1-}M_{1+}^*)$$

can be determined unambiguously by measuring the ϕ and θ dependence of the differential cross section. Figure 13 shows the result of such an analysis.

Experimental data on $Re(E_{1+}M_{1+}^*)/|M_{1+}|^2$ and $Re(S_{1+}M_{1+}^*)/|M_{1+}|^2$ at the $P_{33}(1232)$ resonance mass are shown in Figure 14 and 15, respectively. The results confirm that the ratio E_{1+}/M_{1+} at the resonance mass is quite small, in qualitative agreement with $SU(6)$ and CQM predictions. However, the quality of the data is clearly not sufficient to discriminate against any of the models. For the ratio

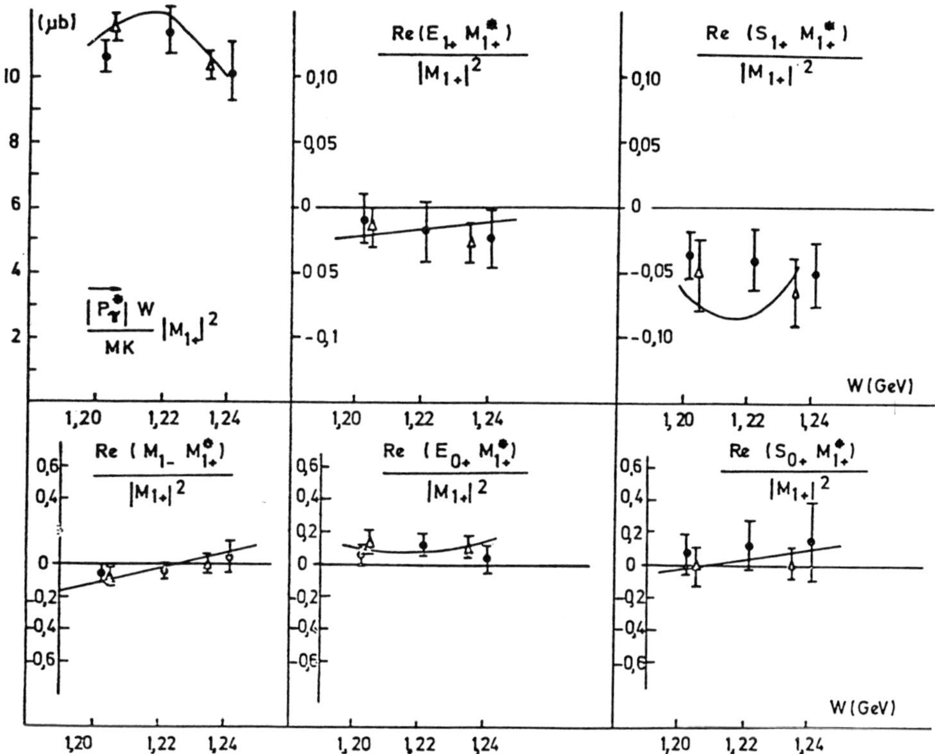

Figure 13. *Multipole analysis of $\gamma_v p \rightarrow \pi^0 p$ in the $P_{33}(1232)$ region. Data from Bonn*[23].

$Re(S_{1+}M_{1+}^*)/|M_{1+}|^2$ there seems to be a discrepancy with the quark model predictions. However, a warning may be in order here: The data in Figure 14 and 15 have not been analyzed to isolate the resonant contributions. They may therefore contain significant non-resonant contributions, whereas the model predictions are for the resonant parts only.

Experiments are in preparation[24],[25] to measure the quadrupole transitions over a large Q^2 range, using polarized electron beams and/or recoil polarimeters. In these experiments one obtains information not only about the terms

$$M_{1+} \ , \ Re(E_{1+}M_{1+}^*) \ , \ Re(S_{1+}M_{1+}^*)$$

but also about the corresponding imaginary parts:

$$Im(E_{1+}M_{1+}^*) \ , \ Im(S_{1+}M_{1+}^*).$$

The imaginary parts of the bilinear terms can be measured only by using polarization degrees of freedom. They are particularly sensitive to phase relations between the multipoles. If the multipoles were strictly in phase, these terms would vanish identically.

Phenomenological Analysis of Pion Production

For the analysis of the resonance region at masses up to 1.8 GeV, the procedure described for the $P_{33}(1232)$ to directly extract the multipoles is not feasible any more as one has to include partial waves up to at least $l_\pi = 3$, and $J \leq 5/2$, and unlike in

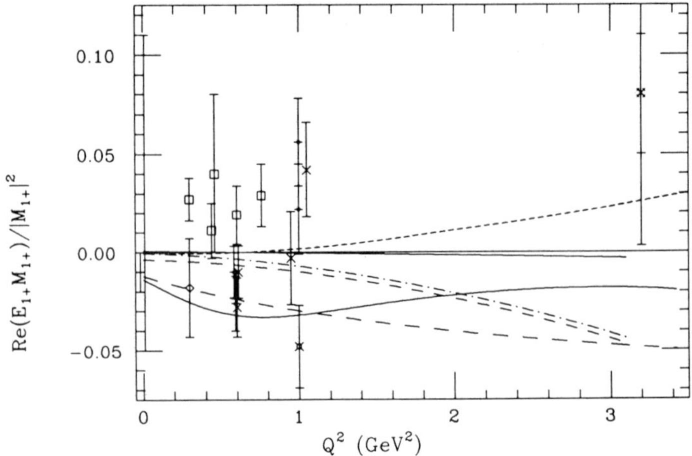

Figure 14. *Ratio $Re(E_{1+}M_{1+}^{*})/|M_{1+}|^{2}$ from electroproduction experiments*[65].

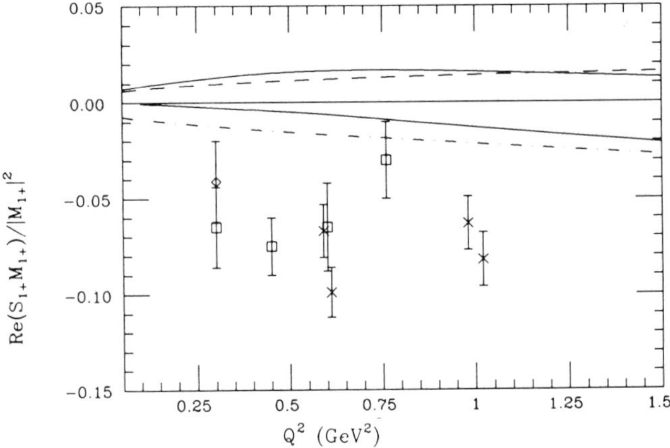

Figure 15. *Ratio $Re(S_{1+}M_{1+}^{*})/|M_{1+}|^{2}$ from electroproduction experiments*[65].

the $\Delta(1332)$ region there is not a single resonant amplitude that completely dominates the differential cross section. Moreover, resonances become broader and overlap, and interferences between resonant amplitudes and between resonant and non-resonant amplitudes become increasingly important. Therefore, alternative analysis methods have been developed that take into account all resonant amplitudes. In addition, a theory or a phenomenological model for the non-resonant amplitudes is needed. To cope with the limited volume of electroproduction data it has been necessary to also make assumptions about the energy dependence of the resonant amplitudes. In photoproduction where sufficient data are available energy-independent analyses have been performed. Two different methods have been employed in electroproduction, the isobar analysis method, and the dispersion relation method at fixed momentum transfer t.

Isobar analysis

In the isobar analysis[26] it is assumed that (1) the resonance masses, spin, isospin, and hadronic decay widths are known from πN scattering analysis, (2) that the nature of the energy dependence of the resonant part is known, and that it e.g. can be

118

described by relativistic Breit-Wigner amplitudes, (3) that pion production can be described by resonant s-channel amplitudes, real Born amplitudes, and additional real background amplitudes with reasonable threshold and high energy behaviour. Obviously, this method will give results that to some degree depend on the specific assumptions about the non-resonant amplitudes. The method works well if resonances constitute the dominant part of the amplitudes, which is largely the case in single pion photo- or electroproduction for masses up to about 1.7 GeV.

Fixed-t dispersion relation analysis

The application of dispersion relations[27],[28] is based on the observation that phenomenological analysis of pion photoproduction data yield good fits assuming real background amplitudes. This suggests that resonances approximately saturate the imaginary parts of the amplitudes, and the real and imaginary parts are related by dispersion relations:

$$ReA_i(s,t) = B_i(s,t) + \int_{(M+m_\pi)^2}^{+\infty} ds' \left[\frac{ImA_i(s',t)}{s-s'} \pm \frac{ImA_i(s',t)}{s'-u} \right] \quad . \tag{23}$$

B_i are the Born amplitudes, and ImA_1 are the imaginary parts of the resonant amplitudes. The advantage of this method is that it implements all general principles such as unitarity, analyticity, crossing symmetry. On the other hand, assumptions about the high energy behavior of the imaginary parts of the invariant amplitudes are necessary, and since integration is performed into the unphysical region questions about the convergence of the multipole series at $|\cos\theta| \geq 1$ remain. In the analysis of pion photoproduction data, this method has been used very successfully. In pion electroproduction, good fits have been obtained for $Q^2 \leq 2\ GeV^2$ (Figure 16).

A realistic analysis must also take into account effects resulting from the opening up of channels other that single pion production. These so-called cusp effects are the result of unitarity constraints on the amplitudes. That these effects can be quite significant is illustrated in Figure 17, which clearly shows the influence of the η threshold on the $\pi^0 p$ and $\pi^+ n$ channels. The very different effect on the charged and neutral pion channels is explained by the larger non-resonant contributions (real part of amplitude) in the $\pi^+ n$ channel while $\pi^0 p$ is dominated by resonant contributions (imaginary part of amplitude). Similar effects are expected to arise from other channels such as $\Delta\pi$.

In the following section I will discuss results of analyses of electroproduction data in terms of resonance couplings.

Electromagnetic Transitions to the $[70,1^-]_1$ and $[56,2^+]_2$

Of the seven non-strange states associated with the $[70, 1^-]_1$ supermultiplet only the $D_{13}(1520)$ and the $S_{11}(1535)$ have been studied in electroproduction experiments in some detail.

The $S_{11}(1535)$ Resonance

The $S_{11}(1535)$ is characterized by a large branching ratio into the ηN channel ($\approx 50\%$). Since the nearby $D_{13}(1520)$ state has a very small decay width into ηN, the $S_{11}(1535)$ can be separated by measuring the differential cross section $ep \to e'p\eta$. This process is dominated by resonant s - wave contributions. The angular distribution is

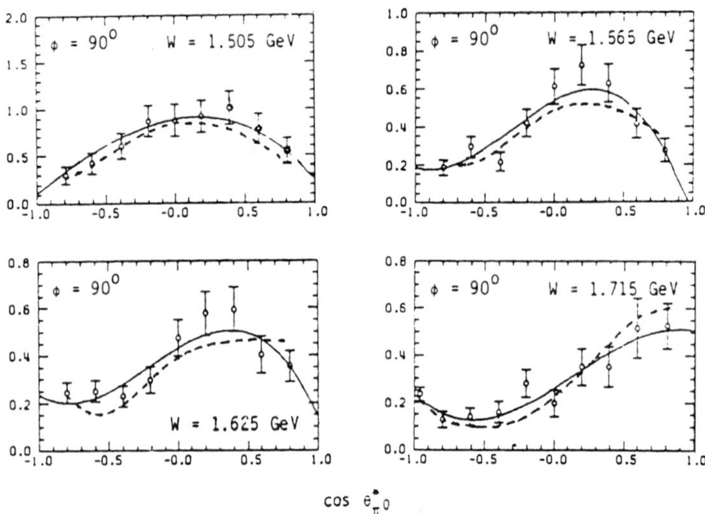

Figure 16. *Example of a dispersion fit (dashed lines) to electroproduction data in the region of the second resonance at $Q^2 = 2\ GeV^2$. The solid lines represent the results of a partial wave analysis fit[29].*

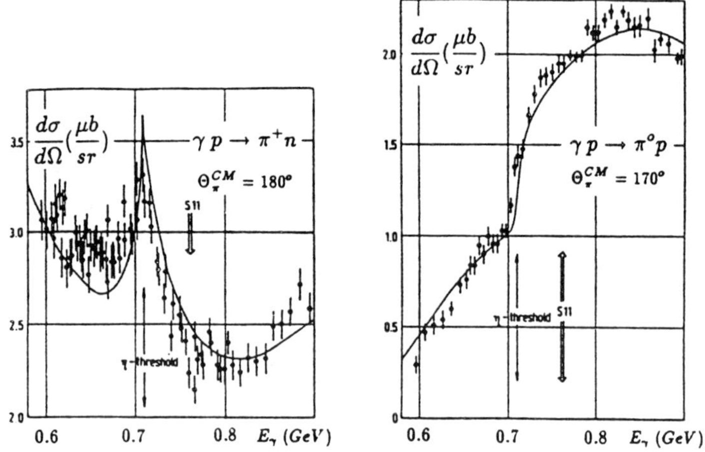

Figure 17. *Differential cross section $\gamma p \rightarrow \pi^+ n$ and $\gamma p \rightarrow \pi^0 p$ at the η threshold[30].*

isotropic, consistent with an s-wave behavior, and the energy dependence is dominated by resonance behavior (Figure 18).

For s-wave dominance the differential cross section can be written as:

$$\frac{d\sigma}{d\Omega^*_\eta} = A(W, Q^2) + \epsilon B(W, Q^2)$$

$$A = \frac{|\vec{p}^{\,*}_\eta| W}{MK} |A_{0+}|^2 \tag{24}$$

$$B = \frac{|\vec{p}^{\,*}_\eta| W}{MK} \frac{Q_0^2}{\vec{Q}^{*2}} |C_{0+}|^2$$

For $Q^2 \leq 1\ GeV^2$, the transverse and the longitudinal contributions have been measured using Rosenbluth separation (Figure 19).

120

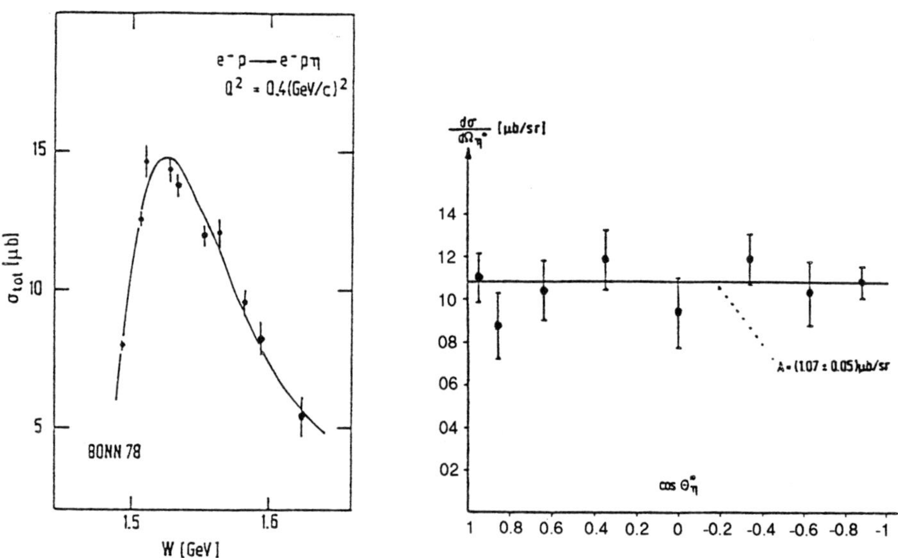

Figure 18. *Energy and angular dependence of η electroproduction* [31],[32].

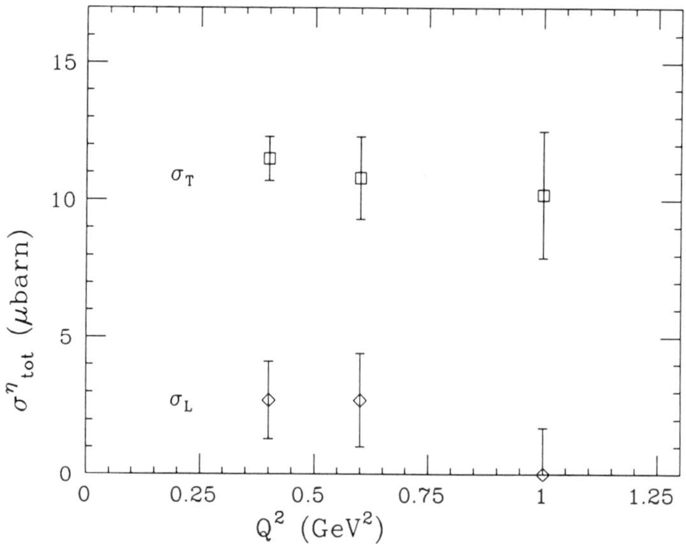

Figure 19. *Transverse and longitudinal photoabsorption cross section for $\gamma_v p \rightarrow \eta p$ at the $S_{11}(1535)$ resonance.*

The transverse photocoupling amplitude can then be determined (Figure 20). It shows an unusually slow falloff with Q^2 which cannot be explained in the simple NRCQM. However, recent extensions of the model to include relativistic effects have been more successful in approximately reproducing this behavior. It is interesting to note that within the framework of a specific model, the absolute normalization and the Q^2 dependence appear to be sensitive to the parameterization of the confinement potential. This lends credibility to the idea that a great deal can be learned about

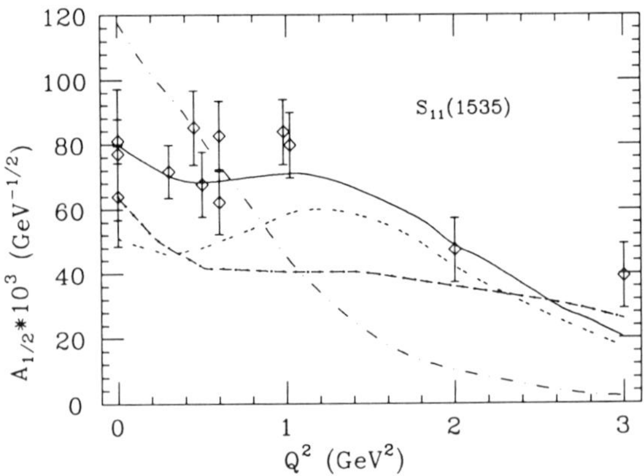

Figure 20. *Transverse amplitude $A_{1/2}$ for the transition $\gamma_v p S_{11}(1535)$. Model calculations by Close and Li[33] (short dashes), Warns et al[18] (solid line, double-dashed line for different confinement potentials), Foster and Hughes[19] (dots), Konen and Weber[34] (long dashes), Forsyth and Babcock [35] (dashed-dotted).*

the properties of the confinement potential by carefully studying many resonance transitions.

Helicity structure of the $D_{13}(1520)$ and the $F_{15}(1680)$

In photoproduction the $D_{13}(1520)$ and the $F_{15}(1680)$ are both excited predominantly by $A_{3/2}$ transitions. This behavior can be accommodated in the harmonic oscillator NRCQM. In this model[37] only single quark spin-flip and single quark orbit-flip amplitudes occur. The ratio of the helicity amplitudes in this model is:

$$\frac{A_{1/2}^{D13}}{A_{3/2}^{D13}} = \frac{1}{\sqrt{3}} \left(\frac{\vec{Q}^2}{\alpha^2} - 1 \right) \tag{25}$$

$$\frac{A_{1/2}^{F15}}{A_{3/2}^{F15}} = -\frac{1}{2\sqrt{2}} \left(\frac{\vec{Q}^2}{\alpha^2} - 2 \right) \ , \tag{26}$$

where the $\vec{Q}^2$ dependent term comes from the spin-flip, and the constant term from the orbit-flip contribution. The ratios can be made to disappear by requiring $\alpha^2 \simeq \vec{Q}^2$ for the $D_{13}(1520)$ and $\alpha^2 \simeq \vec{Q}^2/2$ for the $F_{15}(1680)$, which can be accommodated simultaneously with $\alpha^2 = 0.17 \ GeV^2$ for $\vec{Q}^2$ in the lab frame. With increasing Q^2, $A_{1/2}$ is then predicted to become the dominant contribution in either case[38]. Relativized versions of the NRCQM predict qualitatively the same behavior. This may be best demonstrated by displaying the helicity asymmetry

$$A_{\frac{1}{2},\frac{3}{2}} = \frac{A_{1/2}^2 - A_{3/2}^2}{A_{1/2}^2 + A_{3/2}^2} \ .$$

A compilation of the data[39],[40] is presented in Figure 21 and Figure 24. It is worth noting that effects due to the spatial wave function tend to cancel out in this quantity. The helicity asymmetry is therefore sensitive to the spin-flavor wavefunction. The helicity switch agrees qualitatively with quark model predictions. Calculations for

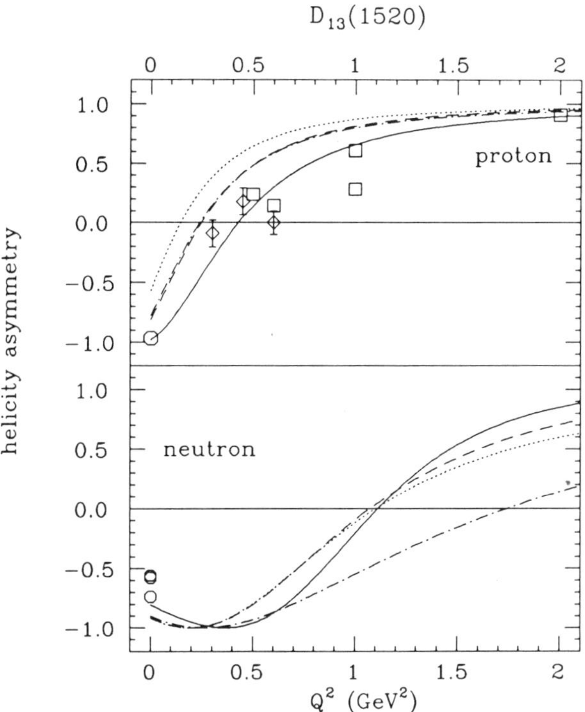

Figure 21. *Helicity asymmetry of the $\gamma_v p D_{13}(1520)$ transition. Quark model calculations by Capstick[3] with different corrections to the NRCQM.*

neutron targets show different sensitivity to ingredients of the model. Excitation of neutron resonances provides independent information about the nucleon structure.

Test of the Single Quark Transition Model

To the degree resonance transitions can be described by a single quark transitions in $SU(6)_W$ symmetric models, radiative transitions between the $[56, 0^+]_0$ and the $[70, 1^-]_1$ multiplet are completely determined by three amplitudes[41], usually called A, B, and C, where A is related to the quark orbit flip current, B to the spin flip current, and C to the combined spin-orbit flip current with $\Delta L_z = 1$. In the SQTM radiative transitions between all states belonging to these multiplets can be expressed in terms of linear combinations of these amplitudes (table 2).

Using the known $A_{1/2}$, $A_{3/2}$ amplitudes for the $S_{11}(1535)$ and the $D_{13}(1520)$ the A, B, C amplitudes can be determined (Figure 22). There is clear evidence for the existence of a non-zero spin-orbit flip amplitude C, which does not exist in the simple NRCQM. The SQT amplitudes can be used to predict transition amplitudes for other states in the same supermultiplet. Unfortunately, information from other states is limited to proton targets and is of poor quality. Current experimental information of the $S_{11}(1650)$, $S_{31}(1620)$, and $D_{33}(1700)$ is summarized in Figure 23. With the possible exception of the $S_{31}(1620)$, the data are not in disagreement with the SQTM predictions, however, they are not accurate enough to test deviations from the SQTM. Whether the deviations seen for the $S_{31}(1620)$ at small Q^2 are significant remains to be seen when more accurate and more complete data will be available.

The most prominent state in $[56, 2^+]_2$ is the $F_{15}(1688)$, and it is the only one

Table 2. Single Quark Transition Amplitudes for $\gamma + [56, 0^+]_0 \rightarrow [70, 1^-]_1$

State	Proton Target	Neutron Target
$S_{11}(1535)$:	$A^+_{\frac{1}{2}} = \frac{1}{6}(A + B - C)\cos\theta$ *	$A^o_{\frac{1}{2}} = -\frac{1}{6}(A + \frac{1}{3}B - \frac{1}{3}C)$
$D_{13}(1520)$:	$A^+_{\frac{1}{2}} = \frac{1}{6\sqrt{2}}(A - 2B - C),$ $A^+_{\frac{3}{2}} = \frac{1}{2\sqrt{6}}(A + C)$	$A^o_{\frac{1}{2}} = -\frac{1}{18\sqrt{2}}(3A - 2B - C)$ $A^o_{\frac{3}{2}} = \frac{1}{6\sqrt{6}}(3A - C)$
$S_{11}(1650)$:	$A^+_{\frac{1}{2}} = \frac{1}{6}(A + B - C)\sin\theta$	$A^o_{\frac{1}{2}} = \frac{1}{18}(B - C)$
$D_{13}(1700)$:	$A^+_{\frac{1}{2}} = A^+_{\frac{3}{2}} = 0$	$A^o_{\frac{1}{2}} = \frac{1}{18\sqrt{5}}(B - 4C)$ $A^o_{\frac{3}{2}} = \frac{1}{6\sqrt{15}}(3B - 2C)$
$D_{15}(1670)$:	$A^+_{\frac{1}{2}} = A^+_{\frac{3}{2}} = 0$	$A^o_{\frac{1}{2}} = -\frac{1}{6\sqrt{5}}(B + C)$ $A^o_{\frac{3}{2}} = -\frac{1}{6}\sqrt{\frac{2}{5}}(B + C)$
$D_{33}(1670)$:	$A^+_{\frac{1}{2}} = \frac{1}{18\sqrt{2}}(3A + 2B + C)$ $A^+_{\frac{3}{2}} = \frac{1}{6\sqrt{6}}(3A - C)$	same same
$S_{31}(1650)$:	$A^+_{\frac{1}{2}} = \frac{1}{18}(3A - B + C)$	same

* $(\theta = $ mixing angle between $^4[8]_{1/2}$ and $^2[8]_{1/2}$ in $\{70, 1^-\}_1)$

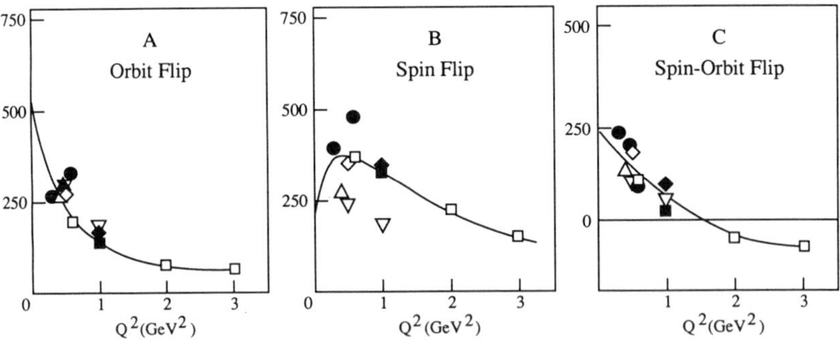

Figure 22. *Single quark transition amplitudes A, B, C in units 10^{-3} GeV$^{-1/2}$*

that has been studied experimentally over an extended Q^2 range. Similar to the $D^+_{13}(1520)$, the photoexcitation is dominantly helicity 3/2 and $A^+_{1/2}(F_{15}) \approx 0$, at $Q^2 = 0$. The data show a rapid change in the helicity structure with rising Q^2 (Figure 24). The switch to helicity 1/2 dominance is qualitatively reproduced by quark model calculations. However, much improved data are needed for a more definite comparison with the theory.

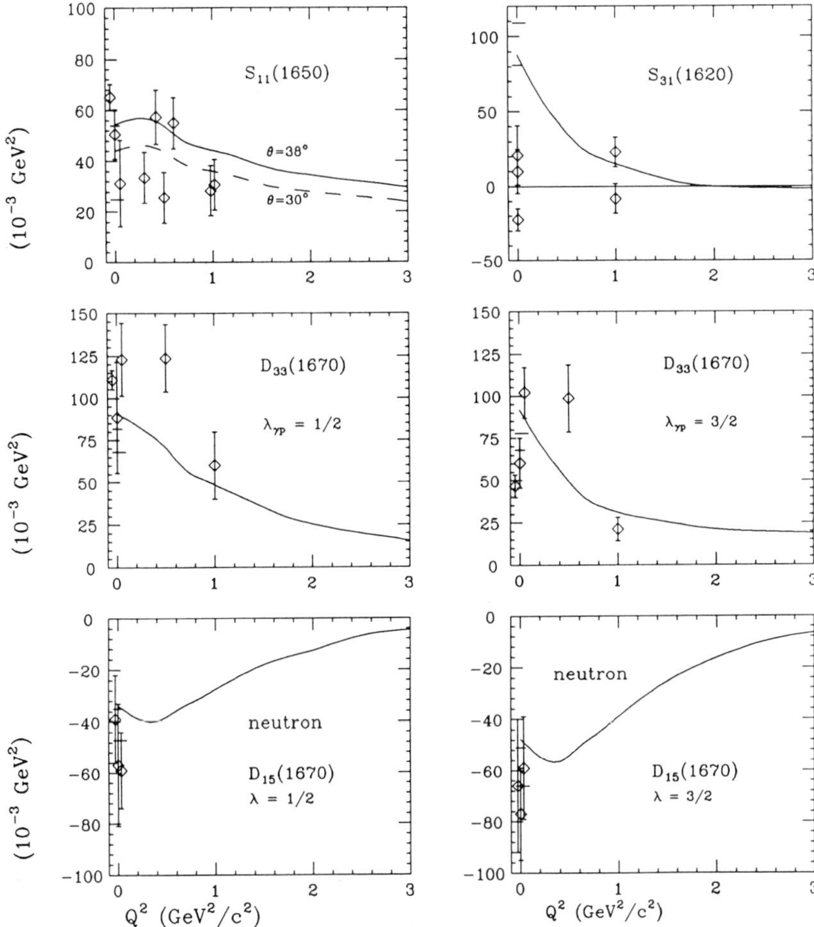

Figure 23. *Transverse photocoupling amplitudes for the the $S_{11}(1650)$, $S_{31}(1620)$, and $D_{33}(1700)$ states. The curves represent SQTM predictions using $S_{11}(1535)$ and $D_{13}(1520)$ data and the algebraic relations of Hey and Weyers[41]. A mixing angle between the $[^48]$ and $[^28]$ quark model states in $[70, 1^-]$ of $\theta = 38°$ was used.*

In the SQTM the transition to the $[56, 2^+]_2$ can be described by four amplitudes A′, B′, C′, and D′. D′ is a spin-orbit flip amplitude with $\Delta L_z = 2$. Without additional assumptions the four contributing SQTM amplitudes cannot presently be determined from available data due to the lack of electroproduction data for a second state in the $[56, 2^+]_2$. The $F_{37}(1950)$ would be a good candidate to obtain additional information as it has a large decay width into the πN channel and can be studied in single pion production.

TOPICS IN NUCLEON RESONANCE PHYSICS

In this section I want to discuss some problems that have generated a great deal of interest in recent years. Their resolution may have a significant impact on our understanding of baryon structure. These issues have either not been addressed at all, or have not been addressed adequately in previous experiments.

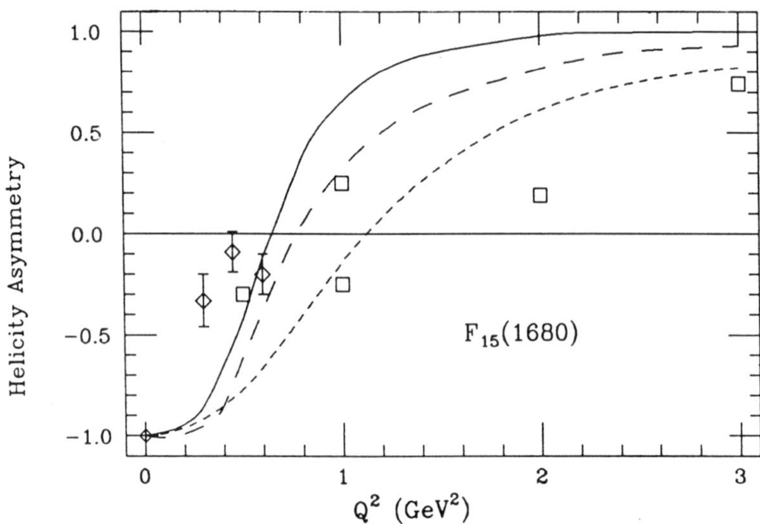

Figure 24. *Helicity asymmetry for the $\gamma_v p F_{15}(1688)$. NRCQM calculation by Ono[36] - long dashed line, Copley[37] - short dashed, Close[33] - solid, with QCD mixing.*

Electroquenching of the $P_{11}(1440)$

In the NRCQM the lowest mass $P_{11}(1440)$ state is assigned to a radially excited q^3 state within the $SU(6) \otimes O(3)$ super-multiplet $[56, 0^+]_2$ (i.e. $L_{3Q} = 0$, $N_{3Q} = 2$). However, the observed low mass of the state, as well as the sign and magnitude of the photocoupling amplitudes have traditionally been difficult to reproduce within the framework of the NRCQM. Moreover, there is experimental evidence that the Q^2 dependence of the photocoupling amplitude $A_{1/2}(Q^2)$ is quite different from what is predicted in the framework of the NRCQM. The data indicate a rapid fall-off of the absolute value of this amplitude with Q^2 whereas the NRCQM, as well as relativized versions, predict a much weaker fall-off or even an initial rise with Q^2. Experimental information about electroproduction amplitudes of the $P_{11}(1440)$ is rather limited, largely due to the complete lack of polarization data, and definite conclusions about the nature of the $P_{11}(1440)$ will have to wait for more accurate data. The non-relativistic quark model predicts for the neutron/proton ratio: $A_{1/2}^n / A_{1/2}^p = -2/3$. In chiral bag model calculations[42] contributions from the pion cloud of the proton bring this ratio closer to -1 at $Q^2 = 0$. With increasing Q^2, however, the role of the pion cloud should be diminished, and the quark composition is expected to dominate the excitation of the $P_{11}(1440)$ at higher Q^2. Precise data on photo- and electro-excitation of the $P_{11}(1440)$ would help to reveal the true nature of this state. An interesting consequence of the $[56, 0^+]_2$ assignment of the $P_{11}(1440)$ is its predicted dominance over the $P_{33}(1232)$ at high Q^2. The NRCQM predicts[37]:

$$\frac{A_{1/2}(P_{11}(1440))}{A_{1/2}(P_{33}(1232))} \propto \vec{Q}^2 . \tag{27}$$

The data are shown in Figure 25. The value -2/3 is preferred for the neutron/proton ratio, although a value closer to -1 is not ruled out. The Q^2 dependence is not well determined, however it is rather obvious that none of the explicit quark models comes even near to describing both the photon point and the Q^2 behaviour suggested by the data. Relativistic corrections give uncomfortably large effects, casting some doubts their convergence. The $P_{11}(1440)$ is predicted to couple rather strongly to longitudinal photons, a feature which is not supported by the data. $S_{1/2}$ is consistent with a small value, or even zero, although significant values at small Q^2 cannot be excluded. Clearly, more precise and more complete data are needed to study the apparently strong Q^2 dependence of $A_{1/2}$ at small Q^2, and to establish more accurate values for the longitudinal coupling.

Gluonic Excitations of the Nucleon

From inclusive lepton scattering experiments we know that in the deep inelastic region about 50% of the proton momentum is carried by gluons. This raises the question of what impact gluons have on baryons spectroscopy. For example, can gluonic degrees of freedom be excited explicitly and generate new spectroscopic states? For some time there have been speculations about the existence of gluonic baryon states $q^3 G$ consisting of three constituent quarks and one constituent gluon[10],[11]. QCD lattice simulations indicate that such configurations may exist for mesons, whereas no such calculations have been performed for baryons. Estimates within the framework of bag models yield masses for the lowest P_{11}^G state around 1.5 GeV.

How can one search for these states? In hadronic production experiments gluonic baryons cannot be distinguished from ordinary q^3 states because they are, unlike gluonic mesons, characterized by quantum numbers which are also possible for the normal q^3 baryon states. However, as their internal structure is quite different from ordinary baryons, electroproduction experiments could be a powerful tool in these studies.

The $P_{11}(1440)$ has been suggested as a candidate for the gluonic partner of the nucleon[43]. The wave function for a nucleon with gluonic degrees of freedom may be written as:

$$|N> = \frac{1}{\sqrt{1+2\delta^2}}[|N_0> -\delta(|^4N_g> +|^2N_g>)] \ ,$$

where $|N_0>$ represents the wavefunction of the 3-quark system, which transforms as a [56] under $SU(6)$ for nucleons, $|^2N_g>$ and $|^4N_g>$ are the wave functions of the $q^3 G$ system with the spin $\frac{1}{2}$ and $\frac{3}{2}$, respectively. The parameter δ is determined by the quark-gluon interaction. The corresponding state orthogonal to the nucleon in spin-flavor space is:

$$|N', J = \tfrac{1}{2}> = \sqrt{\frac{2}{1+2\delta^2}}[\delta|N_0> +\tfrac{1}{2}(|^2N_g> +|^4N_g>)] \ .$$

It turns out that this description allows preservation of the successes of the quark model, most notably the neutron/proton ratio $A_{1/2}^n/A_{1/2}^p = -2/3$. If the $P_{11}(1440)$ were indeed the gluonic partner of the nucleon, a long-standing problem in baryon structure, the strong quenching of the transition formfactor $\gamma_v p P_{11}(1440)$ with Q^2 could be resolved. Obviously, the solution of the puzzle concerning the correct assignment of the $P_{11}(1440)$ could have enormous impact on our understanding of baryon structure and the dynamics of the strong interaction in the non-perturbative regime.

How can we experimentally discriminate between these alternatives? In a model

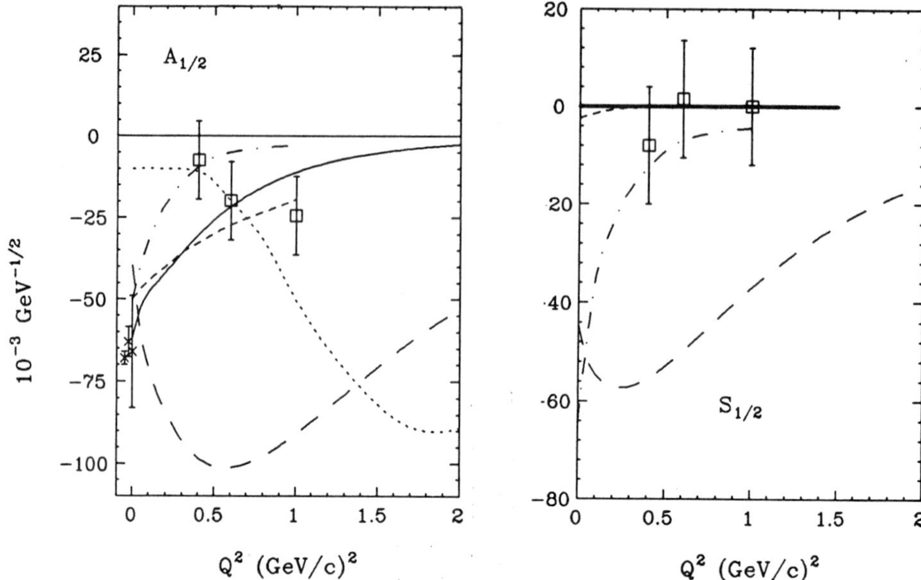

Figure 25. *Transverse (left) and longitudinal (right) photocoupling amplitudes of the $P_{11}(1440)$ for proton targets. The open symbols and the short dashed and dashed-dotted lines represent results of a fixed-t dispersion relation fit by Gerhardt[44]. Long dashes represent calculations for protons within the NRQM with QCD mixing[33]. The dotted line includes higher order relativistic corrections[18]. The solid lines are the result of a calculation assuming the $P_{11}(1440)$ is the gluonic partner of the nucleon[43]. Only calculations that approximately reproduce the photon point have been included.*

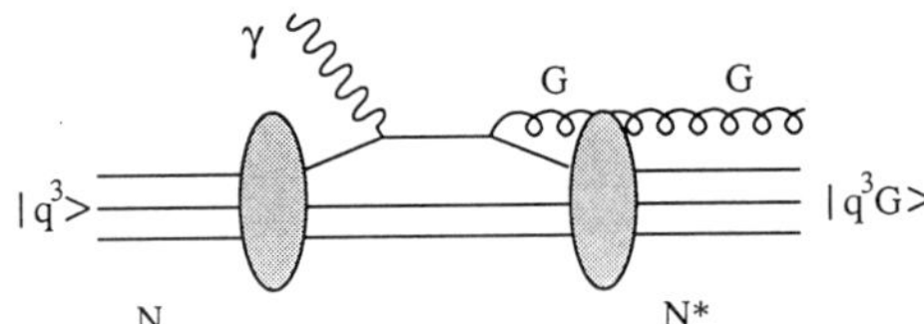

Figure 26. *QCD Compton process in gluonic excitation.*

of the nucleon containing constituent quarks and gluon, a graph that is expected to contribute to gluonic excitations is the QCD Compton process $\gamma q \to G q$ (Figure 26)

The inverse process $g q \to \gamma q$, where g is an elementary gluon, has been studied in detail in hard scattering processes[45] and is well described in perturbative QCD. If the $P_{11}(1440)$ is a $q^3 G$ state, then, because the gluon has only transverse excitation modes the longitudinal coupling is absent, $\gamma_L p \not\to P_{11}$, and

$$S_{1/2}(Q^2) \equiv 0 \ . \tag{28}$$

This is consistent with the analysis of experimental data, although more accurate data especially at small Q^2 are needed for a more definite comparison.

A precise measurement of the Q^2 dependence of the $\gamma_v p P_{11}(1440)$ transverse photocoupling amplitude discriminates between the interpretation of the $P_{11}(1440)$

as a regular $[56, 0^+]_2$ q^3 state, or as a gluonic excitation where the 3-quark system transforms like a $[70]$ under $SU(6)$. The discriminating power is a result of the fact that the respective photocoupling amplitudes are associated with different spin flavor factors for different spectroscopic assignments, so that in the first approximation (if effects from the spatial wavefunction and relativistic corrections are neglected):

$$\frac{A_{1/2}(P_{11}^G)}{A_{1/2}(P_{11})} \sim \frac{1}{\vec{Q}^2} \; . \tag{29}$$

From (27), (28), and (29) one infers that a P_{11}^G state behaves like the $P_{33}(1232)$. Accurate measurement of the Q^2 dependence of the $P_{11}(1440)$ photocoupling amplitude can be used to discriminate between different spectroscopic assignments. The calculation based on the q^3G interpretation is in better agreement with the data than calculations using the non-relativistic or relativized versions of the constituent quark model (Figure 25).

If the $P_{11}(1440)$ is the gluonic partner of the nucleon, then other gluonic states such as P_{13}^G, , P_{31}^G, P_{33}^G must exist as well, with masses around 1.6 - 1.7 GeV. The first two states have no place in the q^3 model at such low masses, whereas the latter one is also expected as a q^3 state. There is weak evidence for $P_{13}(1540)$ and $P_{31}(1550)$ states. Another question is, what would the mass of the lowest q^3 P_{11} be? The $P_{11}(1710)$ might be this state. Its mass is also more in accord with the NRCQM estimate. If this were the case, then the photocoupling amplitudes should exhibit a Q^2 dependence characteristic of a radially excited q^3 state.

Missing q^3 Baryon States

The QCD motivated extensions of conventional quark models predict many states with masses above 1.8 GeV which have not been observed in $\pi N \to \pi N$ reactions. Calculations[7] show that many of the "missing" states tend to decouple from the πN channel due to mixing, however, they may couple with significant strength to channels such as ρN, ωN, or $\pi \Delta$. It is experimentally well established that single pion production decreases with energy, while multi-pion production and vector meson production processes become more important (Figure 27). Electromagnetic production of these channels may therefore be the only way to study the "missing" states. In fact, several of those in the $[56, 2^+]_2$ super multiplet are predicted to couple strongly to photons. For example, the $F_{15}(1955)$, and the $F_{35}(1975)$ should be excited almost as strongly as some of the prominent states at lower masses. Search for these states is important and urgent. There are models, such as the quark cluster model[9] that can accommodate known baryon states, while predicting a fewer number of unobserved states. Future experiments[46],[47] to study electroproduction and photoproduction of vector mesons should provide a definite answer regarding the existence of at least some of these states. Figure 28 shows the predicted effect of one of the "missing" resonances on the differential cross section in $ep \to ep\omega$. At forward angles the process is dominated by diffractive production and the pion exchange diagram, while resonance contributions would dominate at backward angles. Similar behavior is expected for $ep \to ep\rho^0$, while $ep \to en\rho^+$ has no diffractive contributions and is more sensitive to resonant production.

Baryon Resonance Transitions at High Q^2

At high energies, perturbative QCD makes simple predictions about the asymp-

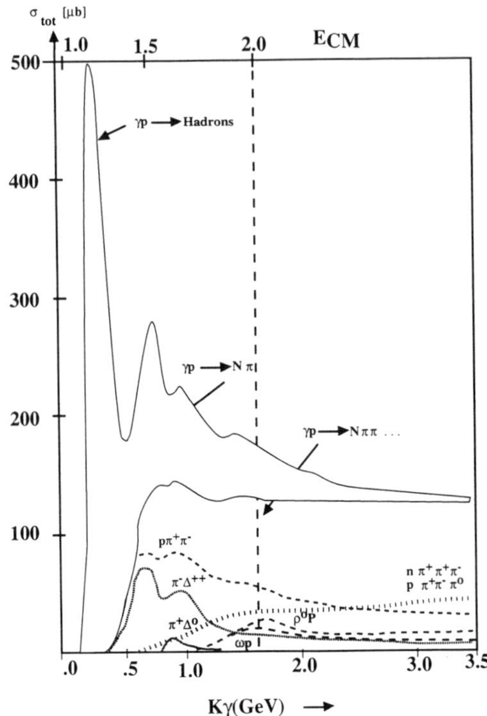

Figure 27. *Photoabsorption and total photoproduction cross section for various exclusive channels.*

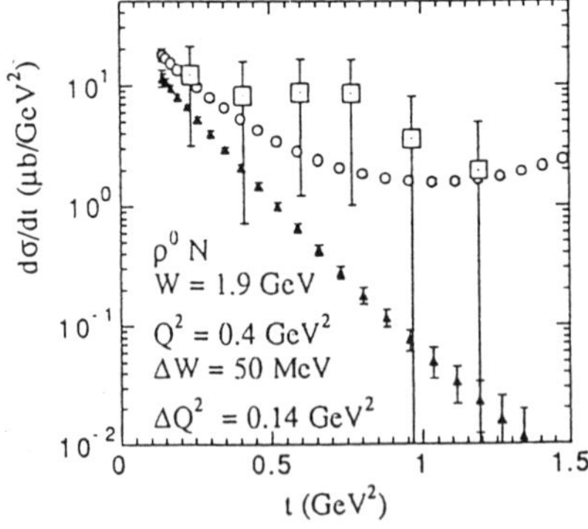

Figure 28. *Differential cross section for $\gamma_v p \to \rho^0 p$. Existing data (large errors), and projected data for a CEBAF experiment[46] are shown; full triangles without, open circles with the "missing" $F_{15}(1950)$ states.*

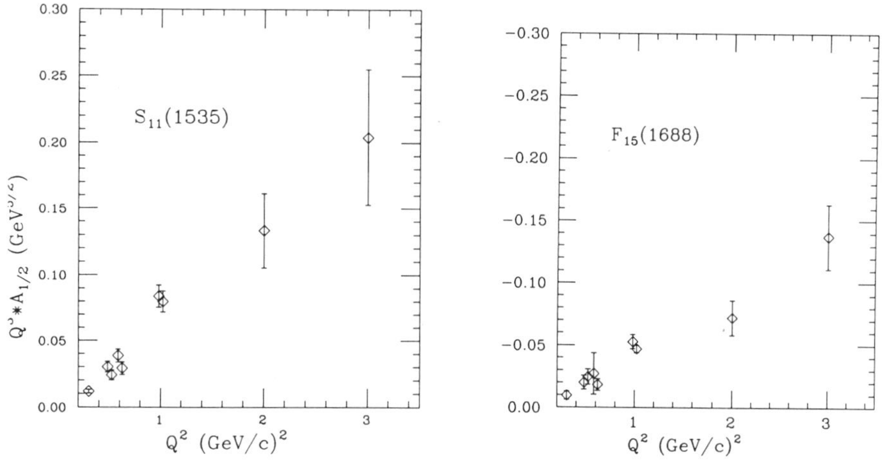

Figure 29. $Q^3 \cdot A_{1/2}$ *for various resonance transitions. The range of normalized QCD sum rule predictions of Carlson[15] for the asymptotic behavior of the $\gamma_v p S_{11}$ transition are indicated as lines labeled QCD SR.*

totic Q^2 behavior of the helicity amplitudes for resonance excitation. Based on the model of Brodsky and Lepage[48], who factorize the process into a hard scattering part and a 'soft' non-perturbative part described by quark distribution functions, it is expected[15] that

$$A_{1/2}(Q^2) = \frac{C_1}{Q^3}, \ A_{3/2}(Q^2) = \frac{C_2}{Q^5}, \ Q^2 \to \infty \ , \tag{30}$$

if logarithmic terms are neglected. Information about the quark distribution functions and the normalization constant C_1 may be obtained from QCD sum rules.

Of course, the first question to address is: At what momentum transfer does this description apply? Some interpretations[15] of the inclusive data have suggested that the asymptotic behavior already is observed at $Q^2 \approx 4$ to $5 \ GeV^2$. However, others[49] have argued that asymptotic behavior will occur only at much higher Q^2. Conclusive tests require exclusive data, where the resonances are uniquely identified, and their respective helicity amplitudes have been separated. Separated data exist for $Q^2 \leq 3 GeV^2$ only, and only for a few states. In Figure 29 the $A_{1/2}$ data are shown, multiplied by Q^3. The onset of the asymptotic regime would be indicated by the Q^2 independence of this quantity. This is obviously not the case for this limited Q^2 regime.

Multiple quark transitions

It is well known[50] that q^3 baryon states belonging to the $[20, 1^+]$ super multiplet with its antisymmetric wave function cannot be excited from the ground state in a single quark transition. Search for direct electromagnetic excitation of these states allows direct tests of the SQT hypothesis. Calculations within the framework of a relativized quark model[18] indicate that multi quark transition (MQT) amplitudes may contribute at a level of 10 to 20% of those for SQT amplitudes of some of the prominent states. Search for these transitions requires high statistics measurements under conditions where interferences of the MQT amplitudes with other dominant amplitudes are important.

Nucleon Resonances and Spin Structure Functions

The results of the EMC measurements[51] on the polarized proton structure functions have prompted numerous speculations about whether or not in the deep-inelastic region the spin of the proton is carried by the quarks. Recent results from the CERN Spin Muon Collaboration (SMC)[52] and SLAC experiment E142[53] on the neutron polarized structure functions added additional speculations including one interpretation the neutron spin is not carried by quarks either. In another interpretation, the (fundamental) Björken sum rule[54] would be violated while leaving the (less fundamental) Ellis-Jaffe sum rule[55] for the neutron intact. The experiments on the neutron use data sets with Q^2 as low as 1 GeV^2. While such low Q^2 values have been used in the analysis of unpolarized lepton scattering, there is a lack of convincing evidence that polarized structure functions exhibit true scaling behavior at such low Q^2. One should also keep in mind that the various spin sum rules are only defined at fixed Q^2 while the experiments integrate over large ranges in Q^2 . Moreover, the W range used in these analyses ($W \geq 2\ GeV$) may overlap part of the resonance region. The general perception is that for $W \geq 2\ GeV$ one probes the deep inelastic and hence scaling regime. However, excited nucleon states with masses as high as 3.0 GeV have been observed, and many states are predicted to exist in the mass region above 2.0 GeV, which raises the interesting question of how to correct for contributions resulting from these states. As the conclusion about the spin of the proton not being carried by quarks rests on relatively small differences between theoretical predictions and the data, it is important to study such contributions before far-reaching conclusions about the origin of the nucleon spin are drawn.

In some interpretations of the EMC results it is assumed that the missing quark spin may be accounted for by the so-called axial anomaly[56]. However, such contributions are rather controversial [57]. Another possibility is that most of the proton spin resides in orbital angular momentum contributions. Such contributions are necessarily associated with extended objects and therefore cannot be probed in deep inelastic scattering, however, they may be accessible at lower energies and momentum transfers. The low Q^2, and low energy loss ν region may therefore contain significant information about the spin structure of the nucleon. This is the kinematical regime where contributions from excited N^* and Δ^* resonances are important.

Polarized Structure Functions of the Proton

The spin-structure of the proton is usually discussed in relation to the deep-inelastic polarized structure functions $g_1(x)$ and the first moment $\Gamma = \int g_1(x)$. In the kinematical regime of resonances, and at low Q^2, use of total helicity $\frac{1}{2}$ and $\frac{3}{2}$ photon-nucleon absorption cross sections is more convenient.

The sum rule by Gerasimov[58], and independently by Drell and Hearn[59] relates the difference in the total photoabsorption cross section on nucleons for $\lambda_{\gamma N} = \frac{1}{2}$ and $\lambda_{\gamma N} = \frac{3}{2}$ to the anomalous magnetic moment of the target nucleon.

$$I_p(0) = \frac{M_p^2}{8\pi^2\alpha} \int_{\nu_{thr.}}^{\infty} \frac{d\nu}{\nu}(\sigma_{1/2}^p(\nu) - \sigma_{3/2}^p(\nu)) = -\frac{1}{4}\kappa^2 \ , \tag{31}$$

where κ is the anomalous magnetic moment of the target nucleon. Assuming scaling behavior the EMC results can be written as:

$$I_p(Q^2) = \frac{2M_p^2}{Q^2}\Gamma_p^{EMC} \simeq \frac{0.222 \pm 0.018 \pm 0.026}{Q^2} \tag{32}$$

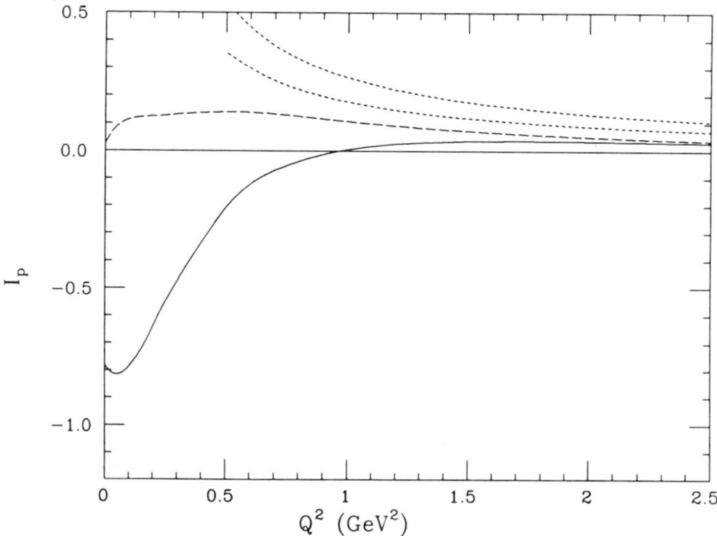

Figure 30. *The integral $I_p(Q^2)$. The solid curve represents the results of the empirical analysis by Burkert and Li[62]. The long dashed line is the result of the same analysis excluding the $P_{33}(1232)$ contribution. The short dashed lines represent the extrapolated EMC results.*

where QCD corrections have been neglected. $I_p(0)$ is large and negative, whereas the EMC data yield a positive $I_p(Q^2)$. In order to reconcile the GDH sum rule with the EMC results, dramatic changes in the helicity structure must occur when going from $Q^2 = 0$ to finite values of Q^2. In an analysis of photoproduction data[60],[61] for energies up to $E_\gamma = 1.7~GeV$ single pion production contributions were found to nearly saturate the sum rule. An analysis of electroproduction data[62] including known resonant channels as well as non-resonant single pion Born terms showed that contributions of the $P_{33}(1232)$ to $I_p(Q^2)$ are dominant at small Q^2. This analysis also showed that contributions from other resonances become significant with increasing Q^2, in fact causing $I_p(Q^2)$ to change its sign at Q^2 between 0.5 to 1.0 GeV^2. Resonance contributions other than the $P_{33}(1232)$ contribute as much as 50% or more of the extrapolated EMC results at $Q^2 = 1~GeV^2$ (Fig. 30), which highlights the importance of the nucleon resonance region for a full understanding of the nucleon spin structure.

AN EXPERIMENTAL STRATEGY

In a discussion of the experimental aspects it is useful to distinguish baryon resonance studies according to the complexity of the final state.

Single Pion and Eta Production

Most of the existing data consist of differential cross sections of single π or η production using unpolarized beams and targets. This channel is particular sensitive to the lower mass resonances ($W \leq 1.7~GeV$) which decay dominantly into the πN channel, or in the case of the $S_{11}(1535)$ into ηN. Much of the theoretical formalism required for data analysis has been worked out for these channels. In single pion electroproduction from nucleons, 11 independent measurements are needed at a given kinematical point (Q^2, W, θ_π) to determine the amplitudes of the process $\gamma_v N \rightarrow N'\pi$ in a model independent fashion. The complete determination of the transition amplitudes in pion and eta production over a large kinematical range is the ultimate goal of nucleon resonance physics with electromagnetic probes. This program requires high statistics measurements of unpolarized cross sections, and detailed measurements of polarization observables using polarized beams, polarized nucleon targets, and the measurement of nucleon recoil polarization.

Measurement of the single pion cross section allows the determination of four response functions σ_T, σ_L, σ_{TT}, σ_{TL} in (13), which are functions of the helicity amplitudes H_i. Measurement of polarization observables yields information on many response functions (table 3). While measurement of the unpolarized response functions allows determination of prominent resonance transitions, the weaker transitions are difficult to determine in such a way.

Table 3. Response functions in pion electroproduction.

Experiment	# Response Functions
$eN \rightarrow eN'\pi$	4
$\vec{e}N \rightarrow eN'\pi$	1
$e\vec{N} \rightarrow eN'\pi$	8
$\vec{e}\vec{N} \rightarrow eN'\pi$	5
$eN \rightarrow e\vec{N}'\pi$	8
$\vec{e}N \rightarrow e\vec{N}'\pi$	5

Since polarization observables contain interference terms between amplitudes they are sensitive to small amplitudes and to relative phases between amplitudes. Even measurements of limited statistical accuracy will be extremely important in determining absolute values and signs of small amplitudes which are otherwise not accessible. Not all the response functions contain independent information. In particular, only four of the response functions measured with a polarized target are different from the ones measured with recoil polarimeters. In many applications the two methods can be quite competitive, which allows one to select the more convenient technique.

Figure 31 shows the sensitivity of the unpolarized cross section for $p(e, e'p)\pi^0$ and the target asymmetry T_{long} to the excitation strength of the $P_{11}(1440)$. As the transition amplitudes of the $P_{11}(1440)$ may be quite small, the sensitivity of polarization experiments is essential in measuring these amplitudes. Similar sensitivities to the $P_{11}(1440)$ amplitudes have been found in measurements of the proton recoil polarization[64] in $p(e, e'\vec{p})\pi^0$.

The main objective is to disentangle the various resonant partial waves. This requires measurement of complete angular distributions with respect to the direction of the virtual photon. Also, measurements in different isospin channels are needed to separate resonant and non-resonant amplitudes with different isospin assignments.

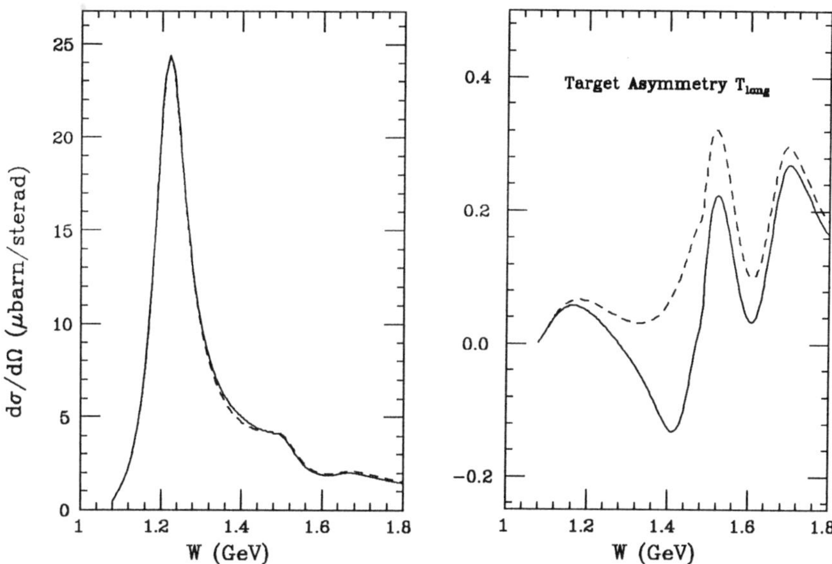

Figure 31. *Predicted cross section (left) for* $\gamma_v p \to p\pi^0$ *at* $Q^2 = 0.25 GeV^2$, $\epsilon = 0.8$, $\theta_\pi^* = 90^o$, $\phi_\pi = 30^o$. *The AO code*[66] *amplitudes were used. Predicted target asymmetry for target polarization along the incident electron beam* T_{long} *for the same kinematics (right). The sensitivity to the amplitudes of the* $P_{11}(1440)$ *resonance is shown: solid line - with* P_{11}, *dashed line - without* P_{11}.

Complete isospin information can be obtained from a study of the reactions

$$\gamma_v + p \to p + \pi^0$$

$$\gamma_v + p \to n + \pi^+$$

$$\gamma_v + n \to p + \pi^- \ .$$

In addition, measurement of

$$\gamma_v + p \to p + \eta$$

selects isospin $1/2$, and is a unique means of tagging the $S_{11}(1535)$ and the $P_{11}(1710)$ resonances which both have a significant coupling to the $N\eta$ channel.

The various experimental requirements call for an experimental setup which allows measurement of complete angular distributions in different isospin channels simultaneously.

Multiple Pion Production

For higher masses, multiple pion production due to decay channels such as $\pi\Delta$, ρN, and ωN becomes the dominant process. For example, the $S_{31}(1650)$ and $D_{33}(1700)$ decay about 70% and 85% of the time, respectively, into the $N\pi\pi$ channel, with dominant contributions coming from the $\pi\Delta$ decay. Obviously, a study of baryon resonance production in this mass region requires measurement of these channels. The information that can be extracted from the two-pion process is potentially very rich since polarization observables can be measured in the final state. For example, the measurement of the $\Delta \to N\pi$ decay allows a determination of the Δ helicity in the process $\gamma p \to \Delta\pi$. This should prove very powerful in determining

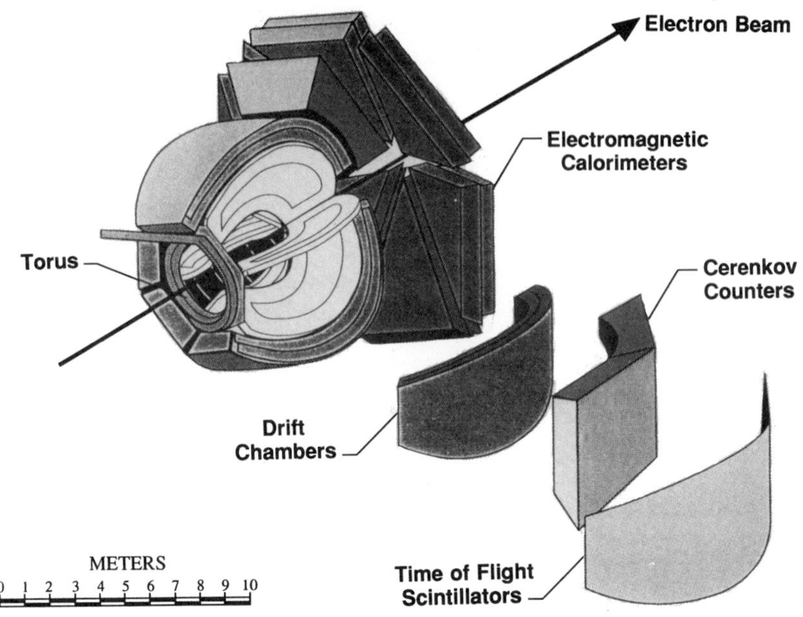

Figure 32. *The CEBAF Large Acceptance Spectrometer (CLAS)*[6]. *Six symmetrically arranged superconducting coils generate an approximate toroidal magnetic field. Drift chambers, time-of-flight counters, gas Cerenkov counters, and an electromagnetic calorimeter provide particle identification, charged particle tracking, and energy measurements for electromagnetic particles. The field free region around the target allows use of polarized solid state targets.*

the resonant contributions to the process. Moreover, clarification of the existence of states such as $P_{13}^G(1540)$ and $P_{31}^G(1550)$ also appears more promising in the multi-pion channel than in single pion production.

Experimental Equipment

A comprehensive and efficient experimental program to study electromagnetic transitions of baryon resonances in a large kinematical region, requires experimental equipment with large solid angle coverage, the capability to measure neutral particles, and compatibility with polarized proton and neutron targets. At CEBAF, a large acceptance spectrometer (CLAS) based on a toroidal magnetic field is under construction[6]. A significant portion of the scientific program for this detector is aimed at studies of baryon resonance excitations using electron and photon beams[41]. Figure 32 shows an artists view of the CLAS spectrometer. As an example, Figure 33 shows a comparison of existing data in the $P_{33}(1232)$ region with projected data expected from measurements with CLAS.

Many details of baryon resonance excitations, in particular at lower masses, may be addressed with magnetic spectrometers, which have small solid angles, but are able to operate at very high luminosities. Single pion production near pion threshold and in the $P_{33}(1232)$ region, or eta production in the $S_{11}(1535)$ region could be measured with high precision. Small solid angle, high rate magnetic spectrometers may also allow accurate measurements of proton recoil polarizations in reactions such as:

$$e + p \rightarrow e + \vec{p} + \pi^0$$

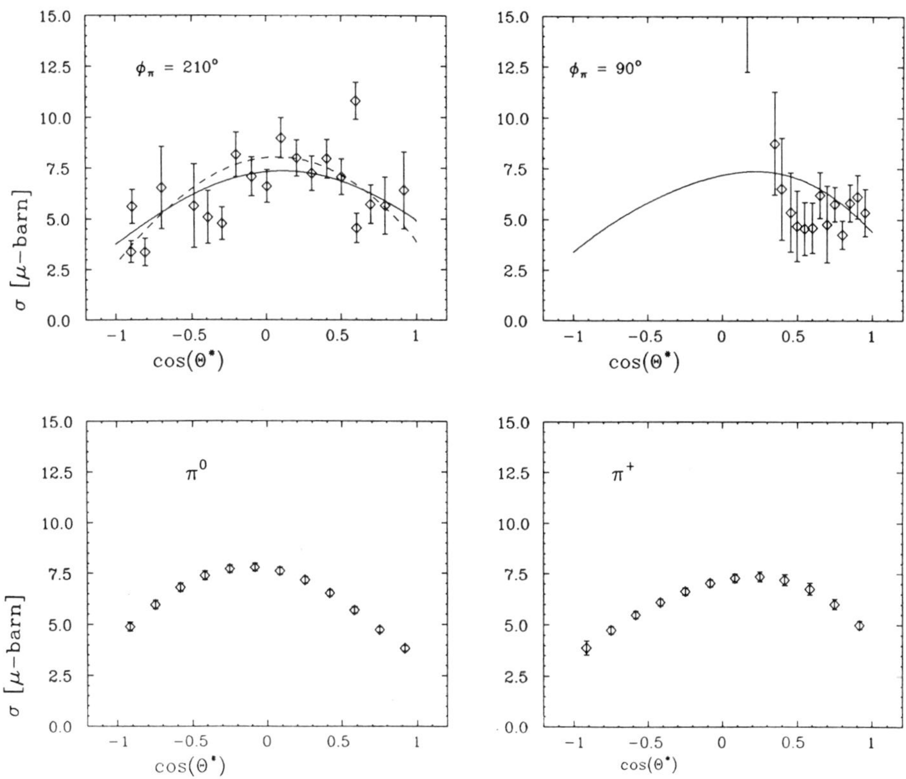

Figure 33. *Differential cross section for π^0 and π^+ production from protons (top), and projected data from CLAS (bottom). A 1000 hour run at a luminosity of 10^{34} $cm^{-2}sec^{-1}$ has been assumed.*

$$e + n \to e + \vec{p} + \pi^-$$

$$e + p \to e + \vec{p} + \eta.$$

Polarization experiments with such spectrometers are in preparation at CEBAF, at MIT-Bates, and at MAMI-B. These are designed to make precise measurements of single π^0 production off protons in the $\Delta(1232)$ region, with the goal of extracting more accurate information about the small E_{1+} and S_{1+} multipoles.

OUTLOOK

In the past, studies of the structure of baryons using electromagnetic excitation of nucleon resonances have suffered from several shortcomings. Firstly, theoretical guidance based on models which have some basis in the theory of strong interaction was established only after the bulk of the experiments had been completed. Secondly, all of the high statistic experiments are single pion production measurements. In view of QCD based quark models, this allows the study of lower mass states with a

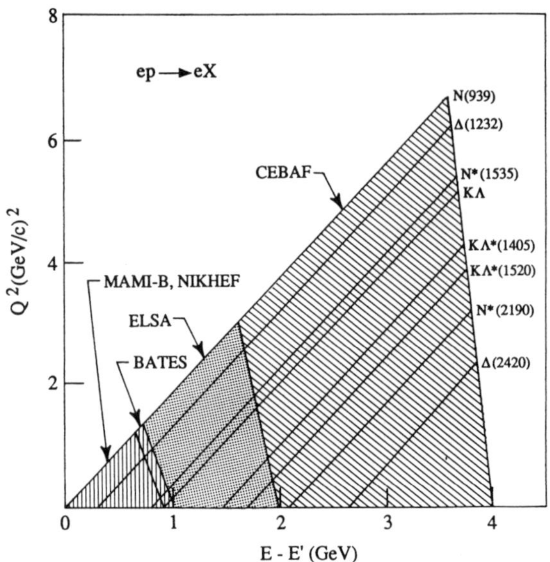

Figure 34. *Kinematical region accessible at various CW electron machines.*

large elasticity. However, single pion production measurements are not suited for the study of most of the higher mass states. These are predicted to largely decouple from the πN channel, which makes this channel insensitive to resonance excitation in the higher mass region. Thirdly, the notorious rate problem in electromagnetic production experiments prevented high statistics measurements to be performed, even in the case of single pion production.

The construction of continuous electron accelerators in the multi-GeV range, and the utilization of large acceptance detectors has opened up the possibility of studying the structure of the nucleon with unprecedented accuracy. The analysis of this data, and the comparison with theoretical descriptions of the nucleon in terms of quarks and gluons will give us detailed insight into the dynamics of hadronic systems and the underlying strong interaction force.

Several new CW electron accelerators in the GeV and multi-GeV range are now under construction. Figure 34 shows the resonance mass range accessible with these machines covering the entire nucleon resonance mass region and a large Q^2 range. Clearly, with these machines, and with the use of modern experimental equipment, the scientific community will have powerful new instruments which will allow for progress on many of the outstanding problems in baryon structure and strong interaction physics.

References

[1] N. Isgur, G. Karl, *Phys. Lett. 72B, 109 (1977); Phys. Rev. D23, 817 (1981)*

[2] S. Capstick, N.Isgur, *Phys. Rev. D34, 2809 (1986); S. Capstick, Phys. Rev. D36, 2800 (1987)*

[3] S. Capstick, *Phys. Rev. D46,2864(1992)*

[4] F. Iachello, *Phys. Rev. Lett.* **62**, *2440 (1989)*
R. Bijker and A. Leviatan, *Proceedings of 6th Workshop on Nuclear Physics at Intermediate Energies, Trieste, May 3 - 7, 1993*

[5] G. Eckart and B. Schwesinger , *Nucl. Phys. A458, 620 (1986)*

[6] V. Burkert and B. Mecking, *Large Acceptance Detectors for Electromagnetic Nuclear Physics,* in: *Modern Topics in Electron Scattering,* eds. B. Frois, I. Sick, *World Scientific, Singapore, 1991*

[7] Particle Data Group, Review of Particle Properties, *Phys. Rev.D45 (1992)*

[8] R. Koniuk, N. Isgur, *Phys. Rev. D21, 1868 (1980)*

[9] K.F. Liu, C.W. Wong, *Phys. Rev. D28, 170 (1983)*

[10] E. Golowich, E. Haqq, G. Karl, *Phys. Rev. D28, 160 (1983)*

[11] T. Barnes, F.E. Close, *Phys. Letts. 123B, 89 (1983)*

[12] R. Longacre, *Nucl. Phys. B211, 1 (1977)*

[13] P. Stoler, *Phys. Rev. Lett. 66, 1003 (1991)*

[14] M. Köbberling, et al., *Nucl. Phys. B82, 201 (1974)*
S. Stein et al., *SLAC-PUB-1518 (1975)*

[15] C.E. Carlson, *Phys. Rev. D34, 2704 (1986)*
C.E. Carlson, J.L. Poor,*Phys. Rev. D38, 2758 (1988)*
C.E. Carlson, *Proc. 12th Internat. Conf. on Few Body Problems*

[16] K. Bätzner et al., *Phys. Lett. 39B, 575 (1972)*

[17] J. Drees et al., *Z. Phys. C7, 183 (1981)*

[18] M. Warns et al., *Z. Phys.C45, 627 (1990)*

[19] F. Foster, G. Hughes, *Z. Phys. C14, 123 (1982)*

[20] H.F. Jones and M.D. Scandron, *Ann. Phys. 81, 1 (1972)*

[21] G. Baum et al.; *Phys. Rev. Lett. 45, 2000 (1980)*

[22] V. Eckardt, et al., *Nucl. Phys. B55, 45(1973)*
K. Wacker et al., *Nucl. Phys. B144, 269 (1978)*

[23] K. Bätzner et al., *Nucl. Phys. B75, 1 (1974)*

[24] R. Lourie, V. Burkert, (co-spokesmen), *Bates Proposal PR 89-03;*
C. Papanicolas et al., *Bates Proposal PR 87-09*

[25] *CEBAF Proposals PR 89-37 and PR 89-42 of the N^* collaboration*

[26] R.L. Walker, *Phys. Rev. 182, 1729 (1969)*

[27] R.G. Moorhouse, in: *Electromagnetic Interaction of Hadrons Vol. 1,*
eds. *A. Donnachie and G. Shaw, Plenum Press, 1978*

[28] D.H. Lyth, in: *Electromagnetic Interaction of Hadrons Vol. 1,*
eds. *A. Donnachie and G. Shaw, Plenum Press, 1978*

[29] R. Haidan, *PhD Thesis, University of Hamburg; DESY Internal report No. F21-79/03 (unpublished)*

[30] K.H. Althoff et al., *Z. Phys. C1, 327 (1979)*

[31] H. Breuker et al., *Phys. Letts. 74B, 409 (1978)*

[32] W. Brasse et al., *Z. Phys. C22, 33 (1984)*

[33] F. Close and Z.P. Li, *Phys.Rev.D42, 2194(1990)*

[34] W. Konen and H.J. Weber, *Phys. Rev. D41, 2201 (1990)*

[35] C.P. Forsyth, J.B. Babcock, *Preprint CMU-HEP 83-4 (1983)*

[36] S. Ono, *Nucl. Phys. B107, 522 (1976)*

[37] L.A. Copley, G. Karl, E. Obryk, *Phys. Letts. 29B, 117 (1969)*

[38] F.E. Close, F.J. Gilman, *Phys. Letts. 38B, 541 (1972)*
F.E. Close, F.J. Gilman, I. Karliner, *Phys. Rev. D6, 2533 (1972)*

[39] H. Breuker et al., *Z. Phys. C17, 121 (1983)*
H. Breuker et al., *Z. Phys. C13, 113 (1982)*

[40] F. Foster and G. Hughes, *Rep. Prog. Phys., Vol. 46, 1445 (1983)*

[41] A.J.G. Hey, J. Weyers, *Phys. Letts. 48B, 69 (1974)*
W.N. Cottingham, I.H. Dunbar, *Z. Phys. C2, 41 (1979)*

[42] K. Bermuth, D. Drechsel, D. Tiator and J.B. Seaborn, *Phys.Rev.D37, 89 (1988)*

[43] Z.P. Li, V. Burkert, Z. Li, *Phys. Rev. D46, 70 (1992)*

[44] C. Gerhardt, *Z. Phys. C4, 311(1980)*

[45] V. Burkert, *Proceedings, 18th Rencontre de Moriond, January 23-29, 1983, La Plagne, pg. 189, Editions Frontières, ed. J. Tran Thanh Van*

[46] H. Funsten et al., *CEBAF Proposal 91-024 (1991)*
M. Ripani et al., CEBAF Proposal 93-06, also: *Proceedings, Workshop on "Exclusive Reactions at High Momentum Transfer ", Elba, Italy, July 1993*

[47] D. Menze, Talk presented at the *Int. Workshop on Baryon Structure and Spectroscopy, Paris, September 22-25, 1991*

[48] S.J. Brodsky, G.P. Lepage, *Phys. Rev. D24, 2848 (1981)*

[49] N. Isgur and C.H. Llewellyn-Smith, *Nucl. Phys. B317 ,526 (1989)*

[50] R.P. Feynman, *Photon Hadron Interactions, Frontiers in Physics, W.A. Benjamin, Inc. (1972), pg.49*

[51] J. Ashman et al., *Phys. Lett. B206,(1988)364; Nucl. Phys.B328 (1989)1*

[52] B. Adeva et al.,*Phys. Lett. B302, 533 (1993)*

[53] P.L. Anthony et al. (E142 Collaboration); Presented at *PANIC '93, Perugia, Italy, June 1993*

[54] J.D. Björken, *Phys.Rev. 148, 1467(1966)*

[55] J. Ellis and R.L. Jaffe, *Phys. Rev. D9, 1444 (1974)*

[56] G. Altarelli, G.G. Ross, *Phys. Lett. B214, 381(1988)*; R. Carlitz, J.C. Collins, A.M. Mueller,*Phys. Lett.B214, 2229 (1988)*

[57] R. Manohar, in: *Polarized Collider Workshop, University Park, Pa. , 1990., AIP Conference Proc. No.223*

[58] S. Gerasimov,*Yad.Fiz.2, 598(1965) [Sov. J. Nucl. Phys. 2 (1966)930]*

[59] S.D. Drell and A.C. Hearn, *Phys. Rev. Lett. 16 (1966) 908*

[60] I. Karliner, *Phys.Rev. D7, 2717 (1973)*

[61] R. Workman and R. Arndt, *Phys. Rev. D45, 1789(1992)*

[62] V. Burkert and Zh. Li, *Phys. Rev.D47, 46(1993)*

[63] D. Drechsel, *6th Int. Workshop on Perspectives in Intermediate Nuclear Physics, Trieste, May 4 - 7, 1993*

[64] R. Lourie, *Z. Phys. C50, 345 (1991)*

[65] W. Albrecht et al., *Nucl. Phys. B27, 615 (1971)*
S. Galster et at., *Phys. Rev. D5, 519 (1972)*
R. Siddle et al., *Nucl. Phys. B35, 93 (1971)*
R. D. Hellings et al., *Nucl. Phys. B32, 179 (1971)*
J.C. Alder et al., *Nucl. Phys. B46, 573 (1972)*
K. Bätzner et al.,*Nucl. Phys. B75, 1 (1974)*

[66] V. Burkert and Zh. Li, *AO code, unpublished*

STRUCTURE FUNCTIONS OF THE NUCLEON

Tjeerd J. Ketel

Department of Physics and Astronomy, Free University
1081 HV Amsterdam, the Netherlands

ABSTRACT

In these lectures recent experiments on deep inelastic scattering are described in which nucleon structure functions were measured. These structure functions are related to quark distributions and therefore yield information on the internal quark structure of the nucleon.
The measured Q^2 dependence of the structure functions represents a beautiful confirmation of the scaling violation predicted by the theory of quantum chromo dynamics (QCD). The results from the New Muon Collaboration allow in addition, by using the QCD description, a precise determination of the gluon distribution inside the nucleon.
Several years ago the European Muon Collaboration (EMC) questioned the influence of the nuclear environment on the internal structure of the nucleon. Later experiments confirmed this so-called EMC effect and also showed nuclear shadowing at small values of the Bjorken x parameter.
Spin dependent structure functions can tell us about the contribution of the quark spins to the nucleon spin. The present situation seems to be confused as both the EMC and the Spin Muon Collaboration results indicated a small contribution of 0.12 ± 0.14 and 0.06 ± 0.25, respectively, while the SLAC experiment E142 claimed a larger quark spin contribution of 0.57 ± 0.11 to the nucleon spin.

STRUCTURE FUNCTIONS AND QUARK DISTRIBUTIONS

Nucleon structure functions have been determined from deep inelastic lepton scattering experiments with increasing detail and kinematical range since the first SLAC experiments by Friedman, Kendall and Taylor [1] in the seventies until the recent experiments at the HERA collider ring in Hamburg.

As an introduction let me recall the cross section for scattering of two spin-1/2 point particles,

$$\frac{d\sigma}{d\Omega} = \sigma_{Ruth}(1 - \beta^2 sin^2\theta/2)(1 - \frac{q^2}{2m^2}tan^2\theta/2), \tag{1}$$

---

[1]Nobel Prize 1990.

with θ being the scattering angle, $\beta = v/c$ the velocity of the projectile, m the mass of the target particle and q the momentum transfer. The first term in the cross section is the Rutherford cross section. The second term is the relativistic correction for the spin-1/2 projectile which observes a magnetic field in its rest frame due to the moving charge of the target particle. These first two terms represent the so-called 'no structure' or Mott cross section σ_{Mott}. The last term is due to the spin 1/2 of the target particle.

For elastic scattering of a spin-1/2 point particle from a particle with an internal structure like the proton we use electric and magnetic form factors G_E and G_M,

$$\frac{d\sigma}{d\Omega} = \sigma_{Mott}\left(\frac{G_E^2 - \tau^2 G_M^2}{1 - \tau^2} - 2\tau^2 G_M^2 tan^2\theta/2\right), \tag{2}$$

where $\tau = q/2m$.

In elastic lepton scattering the proton is barely scratched by the probe. The single experimental variable in the cross section is the scattering angle θ. The outgoing electron energy E is fixed by θ and the incident electron energy.

In deep inelastic lepton scattering the proton is blown apart and the outgoing lepton energy is undetermined. The cross section now depends on two kinematical variables θ and E and two structure functions W_1 and W_2 ;

$$\frac{d^2\sigma}{d\Omega dE} = \sigma_{Mott} cos^2\theta/2 (W_2 + 2W_1 tan^2\theta/2). \tag{3}$$

The surprising experimental result was that $W1$ and $W2$ stay almost constant in the deep inelastic region, while the electromagnetic form factors G_E and G_M, which describe the elastic scattering, decrease rapidly with momentum transfer q (see Fig. 1).

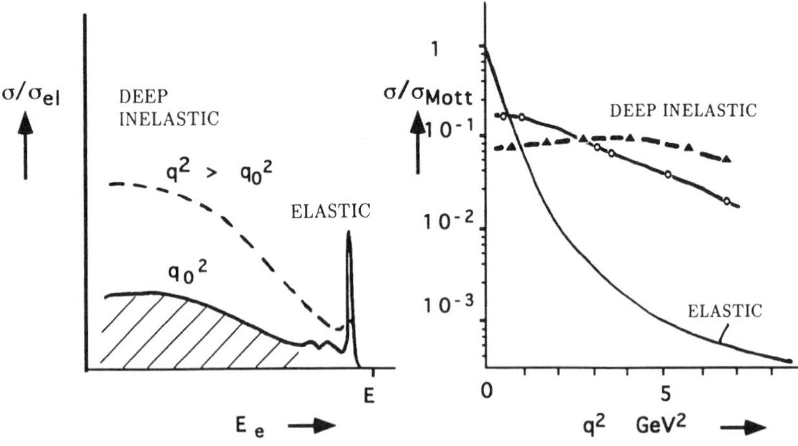

Figure 1. **Comparison of the cross section for elastic scattering and for deep inelastic electron scattering with 4-18 GeV electrons (Ref. [1]).**

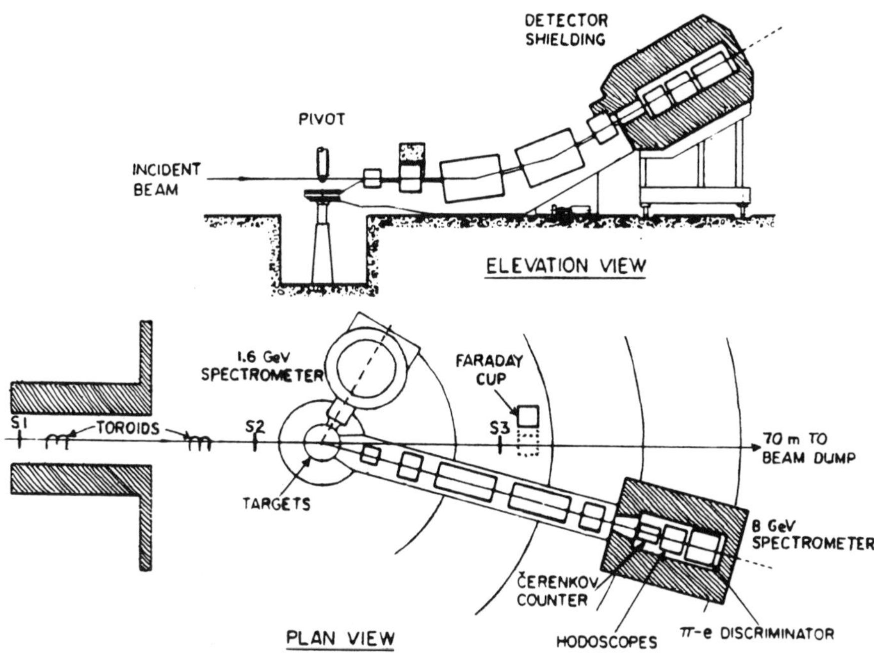

Figure 2. **The 8 GeV spectrometer at SLAC used for the first deep inelastic scattering experiments.**

Experiments

The early SLAC experiments used electrons from the 20 GeV linear accelerator and used an 8 GeV magnetic spectrometer at fixed angles to measure the momentum of the scattered electrons (see Fig. 2). Two 3 meters long bending magnets were needed to get sufficient bending at this high energy. Particle identification was used to separate electrons from charged hadrons in the spectrometer.

At CERN and Fermilab high energy muons (100-500 GeV) from the decay of high energy pions produced by accelerated protons were used. These experiments needed thick targets to compensate for the low beam flux. A forward spectrometer (see Fig. 3a) was used by the subsequent Muon Collaborations (EMC [2], NMC and SMC) at CERN. High kinematical acceptance was obtained due to the strong Lorentz boost of the center of mass system. Scattered muons were cleanly selected behind a calorimeter and a 2 meter thick iron wall which absorbed all hadrons and electromagnetic showers.

Neutrinos of a broad energy band were used at CERN and Fermilab (30-600 GeV). The nuclear targets consisted of thick iron or marble plates interleaved with particle track detectors. Outgoing muons were analysed in toroidal magnetised iron segments (see Fig. 3b).

In the HERA collider ring protons were collided at high energy (820 GeV) with electrons of 30 GeV. In this way the center of mass energy is increased by several orders of magnitude compared to fixed target experiments. Scattered electrons and also target fragments are studied in two large detector set-ups H1 and ZEUS mainly by using modern calorimetry (see Fig. 3c).

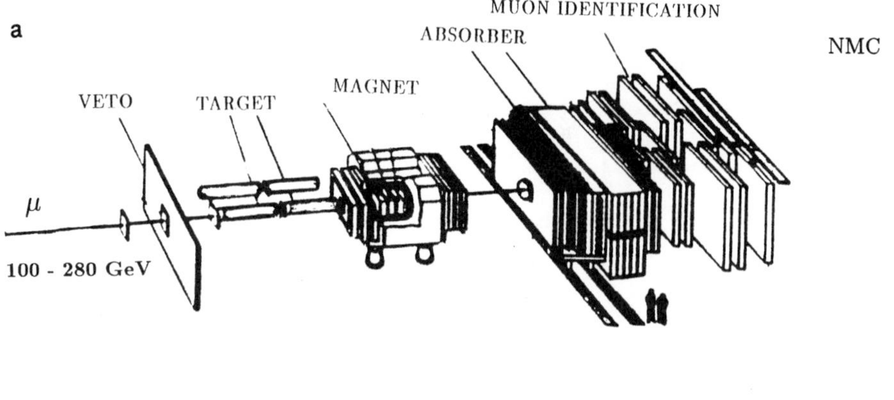

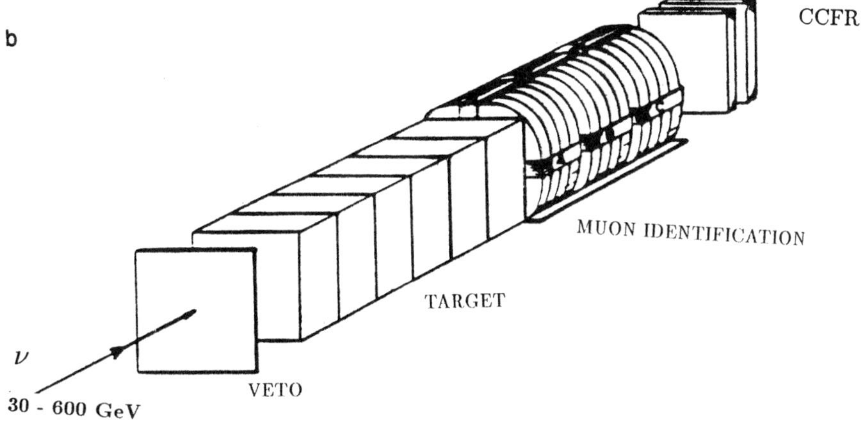

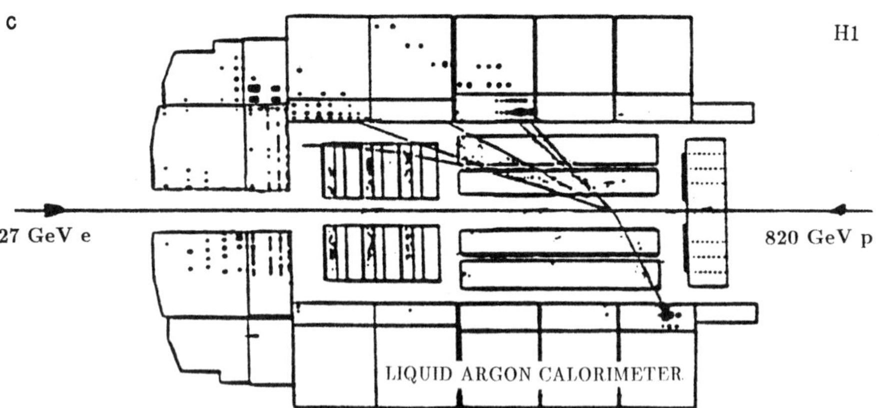

Figure 3. **Experimental set-ups for high energy deep inelastic lepton scattering of (a) NMC at CERN, (b) CCFR at Fermilab and (c) of H1 at DESY.**

Kinematical Variables

In order to compare the results of the different experiments it is clear that we need Lorentz invariant kinematical variables instead of θ and E to describe deep inelastic scattering. The variables Q^2, ν, x and y are defined below as scalar products of four-vectors p, p', P and q and are therefore Lorentz invariant by definition (see Fig. 4).

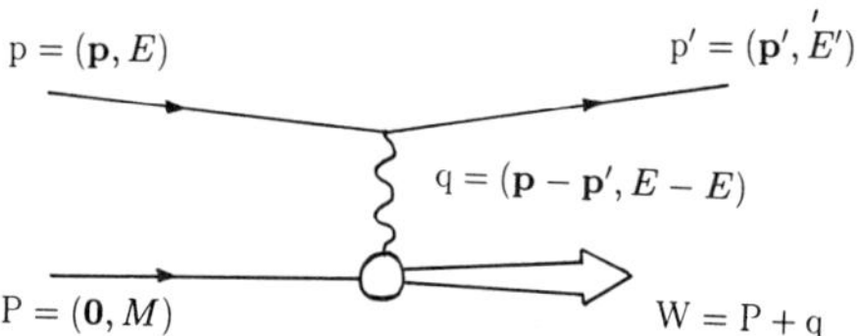

Figure 4. **Feynman diagram for deep inelastic lepton scattering on a nucleon.**

The corresponding expressions for scattering in the fixed target frame is given as well,

$$
\begin{aligned}
Q^2 &= -\text{q.q} = 4EE'sin^2\theta/2 \\
\nu &= \text{P.q}/M = E - E' \\
x &= Q^2/2\text{P.q} = Q^2/2M\nu \\
y &= \text{P.q}/\text{P.p} = \nu/E.
\end{aligned}
\tag{4}
$$

The invariant mass W of the resulting hadronic system is given by

$$
W^2 = (\text{P} + \text{q})^2 = M^2 + 2\text{P.q} - Q^2.
\tag{5}
$$

The proton is broken up in deep inelastic scattering, $W^2 > M^2$, and therefore

$$
x = \frac{Q^2}{2\text{P.q}} < 1.
\tag{6}
$$

Intuitive interpretations of Q^2 and x are discussed in the following (see also [3]). The kinematical variable Q^2 is often referred to as the **resolution** of the virtual photon probe. This can be derived from

$$
\lambda = \frac{h}{p},
\tag{7}
$$

while the virtual photon momentum p is related to Q^2 by

$$
pc = \sqrt{Q^2 + \nu^2}.
\tag{8}
$$

For $Q^2 << \nu^2$, which is always the case for $\nu > 2$ GeV, one has

$$
\lambda \approx \frac{hc}{\nu} \propto \frac{x}{Q^2}.
\tag{9}
$$

This means that, for fixed values of x, Q^2 is inversely proportional to the virtual photon wavelength and hence that Q^2 is a measure for the resolution of the virtual photon probe.

The Bjorken scaling variable x defined in Eq. (4) is referred to as the **momentum fraction** carried by the struck quark, if both Q^2 and ν are large. To illustrate this we use the Breit frame (labeled with an asterix), in which the energy of the incoming and outgoing lepton is equal and thus the virtual photon energy $E_\gamma^* = 0$ (see Fig. 5). Then

$$Q^2 = |\mathbf{q}^*|^2 - 0 = q^2 \tag{10}$$

and therefore

$$\nu = \frac{(\mathbf{P}, E)\cdot(\mathbf{q}, 0)}{M} \longrightarrow Pq = M\nu. \tag{11}$$

At sufficiently high Q^2, transverse momenta can be neglected and the struck quark has equal energy and opposite momentum xP^* after absorption of the virtual photon;

$$q^* = 2xP^*. \tag{12}$$

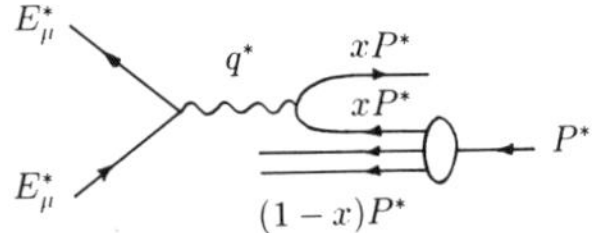

Figure 5. **Absorption of a virtual photon on a quark with momentum fraction x in the Breit system ($\mathbf{E}_\gamma^* = \mathbf{0}$).**

In this frame the Bjorken scaling variable x is equal to the momentum fraction x carried by the struck quark

$$x = \frac{q^2}{2Pq} = \frac{q^*}{2P^*}. \tag{13}$$

The kinematical acceptance of different deep inelastic scattering experiments is shown in Fig. 6. The axes represent the logarithms of x and Q^2. Due to the relation $x = Q^2/2M\nu$ straight lines of constant ν_{max} limit the Q^2 versus x acceptance. The maximum ν for fixed-target experiments is determined by the beam energy; $\nu_{max} = E_\mu = 280$ GeV for NMC. For HERA the maximum ν is

$$\nu_{max} M \;=\; P.q_{max} = \frac{1}{2}W_{max}^2 = 1/2(P+q)_{max}^2 \sim 4 \times 10^4 \; GeV^2, \tag{14}$$

where $(P+q)_{max} = (30 - 820, 0, 0, 30 + 820)$ GeV.

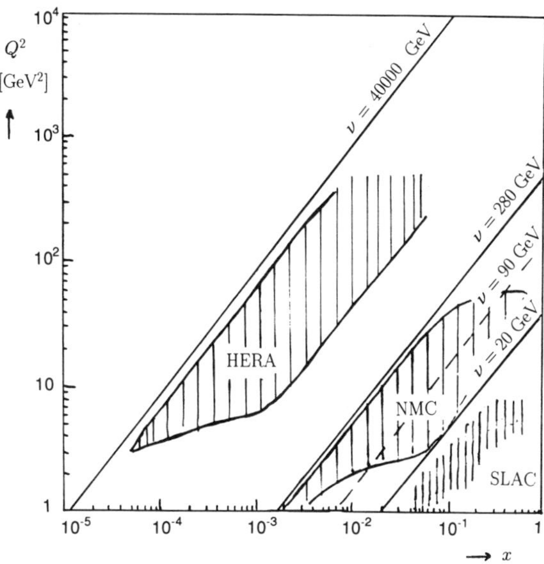

Figure 6. **Kinematical acceptance in x and Q^2 of SLAC, CERN and HERA experiments.**

Cross Sections, Structure Functions and Quarks

The cross sections for deep inelastic scattering can be expressed as functions of two independent kinematical variables (for example x and Q^2) and structure functions, F_1, F_2 and F_3, which are different for protons and neutrons [3, 4].

For polarisation-averaged cross sections we find for $e(\mu) + N \rightarrow e'(\mu') + X$

$$\frac{d^2\sigma}{dx\,dQ^2} = \frac{4\pi\alpha^2}{Q^4}\frac{1}{x}[(1-y)F_2(x,Q^2) + xy^2F_1(x,Q^2)] \tag{15}$$

and for neutrino interactions $\nu_\mu + N \rightarrow \mu + X$

$$\frac{d^2\sigma}{dx\,dQ^2} = \frac{G_F^2}{2\pi}\frac{1}{x}[(1-y)F_2^\nu(x,Q^2) + xy^2F_1^\nu(x,Q^2) \pm y(1-\frac{y}{2})xF_3^\nu(x,Q^2)]. \tag{16}$$

where the $+$ sign in front of F_3 applies for neutrinos and the $-$ sign for antineutrinos.

The experimental cross section for deep inelastic electron scattering was found to decrease as Q^{-4} like the Mott cross section for point particles. This is called scaling, as the structure functions F_1, F_2 and F_3 only depend on x and not on Q^2. Scaling was predicted by Bjorken in the limit of $Q^2 \rightarrow \infty$ for elastic lepton scattering on point charges of partons inside the nucleon.

In the following approximation we relate the structure functions to quark distributions and we assume massless quarks. A virtual photon with helicity $+1$ is absorbed on a free quark with helicity $-1/2$ with a cross section $\sigma_i^{\rightleftarrows} \propto e_i^2$ (see Fig. 7). The charge

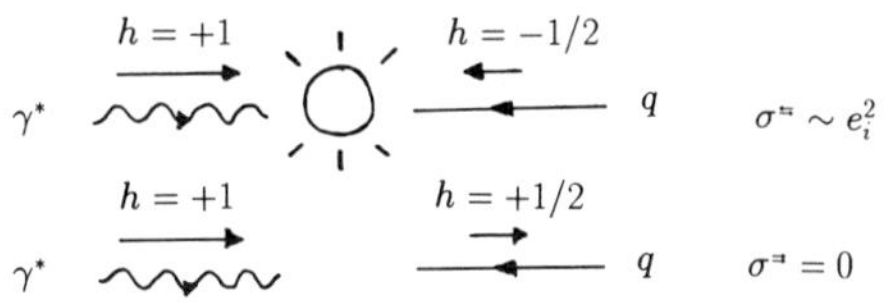

$$\gamma^* \quad \sim\!\!\sim\!\!\sim \quad \bigcirc \quad \longrightarrow q \qquad \sigma^{\rightleftarrows} \sim e_i^2$$

Figure 7. Absorption of a virtual photon with helicity +1 on quarks with helicity −1/2. There is no absorption on quarks with helicity +1/2.

e_i is 2/3 for up quarks and −1/3 for down quarks. No absorption of this photon can take place when the quark has helicity +1/2; thus $\sigma_i^{\rightrightarrows} = 0$.

In a nucleon we have a probability $q_i^+(x)$ to find quarks with flavour i at a certain momentum fraction x with their helicity along the nucleon spin. This gives a cross section $\sigma^{\rightleftarrows}(x)$ for the absorption of a polarised virtual photon on a nucleon with its spin opposite to that of the photon,

$$\sigma^{\rightleftarrows}(x) \propto \sum_i e_i^2 q_i^+(x). \tag{17}$$

Quarks with opposite helicity to the nucleon spin have a probability $q_i^-(x)$. They contribute to the cross section when the nucleon has its spin parallel to that of the photon,

$$\sigma^{\rightrightarrows}(x) \propto \sum_i e_i^2 q_i^-(x). \tag{18}$$

The polarisation-averaged cross section can be then written as

$$\sigma(x) = \frac{1}{2}(\sigma^{\rightleftarrows}(x) + \sigma^{\rightrightarrows}(x)) \propto \frac{1}{2}\sum_i e_i^2 q_i(x) = F_1(x) \tag{19}$$

with $q_i = q_i^+ + q_i^-$ being the probability to find a quark with flavour i at momentum fraction x. Consequently $x q_i(x)$ is the momentum distribution of quarks with flavour i.

The total momentum carried by all quarks was found to be only a part of the nucleon momentum;

$$\int_0^1 \sum_i x q_i(x)\,dx = 0.48 \pm 0.03. \tag{20}$$

The explanation of this result is that gluons carry the rest of the nucleon momentum.

The results of several early deep inelastic scattering experiments showed good agreement between high energy muon and neutrino experiments [4], which are summarised in Fig. 8. For comparison, the structure function F_2^μ determined from muon scattering has been multiplied with a factor 18/5 to take into account the factor e_i^2 in virtual photon absorption compared to W boson exchange in neutrino interactions.

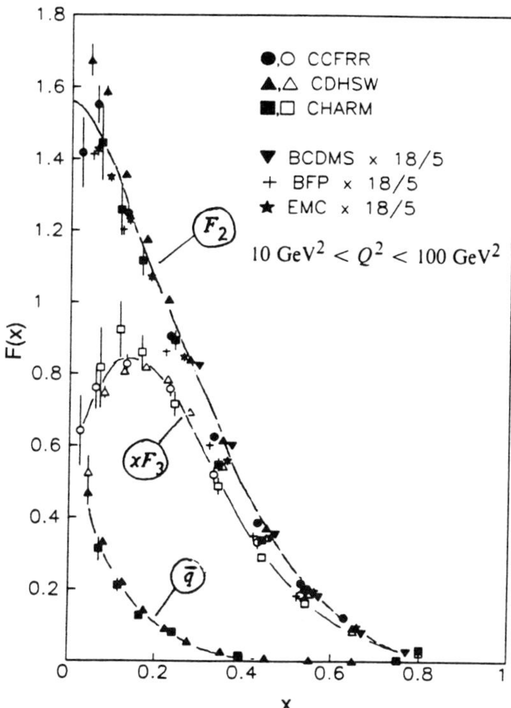

Figure 8. **F_2, F_3 and $\overline{q}$ from several high energy muon and neutrino experiments.**

The factor 18/5 is obtained in the following way. For elastic scattering in the limit of asymptotically free quarks, where from now on we will distinguish between quarks q and antiquarks $\overline{q}$, one has

$$
\begin{aligned}
F_2^\mu = 2x F_1^\mu &= \sum_i e_i^2 x[q_i(x) + \overline{q}_i(x)] \\
F_2^\nu = 2x F_1^\nu &= \sum_i x[q_i(x) + \overline{q}_i(x)] \\
x F_3^\nu &= \sum_i x[q_i(x) - \overline{q}_i(x)].
\end{aligned}
\tag{21}
$$

and F_2^ν is the sum of structure functions for ν and $\overline{\nu}$.

Isospin symmetry between proton and neutron means that the up quark distribution in the proton is the same as the down quark distribution in the neutron, $u = q_u^p = q_d^n$. Similarly for the down quark distribution in the proton one has $d = q_d^p = q_u^n$. For isoscalar nuclei (N=Z) we can show that $F_2^\nu = 18/5 F_2^\mu$ when we ignore the contribution from strange quarks;

$$
\begin{aligned}
F_2^{\mu p} + F_2^{\mu n} &= x[4/9(u + \overline{u}) + 1/9(d + \overline{d}) + 4/9(d + \overline{d}) + 1/9(u + \overline{u})] = \\
&= x[5/9(u + \overline{u}) + 5/9(d + \overline{d})] \\
F_2^{\nu p} + F_2^{\nu n} &= x[2(u + \overline{u}) + 2(d + \overline{d})].
\end{aligned}
\tag{22}
$$

Sum Rules

The structure function $F_3(x)$ is extracted from neutrino and antineutrino deep inelastic experiments. The integral $\int F_3 dx$ represents the probability to find valence quarks in a nucleon, since

$$F_3(x) = \sum_i [q_i(x) - \overline{q}_i(x)] = \sum_i q_i^{val}(x). \tag{23}$$

Assuming three valence quarks, Gross and Llewellyn Smith were able to calculate this integral first [5]. Including up to third order QCD corrections, one finds:

$$\int F_3 dx = 3[1 - \frac{\alpha_S}{\pi} + O(\alpha_s^2) + O'(\alpha_s^3)] = 2.55 \pm 0.05. \tag{24}$$

The most recent data from the CCFR collaboration at Fermilab [6] give an experimental value which is in perfect agreement with this prediction (see Fig. 9). In this figure the integral $\int F_3 dx = \int x F_3 d(log x)$ corresponds to the area under $x F_3$ when x is plotted logarithmically along the horizontal x-axis.

Another sum rule can be defined for $F_2^p - F_2^n$:

$$S_G = \int_0^1 (F_2^p - F_2^n) \frac{dx}{x} = \frac{1}{3}. \tag{25}$$

This so-called Gottfried sum rule [7] assumes symmetry between up and down sea-quark distributions, as is discussed below.

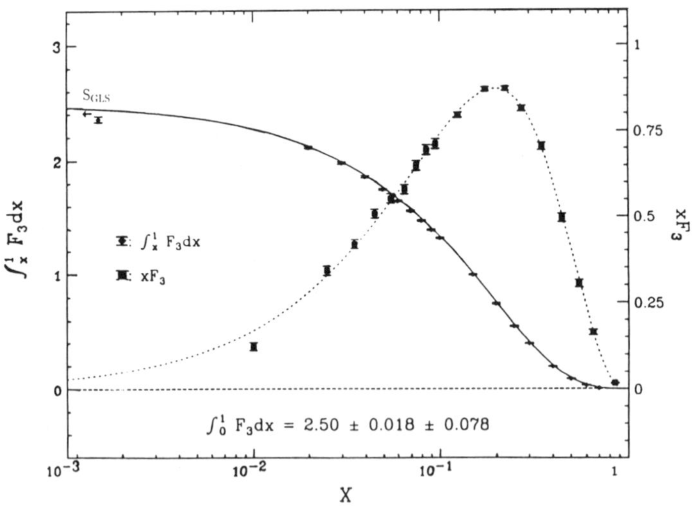

Figure 9. Comparison of the Gross-Llewellyn Smith sum rule with CCFR data at $Q^2 = 3$ GeV^2.

In the quark parton model using Eq. (21) one has

$$S_G = \int_0^1 \sum_i e_i^2 [q_i^p + \overline{q}_i^p - q_i^n - \overline{q}_i^n] dx. \tag{26}$$

By definition $q_i = q_i^{val} + q_i^{sea}$ and $\overline{q}_i = q_i^{sea}$, and hence

$$S_G = \int_0^1 \sum_i e_i^2 [q_i^{p,val} - q_i^{n,val} + 2\overline{q}_i^p - 2\overline{q}_i^n] dx. \tag{27}$$

Again by applying isospin symmetry between the proton and the neutron, we find for the Gottfried sum

$$S_G = \left(\frac{4}{9} - \frac{1}{9}\right) \int_0^1 (u^{val} - d^{val} + 2\overline{u} - 2\overline{d}) dx = \frac{1}{3} + \frac{2}{3} \int_0^1 (\overline{u} - \overline{d}) dx. \tag{28}$$

When $\overline{u} - \overline{d} = 0$ the Gottfried sum is $1/3$. Therefore, a measurement of $S_G = 1/3$ is a test for flavour symmetry of the quark sea.

NMC has determined the Gottfried sum experimentally (see Fig. 10). Its value deviates significantly from $1/3$ [8],

$$S_G(NMC) = 0.258 \pm 0.010(stat.) \pm 0.015(syst.). \tag{29}$$

One of the explanations put forward is that pair creation of up and anti-up quarks is more inhibited in the proton than in the neutron due to the presence of two valence up quarks in the proton.

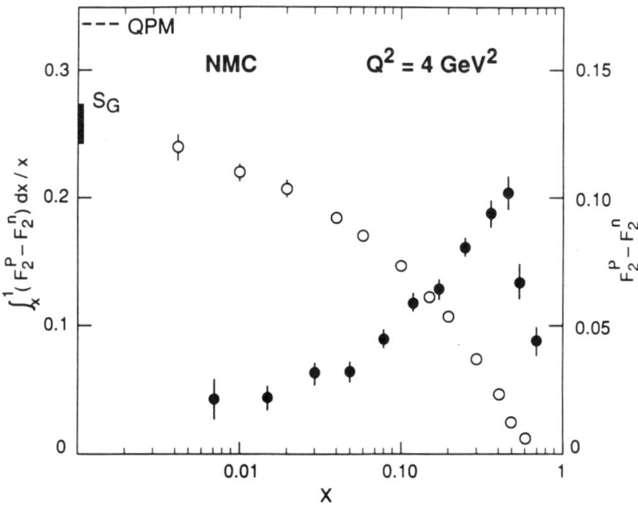

Figure 10. **The Gottfried sum measured by NMC. $F_2^p - F_2^n$ (full circles) and $\int_x^1 (F_2^p - F_2^n)/x'\, dx'$ (open circles) as a function of x at $Q^2 = 4$ GeV^2.**

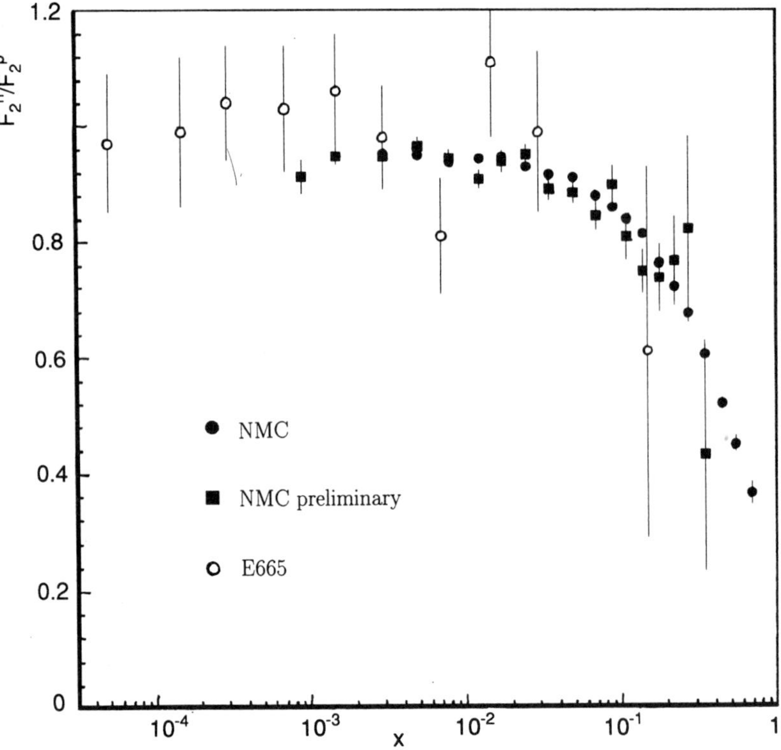

Figure 11. $\mathbf{F_2^n/F_2^p}$ as a function of x.

The Gottfried sum was obtained by NMC from the measurements of the structure function of the deuteron F_2^d and of the ratio F_2^n/F_2^p,

$$F_2^p - F_2^n = 2F_2^d \frac{1 - F_2^n/F_2^p}{1 + F_2^n/F_2^p}, \tag{30}$$

using $2F_2^d = F_2^p + F_2^n$. The advantage of this method is that the systematic error is small, since the ratio F_2^n/F_2^p was determined very accurately. New preliminary data on the ratio F_2^n/F_2^p from NMC extend to smaller values of x. They are shown in Fig. 11. The ratio seems to remain slightly below unity at small x, which may indicate shadowing in the deuteron. In the lecture on nuclear effects, shadowing is discussed for nuclear targets.

Scaling Violation

Scaling of the deep inelastic cross section, which originally suggested elastic scattering on asymptotically free point charges, is found in more extensive experiments to be violated. This is understood to be the result of the interaction of quarks with other quarks and gluons in the nucleon. In fact scaling violation has been accurately determined and is in agreement with predictions from quantum chromodynamics (QCD), the theory of strong interactions, which describes the interaction between quarks.

Fig. 12 represents the highest-accuracy data on deep inelastic $\nu \rightarrow \mu$ scattering over a large range of Q^2 obtained by the CCFR collaboration at Fermilab. Both $F_3^\nu(x, Q^2)$ and $F_2^\nu(x, Q^2)$ show changes with Q^2, almost linear in $\log(Q^2)$, for different values of x.

152

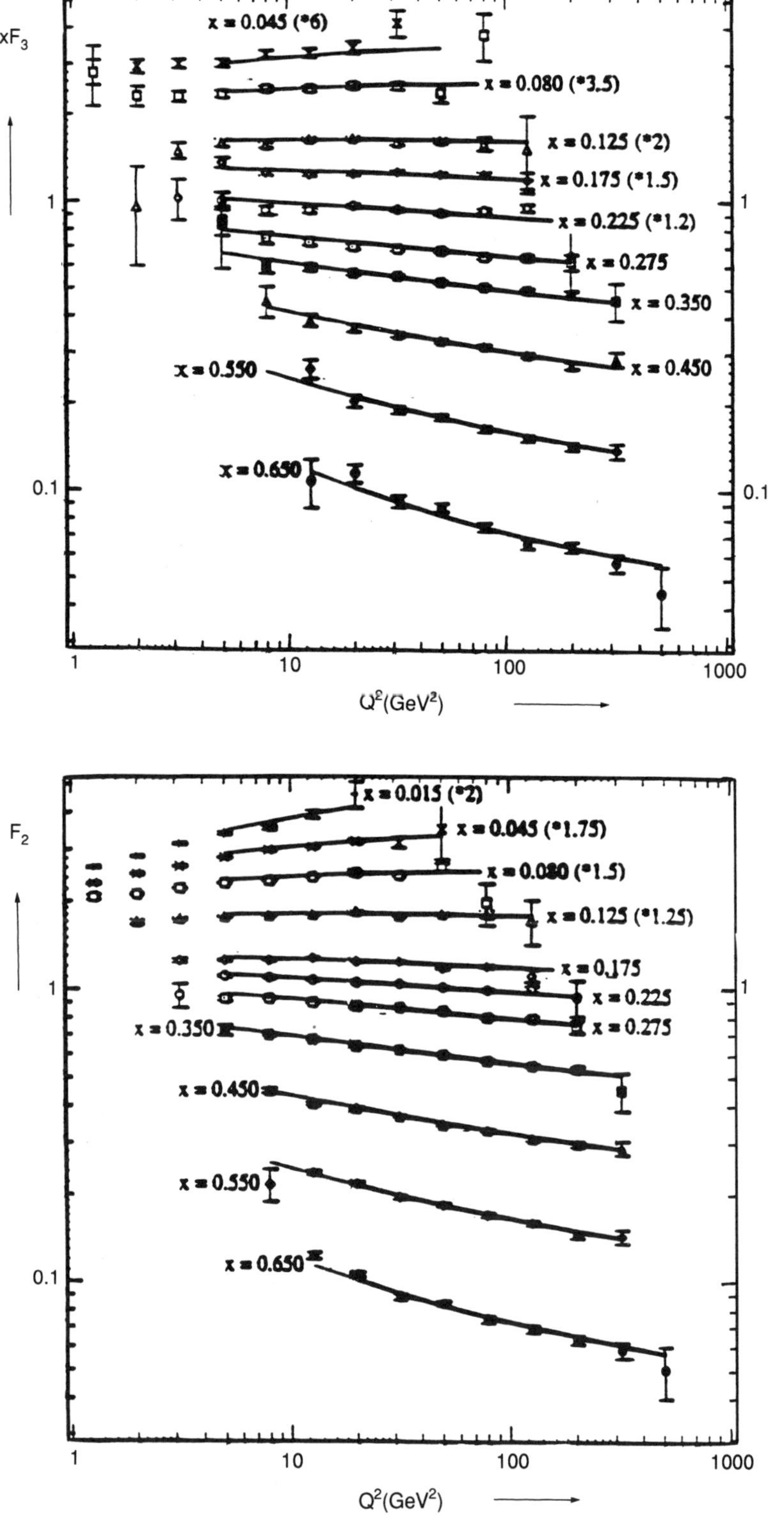

Figure 12. Scaling violation of $F_3(x)$ and $F_2(x)$ in CCFR neutrino data.

QCD Analysis

The scaling violation can be understood from the increased resolution of the virtual photon at higher Q^2. Due to the strong interaction the quarks emit virtual gluons and the gluons fluctuate into quark-antiquark pairs. At sufficiently high Q^2 the quarks within these processes are resolved by the virtual photon (see Fig. 13a).

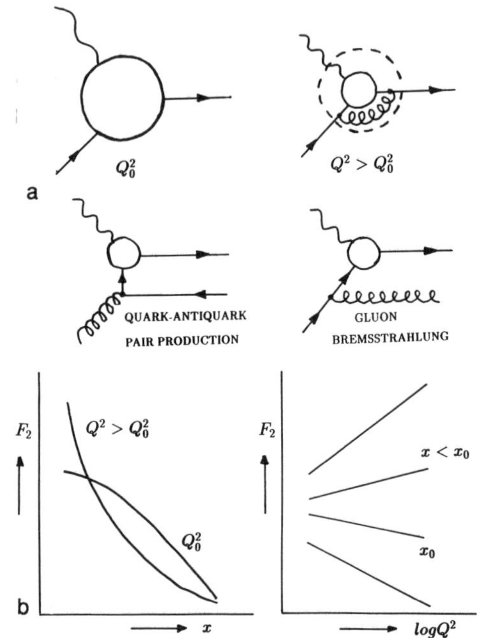

Figure 13. $\mathbf{Q^2}$ **resolution and scaling violation.**

In the QCD-corrected quark parton model the structure function

$$F_2(x, Q^2) = \sum_i e_i^2 x q_i(x, Q^2) \tag{31}$$

is given by quark momentum distributions $x q_i(x, Q^2)$ which depend on Q^2. The Q^2 dependence of the probability to find a quark at a certain value of x is described by the Altarelli-Parisi equations [9].

$$Q^2 \frac{d}{dQ^2} q(x, Q^2) = \frac{\alpha_s(Q^2)}{2\pi} \int_x^1 \frac{dy}{y} [q(y, Q^2) P^q(\frac{x}{y}) + G(y, Q^2) P^g(\frac{x}{y})], \tag{32}$$

where $G(x, Q^2)$ is the gluon distribution and P^q, P^g are so-called QCD splitting functions. In Fig. 13b the x dependence of F_2 for different Q^2 and the Q^2 dependence for different values of x are shown. At higher Q^2, the quarks share part of their momentum with emitted gluons, thus the quark momentum distribution shifts to smaller x. At small x the number of quarks increases also due to pair production from gluons.

A compilation of recent μ and ν data given in Fig. 14 shows the high accuracy and the degree of agreement of different deep inelastic scattering experiments on isoscalar

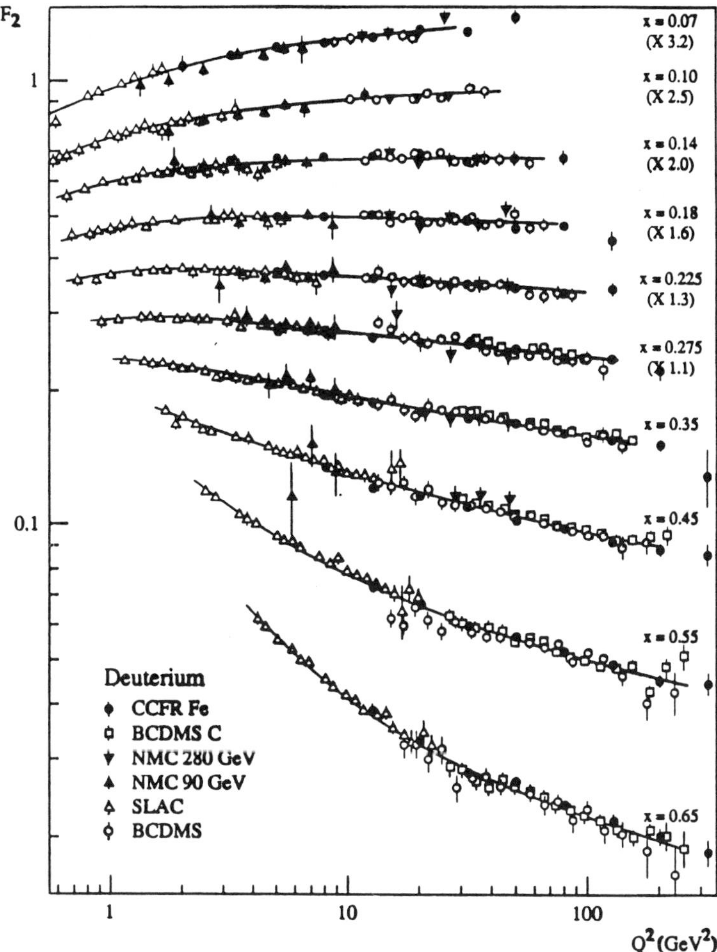

Figure 14. $\mathbf{F_2(x, Q^2)}$ **obtained with high accuracy on isoscalar targets in different experiments.**

(N=Z) targets. The increase of F_2 at small x and the decrease at large x with increasing Q^2 are clearly illustrated by the data.

Parton distributions can be obtained by a QCD analysis of the structure functions $F_2^p(x, Q^2)$ and $F_2^d(x, Q^2)$; this has recently been carried out by NMC [10]. To perform such an analysis a parametrisation for the flavour singlet quark distribution $q^{SI}(x, Q_0^2)$, the flavour non-singlet quark distribution $q^{NS}(x, Q_0^2)$, and the gluon distribution $G(x, Q_0^2)$ was chosen at a fixed Q_0^2.

For non-singlet quark distributions, like xF_3 and $F_2^p - F_2^n$ there is no gluon contribution to the evolution, as the quark and antiquark contributions from pair creation cancel. Also other sources of Q^2 dependence were considered:

(a) Threshold effects for heavy quarks, $g \rightarrow q + \overline{q}$.

(b) Target-mass correction, $\zeta = 2x/(1 + \sqrt{1 + 4M^2x^2/Q^2})$.

(c) Higher-twist effects due to secondary interactions of the struck quark,

$$F_2 = F_2^{LT}(1 + \frac{H(x)}{Q^2}).$$

$$(33)$$

155

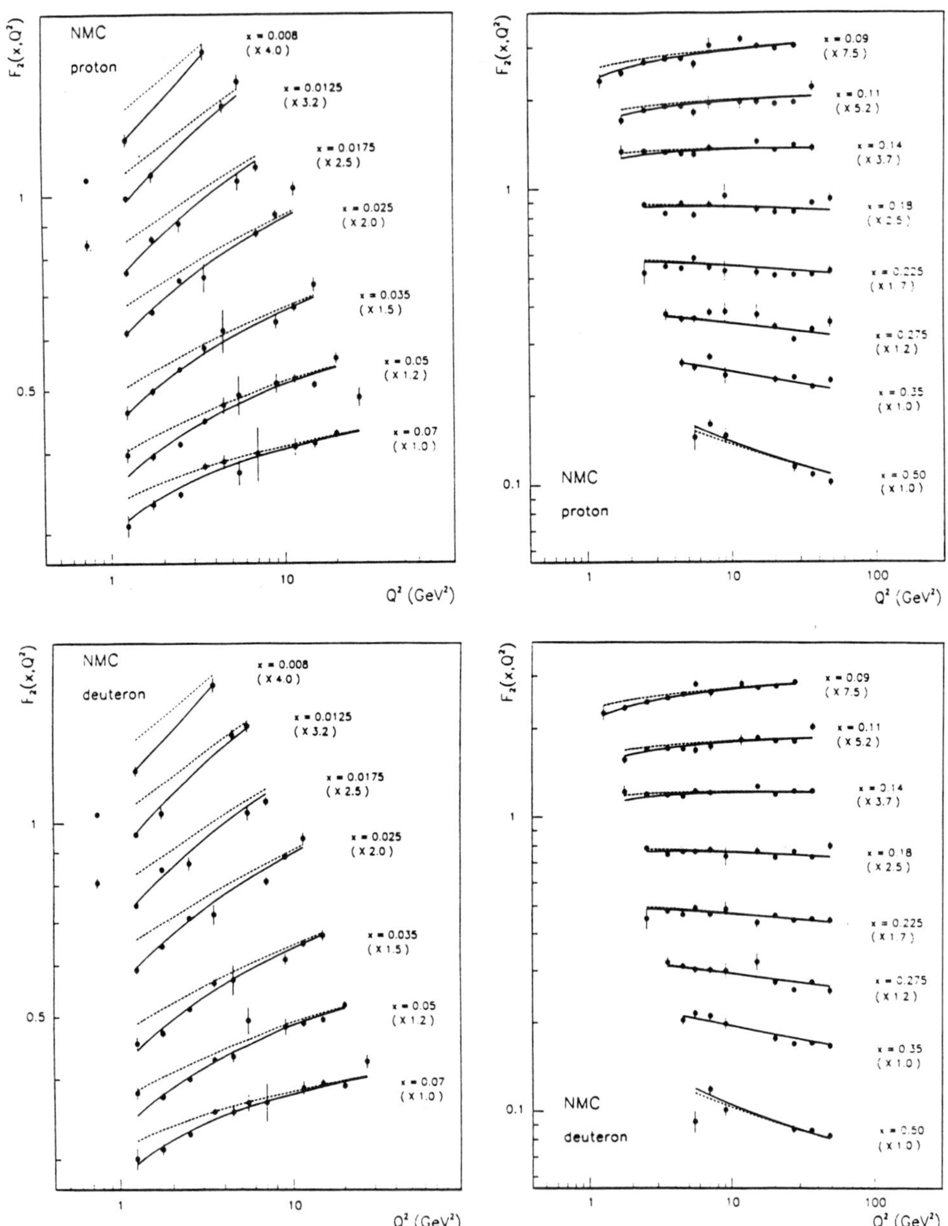

Figure 15. **Results of the QCD fits to $\mathbf{F_2^p(x, Q^2)}$ and $\mathbf{F_2^d(x, Q^2)}$ by NMC including higher twist (solid curves) and with the same parameters but no higher-twist contribution (dotted curves).**

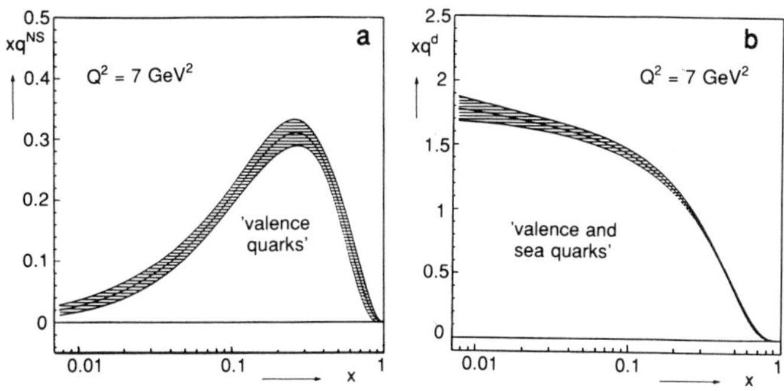

Figure 16. **Parton distributions from the QCD analysis by NMC on** $\mathbf{F_2^p(x, Q^2)}$ **and** $\mathbf{F_2^d(x, Q^2)}$. **The result of the fit with the error band is given for (a) the non-singlet distribution** $\mathbf{xq^{NS}(x)}$ **and (b) the singlet distribution** $\mathbf{xq^d(x)}$.

The higher-twist effects were parametrised with a single parameter H for small values of x and for larger values of x the results of a fit to high statistics data from SLAC and BCDMS were used. From the parton parametrisation at Q_0^2 the structure function $F_2(x)$ can be calculated at any Q^2 using the QCD splitting functions. In the NMC analysis [10] in total 12 parameters were fitted to the data shown in Fig. 15.

The results for the parton distributions at $Q^2 = 7$ GeV^2 are shown in Fig. 16. The non-singlet distribution relates to valence quarks, while the singlet distribution combines valence and sea quarks, which emerge at small x.

Also a remarkably accurate gluon distribution is obtained, although the muons do not scatter from gluons which carry no electric charge. The gluon distribution enters in the Q^2 dependence indirectly through quark-antiquark pair creation. In Fig. 17 the new result for the gluon distribution is compared with results obtained from similar fits to other data sets.

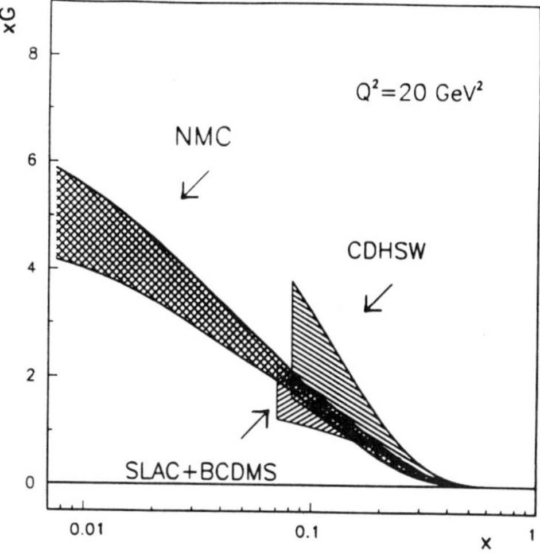

Figure 17. **The gluon distribution** $\mathbf{xG(x)}$ **from QCD analyses of different data sets.**

The running coupling constant $\alpha_S(Q^2)$ has also been determined in the QCD fit by NMC,

$$\alpha_S(7\ GeV^2) = 0.264 \pm 0.018(stat.) \pm 0.070(syst.) \pm 0.013(h.t.), \qquad (34)$$

where the systematic uncertainty due to the higher-twist term has been estimated to be 0.013. At the mass of the Z^0 boson, the extrapolated value of $\alpha_S(Z^0) = 0.117 + 0.011/ - 0.016$. This value agrees well with the more precise results obtained from Z^0 decays at LEP.

Future Experiments

In the region of very small x and high Q^2 the first data on F_2 from ZEUS and H1 [22] at HERA show an enhancement of F_2. The H1 data shown in Fig. 18 agree well with the NMC data at larger values of x. They indicate a stronger increase of the sea quarks at small x. More new data are expected soon in this interesting new region of high Q^2 and very small x.

NUCLEAR EFFECTS

Deep inelastic scattering has not only been studied on protons and deuterons, but also on heavier targets like iron. The use of heavy and dense targets has been imperative

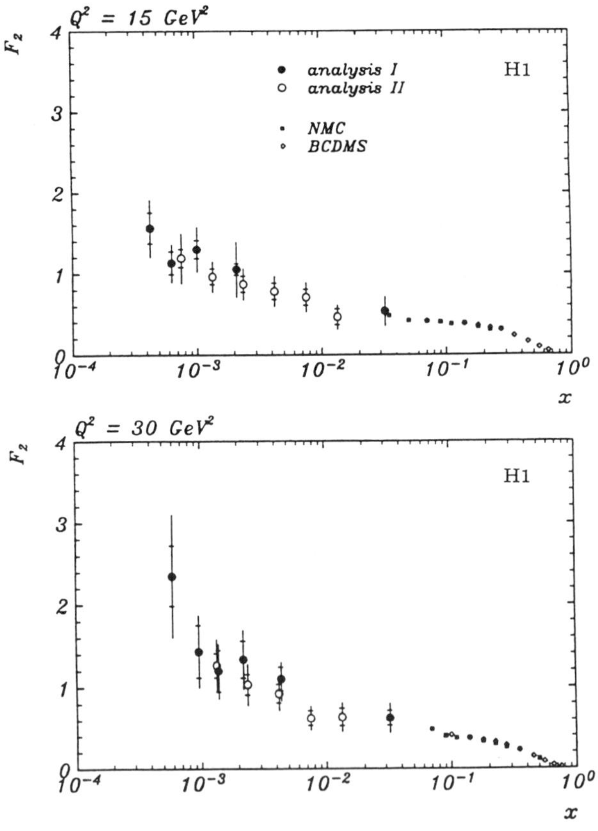

Figure 18. $\mathbf{F_2^p(x)}$ at $\mathbf{Q^2 = 15\ GeV}$ measured at **HERA** and at **CERN**.

for most neutrino experiments to compensate for the small cross section. The structure function F_2 was supposed not to depend on the target material. As the reaction was shown to take place on the quarks inside the nucleon, this was considered to be a good approximation as long as the difference between the number of protons and neutrons was taken into account.

EMC Effect

In a comparison by EMC of the nucleon structure function F_2^{Fe} obtained on an iron target and of F_2^d on a liquid deuterium target this assumption was shown to be invalid [11]. This so-called EMC effect was confirmed by other experiments [4] (see Fig. 19).

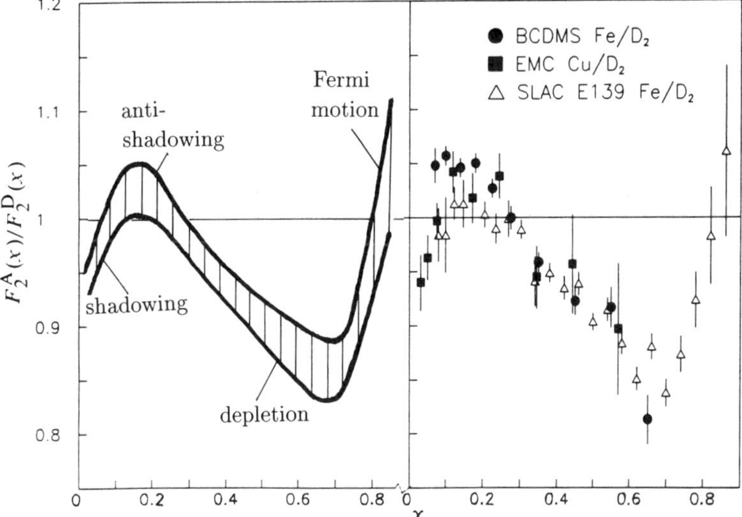

Figure 19. **(a) Schematic representation of $\mathbf{F_2^A/F_2^d}$ indicating shadowing, anti-shadowing, depletion and Fermi motion. (b) Ratio of nucleon structure functions $\mathbf{F_2^A(x)}$ on Fe and Cu to $\mathbf{F_2^d(x)}$ on D in different experiments.**

The ratio of the nucleon structure functions F_2^A determined on a nuclear target with mass number A to F_2^d of the deuteron shows the expected rise for $0.7 < x < 1$ due to Fermi motion of nucleons inside the nucleus. For $0.3 < x < 0.6$ an unexpected depletion was observed, the EMC effect, which caused quite some excitement. This depletion increases for heavier nuclei.

Between the many models which were proposed to explain the EMC effect two directions can be distinguished:

(a) Q^2 rescaling models, which attribute the EMC effect to a higher effective Q^2 in nuclei. An increase of confinement radius for nucleons in a nuclear environment was postulated to explain this effect [12].

(b) Bjorken x rescaling models, which take into account a lower effective nucleon mass $M^{eff} \leq 0.9M$ due to nuclear binding effects.

Shadowing

For $x < 0.1$ also depletion of F_2^A was observed, which is called shadowing. The small enhancement of F_2^A around $x = 0.2$ is called anti-shadowing. The effect of shadowing is stronger for heavier nuclei (see Fig. 20).

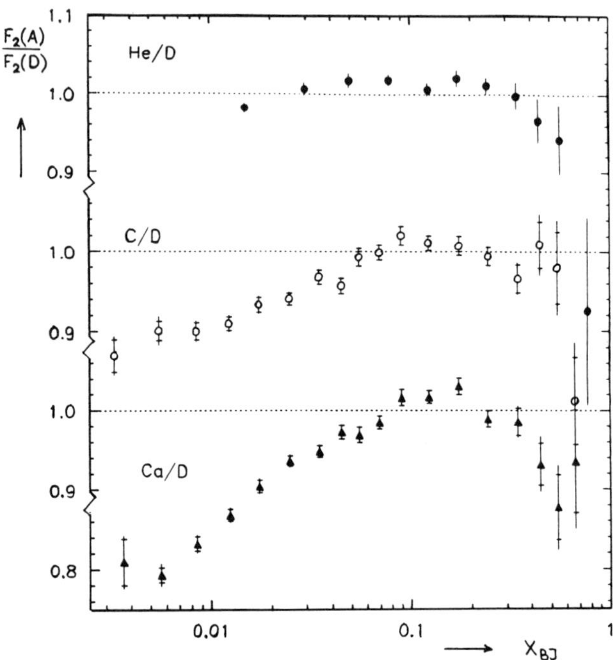

Figure 20. **Ratio of nucleon structure functions F_2^A/F_2^d for He/D, C/D and Ca/D from the NMC experiment. Shadowing for x < 0.05 is stronger at small x and high A.**

Several models were proposed to explain shadowing in deep inelastic scattering due to either the virtual photon probe or the nuclear environment:

(a) Vector-meson dominance models, which attribute the shadowing to hadronic components in the virtual photon.

(b) Parton fusion models, which predict a change in the number of quarks at small x where quarks from different nucleons start to overlap.

In the **vector-meson dominance** model the wave function of the virtual photon contains the usual bare photon and vacuum fluctuations of e^+e^- pairs and in addition $q\bar{q}$ pairs representing virtual vector mesons like ρ^0 and ϕ, which have the same quantum numbers as the photon [13].

We can define a hadron absorption length l, which determines the A dependence of the absorption cross section, like in hadron-nucleus absorption;

$$\begin{aligned}
\sigma_{hA} &= A\sigma_{hN} \quad for \ l >> 2R \\
\sigma_{hA} &= A^{2/3}\sigma_{hN} \ for \ l << 2R,
\end{aligned} \tag{35}$$

where R is the nuclear radius.

For example, shadowing is observed in the absorption cross section for pions on copper nuclei, $\sigma_{\pi Cu} = A_{Cu}^{0.85}\sigma_{\pi N}$, with a pion absorption length of $l \sim 2.4$ fm, which is slightly smaller than the nuclear radius of $R = 4.5$ fm. Similarly, shadowing in deep inelastic lepton scattering can be expected from the absorption of virtual vector mesons.

A fluctuation length d can be defined, which depends on the four-momentum of the virtual photon and on the mass m_h of the virtual vector meson;

$$d(m_h, Q^2) \approx \frac{2\nu}{m_h^2 + Q^2}. \tag{36}$$

For real photons $d \approx 2\nu/m_h^2$ and for virtual photons $d \approx 2\nu/Q^2 = 1/Mx$.

In shadowing models the cross section is not proportional to A, when the absorption length l is smaller than the nuclear size and the fluctuation length d of the hadronic state is larger than the nuclear size, so that no virtual vector mesons are regenerated inside the nucleus,

$$d < 2R < l. \tag{37}$$

In the **parton fusion** models both shadowing and anti-shadowing are predicted due to the change of the quark momentum distribution, and thus of $F_2(x)$, when the wave functions of quarks from different nucleons start to overlap [14].

Let me explain the x dependence of parton fusion by considering the internucleon distance z in the Breit frame,

$$z = \frac{h}{\gamma m_\pi} = \frac{Mh}{E_N m_\pi} \approx \frac{Mh}{P_N m_\pi}. \tag{38}$$

The position resolution of a parton with momentum fraction x is

$$dz = \frac{h}{x P_N}. \tag{39}$$

The wave functions of partons overlap with other nucleons when $z < dz$, thus

$$\frac{M}{P_N m_\pi} < \frac{1}{x P_N}. \tag{40}$$

This occurs at

$$x < x_0 = \frac{m_\pi}{M} \sim 0.15. \tag{41}$$

When partons from different nucleons fuse the momentum distribution will be enhanced for $x \sim x_0$ and depleted for $x \sim x_0 A^{-1/3}$ (see Fig. 21).

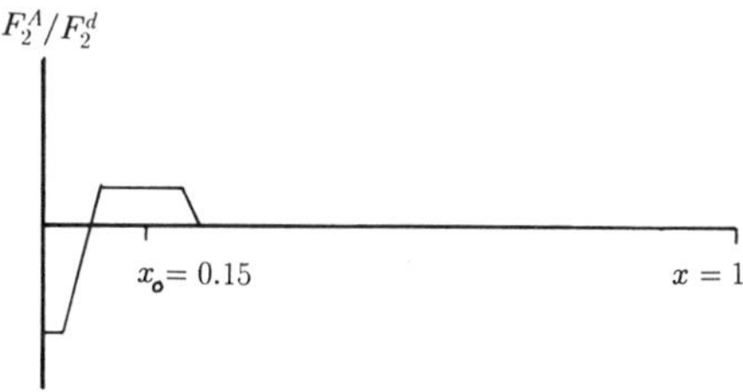

Figure 21. **Shadowing and anti-shadowing in parton fusion models.**

New data from E665 [15] are shown in Fig. 22. They show again the increase with
A of the shadowing effect and indicate at very small x saturation of shadowing.

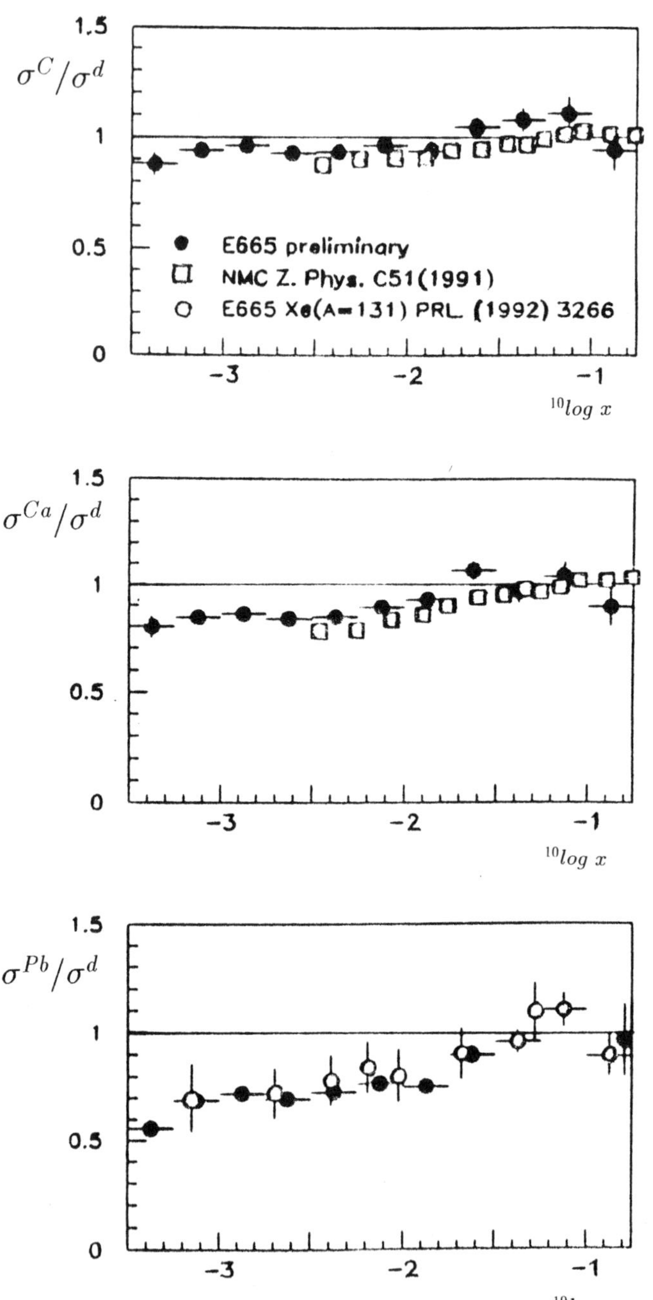

Figure 22. **Preliminary results from E665 together with those of NMC on
shadowing for heavy targets show saturation at small x.**

THE SPIN PUZZLE

The spin of the nucleon has long been considered to be the result of quark spins coupled to $s = 1/2$ in the lowest energy state without orbital momentum. Measurements on spin structure functions seem now to contradict this simple picture.

The deep inelastic cross section for longitudinally polarised electrons or muons on longitudinally polarised nucleons depends on the relative orientation of the spins. This dependence is described by the spin structure functions G_1 and G_2;

$$
\begin{aligned}
\frac{d^2\sigma}{dQ^2 d\nu} = {} & \frac{4\pi\alpha^2}{Q^4}\frac{E}{E'}[2W_1 sin^2\theta/2 + W_2 cos^2\theta/2] \\
& \pm \frac{2\pi\alpha^2}{Q^2 E^2}[MG_1(E - E'cos\theta) - Q^2 G_2],
\end{aligned}
\tag{42}
$$

where the sign $(\pm)$ depends on the relative longitudinal polarisation of the incoming lepton and the target nucleon. Therefore G_1 and G_2 can experimentally be determined from the difference in the cross section for opposite relative polarisation directions. These structure functions can be related to the more frequently used dimensionless structure functions,

$$
\begin{aligned}
F_1 &= MW_1 \\
F_2 &= \nu W_2 \\
g_1 &= M^2\nu G_1 \\
g_2 &= M\nu^2 G_2.
\end{aligned}
\tag{43}
$$

In the approximation of asymptotically free quarks the absorption of a virtual photon with helicity $+1$ depends on the spin direction of the nucleon and the probabilities $q_i^+(x)$ and $q_i^-(x)$ for quarks of flavour i with momentum fraction x (Eq. 17). The difference in the cross section for opposite nucleon polarisations is related to the spin structure function $g_1(x)$ as follows:

$$
\sigma^{\rightleftarrows}(x) - \sigma^{\rightrightarrows}(x) \sim \sum_i e_i^2[q_i^+(x) - q_i^-(x)] = 2g_1(x).
\tag{44}
$$

Here $\Delta q_i(x) = q_i^+(x) - q_i^-(x)$ can be considered as the spin distribution of quarks with flavour i as a function of x.

The cross section asymmetry A_1 for a virtual photon with helicity $+1$ on a nucleon with its polarisation antiparallel and parallel to the virtual photon direction is directly related to the structure functions g_1 and F_1 by

$$
A_1(x) = \frac{\sigma^{\rightleftarrows}(x) - \sigma^{\rightrightarrows}(x)}{\sigma^{\rightleftarrows}(x) + \sigma^{\rightrightarrows}(x)} = \frac{g_1(x)}{F_1(x)}.
\tag{45}
$$

SLAC and EMC Experiments

In deep inelastic spin-dependent scattering experiments A_1 is determined from count-rate asymmetries

$$
\frac{N^{\rightleftarrows}(x) - N^{\rightrightarrows}(x)}{N^{\rightleftarrows}(x) + N^{\rightrightarrows}(x)} \sim P_b P_t f D A_1,
\tag{46}
$$

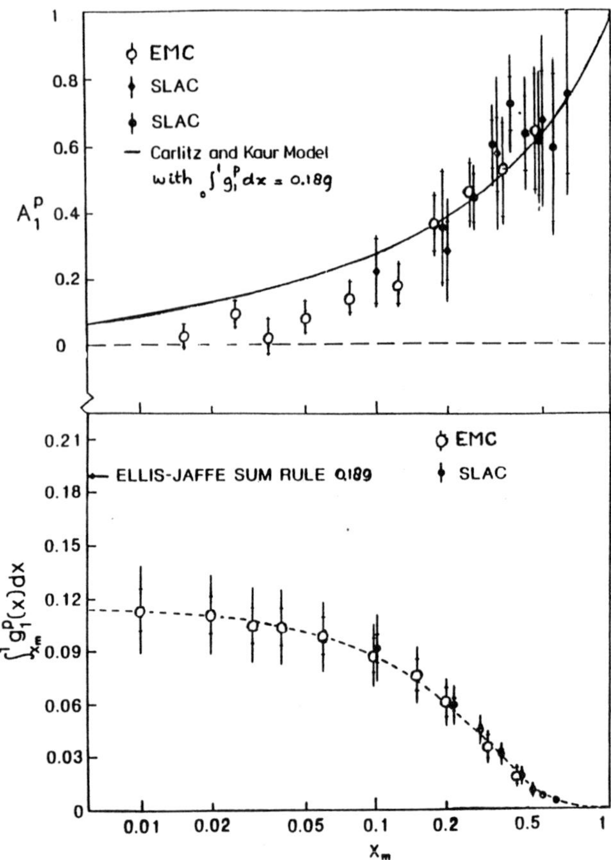

Figure 23. **Asymmetry $A_1^P(x)$ and spin structure function $g_1^P(x)$ of the proton measured in SLAC and EMC experiments.**

which also depend on the longitudinal beam polarisation P_b, the target polarisation P_t, a dilution factor f due to nucleons from carbon and other materials in the target which are not considered in the target polarisation. The depolarisation factor D takes into account how much of the electron or muon polarisation is carried by the virtual photon [16]. For frozen ammonia and butanol targets, which were used in these spin experiments, the count-rate asymmetries were very small, in the order of 10^{-3}. Therefore, these experiments needed high statistics and were difficult to carry out.

The spin structure function $g_1(x)$ of the proton was measured in SLAC experiments and by EMC at CERN [17]. The results on $A_1^p(x)$ and $g_1^p(x)$ are shown in Fig. 23. The first moment of g_1^p including extrapolations to $x = 0$ and to $x = 1$ was found to be

$$\Gamma_1^p = \int_0^1 g_1^p(x)dx = 0.126 \pm 0.010(stat.) \pm 0.015(syst.). \tag{47}$$

This result was compared with the prediction by Ellis and Jaffe [18],

$$\Gamma_1^p(EJ) = 0.189 \pm 0.005. \tag{48}$$

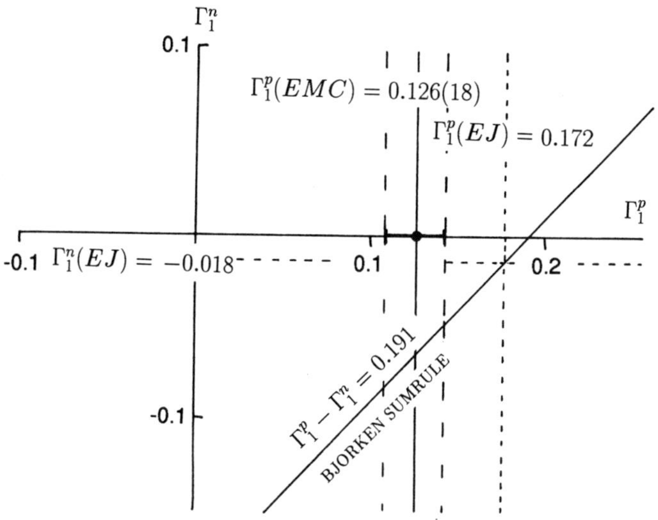

Figure 24. **First moments $\Gamma = \int_0^1 g_1(x)dx$ for proton and neutron compared with the Bjorken sum rule $\Gamma_1^p - \Gamma_1^n = 1.91$.**

The deviation from the Ellis-Jaffe prediction initiated new discussions on the origin of the nucleon spin. We have seen in Eq. (44) that g_1 contains information on the probabilities $q_i^\pm$ to find quarks with one or the other spin direction inside the nucleon.

The first moment of the neutron spin structure function g_1^n is related to that of the proton via the fundamental Bjorken sum rule

$$\Gamma_1^p - \Gamma_1^n = \frac{1}{6}\left|\frac{g_A}{g_V}\right|(1 - \frac{\alpha_S}{\pi}) = 0.191 \pm 0.002, \qquad (49)$$

which relates the difference in spin structure functions of proton and neutron to the ratio of weak axial-vector and vector coupling constants g_A/g_V observed in neutron β-decay. A deviation from the Ellis-Jaffe prediction for the proton then requires a similar deviation for Γ_1^n of the neutron according to the Bjorken sum rule. This is illustrated in Fig. 24, where a more recent value of $\Gamma_1^p(EJ) = 0.172 \pm 0.009$ is used from new data on hyperon decays [19].

New experiments to measure the neutron spin structure function and to remeasure the proton spin structure function were therefore proposed on polarised proton, deuteron and helium-3 targets at CERN, SLAC and DESY (see Table 1).

The assumption made by Ellis and Jaffe is that the contribution from the strange sea to the nucleon spin is zero;

$$\Delta s = \int_0^1 (\Delta s(x) + \Delta \bar{s}(x))dx = 0. \qquad (50)$$

In this case the weak coupling constants from hyperon decay and from neutron decay alone determine the values of Γ_1^p and Γ_1^n.

If we allow the strange sea to be polarised, $\Delta s \neq 0$, the experimental result on Γ_1^p yields the following flavour contributions of the quark spins to the nucleon spin [17]:

$$\Delta u = +0.78 \pm 0.03 \pm 0.05$$
$$\Delta d = -0.47 \pm 0.03 \pm 0.05$$
$$\Delta s = -0.19 \pm 0.03 \pm 0.05.$$

This means that the total fraction of the spin carried by the quark spins is only

$$\Delta \Sigma = \int_0^1 \sum_i \Delta q_i(x) dx = 0.12 \pm 0.10(stat.) \pm 0.14(syst.). \tag{51}$$

This is called the spin puzzle. The probabilities for all quarks to carry spin parallel or antiparallel to the nucleon spin adds up to a small fraction, consistent with no preferred spin direction at all. The question arises: Where is the nucleon spin ?

Other contributions are now considered as well; angular momentum and also polarisation of gluons which can contribute to deep inelastic lepton scattering asymmetry via a process which is known as the axial anomaly;

$$s_N = \frac{1}{2}\Delta\Sigma + < L_z > + \Delta G. \tag{52}$$

It will be the task of the new experiments to confirm these results and to map the contributions to the nucleon spin.

Table 1. New experiments on spin structure functions.

Experiment	Beam		Target		x-range	L $1/cm^2 s^1$
SMC-CERN 1991-1995 p,n	$\vec{\mu}$ I P_b	100-200 GeV 0.05 pA 0.80	$C_4H_{10}O$ $C_4D_{10}O$ $P_p \sim 0.85$ $P_d \sim 0.40$	$3.10^{25}/cm^2$ $f \sim 0.13$ $f \sim 0.24$	0.005 $\downarrow$ 0.60	10^{32}
E142/SLAC 1993 n	$\vec{e}$ I P_b	23 GeV 6 μA 0.40	$^3He(10\ atm)$ $P_{^3He} \sim 0.50$	$10^{21}/cm^2$ $f \sim 0.12$	0.04 $\downarrow$ 0.60	3.10^{35}
E143/SLAC 1994 p,n	$\vec{e}$ I P_b	30 GeV 10-100 nA 0.80	NH_3, ND_3 $P_p \sim 0.85$ $P_d \sim 0.40$	$\sim 10^{24}/cm^2$ $f \sim 0.17$ $f \sim 0.30$	0.04 $\downarrow$ 0.60	1.5×10^{35}
E154/SLAC 1995 n	$\vec{e}$ I P_b	50 GeV 6 μA 0.80	$^3He(10\ atm)$ $P_{^3He} \sim 0.50$	$10^{21}/cm^2$ $f \sim 0.12$	0.02 $\downarrow$ 0.60	3.10^{35}
E155/SLAC 1995 p,n	$\vec{e}$ I P_b	50 GeV 10-100 nA 0.80	NH_3, ND_3 $P_p \sim 0.85$ $P_d \sim 0.40$	$\sim 10^{24}/cm^2$ $f \sim 0.17$ $f \sim 0.30$	0.02 $\downarrow$ 0.60	1.5×10^{35}
HERMES/DESY $\geq$ 1995 p,n	$\vec{e}$ I P_b	35 GeV 60 mA 0.50	$H,D,^3He$ $P_p \sim P_d \sim 0.8$ $P_{^3He} \sim 0.6$	$\sim 10^{14}/cm^2$ $f \sim 1$ $f \sim 1/3$	0.02 $\downarrow$ 0.80	3.5×10^{32}

Recent Results

First results on the spin-dependent structure functions g_1^d of the deuteron and g_1^n of the neutron were reported this year. In Fig. 25 the present world data on spin structure functions is shown.

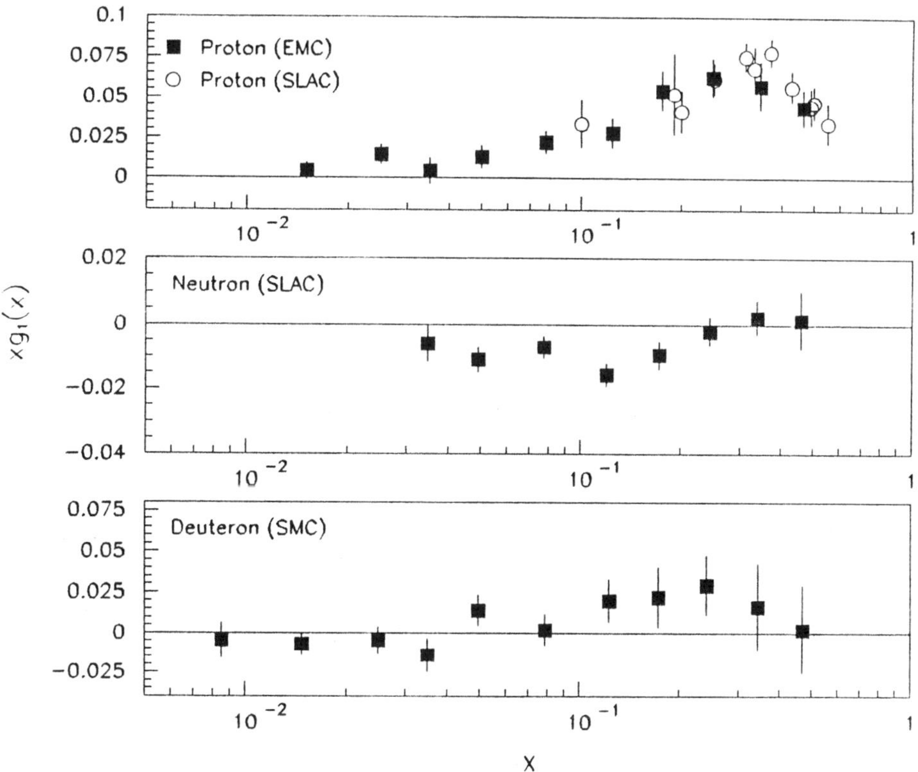

Figure 25. **Spin-dependent structure functions $g_1^P(x)$ from E80, E130 and EMC, $g_1^n(x)$ from E142 and $g_1^d(x)$ from SMC.**

The Spin Muon Collaboration (SMC) at CERN measured $g_1^d(x)$ over the kinematical range $0.006 < x < 0.6$ for $Q^2 > 1$ GeV^2 using a polarised μ^+ beam of 100 GeV and a polarised deuteron target [20].

The target material consisted of frozen deuterated butanol beads cooled to 0.5 K in a helium-3 dilution refrigerator. A longitudinal magnetic field of 2.5 T was used to polarise electrons introduced by a paramagnetic dopant.

The method to polarise the deuterons is dynamical nuclear polarisation (DNP), which uses microwaves to induce second order transitions between electron states in which also the nucleon spin is flipped. By tuning the microwave frequency the resulting deuteron polarisation can be either parallel or opposite to the direction of the magnetic field.

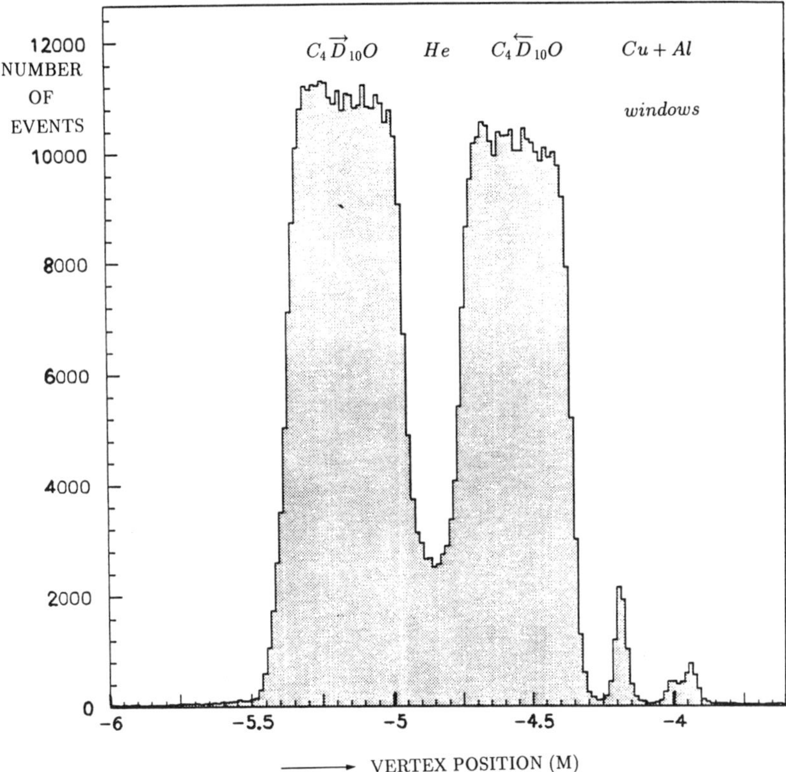

Figure 26. Vertex reconstruction of the interaction point inside the target. Interactions in the first target half and the second oppositely polarised target half are clearly separated. Interactions in the helium mixture and several vacuum windows can be distinguished as well.

To reduce the systematic error in the cross section asymmetry the target was split into two halves of 40 cm length each separated by 20 cm along the beam direction. The target halves were oppositely polarised using different DNP frequencies. Also the polarisation direction with respect to the beam in each target half was regularly reversed by rotating the magnetic field using a transverse field. Reconstruction of the interaction vertex was used to determine in which target half the deep inelastic scattering took place (see Fig. 26).

The result from SMC [20] for the integral of g_1^d was

$$\Gamma_1^d = 0.023 \pm 0.020(stat.) \pm 0.015(syst.),\tag{53}$$

including a total contribution from the unmeasured regions $x < 0.006$ and $x > 0.6$ of -0.001 ± 0.005. This value also deviates from the Ellis-Jaffe prediction for proton and neutron. This again indicates a finite negative spin contribution from the strange sea. Also the total contribution of the quark spins to the nucleon spin is small,

$$\Delta\Sigma(SMC) = 0.06 \pm 0.25.\tag{54}$$

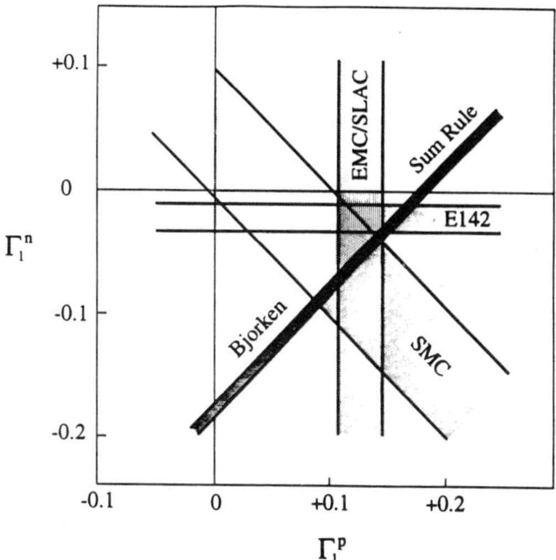

Figure 27. **First moments $\Gamma = \int_0^1 g_1(x)dx$ for proton, neutron and $\Gamma_1^p + \Gamma_1^n$ from the deuteron compared with the Bjorken sum rule.**

The SLAC experiment E142 measured $g_1^n(x)$ over the kinematical range $0.03 < x < 0.06$ for $Q^2 > 1$ GeV^2 using a longitudinally polarised electron beam of $19 - 26$ GeV and a polarised ^{3}He gas target [21].

The ^{3}He nuclei were polarised through spin exchange with optically pumped rubidium vapour. The target was 30 cm long and contained 69×10^{20} atoms/cm^2 (8.6 atm at 0 °C). The ^{3}He wave function is primarily in a S-state in which the two protons pair with opposite spins due to the Pauli exclusion principle leaving the neutron spin as the dominant contribution to the spin-dependent cross section. Corrections due to the other parts of the ^{3}He wave function were applied in the extraction of A_1^n. The result for the integral of $g_1^n(x)$ was

$$\Gamma_1^n(E142) = -0.022 \pm 0.007(stat.) \pm 0.011(syst.), \tag{55}$$

including a total contribution from the regions $x < 0.03$ and $x > 0.6$ of -0.003 ± 0.007. This value agrees with the Ellis-Jaffe prediction for the neutron $\Gamma_1^n(EJ) = -0.018 \pm 0.009$. Therefore, E142 concludes that the spin contribution of the strange sea is zero and the total contribution of the quark spins to the nucleon spin is large,

$$\Delta\Sigma(E142) = 0.57 \pm 0.11. \tag{56}$$

The conclusions by SMC and E142 on the Bjorken sum rule and the spin contribution $\Delta\Sigma$ are different. This is illustrated in Fig. 27. The combination of Γ_1^d of SMC and Γ_1^p of EMC gives good agreement with the Bjorken sum rule

$$\Gamma_1^p - \Gamma_1^n = 0.20 \pm 0.04, \tag{57}$$

while the combination of Γ_1^p of EMC and Γ_1^n of E142 gives a deviation of almost two σ with the Bjorken sum rule

$$\Gamma_1^p - \Gamma_1^n = 0.15 \pm 0.02. \tag{58}$$

Combination of E142 and SMC yields an even smaller value for the Bjorken sum.

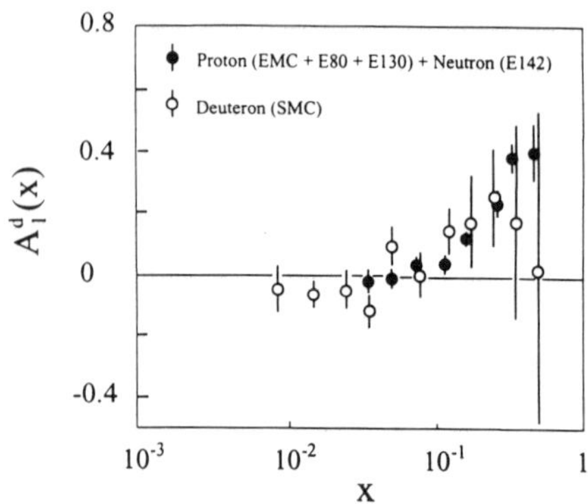

Figure 28. Comparison of $A_1^d(x)$ measured by SMC and calculated from $A_1^p(x)$ and $A_1^n(x)$ measured by EMC+SLAC and E142, respectively.

However, if we compare the measurements in the region of x where the two measurements overlap the data agree well within their errors, as can be seen in Fig. 28. It is only in the small x region where the E142 extrapolation and the SMC data do not agree. This can best be illustrated with the value for Γ_1^n from the combined data sets of proton, neutron and deuteron spin measurements

$$\Gamma_1^n(E142 + EMC + SMC) = -0.055 \pm 0.026, \tag{59}$$

for which the change with respect to Eq. (55) is almost completely due to the small x region, where no E142 data exists at $Q^2 > 1$ GeV^2.

A combined analysis of all available data at present gives a consistent result in agreement with the Bjorken sum rule and indicating only a small contribution of the quark spins to the spin of the nucleon.

The measurements at small values of x can only be done at high beam energies. SMC is measuring g_1^p at 190 GeV beam energy in 1993 and continues with g_1^d in the year after. High-statistics experiments at 30 GeV will be performed at SLAC on g_1^p and g_1^d starting end of 1993. Running with 50 GeV electrons at SLAC is foreseen for 1995. At DESY experiments at 30 GeV will start in 1995 with internal targets of p, d and ^{3}He. The features of the new experiments are summarised in Table 1. Also measurements will be done on the spin structure function g_2 which was not discussed in these lectures.

Altogether a broad programme of various spin-structure experiments at several laboratories will be realised in the near future to provide more insight into the origin of the nucleon spin.

EXERCISES

- Show that elastic scattering on a proton corresponds to $x = 1$.

- Give a naive quark model expectation for the ratio F_2^n/F_2^p, (a) at $x \to 0$ where the sea quarks dominate and (b) at $x \sim 1/3$ with only valence up and down quarks.

- Explain how Fermi motion causes a rise in F_2^A/F_2^d at $x \to 1$.

- Propose a deep inelastic scattering experiment which may attribute shadowing in deep inelastic muon scattering to either vector-meson dominance or to parton fusion.

REFERENCES

[1] M. Breidenbach et al., Phys. Rev. Lett. 23: 935 (1969).

[2] O.C. Allkofer et al., Nucl. Instr. Meth. 179: 445 (1981).

[3] F. Halzen and A.D. Martin, Quarks and Leptons, New York: Wiley, 1984.

[4] Particle Data Group, Phys. Lett. B204: 1 (1988).

[5] D.J. Gross and C.H. Llewellyn Smith, Nucl. Phys. B 14: 337 (1969).

[6] CCFR, Rencontres de Moriond (1993).

[7] K. Gottfried, Phys. Lett. 18: 1174 (1967).

[8] NMC, P. Amaudruz et al., Phys. Rev. Lett. 66: 2712 (1991);
CERN PPE/93-117 (1993).

[9] G. Altarelli and G. Parisi, Nucl. Phys. B 126: 288 (1977).

[10] NMC, M. Arneodo et al., Phys. Lett. B 309: 222 (1993).

[11] EMC, J.J. Aubert et al., Phys. Lett. B 123: 275 (1983),
EMC, J.J. Aubert et al., Nucl. Phys. B 293: 740 (1987).

[12] F.E. Close, R.L. Jaffe, R.G. Roberts and G.G. Ross, Phys. Rev D31: 1004 (1985).

[13] L.V. Gribov, E.M. Levin and M.G. Ryskin, Phys. Rep. 100: 1 (1983).

[14] N.N. Nicolaev and V.I. Zakharov, Phys. Lett. 55 B: 397 (1975),
J. Qui, Nucl. Phys. B 291: 746 (1987).

[15] E665, Rencontres de Moriond (1993).

[16] V.W. Hughes and J. Kuti, Ann. Rev. Nucl. Part. Sci. 33: 611 (1983).

[17] EMC, J. Ashman et al., Nucl. Phys. B 328: 1 (1989).

[18] J. Ellis and R.L. Jaffe, Phys. Rev. D 9: 1444 (1974); D 10: 1669 (1974).

[19] F.E. Close and R.G. Roberts, Phys. Lett. B 316: 165 (1993).

[20] SMC, B. Adeva et al., Phys. Lett. B 302: 533 (1993).

[21] P.L.Anthony et al., Phys. Rev. Lett. 71: 959 (1993).

[22] HERA, Rencontres de Moriond (1993).

NUCLEAR FILTERING AND QUANTUM COLOR TRANSPARENCY: AN INTRODUCTORY REVIEW

John P. Ralston

Department of Physics and Astronomy
University of Kansas
Lawrence, KS 66045 USA

ABSTRACT

Color transparency is the proposal that under certain circumstances the observed strong interactions can be controlled and in fact reduced in magnitude. I give a comprehensive review of our approach, which is based on the perturbative QCD study of hard exclusive reactions in free space and in nuclear targets. Topics considered include quasi-exclusive reactions with either electron or hadron beams in the initial state, and many kinds of outgoing particles in the final state. The description of color transparency in terms of light-cone matrix elements, and its interesting character as a two-scale pQCD process, is presented as a foundation for theoretical discussion. The phenomenon of *nuclear filtering*, which involves the modification of quark wave functions in hadrons to smaller transverse space dimensions, leads to a broad new program to study the strong interactions and hadron structure. I review existing experimental data and certain experiments planned at various facilities. A recently formulated systematic method to define color transparency directly in terms of experimental data and with minimal theoretical model dependence plays an important role. The analysis shows evidence that color transparency may already have been observed in the pioneering experiment of Carroll *et al.* I will also discuss the role of spin and color transparency as powerful tools for learning more about the internal quark configurations of hadrons in future experiments. The subject has the potential for strong scientific complementarity and progress in exploring hadron physics at Brookhaven, SLAC and CEBAF as well as future facilities.

I. OVERVIEW

There is a wonderful short story called *Le Passe-muraille* by the French author Marcel Aymé[1]. The story is about a man who, one day, suddenly realizes that he can pass through solid walls. Initially he is quite upset about this. Instead of taking advantage of the opportunity to rob banks, visit his girlfriend, etc, he goes to a physician to see what is wrong with him. But after a while, he learns to enjoy the new situation, and begins a new life of robbing banks, walking out of locked jail cells, and so on.

Color transparency is the theoretical story that protons and other hadrons can do the incredible - that under certain circumstances they can pass through solid nuclear matter. Like the Aymé story, this contradicts certain rules of physics, which is one reason why it is so interesting. The rules which are contradicted are those of conventional strong interaction physics, which would say that for 1-10 GeV of beam energy, the proton-nucleon inelastic cross section is about 30-35 mb. In nuclear matter with a density of about 1/6 nucleon/fm^3, the mean free path before having an inelastic collision should be about 2 fm.

But the conventional rules may not be right; if color transparency is correct, then under certain circumstances we can prepare a proton or other hadron to make it pass through many Fermi of nuclear matter. For all practical purposes, sometimes the observable consequences of the strong interactions can be turned off.

Theorists believe the color transparency story because it is based on another set of beliefs, perturbative QCD. However, the experimental situation is not settled; there is much discussion at this time whether color transparency has actually been observed, or will be observed at laboratory energies. So the whole subject is quite lively and controversial; no matter which way it is presented it retains elements of the fantastic and unbelievable. The strength of theoretical faith is significant, and is an interesting phenomenon itself, because pQCD has never explained the "real" strong interactions, as known and loved by nuclear physicists. The energy region of interest is the relativistic one because we still do not know how QCD acts in the low-energy region. However, we do not want extremely high energies: the region of 1-30 GeV is best. The goal is not to learn more about pQCD, but more about the real strong interactions using pQCD as a tool.

le passe–muraille in Dronten

We hope, in addition, to learn fundamentals of the strong interactions by having a second universe, namely reactions inside nuclear targets, to compare with the usual one. I will argue that this technique, of using the nucleus as a test medium to compare and contrast with the "control" reactions of hadrons in free space, may eventually be one of the most significant advances in understanding strong interaction physics ever. At the present time, color transparency plays a central role in the growing, broad subject of the interface of pQCD with the non-perturbative structure and interactions of hadrons. Ultimately the goal of this and other current study in this area is to solve the problem of the strong interactions. This is an ambitious goal, and one worth doing well.

The Basic Idea: Survival of the Smallest

Hadrons are made of quarks and gluons. The hadrons are asymptotic energy eigenstates, while the quarks and gluons are not. It follows that hadrons fluctuate between different quark and gluon configurations when we study them. To observe color transparency we need to measure a hadron when it has fluctuated to a configuration which will have small interactions.

Perturbation theory can tell us when the interactions do become small. It is important to note that this is self-consistent. Within perturbation theory, there are calculations[2] by Low and Nussinov (in QED) and Gunion and Soper (in QCD) predicting that a spatially small, zero-total charge system will have a small interaction cross section. The reason that the cross section is small is that the dipole moment is small, cutting off the amount of radiation exchanged. The scattering cross section $\sigma(b)$ for the region where two Fermions in neutral "atoms" are separated by a transverse position separation b is of order:

$$\sigma(b) \approx \sigma_{tot} b^2 / <b^2> , \qquad (1.1)$$

where σ_{tot} is the total cross section, and $<b^2>$ is an average b^2 in the states. Note that the transverse separation is the coordinate that matters. It is invariant under boosts along the direction of motion of a particle. So, if this picture is correct, for color transparency we need to somehow select configurations of quarks inside hadrons which have small transverse separation. From other phenomenology we know that such configurations are rare.

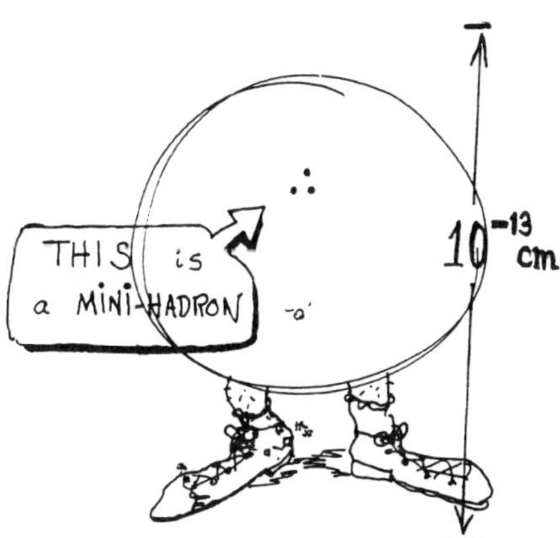

inside Mr. (or Mrs.) Hadron)

But (the theory story goes) such configurations are the important ones when we study hard exclusive reactions. ("Hard" means more than high energy; it means that every invariant momentum transfer Q^2, such as the Mandlestam variables s, t and u for $2 \rightarrow 2$ reactions, is big compared to a GeV^2.) "Exclusive" means that all the momenta of all participating hadrons are measured.) Again from pQCD, one can show that at large Q^2 the biggest contributions come from the minimal Fock space number projection of the hadrons: 3 quarks in a proton, 2 in a meson. Moreover, the typical quarks selected are separated by a transverse distance b that gets smaller as Q^2 increases, with the important integration region being $b^2 \leq 1/Q^2$. Since these are statements about short distance where pQCD works, this again is self-consistent. (Important caveats and details will be brought up later). Finally- and this is the idea known as "nuclear filtering" - if a configuration happens to be spatially large, it should have large inelastic interactions, and we will just reject it when we select the quasi-elastic data. So with large Q^2 and large A we have two excellent ways to select the rare, short distance components of quark wave functions.

A nuclear target now provides the ideal detector. Let us have a hard exclusive reaction right inside a nucleus, which will be standing there to attenuate the participating particles. It is "passive" if we think of it as measuring the event; it is "active" if it plays a role in giving us the configurations that we want. An example would be knocking a proton out of a nucleus with a fast incoming proton, the reaction $pA \to p'p''(A-1)$. A signal of color transparency is a reduced attenuation cross section in the nuclear target as Q^2 increases. Ideally, we would measure the cross section $d\sigma/dt$ at fixed cm angle as a function of Q^2, extract the nuclear attenuation rate, and see the typical "size" of the quark separations decreasing rapidly with Q^2. This summarizes the basic idea, as originally expressed by Brodsky and Mueller[3].

Backup!

In presenting the basic idea, I skipped over many essential details. In fact, there may be some details in the story on which not everyone agrees! Let me dig into some interesting questions:

What is meant by a hadron fluctuating in its configurations?

First let us define the configurations. A hadron state $|\, p,s >$ can be expressed as a superposition over a (gauge invariant) Fock space of quarks and gluons:

$$|\, p,s >=|\, qqq >< qqq \,|\, p,s > + \,|\, qqq\bar{q}q >< qqq\bar{q}q \,|\, p,s,> + \,|\, qqqg >< qqqg \,|\, p,s > +...$$

There is no pretense that any particular components of the Fock space are dominant here: we are just using a basis which is natural for the perturbative discussion. Note that the expansion is over all numbers of quarks and gluons and all their momenta.

Since a hadron is an energy eigenstate, its probability to look like any configuration is time independent:

$$|< qqq \,|\, e^{-iHt} \,|\, p,s >|^2 = |< qqq \,|\, ps >|^2$$

So the concept of a hadron fluctuating in free space does not apply. However, this is a meaningless probability, because we have not said how it was measured. When we look inside a proton to see what it is made of, we couple the proton to a perturbation. Then the time dependence of fluctuations between internal constituents becomes meaningful. The time scale of these fluctuations is unknown but it is determined by both the state and the interaction Hamiltonian. One can say then, maybe we are not looking only at the proton, but also at the process of measuring the proton. This is correct but also just the way quantum mechanics works. As long as we can arrange to understand how a proton can pass through a nucleus that it OK! It is a hint that quantum mechanics, and the selection of what will be measured, plays a big role in color transparency.

So, too, does the quantum mechanical basis play a role: theorists do not agree on which basis is the best to use. In perturbing the system to measure color transparency, it is quite sensible to consider the resulting system as a superposition over asymptotic hadronic states rather than the fundamental Fock space. In this basis, pQCD apparently predicts that some superposition of hadrons is made which has small strong interactions due to destructive interference. This is neither more nor less rigorous than using the Fock space, since in the hadronic basis one has to rely on a postulate of completeness. This postulate goes back to the days of the S-matrix; it has never been straightfoward how the collective coordinates of the hadrons are to reproduce

the internal coordinates, the quarks, in the system with confinement. The hadronic
viewpoint's advantage is that one has directly measured information about the interactions of hadrons. Its disadvantage is that the system has to be "tuned" into the actual
superposition that Nature chooses, a task which is not at all systematic.

*Why are exclusive processes emphasized? Couldn't we observe color transparency
in all kinds of reactions?*

Maybe, but probably not. Inclusive reactions sum over all kinds of channels and
all kinds of configurations of the quarks inside the hadrons. Inclusive processes are ideal
when we know nothing at all about hadrons, which was the situation in establishing
QCD. Theory work is trying hard to push the ideas of color transparency into inclusive
reactions. Perhaps some of the configurations are "transparent", but these are probably
an infinitesimal part of most cross sections.

Besides, exclusive processes are a wonderful frontier. At large momentum transfer, they involve projections of hadrons onto just a few quarks. They are difficult to
measure, but this is because there are many, many channels. These are exactly the
projections we are interested in learning about on the interface of non-perturbative
QCD physics.

*Just because the perturbative treatment becomes self consistent, does that mean
that one has to believe it?*

Certainly not. Many interesting questions - such as what happens to a hadron's
"cloud" of soft pions and/or soft gluons, the way it sits in the chirally broken vacuum, and the way a hadron reforms after being struck hard - are all questions beyond
perturbation theory. It is conceivable that perturbation theory will fail. But if color
transparency does not work, then we will have observed where perturbation theory fails
in a region where it might have worked, and learned something fundamental.

*Why are small transverse separations between quarks involved in hard scattering
reactions?*

This is a straightforward technical question, which will take many pages not to
answer simply. The true answer is: it is not always true. One must check on a
reaction-by reaction basis in perturbation theory to know when it is true. The best
known exceptions to having short distance at large Q^2 are the "independent scattering"
configurations, introduced by Landshoff[4]. These configurations play a major role in our
interpretation of existing data, but we believe from studying them that the "basic idea"
remains correct for the case of large enough nuclei with $A >> 1$.

Understanding when short distance occurs between quarks is a current research
problem. By studying it, we hope to learn more about the way the quarks themselves
are arranged. We do not really want always to have short distance. If different distances
are involved in different reactions, we can effectively take "snapshots" of the quark wave
functions to see how the hadrons are made.

Experimental Status

Before going into more technical matters, it is best to understand the experimental observables and what has already been measured. In this part, I will concentrate
on the experiment which has published its data. (The experiment (NE-18) performed
at SLAC has already measured proton knockout at the quasi-elastic peak in electroproduction, $eA \rightarrow e'p(A - 1)$, but the data has not yet been officially released. Many
other experiments are planned or are in progress at BNL, SLAC, CEBAF, Fermilab
and elsewhere. Color transparency plays a major role in discussions for new facilities
such as the European Electron Accelerator ELFE.) Skipping the details and history of
theoretical work for now, let us go straight to the nitty-gritty.

The first color transparency experiment was done at BNL by the group of Carroll,
et al.[5]. It is considered a pioneering experiment, which is being repeated this year with
much higher statistics and better instruments. A proton beam was used to study
$pA \rightarrow p'p''(A - 1)$. Data was taken simultaneously on 5 targets, namely ^{1}H (in the

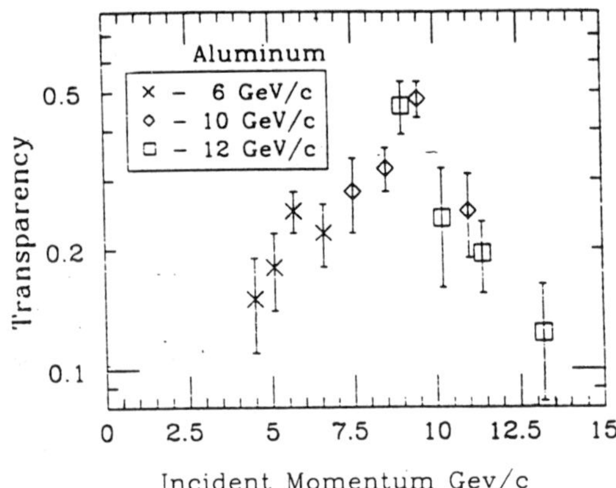

Incident Momentum Gev/c

Fig (1). The transparency ratio (Ref (5)) for the Aluminum target as a function of effective beam energy

form of polyethylene,), ^{12}C, ^{27}Al, ^{65}Cu and ^{207}Pb. The beam energies were 6, 10 and 12 GeV, although data was reported at 12 GeV only for Al and Cu . The data was measured at a quasi-elastic point of $90°$ cm scattering. (At this point, the data just depends on one scale $Q^2 = -t = -u = s/2$.) That is, the spectrometers were set at the correct angle and energy for an elastic collision, within the resolution of the instruments, and a veto was also placed on the nuclear targets to exclude events where a large amount of inelastic energy was deposited. In practice one is sure that no pions were produced but not sure what the excited nuclear state left behind is. The experiment[5] reported a ratio $T(Q^2, A)$ called the "transparency ratio", defined by

$$T(Q^2, A) = \frac{1}{Z} \frac{d\sigma(pA \to p'p''(A-1))/dt}{d\sigma(pp \to p'p'')/dt} \Bigg|_{\theta_{cm}=90°}$$

One reason for reporting the transparency ratio is that certain systematic experimental uncertainties in the data cancel out. It is often loosely said that the free space $pp \to pp$ scattering in the denominator scales like s^{-10}, providing evidence that the scattering of a limited number of point-like quark constituents (the "quark-counting" rules[22]) is taking place. Although this is not exactly true, the transparency ratio conveniently takes out a very rapidly varying function of the energy. The factor of $1/Z$ represents the expectation that the hard scattering is *incoherent*: the only interference that should occur at large momentum transfer comes from one proton being struck at a time. The transparency ratio is an observable quantity which takes into account a basic idea of *factorization*, which roughly means that the hard scattering effects multiply the soft nuclear interaction effects. A working, but not very precise theoretical expression of this is

$$measured\,rate(Q^2, A) = [hard\,scattering\,rate(Q^2)][survival\,probability\,(\sigma_{eff}, A)]$$

On the basis of factorization of the hard scattering from the effects of propagation in the nuclear medium, one *might* expect that the hard scattering rate would cancel out in the transparency ratio. *With this assumption* the transparency ratio would measure the survival of particles passing through the medium and should rise with increasing Q^2. *This is not a safe assumption*, and in fact the naive expectation did not occur; instead the published data show a rise, and then a decrease of the ratio for beam momenta above 10 GeV. This is shown in Fig (1).

Remark: The role of Fermi motion in the experiment is significant. We must allow for events at shifted kinematic points in the experimental definition of "quasi-elastic". Fermi motion can also create a sizable shift in the effective cm energy-squared s. Let p' be the 4-momentum of a proton at rest. Including Fermi momentum k_F, the initial momentum of a struck proton is $p' + k_F$. Calculate the cm energy-squared s for a collision of this object with the beam carrying momentum p_b:

$$s = (p_b + p' + k_F)^2 \cong 2m^2 + 2E_b m + 2p_b \cdot k_f.$$

For p_b in the region of 10 GeV and k_F about 0.3 GeV! the change in s from its value if the target were at rest is about 3 GeV! A small change in the value of s can make a big difference in the cross section. An experimental cure for this problem is to measure enough kinematic variables to determine the Fermi momentum of each struck proton, which was done in the BNL experiment. Thus Fig (1) shows data for values of the effective beam energy which range around the values actually used, as the Fermi motion effects were incorporated by the experimentalists. A weakness in the procedure is that one must know the Fermi momentum distribution: the experiment used its own measured out-of plane momentum distributions as well as standard models for this. Needless to say, the role of the Fermi motion may be more subtle than this and it has received much theoretical attention.

The data, continued: the rise and fall of the transparency ratio with increasing energy at first caused a great stir. By now it is clear that the structure of the ratio with energy is due to the process of making a division by the free-space $pp \to pp$ data. That data fall roughly like $s^{-9.7}$ but also are modulated by oscillations with the logarithm of the energy (Fig (2)).

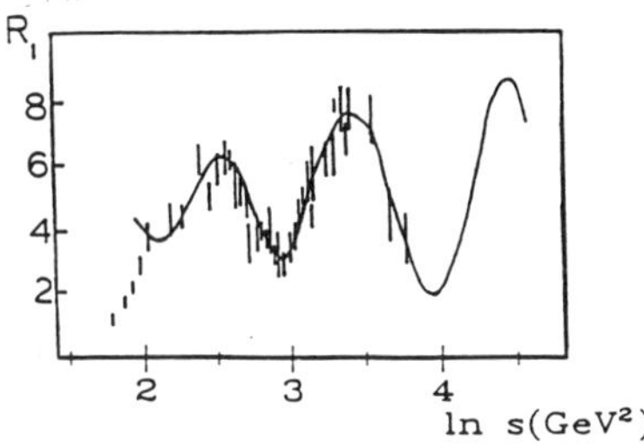

Fig (2). Oscillations in the elastic pp scattering cross section $R = s^{10} d\sigma/dt$. From Ref. (6).

The oscillations are rather clear evidence for quantum mechanical interference of amplitudes. Apparently the numerator of the transparency ratio - which is the nuclear scattering - has little if any oscillations, while the denominator has oscillations, so that the *ratio* must oscillate. Here is the evidence for this: consulting Fig. (3), the rise in the transparency ratio is exactly correlated with a falling oscillation of the free space data, indicating that the nuclear target is showing reduced oscillations.

Making a plot of the nuclear differential cross section by itself, *with no division by the free space denominator* but taking out a factor of s^{-10}, one sees that the oscillations, if present at all, are much reduced in the nuclear target (Fig (4)).

Evidence for Filtering

We have strong reasons for believing we understand the oscillations in the free space data, as part of a QCD phenomenon we have studied for more than 10 years[7,8]. But temporarily setting aside the explanations proposed for the oscillations, *if* there is interference *then* the free-space $pp \rightarrow pp$ amplitude has not settled into its asymptotic form. It must be different from the processes occurring inside the nucleus, which do not show the same oscillations. Something interfering in the free space $pp \rightarrow pp$ process has apparently been killed in the nuclear target. There are two explanations proposed for this:

Brodsky and de Terramond[9] propose that there are several charmed dibaryon resonances in the region of interest, which are eliminated in the nuclear case. The states have not been observed elsewhere but the value of the threshold kinematics

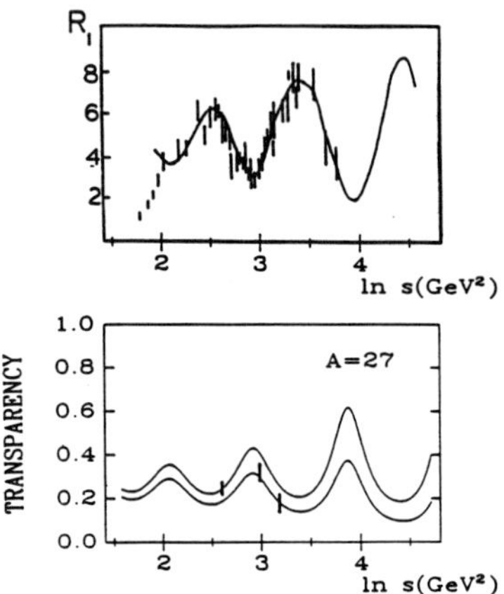

Fig (3). Comparison of oscillations in the pp elastic differential cross section (top) with the transparency ratio of Al (bottom) on the same energy scale. Data points re-binned in Ref. (5). Curves from Ref. (10).

do coincide with charmed quark masses. By adjusting several parameters the shape of the differential cross section and the spin analyzing power A can be reproduced. Presumably the charmed states could be observed to test the idea. It could also be applied to other reactions.

Ralston and Pire[10] propose that the free space $pp \rightarrow pp$ scattering consists of roughly two perturbative QCD regions, which have an energy dependent "chromo-coulomb" phase difference[7,8]. One of the regions corresponds to "large"quark configurations which would be filtered away in the nuclear medium, leading to disappearance of the oscillations in the nucleus. By proposing that the oscillations are reduced for A>>1 the energy dependence of the BNL transparency ratio is reproduced with no free parameters. The fact that such energy dependent phases occur in QCD has been confirmed by Sen[11] and Botts and Sterman[12]. Spin observables in this model of the free-space scattering have been fit by Ramsay and Sivers[13], and Carlson, Myrher and Chashkuhnashvilli[14]. However, the fundamental parameters needed to fit the data have not yet been obtained from the (quite complicated) pQCD calculations.

The fact that there are "large" configurations in the hard scattering description of pQCD was not widely appreciated until recently. Such processes, which involve independent scattering of quarks at separated scattering centers, simply do not obey the assumptions of the "basic idea" presented earlier. Nevertheless, independent scattering occurs in the perturbation theory and must be dealt with to understand color transparency. Both of the explanations above agree on the proposal that the structure in the transparency ratio occurs due to dividing a smoothly varying, somehow filtered numerator by an oscillating denominator. Early fits to the data involving an expanding, time-evolving system by Farrar, Liu, Frankfurt, and Strikman (FLFS)[15] not incorporating the structure of the denominator have been ruled out - assuming the data is correct. Nevertheless, that paper introduced ideas which have had substantial impact, and will be discussed below.

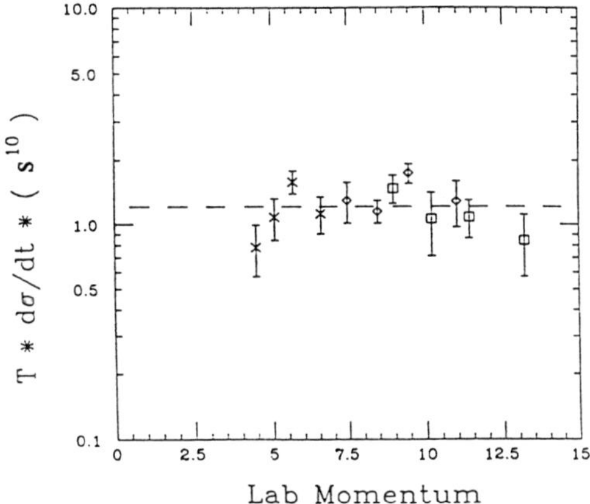

Fig (4). The nuclear differential cross section obtained from $s^{10}T(Q^2, 27)\, d\sigma/dt(pp \to pp)$ for the Aluminum target. From Ref.(6).

These interpretations of the data have been with us for several years and have certain modifications and refinements have been studied. In a series of papers with various assumptions, Jennings and Miller[16] and also Kopeliovich[17] and have studied both the idea that the system can be described by interfering hadronic states and hybrids of combining the Ralston and Pire and the FLFS idea. N. Nikolaev and collaborators have made great efforts to incorporate color transparency into the traditional theory of diffractive processes. In a recent comprehensive review to the diffractive approach, Nikolaev[18] favors the Pire–Ralston explanation but argues that the experiment is kinematically flawed in coupling Q^2 and energy at fixed–angle kinematics. Anisovich et al[18] have made a careful study of the relative importance of filtering incorporating the details of the small size configurations. Miller and Lee[19] and Benhar et al[20] examined the effects of nuclear correlations. Some studies conclude that the approach of Ref. [10] describes the data but seems too simple; adding more complicated nuclear effects can make the agreement with data worse. Generally the studies agree that if an interfering component is blocked in the nuclear target, we should expect oscillating transparency.

An Impasse. Conceptually the appearance of an oscillating transparency ratio caused somewhat of an impasse. One lost confidence in the interpretation of the ratio. We know that the experiment on the nuclear target measures a combination of the hard scattering rate and the nuclear attenuation. If we honestly admit that the nuclear hard scattering rates are unknown, how can we extract a signal of color transparency from the nuclear attenuation? Suppose, for example, that one overestimated the nuclear hard scattering rate- could this simply be compensated by an corresponding underestimate of the nuclear transparency?

Experimentally Defining Color Transparency: Extracting an Attenuation Cross Section

The problem of measuring the nuclear attenuation is similar to extracting the total inelastic cross section from an experiment on a target when the "beam flux" is unknown. It can be done in an almost model-independent manner, but this was only realized recently. We turn now from discussing the data itself to a new, systematic procedure to interpret the data[21]. We believe that it will allow color transparency to be defined as a relation between observables in an almost model-independent manner.

Consider $A >> 1$, so that whatever filtering of the hard scattering to short distance that is occurring has set in. (The radius of a nucleus is about 1.2 $A^{1/3}$ fm. If the cross section is incoherent across the nuclear volume, then most events will have gone through quite a few fermi with soft interactions before having a hard scattering.) For experimental purposes, the impulse approximation can be used to separate mathematically the hard scattering and the process of propagation through the nuclear medium:

$$measured\ rate(Q^2, A) = [scattering\ rate(Q^2)][survival\ probability\ (\sigma_{eff}, A)] \quad (1.2)$$

Here by "measured rate" we mean data for either a cross section or a transparency ratio. If one studies the transparency ratio, then "scattering rate" is actually the *ratio* of the hard scattering rates in the nuclear target to the analogous free space scattering. On the other hand if (1.2) is applied to a cross section on a nucleus then "scattering rate" actually equals the hard scattering rate in the nuclear target.

We observe that the A dependence can be experimentally isolated by taking the logarithm of (1.2) at fixed $Q^2 = Q_o^2$:

$$log\{measured\ rate\ (Q_o^2, A)\} = log\{survival\ probability\ (\sigma_{eff}, A)\} + log\{N(Q_o^2)\}.$$

Here $N(Q_o^2)$ is a term independent of A: it is the log of some hard scattering rates. Whatever this constant is, the attenuation cross section σ_{eff} is the parameter which determines the shape of the survival probability as A is varied. *Thus to determine σ_{eff} we want a fit to the shape of the A dependence of the right hand side, but a fit only up to an additive constant.* In practice this is quite easy. For example, plot the measured rate as a function of A on log-scale paper: adding an overall constant is just a rigid translation of the fit on the paper. After this is done, σ_{eff} is known at $Q^2 = Q_0^2$.

Relating the shape of the survival probability to σ_{eff} requires a theoretical model. This model dependence has nothing to do with color transparency: it is simply a question of defining how a cross section is used. On the other hand, in certain models of color transparency even the very general factorization (1.2) does not follow - more on these later. We are interested in an experimental procedure here, and will continue with how a cross section is extracted.

After determining the survival probability, the $N(Q^2)$ values contain the information about the hard scattering rate. The interpretation depends on whether (1.2) is applied to the transparency ratio or a cross section on the nuclear target. If "measured data" means the transparency ratio then one extracts the ratio of scattering in the nuclear target to the free space rate. If "measured data" is a cross section then

one finds the scattering rate in the nuclear target directly. No normalization needs
to be set using isolated hadron events, nor is color transparency being determined by
comparison with a theory. This is progress: we can use the data itself to learn about
the differences between the nuclear and free space cases.

The procedure is very general, and we expect it to be very useful. Now let us
apply it to the data of Carroll *et al*[5] as an example. What is found is shown in Fig (5).

The plots represent the fit of the BNL experimental data to unknown functions
$N(Q^2)$, $\sigma_{eff}(Q^2)$ with:

$$s^{10}\frac{d\sigma_A/dt}{Z} = N(Q^2)P_{ppp}(\sigma_{eff}(Q^2), A) \tag{1.3}$$

The factors of s^{10} and $1/Z$ are simply definitions to take out some typical orders of
magnitude - their usage introduces no bias. Since $\sigma_{eff}(Q^2)$ and the normalization
$N(Q^2)$ are free, the fit to the data may or may not show that these parameters vary
with Q^2. Of course, finding a variation with energy of σ_{eff} was the goal of color
transparency, but at this stage the fitting is an objective, empirical procedure.

Disregarding the overall normalization, the attenuation cross sections extracted
from the *shape of the data as A increases* reveal a definite and statistically significant
Q^2 dependence, shown in the table below. The two beam energies of 6 and 10 GeV^2
correspond to $Q^2 = -t = 4.8 GeV^2$ and $8.5 GeV^2$, respectively. The intermediate values
of Q^2 were provided by the Fermi motion variation reported for the Al target. The
best χ^2 fit to the shape of the nuclear cross section is obtained with:

Table 1

$$\sigma_{eff}(E = 6\ GeV) = 17 \pm 2mb; \quad N(E = 6\ GeV) = (5.4 \pm 0.4)\zeta: \chi^2 = 0.28$$

$$\sigma_{eff}(E = 10\ GeV) = 12 \pm 2mb; \quad N(E = 6\ GeV) = (3.3 \pm 0.4)\zeta: \chi^2 = 0.53$$

$$\sigma_{eff}(Q^2) \cong 40mb(2.2GeV^2/Q^2) \pm 2mb \tag{1.4}$$

where $\chi^2 = \Sigma[(y_i - d_i)/\Delta d_i]^2$, d_i are the data points, Δd_i is the error in d_i and y_i
are the theoretical values, and $\zeta = 5.2 \times 10^7\ mb\ GeV^{18}$ is a constant containing the
overall normalization of the cross section. The low χ^2 values indicate rather good fits
to five data points. Corresponding to these values are the survival probabilities shown
in Fig (6). (The fit does not change significantly if the data point of Lithium is deleted:
$\sigma_{eff}(E = 6\ GeV) = (19c \pm 2)\ mb$ and $\sigma_{eff}(E = 10\ GeV) = (13 \pm 2)\ mb$ in that case.
The two targets measured at 12 GeV were not sufficient to be useful.)

The rise with Q^2 is clear evidence for observation of color transparency in the
experiment.

Hard Scattering in the Nuclear Medium

If the survival probability shows such an increase with Q^2, why was it not immediately obvious in the nuclear data itself, as illustrated in Fig (4)? The answer is that
this data is the combination of the survival probability and the hard scattering rate,
and we did not know the hard scattering rate. If anything there was a prejudice that
it should fall like s^{-10}, because this was (sort of) expected in pQCD from the quark-
counting rules (reviewed below). The hard scattering rate found from the fit falls faster
than s^{-10} and at first seemed impossible. After puzzling over this, we realized that one
should really compare the nuclear hard scattering rate to the pQCD prediction,

$$s^{10}ds/dt_{pQCD} = (\alpha_s(\mu^2/\Lambda_{qcd}^2))^{10}f(t/s) \tag{1.5}$$

Although α_s varies only logarithmically, the factor of α_s^{10} is numerically equivalent to
about 1-1.5 inverse powers of s in this energy region. The scale μ^2 of α_s should be of

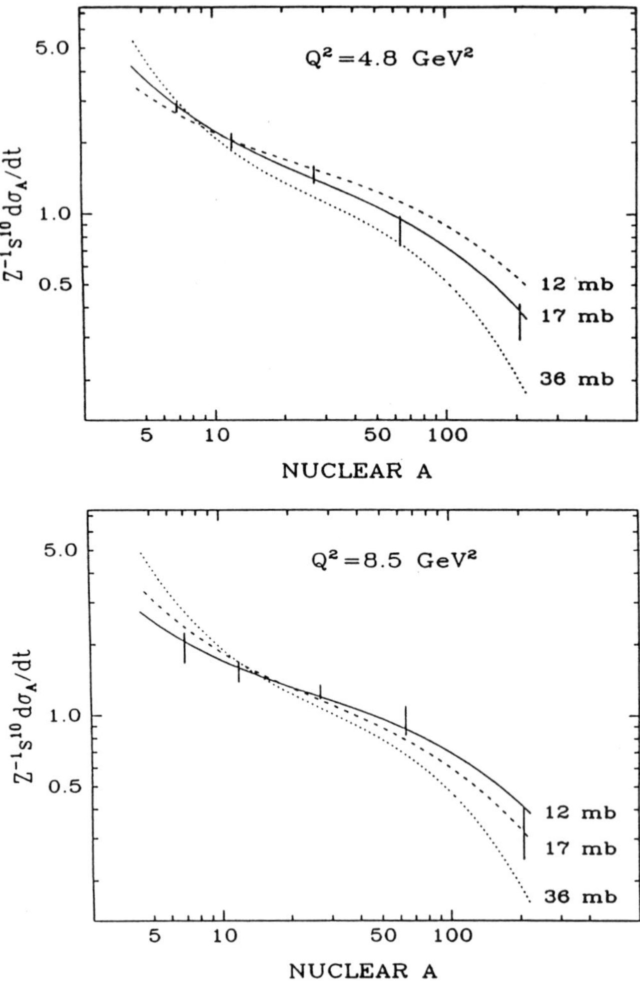

Fig (5). Fitting the A dependence of the nuclear cross section with attenuation cross sections σ_{eff} = 12, 17, and 36 mb at fixed momentum transfer. (a) $Q^2 = 4.8$ GeV^2. (b) $Q^2 = 8.5$ GeV^2. The solid line shows the best fit. The dashed lines show constrained fits using permutations of σ_{eff} and best normalizations.

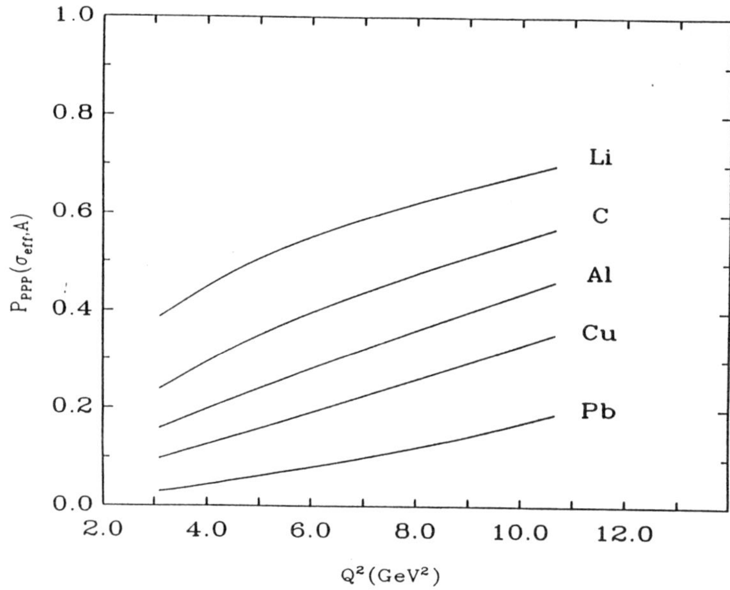

Fig (6). The survival probabilities P_{ppp} extracted from the fit to the BNL data.

order Q^2 but can never be determined until next-to-leading order calculations are done. Thus we used $\mu^2 = 0.5\ Q^2$ and $\mu^2 = 1.5\ Q^2$, producing two curves for comparison: a reasonable value of μ^2 should be somewhere in between. The results are shown in Fig. (7).

No editing of the data has been done: it is clear that if some low-Q^2 data were removed, the lines would be quite flat, indicating beautiful agreement of the data with the pQCD expectation listed above. This is quite a spectacular result: although the free space data has complicated oscillations, the data after "going through a nuclear filter", seems to obey the simplest theoretical expectations.

Preliminary Results on Electroproduction

Before leaving the "experimental status" section, electroproduction and the SLAC experiment should be mentioned. The experiment will report measurements of $eA \rightarrow e'p(A-1)$ on the nuclear targets ^{1}H, ^{2}D, ^{12}C, ^{56}Fe and ^{197}Au, at Q^2 values of 1, 3, 5, and 6.8 GeV^2. The microscopic process is the absorption of the virtual photon by a proton, which subsequently recoils, interacts with the nuclear medium, and is then detected at the quasi-elastic point. If the experiment measures $d^6\sigma/d^3p\,d^3k$, where p and k are the outgoing proton and electron momentum, then the Fermi motion of the struck particles (modulo interactions) can be divided out by knowing the nuclear spectral function.*

This kind of experiment is beautifully complementary to the hadron beam experiments. The initial state of the large Q^2 virtual photon is very clean and well understood: one is measuring the photon's absorption and final state interactions in

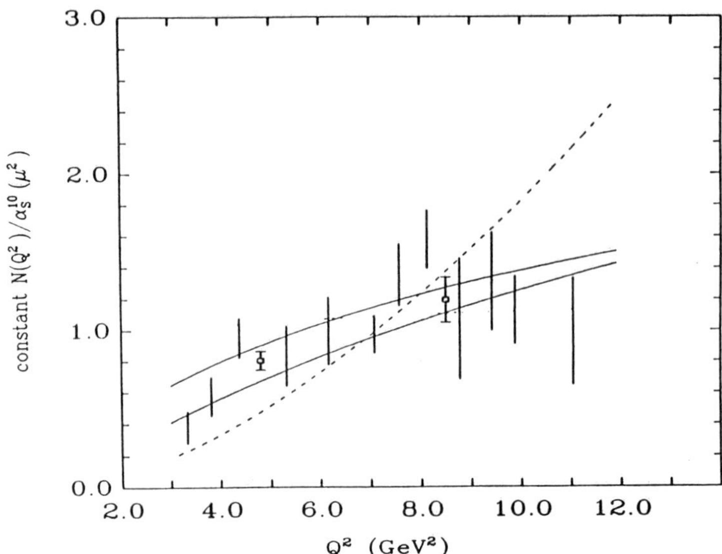

Fig (7). Comparison of the hard scattering rate to pQCD including effects of $\alpha_s(\mu^2)$; flat indicates agreement. The normalization factor $N(Q^2)/(d\sigma/dt_{pQCD}$ (Eq 1.5) is plotted versus Q^2. The upper (or lower) curve uses $\mu^2 = 1.5\,Q^2$ (or $0.5Q^2$). The dashed line indicates disagreement of the nuclear hard scattering rate with $s^{-9.7}$.

the nuclear target. There is tremendous interest in this and other experiments which could measure meson knockout, transition form factors, etc. A clean theoretical matrix element is measured in the experiment:

$$< p, A - 1 \mid J_{em}^\mu \mid A >$$

However, the experiment is somewhat insensitive to the attenuation cross section of the system crossing the nucleus, so that comparatively high precision data may be needed to extract an attenuation cross section. This is shown in Fig. (8), which shows some predicted survival probabilities for several different attenuation cross sections σ_{eff}. Even with cross sections varying by a factor of about three, the curves on the log-plot are almost parallel. Even if there is a decreased σ_{eff}, the effects of color transparency (if any) might be masked by a simple change in the normalization of the event.

We have already seen that such a normalization is equivalent to the hard scattering rate, so it is tricky in electroproduction to separate the ambiguities of hard scattering inside the nucleus from the extraction of σ_{eff}.

To understand this in more detail one can make an analytic calculation. Assuming a uniform nucleon number density $n = 1/6fm^3$ and nuclear radius $R = 1.2fmA^{1/3}$, one can find the survival probability P_p of one proton taken at random over the volume of a nucleus and crossing on a straight line with exponential attenuation. It is:.

$$P_p(\sigma_{eff}, A) = P_p(A^{1/3}\sigma_{eff}) = \frac{3}{4y} + \frac{3}{4y^2}e^{-2y} - \frac{3}{8y^3}(1 - e^{-2y}) \; ; \; y = \frac{R_A}{\ell} \cong 1.2(A^{1/3}fm)$$

$$(1.6)$$

---

* This is a legitimate worry. The spectral function may depend on Q^2, and its normalization has historically been a sticky point of nuclear theory. Its value at larger values of k_f is not well established.

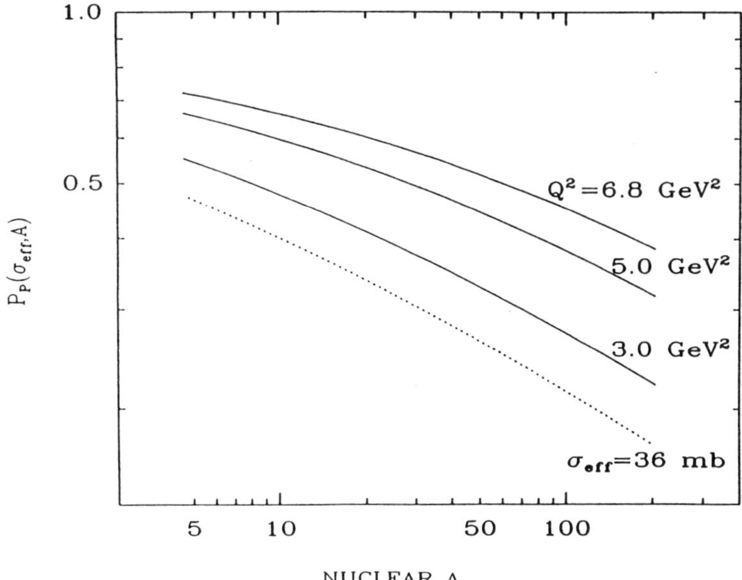

Fig. 8. Plots of P_p, the survival probability in electroproduction, as a function of A for various Q^2. The plots also show the effects of using different values of $\sigma_{eff}(Q^2)$, which are evaluated using Table 1.

(For an exercise, show this.) This formula was used to generate Fig. (8). Even though there is an integration over a sphere, this formula depends on only one variable $y = R_A/\ell$, which is the nuclear radius $R_A = 1.2 fm A^{1/3}$ in units of the mean free path $\ell = 1/n\sigma$. In the limit $\ell >> R_A$ we have 100% survival. In the limit of $\ell << R_A$ the problem is quite simple. Particles survive if they come from a thin skin of depth ℓ on the "back side" of the struck nucleus. The volume containing such survivors is $\pi R_A^2 \ell$; the area πR_A^2 occurs because we want the projected area along the beam direction. The survival probability, in this limit, should be the ratio of the volume containing survivors to the total volume:

$$P_p(\ell << R_A) = \frac{\pi R_A^2 \ell}{4\pi R_A^3/3} = \frac{3}{4n\sigma_{eff}A^{1/3}}$$

Now the ambiguity in examing data in this case is brought out by considering a transparency ratio

$$T(Q^2, A) \sim \frac{N_A(Q^2)}{N_1(Q^2)} P_p = \frac{3N_A(Q^2)}{4n N_1(Q^2)} \frac{1}{A^{1/3}\sigma_{eff}(Q^2)} ,$$

where $N_A(Q^2)$ and $N_1(Q^2)$ are nuclear and free–space hard scattering rates, and we indicate the possible Q^2 dependence of $\sigma_{eff}(Q^2)$ explicity. The left–hand side of the equation above is measured. But in the limit chosen there is no model–independent way to extract $\sigma_{eff}(Q^2)$! That is because there are two unknowns, and we can let

$$N_A(Q^2) \to \lambda(Q^2)N_A(Q^2) \; ; \; \sigma_{eff}(Q^2) \to \lambda(Q^2)\sigma_{eff}(Q^2)$$

where $\lambda(Q^2)$ is any function, and still fit the data.

The way out is to use the A dependence to vary the target thickness over a reasonable range of ℓ/R_A. With such a procedure the BNL data was fit using the shape of the

A–dependence, and resolving the ambiguity above. The same procedure can be done in electroproduction, but it is more difficult. One can compare Fig. (8) to the survival P_{ppp} of *three* particles in a nuclear target, all having the same exponential attenuation law, averaging over the volume, and with the 90° kinematics of the BNL experiment. To a surprisingly good accuracy, the survival of three particles is proportional to the cube of the survival of one, for $A \gtrsim 7$. (The empirical proportionality constant is given by $P_{ppp} \cong 1.9 P_p^3$.) Because of this cube effect, *the curvature in the A-dependence* in the (p, 2p) experiment is easy to pick out. But in electroproduction a high probability of survival occurs because the attenuated "particle" only has to cross a small part of the nucleus once. And this means that the experiment does not necessarily have a high discriminating power in measuring the attenuation cross section, unless the data is of comparatively high precision.

Several conference presentations of the preliminary data[23] report that the ratio of the SLAC NE-18 data as a function with Q^2 to a plane wave impulse approximation show only a very mild increase with Q^2. Plots of the A dependence of fixed Q^2 do seem to rise systematically with increasing Q^2, but it is not a very big effect. *If the transparency ratio used is interpreted to measure the survival probability, then the experiment shows only a weak signal for color transparency. However, we have already seen that the transparency ratio cannot generally be interpreted as a survival probability.* If the SLAC data becomes good enough to extract an attenuation cross section along with a hard scattering rate, then we suggest that would be the preferred procedure. But it is not yet clear whether the data will be of this quality. Without this, the experiment can only determine a curve relating the hard scattering rate (which is unknown) and the survival probability (which is unknown). This is useful information, but does not say whether or not color transparency is being observed.

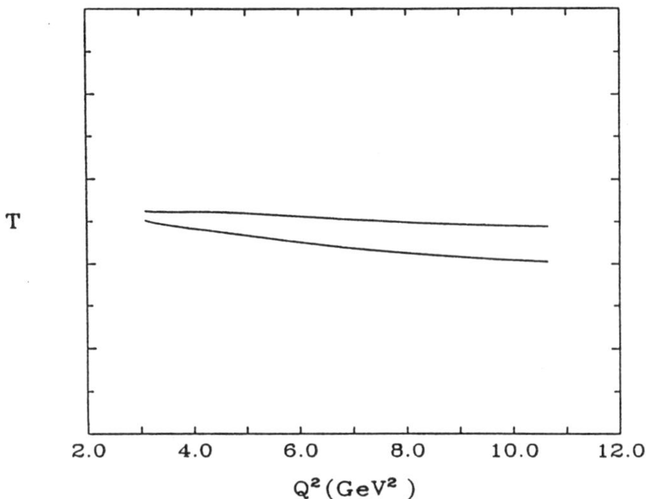

Fig (9). A transparency ratio for electroproduction made with $\alpha_s(\mu^2)^4 P_p(\sigma_{eff}(Q^2), A)$ evaluated for the Iron target and using $\sigma_{eff}(Q^2)$, from the BNL data. The range between the lower and upper curve uses $0.5Q^2 < \mu^2 < 1.5Q^2$. Vertical scale depends on overall normalization of hard scattering rate in nuclear medium which has not been specified.

Remark: Theoretically, the description of the hard quasi-elastic scattering in electroproduction begins with exclusive form factors: the elastic form factor, and transition form factors to whatever states end up converting to protons after travelling through the nucleus. There are plenty of reasons to believe that the electromagnetic form factors used to describe the process in a nuclear targets may not be the same ones as in free space. The reason: the free space processes at laboratory values of Q^2 are

not truly short distance dominated. Isgur and Lewellyn-Smith (ILLS)[24] have made
the case for this, claiming that *none of the perturbative arguments apply at all.* But
if this is so, how can one explain the many regularities of power law behavior which
are observed in many experiments? It would be unscientific to ignore this qualitative
success! More recently, Li and Sterman and Li[25] have shown that the ILLS arguments
focus on an integration region that does not occur. In the new estimates, less than
about one-half of the amplitudes in free space scattering could come from dangerously
large quark separations.*The new ingredient is the appearance of Sudakov effects in the
hard scattering form factor. Would these offensive regions be filtered away in a nuclear
target? Nobody yet knows, but our calculations[6,7,21] indicate that it should happen.
We would then see a repeat of the previous phenomenon: for experiments inside the
nuclear target there would be a more pure, short distance form factor, and α_s should
make its appearance in the naive way for the hard scattering.

This is an audacious speculation but we tried it anyway[21]. From the elastic form
factor and quark-counting diagrams the hard scattering scales like α_s^4/Q^4. The free
space form factor scales like $1/Q^4$. Taking the attenuation cross sections (Table 1)
extracted from the BNL data, calculating the P_p survival factors with the exercise
above, and multiplying by $\alpha_s^4(\mu^2)$, we produced the plot in Fig (9) for the Fe target.

This is not a real transparency ratio - for example, we are not sure which observable ratio the experiment will report until the experimental report is released! But it
shows that the reports of the NE-18 data are quite consistent with the overall picture.
Optimistically, maybe we have another indication that QCD has become cleaner when
being observed after nuclear filtering.

Mini-Summary

I have presented the basic idea of color transparency, and the experimental situation. Experimentally the subject is becoming well defined. Color transparency is likely
to remain with us for a long time as relations between observables, and in hadron–
initiated reactions as a big enough effect to be robust when examined in different
theoretical models. With the systematic data analysis procedure, one can use all of the
information in the A and the Q^2 dependence to extract an attenuation cross section.
Nuclear attenuation effects can be conceptually separated from the effects of the hard
scattering rate. The BNL experiment has been subtle to interpret, but it definitely
shows evidence for the observation of color transparency.

I have not yet presented the theoretical basis of color transparency in any detail,
but in the next section I will show that its foundations are as good or better than the existing pQCD treatments of hard exclusive processes. The worries there are quantitative:
are the energy and Q^2 scales in the laboratory large enough so that the approximations
apply? Do we understand soft quark scattering well enough to make decent approximations? What quantum mechanical basis is the best for the non-perturbative part of
the calculation? What are we supposed to use for quark wave functions in hadrons?
These are interesting and reasonably fundamental questions.

Foundations have been put down to make color transparency both experimentally
and theoretically well defined and promising. Nevertheless, there is still much work to
do in both areas!

II. THE THEORY OF QUANTUM COLOR TRANSPARENCY

The Impulse Approximation for the Hard Scattering

We will now go into greater theoretical detail in the theory of color transparency.
The original arguments are based on the physics of factorization, which is a mathe-
matical way of imposing the impulse approximation. I want to separate the way this is

---

* This may sound like a negative conclusion but it is an improvement on ILLS who claim that
not even a tad of the amplitude is ever short distance.

used in QCD from some other uses. We view collisions as a three step process: the soft, long-time dynamics before the hard scattering, the hard scattering itself, and the long-time dynamics after the hard scattering. Perturbative QCD is a model for the hard scattering of the quarks. The impulse approximation separates this from the other dynamics, without saying what those other processes are. This makes color transparency interesting: it combines the perturbative and the non-perturbative. And, by no means are we going to claim that the interaction is so fast that a particle automatically escapes the nucleus! *What we want is to separate the nuclear effects from the perturbative ones.*

The impulse approximation is actually a feature of classical Hamiltonian dynamics. A classical particle's state is represented as a point in the classical phase space $(q(t), p(t))$. If at a time $t = 0$ the system's Hamiltonian changes very suddenly, then the particle starts at $(q(0), p(0))$ and just time evolves according to the new Hamiltonian. In quantum mechanics in the Schroedinger picture the dynamical phase space coordinates of the system are given by the wave function.* So if the old wave function for time $t < 0$ is given by $| s(0) >$, and if for $t > 0$ we suddenly turn on a perturbation, then the time evolution of the new state is given by

$$| s(t) >= \sum_n e^{-iE_n t} | E_n >< E_n | s(0) >$$

where $| E_n >$ are the energy eigenstates of the new Hamiltonian. The probability that a particular transition occurs in the impulse approximation is simply given by wave function overlaps from the initial to the suddenly perturbed system. Note that we do not have to follow the details of time evolution as the sudden perturbation is switched on, so long as this change is much faster than the time scales already present. It would be a mistake, for example, to assert that if the perturbation is turned on in a short

our ancestors

time Δt then the system after the perturbation will have energies ranging up to $1/\Delta t$ by the uncertainty principle. (This would be a misapplication, because if we really looked at the wave packet we would see almost no high energy components since it did not have time to change at all.)

---

* This is not just an analogy, but literally true. See Ref. 26.

This wonderful trick from our ancestors is a foundation[†] of perturbative QCD. It allows theorists to be proud of their ignorance about a hadron's quark–basis wave functions. Instead, one can just make a list of what the wave functions might be, and arrange to have the experiment *measure* the wave functions. Or, if useful, one can model the wave function, but in a way that has been separated from the more fundamental issue of factorization itself.

Wave Functions

For example, let us make a list of the minimal quark wave functions for the simplest case, a pion. The wave function to find a quark-antiquark pair in a pion state $| \pi(p) >$ with momentum p can be expressed thus:

$$< 0 | T[\psi_\alpha(b)\bar\psi_\beta(0)] | \pi(p) >= \{A\rlap{/}{p}\gamma_5 + B[p,\rlap{/}{b}]\gamma_5 + C\rlap{/}{b}\gamma_5 + D\gamma_5\}_{\alpha\beta} \qquad (2.1)$$

Technically one can write a path ordered exponential "string" of gluon fields between the quarks to make this gauge invariant; in a light cone gauge, this string can be arranged to be the unit operator. The list (2.1) is trivial: nothing was used except Lorentz invariance, parity, etc. Now assume that the pion is moving fast, namely its momentum $p >> GeV$. The first two wave functions (A, B) are the big ones by "power counting" - namely, they come from the parts of the quark Dirac operators which get big in a boost, and scale with the big number p. The list is also organized by powers of b; it makes it easier to pick out the short distance components. The C and D wave functions are kinematically sub-leading. (For an exercise, build your own vector meson such as a ρ: make a list for the (8) Lorentz covariant vector meson wave functions. For overkill, try it for the proton. Another exercise: classify the wave functions under spin and orbital quantum numbers.)

Momentum Flow

Usually calculations are done in momentum space. One sets up a coordinate system oriented locally along each particle's momentum, which is large in the "+" direction. My convention is $p^\pm = (p^0 \pm p^3)/\sqrt{2}$; then the invariant dot product is $A \cdot B = A^+ B^- + A^- B^+ - \vec{A} \cdot \vec{B}$. With these variables the space coordinate $x^+ = (x^0 + x^3)\sqrt{2}$ acts like "time", which is conjugate to the momentum variable k^-. The Feynman x variable measures the + component of momenta with quark $k^+ = xp^+$. The relation of the wave function in transverse momentum k_T and transverse separation space b is simply a Fourier transform:

$$\psi(k_T, x) = \int d^2 b \, e^{ib \cdot k_T} \tilde\psi(b_T, x) = \int d^2 b \, e^{ib \cdot k_T} \int dx^- e^{ixp^+ x^-} \tilde\psi(b, x^-, x^+ = 0)$$

One can also make an expansion[*] in L_z, the orbital angular momentum of the quarks about the particle's momentum direction, according to

$$\tilde\psi(b, x) = \sum_m e^{im\varphi} \tilde\psi_m(| b |, x) \qquad (2.1\,1/2)$$

Note that by continuity and differentiability a wave function with cylindrical orbital angular momentum m must go like b^m as $b \to 0$; this will be important later. A priori, in the high energy limit we know nothing about the wave functions except symmetry properties, so the A and B terms are on an equal footing. For instance, it is entirely unjustified to assume from non-relativistic quark models that the pion would

---

[†] Demonstrating that the impulse approximation is self consistent in QCD is far from trivial, due to the presence of almost every kind of mixed ultraviolet and infrared divergence, but it has been done in "factorization" proofs.

[*] While expanding in the cylindrical orbital angular momentum about the z-axis is useful, it would be a bad idea to expand in full spherical harmonics, because they are not invariant under Lorentz transformations along the particle's direction of motion. Some good quantum numbers are Feynman x, the L_z projection, and the quark helicities.

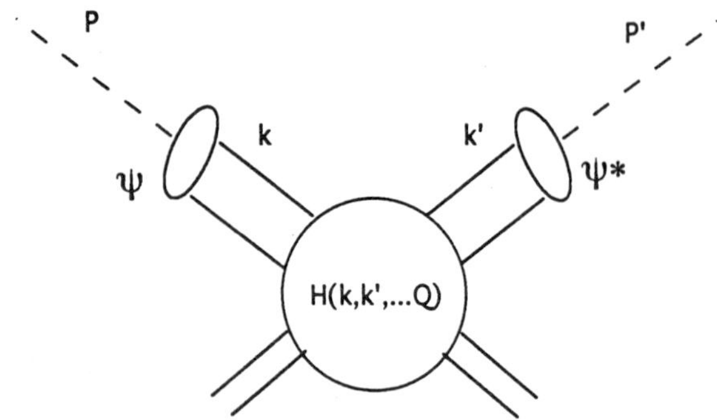

Fig (10). Schematic Feynman diagram showing separation of hard scattering and wave functions

be dominated by the "cylindrical s-wave" ($m = 0$) component, since non-relativistic quarks are some kind of quasi-particles and not the same as light-cone quarks. The normalization of each component of the wave function is also a dynamical question. The total of the probabilities to find a $q\bar{q}$ pair is supposed to be less than unity, so as to leave room for all the other multi-quark and gluon components! One thing from experiment that is known is the normalization of the pion wave function from pion decay:

$$ip^\mu f_\pi = <0 \mid \bar{\psi}(0)\gamma^\mu\gamma_5\tau_+\psi(0) \mid \pi(p) > = \frac{1}{4}Tr[(\gamma^\mu\gamma_5) < 0 \mid \psi(0)\tau_+\bar{\psi}(0) \mid (p) >]$$

This is a solid number because to a very good approximation the W-boson only couples to a pair of quarks.

Let us now see how the impulse approximation separates the hard scattering from the rest of the dynamics. In Feynman diagrams the configurations of the system to be summed over are represented by momentum space integrals. A step where the impulse approximation is used is in the integration over each particles respective k^- when it occurs in loops. Since the light-cone "time" coordinate is x^+, then integrating over k^- sets the internal times of the system to a value, namely *zero*, as previously indicated.

With this in mind, consider the diagrams for a hard exclusive reaction shown in Fig. (10). Let the scattering amplitude[*] be called M. Following the Feynman rules, we have a $d^4k = dk^+dk^-d^2k_T$ integral to do for every loop. The momentum k flows through the wave functions, and into the perturbative part of the diagram (H) where hard gluons are exchanged.

Schematically the diagram can be written

$$M = \int d^4k_1 \ldots d^4k_f \psi^*(k_f) \ldots H(k_1 \ldots k_f, Q)\psi(k_i) \tag{2.2}$$

Now, if the impulse approximation can be used we can make a power series expansion of the hard scattering kernel H in powers of the $\psi(k_1)k^-$ and integrate each wave function over its k^-. This leaves the transverse momentum integrals. *IF* the hard scattering kernel allows a power series expansion to be made in k_T, then we can again de-couple

---

[*] Our normalization convention is that of the text of Itzykson and Zuber.

most of the integrals, obtaining

$$M \simeq \int dx_i \ldots dx_f (P_1^+ \ldots P_f^+) \times$$

$$\prod_{i,f} \int_0^Q d^2 k_T \psi_f^*(k_{T_f}, x_f) H(x_1 \ldots x_N; k_{T_1} = 0 \ldots k_{TN} = 0; Q) \int_0^Q dk_T^i \psi_i(k_{T,i}, x_i)$$

$$(2.3)$$

Power Behavior

The H part of the diagram is now evaluated at the on-shell point $k^- = k_T = 0$. This is gauge invariant in perturbation theory. If we count the dimensions of the diagram and the number of far off shell gluons and fermions, we obtain the "quark-counting" results for the scaling of the diagrams with Q^2. This power counting is designed to be simple and crude; we are ignoring logarithmic effects, such as α_s. Power counting of the hard scattering works this way[27,22]: the amplitude M for n total constituents to participate in hard scattering scales like $s^{2-n/2}$. The differential cross section $d\sigma/dt$ goes like M^2/s^2, or like s^{-n+2}. For proton proton scattering, $n = 12$ and we get the s^{-10} dependence quoted earlier. The overall normalization is not determined; it is an important, but extremely complicated quantity.

Remark on Independent Scattering: Before leaving this topic, I must warn the reader that the exception to the argument above is just as interesting as the argument. There can be a singularity as one tries to expand the hard kernel H. By the general Landau theorem[28], such singularities in perturbation theory are associated with kinematics in which the process can occur on-shell classically. A rather big, power law singularity is known to happen in pQCD when the momentum regions of the diagram correspond to *independent scattering*: quark beams collide at isolated centers. For some particular "quark counting" diagram there may be an integration region in which all the internal momenta obey the quark counting assumptions. The *same diagram* with the same external kinematics may have a region where these assumptions are violated-*this usually happens*, in fact, *if the diagram has a region where an internal gluon can carry small momentum*[29]. For this reason the term "quark-counting diagrams" is a misnomer - one should speak of the "quark counting regions" and the "independent scattering regions". We will set aside this important problem for now and come back to it later. The argument is correct for *a particular region*.

The Distribution Amplitude

The object

$$\phi(Q^2, x) = \int_0^Q d^2 k_T \psi(k_T, x) \tag{2.4}$$

is called the distribution amplitude, and was invented by Brodsky and Lepage*.[21] As $Q^2 \to \infty$, the distribution amplitude tends to evaluate the wave function at the point of zero transverse space separation, by Fourier analysis. One can see this explicitly[30] by transforming the integrals to obtain

$$\phi(Q^2, x) = 2\pi \int_0^Q d \mid b \mid Q J_1(Q \mid b \mid) \tilde{\psi}_0(\mid b \mid, x) \tag{2.5}$$

One sees that the Q^2 dependence of $\phi(x, Q^2)$ "maps out" the b-dependence of the wave function in the region $b^2 \leq 1/Q^2$. (Amusingly, the Bessel transform in (2.5) is

---

* Our convention does not include a logarithmically varying factor of Q^2 introduced for renormalization group analysis in Ref. (22). Either convention can be used without affecting our argument.

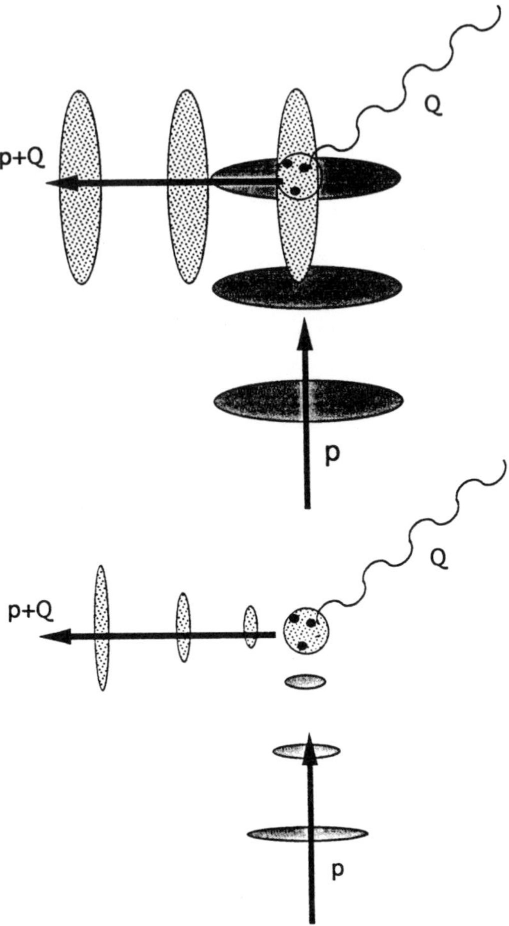

Fig (11). Cartoons of different approximation to the hard scattering. In the impulse approximation (top) we need the overlap of a proton onto the struck quarks and vice-versa. In an "adiabatic" picture (bottom), the system would be quantum mechanically prepared into a small size and coherently scattered.

invertible; one could actually solve for the b-space wave function by measuring $\phi(Q^2, x)$ at all Q^2). The wave function in question has had its hard perturbative "tail" at large k_T^2 removed, and put into the hard scattering. The working assumption is that the wave function is unknown but smooth,[†] characterized by hadronic physics scales of a Fermi or so. There is also information available from the renormalization group, which can organize the limit of extremely short distances in powers of $\log(b^2)$. For laboratory values of Q^2 it is unlikely that this is enough and it is necessary to model the wave function. Finally one is left with the x-integrals, which flow through the diagrams and cannot be eliminated.

We have obtained a main result: the quarks participating have been shown to be separated by a transverse distance $b^2 \leq 1/Q^2$. *Note that this is a statement about a dominant integration region, not about the proton itself.* In free space scattering, for example, the cartoon picture is shown in Fig.(11).

[†] For the approach to very large Q^2, ϕ satisfies a renormalization group equation, by which one can show that the general behavior is a sum of polynomials in x and powers of $\log(Q^2/\Lambda_{QCD})$.

As the cartoon shows, the proton as a whole does not have to shrink down to participate in a hard scattering it does not know is coming. Instead, the quark wave function in the proton is whatever it is; the perturbative impulse occurs quickly and selects a part of its wave function, makes it "turn the corner" with the hard gluon exchanges, and then sends it on its way. From quantum mechanics, we then find the overlap of these quarks onto an outgoing proton, and we have no obligation to follow their other projections if they are not measured. So, even if a normal proton has a soft gluon cloud, we do not necessarily have to understand it to make the perturbative arguments. Whatever amplitude there is to find an outgoing proton after the hard scattering is contained in the overlap coefficients of the impulse approximation, in this case the distribution amplitudes. These amplitudes are small numbers that do not depend strongly on Q^2.

Hadron Helicity Conservation

Note that a particular orbital angular momentum component of the wave function is selected by (2.4, 2.5). If we expand the wave function according to (2.1 1/2), then only the $m = 0$ term contributes in (2.5):

$$\int_0^Q d^2 k_T \psi(k_T, x) = \int_0^Q d^2 k_T e^{ik_T \cdot b} \int d^2 b \sum_m e^{im\varphi} \tilde{\psi}_m(|\, b\,|, x) \sim \sum_m \delta_{mo} \tilde{\psi}_m(b \cong 1/Q, x)$$

The wave functions of interest for this argument therefore give the hadron helicities as the sum of the quark helicities.

In the high energy limit the quark helicities do not flip when gluons are exchanged: this is because pQCD is almost perfectly chirally symmetric. It is quite easy to see this. The rule for Dirac quark spinors is

$$\gamma_5 u(k, \lambda) = \lambda u(k, \lambda)$$

$$\gamma_5 v(k, \lambda) = -\lambda v(k, \lambda)$$

for spinors of helicity λ. By permuting γ_5 through a diagram and using $\{\gamma_5, \not{k}\} = 0$ for every propagators carrying momentum k, the helicity is seen to be invariant along a quark line, no matter how many gluons are exchanged. (If in a calculation you retain a term going like a light quark mass, it will be suppressed by a power of $m/p << 1$.) Putting these results together gives the famous hadron helicity conservation rule due to Brodsky and Lepage.[22] In a reaction of the form $A + B \to C + D$, then the sum of the quark helicities going in to the reaction is the same as the sum going out:

$$\lambda_A + \lambda_B = \lambda_C + \lambda_D \tag{2.6}$$

This rule is evidently as general as the factorization; *it is an exact dynamical symmetry of the quark-counting model.* It is, however, violated by experiments in almost every case of scattering with isolated hadrons tested! Nevertheless, we have reasons to believe that it will be important in the study of color transparency, so we continue.

Quantum Color Transparency

Now let us formalize the ideas of color transparency, following Ref. (30). In a large nuclear target, the ideas of factorization are the same as in free space. The separation between the hard scattering and the nuclear target is shown in Fig (12).

There are therefore nuclear distribution amplitudes, which are Green functions for quarks to propagate from a given small transverse separation through a nucleus and to a given final state hadron. We could just stop here, since these are non-perturbative objects and unknown. However, we can try to relate the nuclear distribution amplitude to the free space one. For this, we must model the long-time, soft interactions of the quark system passing through the nuclear medium.

Consider the elastic scattering of one of the quarks on the nucleus. Let it be a plane wave in the z-direction, with wave function $\psi(x)\delta^2(k_T)$. Let the scattering

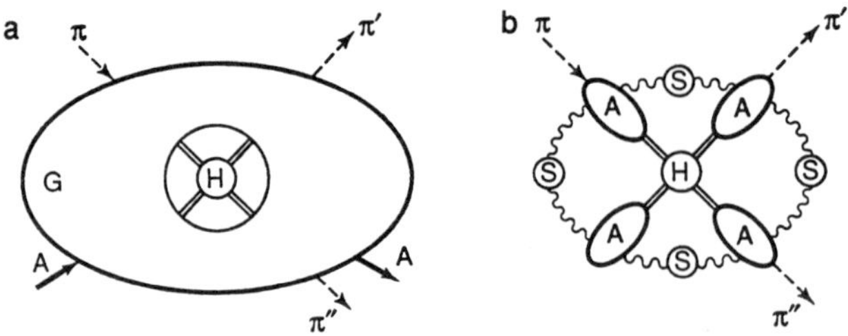

Fig (12). a) Separation of the hard scattering kernel in the nucleus A. (b) Factorization of the hard scattering into nuclear distribution amplitudes.

amplitude for a quark on the nuclear target at cm energy-squared s and invariant squared momentum transfer t be $F_A(s,t)$. If x comes in and x', k'_T go out, then $-t = (x/x')k_T^{2'}$.

For a wave packet given by amplitude $\psi_o(x, k_T^2)$ coming in, by superposition we have a scattered wave

$$\psi'(k_T, x) = \int d^2 k'_T \ F_A(s, \ (k_T - k'_T)^2 x/x')\psi_o(k'_T, x)$$

Except for possible endpoint singularities the x-dependence is inessential. Moreover, since the x-dependence of wave functions is not well known and model dependent, we suppress it. We are also not explicitly writing the location of the struck hadron in the nucleus, which is integrated over. Going to transverse separation (b) space, with Fourier transformed quantities indicated by an tilde, we have a transmitted wave $\tilde{\psi}_A(x, b)$ which is the original wave minus the scattered wave

$$\tilde{\psi}_A(\tilde{x}, b) = \tilde{f}_A(s, b)\tilde{\psi}_o(b, \tilde{x}), \tag{2.7}$$

where $\tilde{f}_A(\tilde{s}, b) = 1 - \tilde{F}(s, b)$ will be denoted as the nuclear filtering amplitude.

It is conventional to write $\tilde{f}_A$ in terms of an eikonal function $\chi(s, b)$: $\tilde{f}(s, b) = 1 - \exp(i\chi(s, b))$. This is very general, and applies whatever the model for the scattering. In hadronic physics $\chi(s, b)$ typically goes like $\chi(s, b) = is^\alpha \exp(-b^2/2b_0^2)$. Expanding the eikonal to lowest order in χ reproduces the naive Born–level picture, which is instructive:

$$\tilde{f}_A(s, b) \approx s^\alpha e^{-b^2/2b_A^2}$$

Here the exponential in b^2 represents cutting off of large impact parameter quarks beyond a scale b_A^2. Evidently b_A^2 is of order $A^{-1/3}$ fm for large A, since we could write

$$e^{-b^2/2b_A^2} \approx e^{-Z(A)/2\lambda} \ : \ \lambda \cong \frac{1}{n\sigma(b^2)} \ , \ \sigma(b^2) \sim b^2 \ , \tag{2.8}$$

with n the average nuclear density seen in the problem and $Z(A) \approx A^{1/3} fm$, a distance scale.

The main effect of traveling through the nucleus should be filtering, i.e., the cutting off of quarks separated by a large impact parameter.[8] Let us examine the general effect of this on $\phi_A(x, Q^2)$. We now have

$$\phi_A(Q^2, x) = 2\pi Q \int_0^\infty db \ J_1(Qb)\tilde{f}_A(s, b^2)\tilde{\psi}(b, x) \ , \tag{2.9}$$

having done the integrals to find a Bessel function $J_1(Qb)$. The content of this formula is in its Q dependence, which is entirely in the $QJ_1(Qb)$ combination. This just comes from Q in the upper k_T limits, i.e., factorization. The remaining content is in the A dependence, entirely from the filtering amplitude $\tilde{f}_A$. Even assuming the full $\tilde{f}_A$ is complicated and model dependent, we have the power-series expansion around $b = 0$ of

$$\tilde{f}_A(s, b^2) \sim 1 - A^{1/3} n b^2 \sigma'_{eff} + \ldots$$

where $b^2 \sigma'_{eff}$, by definition, is an effective cross section. We are assuming only that, at $b = 0$, cancellation of the color dipole moment of the singlet state occurs, so that $\tilde{f}_A(b = 0) = 1$. (Actually, we do not have to limit ourselves to a filtering amplitude that is analytic at $b = 0$. There is a negligible perturbative component to the scattering that should go like $\alpha_s(1/b^2)$.)

The only physical input so far has been to propose factorization and incorporate filtering in the nucleus. Nevertheless, transparency is so general that this is all we need to assume. The point is that even though $\tilde{f}_A$ has a filtering effect with a large cross section, say, $\sigma'_{eff} \sim 1 fm^2$, at large enough Q the integral in (4) is independent of A; in fact, it gives the free-space value:

$$\lim Q \to \infty \ \ \phi_A(Q^2, x) = \phi_0(Q^2, x). \tag{2.10}$$

This comes about because the integral as $Q \to \infty$ over the Bessel function $J_1(Qb)$ only requires the value of $\tilde{\psi}_A(b, x)$ as $b- \to 0$. This is transparency. It is remarkably general in the asymptotic limit $Q^2 \to \infty$.

The Oscillating Transparency Ratio

What if there are independent scattering regions? These are configurations which are not transversely small, and yet can cause a hard scattering in free space. But if these are not small, they should not survive the trip through the nuclear medium: they will die for $A >> 1$.

Recall the oscillations in the pp→pp data. They are not unique to pp scattering: somewhat the same oscillations should occur, and do occur, in $\pi p \to \pi p$ scattering. Many years ago we interpreted these oscillations as evidence for independent scattering[7]. In a simple two-component model we obtained the oscillations via quantum mechanical interference. Let the full amplitude be called M, and let it be made of two pieces, M_{qc} (the quark-counting part) and M_{is} (the independent scattering part):

$$M = M_{qc} + M_{is} e^{i\phi(Q^2)} \tag{2.11}$$

The independent scattering part has an energy dependent relative phase in pQCD,

$$\phi(Q^2) = const \ \pi ln \ ln \ (Q^2/\Lambda^2_{QCD}) \tag{2.12}$$

The $const$ is unknown but calculable in pQCD: a reasonable guess is that it should be between C_F^3 and N_c^3. Our old fit to the free space oscillations found $const. \cong 16$. Recently Carlson et $al.$[14] refit more invariant spin amplitudes with more parameters and found $const. \cong 4$. (By arranging to have "beats" between amplitudes this works well enough.)

The nuclear target should filter away M_{is}. Let us do this in the crudest, most direct way possible by multiplying M_{is} by a factor of $\exp(-n\sigma A^{1/3}/2)$ with σ a "big" attenuation cross section of 30 mb and n the nucleon number density in the target. Then we have a model for the transparency ratio:

$$T(Q^2, A) \sim \frac{1}{Z} \frac{\mid M_{qc} + e^{-n\sigma A^{1/3}/2} M_{is} e^{i\phi(Q^2)+i\delta_A} \mid^2}{\mid M_{qc} + M_{is} e^{i\phi(Q^2)} \mid^2} \qquad (2.13)$$

This should be adequate if the filtering of the *is* part is much larger than the *qc* part. In our original work we were not sure what the effects of nuclear phases might be, so we introduced a relative phase δ_A from nuclear effects. We let δ_A vary over its full range, $0 < \delta_A < 2\pi$, at each energy, to generate a range of theoretical uncertainty. The nuclear phase effects do not matter, however, once a single component of the amplitude dominates.[10] The results of the oscillating transparency ratio above were already shown in Fig (3).

The idea reproduces the structure in energy dependence of the BNL data with no free parameters. Of course, this does not prove that the oscillations were caused by independent scattering - it only shows that the filtering away of a spatially large configuration is beautifully consistent in QCD. Below we will see the reason for the energy dependent phase, and we will consider further evidence for the presence for independent free space scattering when we study hadron helicity violation.

A Scaling Law

Is there another way to check these ideas experimentally? The analysis above would allow us to adjust a number of parameters to fit the data. We want a check of the general formalism which is *as model independent as possible.*[*] We will follow the treatment of Ref. (31).

Here is a simple way to think about the mathematics of quantum color transparency: there are two "infrared" cutoffs of the quark transverse sizes in nuclear targets: large Q^2, and large A. Which cutoff wins? Color transparency would increase with Q^2 at fixed A, but has to decrease with large enough A at fixed Q^2. In the "basic idea" we completely glossed over this point: the limit of large Q^2 is not well defined until we say how the limit is taken in A! So, this is a two-scale problem, and we should seek the controlling combination of variables.

Recalling (2.9), let us suppose that the wave functions are slowly varying because Q^2 is already big enough so that we are close to their "centers". Then we find

$$\phi_A(Q^2, x) \approx 2\pi\tilde{\psi}(b = 0, x) \int_0^\infty db\, Q\, J_1(bQ)\tilde{f}(s, b^2 A^{1/3})$$

We can make a simple change of variables in this expression to see that

$$\phi_A(Q^2, x) \approx \phi(Q^2/A^{1/3}, x) = 2\pi\tilde{\psi}(0, x) \int_0^\infty dy\, J_1(y)\tilde{f}(y^2 A^{1/3}/Q^2) \qquad (2.14)$$

That is, the survival probability for large A and large Q^2 should be a function of a special combination $Q^2/A^{1/3}$ rather than A and Q^2 separately. Very dissimilar events - one at $A = 200$, $Q^2 = 20$ GeV^2 and another at $A = 50$, $Q^2 = 12$ GeV^2- would have the same survival probability. We call this *transparency scaling*. The idea is to isolate the big dimensionful variables in a problem as a test of the general dynamical scheme. One is evidently on much safer theoretical ground in identifying such a general feature of the theory than in comparing calculated numbers with data in a detailed model.

In organizing large dimensionful variables, transparency scaling is similar to Bjorken scaling. Recall that Bjorken scaling tells us we are probing nearly on-shell partons,

---

[*] Otherwise the temptation to a theorist to cheat sometimes becomes overwhelming.

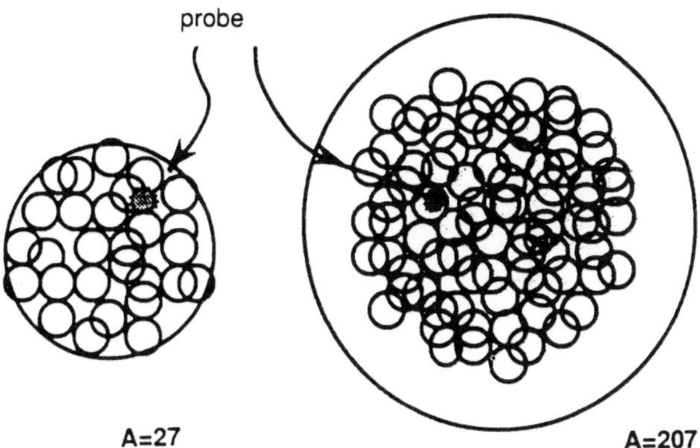

Fig (13). Interpretation of transparency scaling. A small probe zooming through a big target has the same survival probability as a big probe in a small target.

without saying anything about the parton distributions. Is there a physical picture of transparency scaling?

It is quite simple (Fig 13). The quark integration regions which survive to be observed must have

$$\sigma_{eff}(b^2)nA^{1/3} < 1.$$

Since $\sigma_{eff} \approx b^2$ and $b^2 \leq 1/Q^2$, survival probabilities should be a function of $Q^2/A^{1/3}$. Events with the same total amount of interaction can come from compensation between the size of the cross section and the target length. If we see this in data, it means that Q^2 sets the inverse size of the quark regions selected over a range of nuclear sizes.

A Successful Check

To check this in an unprejudiced way, we could fit the differential cross section or transparency ratio data to an unknown function of Q^2 (the hard scattering rate) times another unknown function of Q^2/A^α. The prediction is that $\alpha = 1/3$ should bring dissimilar data into line on some universal curve. Unfortunately there is not enough data in the BNL set to do this at present. To continue we can assume[32] that the hard scattering rate is the short distance pQCD form (1.5). (We already know that there are no oscillations in the data, and we have the previous fit in extracting σ_{eff}. That fit used an exponential attenuation model for the survival probability. Here we will make no model whatsoever for it, so the check is independent.) We simply define the survival probability to be the nuclear cross section $d\sigma/dt$ divided by $\alpha^{10}\sigma^{(s)} s^{-10}$.

The survival probabilities thus defined are plotted versus Q^2/A^α in Fig (14) for $\alpha = 0.2, 1/3$, and 0.5. The data points from BNL are circles ($Q^2 = 4.8\ GeV^2$), squares ($Q^2 = 8.5\ GeV^2$), and crosses* ($Q^2 = 10.4\ GeV^2$). It is clear form the figure that $\alpha = 1/3$ is favored over the other values of α; for $\alpha = 1/3$ one curve could describe all the data. This is quite a gratifying agreement with a theoretical prediction[31].

There remains a minor point: in the factorization of survival probability from hard scattering, one relative normalization is always an ambiguity. We can always scale $P(Q^2/A^\alpha)$ by a constant, and absorb this into the hard scattering rate $N(Q^2)$.

---

* The two data points reported by the experiment at beam energy of 12 GeV are useful for the scaling analysis. We also studied these using the global fit procedure extracting σ_{eff} but in this case the size of the experimental errors and the lack of enough A values prevented extracting a statistically meaningful σ_{eff}.

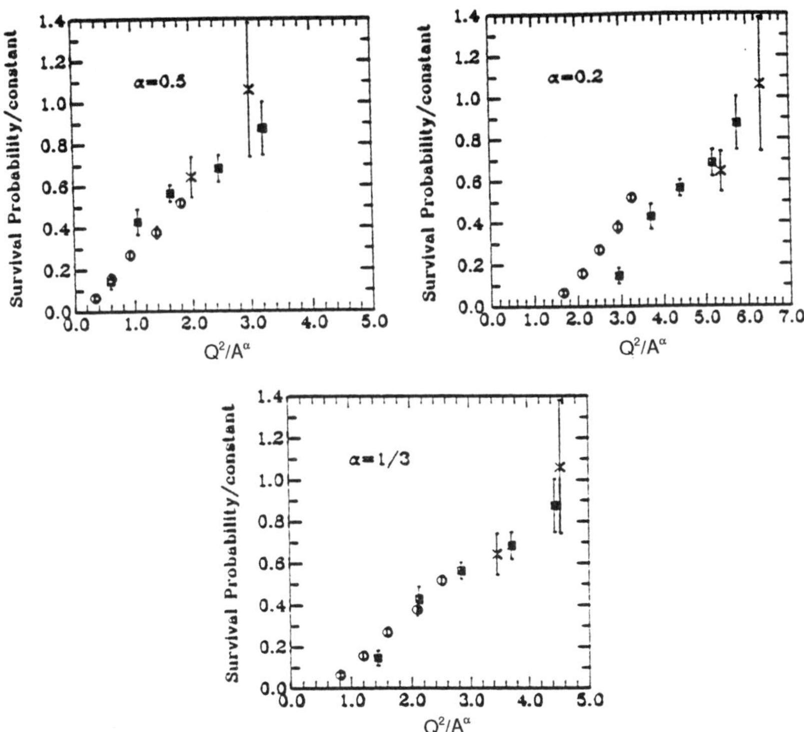

Fig (14). Scaling dependence. The survival probability/constant as a function of Q^2/A^α for the BNL data for three different values of α. One curve could fit all the data for $\alpha = 1/3$.

The absolute magnitudes of the hard scattering rate or survival probability thus cannot be determined by this particular procedure.

Independent Scattering

We have seen that the theory and the experimental checks are nicely consistent for the nuclear target. In that case, we allowed ourselves to ignore independent scattering because of nuclear filtering. However, we still want to understand the free space scattering too - it is still important!

Consider a hard quark-quark scattering in its cm system. Suppose one of the final state quarks goes off in a direction given by p'. For a second pair of quarks there is nothing to prevent a hard scattering that accidentally sends its quark in the same direction as the first. This is independent scattering. It is evidently an unlikely process, because the phase space for two hard scatterings to coincide in direction and center of mass motion is small. However, all elastic hard scattering processes are rare: we will compare this process to the quark counting process, in which all of the hard scatterings are connected by a chain of gluons or far-off shell quark propagators.

The power counting of independent scattering goes this way. Sit in a frame such as the hadron cm frame where all the hadron momenta are big, of order $\sqrt{s}$. The outgoing beams of quarks must coincide in direction well enough to make hadrons in the final state; any discrepancy allowed is set by the wave functions, which have small relative k_T. This of course is much smaller than the beam energies, so we can pretend it is almost zero. The desired independent scattering integration region is thus equivalent to certain delta functions of 4-momenta. Conserving 4-momenta, for each pair of scatterings there are three large momenta for each scattering, and one out-of plane transverse momentum. This component of the transverse momentum is not as big as $\sqrt{s}$ but instead depends sensitively on what the hadronic wave function will allow. It should be of order $\sqrt{<k_T^2>}$ in the state, which for purposes of counting is the same as $1/\sqrt{<b^2>}$.

On–shell quark-quark scattering is dimensionless and scale invariant (up to logarithmic corrections) in QCD. Each delta function of a big momentum counts as $1/\sqrt{s}$ because we can write $\delta(k - k') = (1/\sqrt{s})\delta(x - x')$. The overall amplitude M for a pair of quarks to coincide in direction to make a hadron therefore scales like the product of the delta functions of momentum, namely like $\sqrt{<b^2>}(1\sqrt{s})^3$.

Now let us calculate the cross section. In the high energy limit, $d\sigma/dt = $ const. $|M|^2/s^2$, giving $|M|^2 \sim <b^2> s^{-5}$ for meson-meson scattering . This beats the quark-counting process, which for $n = 8$ goes like s^{-6}. Suppose we go to proton-proton scattering. We have the same argument, but have to add another quark-quark scattering to coincide with the first ones. Counting three more delta functions of big momentum the amplitude-squared has to be smaller by $\sqrt{s}^{-3}$. Independent $pp \to pp$ scattering thus has $d\sigma/dt = $ const. $(<b^2>)^2 s^{-8}$. This beats the quark-counting process, which goes like s^{-10}. (For an exercise, find the lowest order independent scattering and power counting for the meson-proton case. It is fairly tricky).

While power counting is adequate for the overall scaling, one has to make a real calculation for the dependence on s, t, u, namely the angular distribution. This is a way of separating some kinds of independent scattering, such as t-channel gluon exchange, from others involving quark interchange. Now recall from the earlier discussion that independent scattering refers to *internal kinematic regions*. Since there are a huge number of independent scattering regions in higher order Feynman diagrams, the angular distribution is not a test for the presence of independent scattering. Some detailed calculations at lowest order for t–channel gluon exchange are available in the literature[34].

The "out-of plane" transverse momentum is the main focus in studying independent scattering. To control this variable, it is convenient to make a Fourier transform to conjugate space variables b. We let the $\hat{x}$-direction be out of plane; this is the same for all the hadrons. We let the y_i variables be chosen relative to the hadron momenta directions, i.e., $\vec{b}_y$ is in the (vector) $\vec{b}_x \times \vec{p}_i$ direction for each momentum p_i. Then there is a useful formula for the independent scattering amplitude:

$$M(Q) = Q^{-P} \int db_x \prod_{i,f} dx_i dx_f \tilde{\psi}^{*f}(x_i. b_x; b_y^f \approx 1/Q) \bar{H} \tilde{\psi}^i(x^i, b_x; b_y^i \approx 1/Q) \qquad (2.15)$$

This was first given by Botts and Sterman[12].

Sudakov Effects

In the 1950's, V. Sudakov[33] studied the large Q^2 dependence of the electromagnetic form factor of an electron. He observed that at one-loop order, the form factor received corrections going like $\alpha \, log^2(Q^2/\Lambda^2)$ from the integration region where the internal photon loop momenta are "soft", meaning much less than Q. Here $\Lambda << Q$ is any other scale in the problem that acts like an infrared cutoff. The soft photon region can be calculated to all orders in perturbation theory, and moreover, the argument can be made that this region is the only one giving terms of order $[\alpha \, log^2(Q^2/\Lambda^2)]^J$. The result of summing terms of this kind at all orders is a multiplicative factor which is the exponential of the one-loop graph: a form factor going like $\exp(-\alpha \, $ const $\, log^2(Q^2/\Lambda^2))$. For large Q^2 this goes to zero faster than any power of Q^2! Sudakov correctly interpreted this as follows: at large Q^2, a struck electron has many chances to radiate an enormous number of photons into a very large phase space. These inelastic events dominate the probability. This leaves very little probability for an elastic event, which explains why the elastic form factor must be small. Calculating the soft radiation, a separate task, confirms this.

This phenomenon is a universal one and occurs in all elastic scattering calculations in gauge theories[29]. There is a fast, heuristic way to understand the functional form.* Consider an e^+e^- collision which is going to produce a time-like photon. The electron and positron are never really free particles but are accompanied by clouds of soft photons. These can be exchanged before the event, producing a phase-shift on the particles' wave functions. That phase should be given by an eikonal, namely the line integral $e \int A \cdot dx$ where A is the vector potential. By dimensional analysis (or inserting the Coulomb potential) this goes like the logarithm of the ratio of the endpoints. The closest approach is of order $1/Q$, and the furthest is of order $1/\Lambda$. Thus there is a phase proportional to $\alpha \, log(Q^2/\Lambda^2)$ for $Q^2 > 0$. The analytic function of Q^2 having this phase is the exponentially dependent Sudakov form factor already quoted. (For an exercise show this).

In QCD the same phenomenon has to occur for quarks[8]. The Coulomb field in coordinate space x is modified by the running coupling, $\alpha_s(x) = const/ \, log(1/x^2\Lambda^2_{qcd})$. Repeating the calculation above, the QCD Coulomb phase goes like

$$i\int_{1/Q}^{1/\Lambda} dx \frac{1}{x \, log(1/x^2\Lambda^2_{QCD})} = i \, log\left(\frac{log(Q^2/\Lambda^2_{QCD})}{log(\Lambda^2/\Lambda^2_{QCD})}\right) \equiv i\phi(Q^2, \Lambda^2_{QCD}, \Lambda^2) \quad (2.16)$$

This is a wonderful result: the appearance of Λ_{QCD} in the Q^2 dependent part is an indication that this phase is perturbatively calculable, and its Q^2 dependence is not too sensitive to the "infrared region". (In fact, in certain gauges the phase appears as a *purely imaginary anomalous dimension*.[8,11,12]) Now if we want an analytic function to have this phase, then to leading log order it has to be

$$S(Q^2, \Lambda_{qcd}, \Lambda^2) = exp[-\frac{const.}{\pi} log(Q^2/\Lambda^2) \, log\left(\frac{log(Q^2/\Lambda^2_{QCD})}{log(\Lambda^2/\Lambda^2_{QCD})}\right) - i\phi(Q^2, \Lambda^2_{QCD}, \Lambda^2)]$$

There is no true infrared cutoff in QCD except bound state and confinement effects. But if we have a color neutral "atom" of quarks and antiquarks, then there is destructive interference of radiation from the various color charges inside, just as in QED. From translational symmetry, radiation of a gluon of momentum q from an emitter at the origin and an opposite transverse emitter at a transverse position b are added like $1 - e^{-ibq}$. This gives a cutoff on radiation exactly like an infrared cutoff, namely $q \lesssim \Lambda \to 1/b$, giving

$$S(Q^2, \Lambda_{qcd}, b^2) = exp(-const \, log(Q^2/\Lambda^2) log\left(\frac{log(Q^2/\Lambda^2_{QCD})}{log(1/b^2\Lambda^2_{QCD})}\right) - i\phi(Q^2, \Lambda^2_{QCD}, 1/b^2)$$

$$(2.17)$$

There is another way to think of all this. Big b-zones of the wave function radiate; small b zones survive. Consequently, in any exclusive perturbative problem we can simply adopt the b-dependence of (2.17) as a model of the b-space region that will be relevant: (2.17) is an effective perturbative model of the wave function. It is in fact a *confined* wave function, suggesting a bound–state character, obtained from perturbation theory! The identification with a wave function can be made to be literally true in an axial gauge; resummation of leading logarithms produces a Sudakov factor in the wave function. The big scale Q comes from the dependence on the gauge vector n which generates logs of $p \cdot n$. (For an exercise, show that this bound state–type effect would not occur with the running coupling of QED.)

Returning to independent scattering, there are two effects of all this work:

---

* One is allowed to be cavalier when working with leading log approximations because all kinds of sub–leading effects drop out. The virtue of the approach presented here is that it saves many pages of calculation.

(1) Whatever the b-space wave functions, as $Q^2 \to \infty$ the system will radiate (giving no elastic amplitude) except for the region $b \to 0$. It follows that the dominant amplitudes that actually give an elastic scattering will only sample $b \to 0$. To see this quantitatively, consider the region where the Sudakov exponent is of order unity and stationary. By inspection, this region is $b \sim Q^{-A}$, where A is a positive (but not necessarily integer) power. This will be called the *asymptotic* region. On the basis of such an argument Mueller[29] concluded that for $Q^2 \to \infty$ the asymptotic region dominates and the independent scattering becomes a short-distance process. However, with numerical studies, Botts[37] showed that the asymptotic region dominance sets in at about $\sqrt{Q^2} \gtrsim 1 TeV$, which is far too high to be relevant to experiments. Below such large Q^2 values, the Sudakov effects are not strong enough to eliminate contributions from larger regions of b.

(2) For large $Q^2 > GeV^2$ the dominant amplitudes apparently come from the region of the original independent scattering argument, $b \cong const$. In the large Q^2 region, elastic scattering should be sensitive to the phase of the wave function, which we have already seen goes like $log(log(Q^2/\Lambda^2_{qcd}))$.

As a test of this idea, we had fit the oscillations in the $pp \to pp$ data to this functional form[7]. Such an interpretation is supported by the work of Botts and Sterman[12]. Nevertheless, this interpretation of the data's oscillations is not yet established: the diagrams chosen for the calculation of Ref.(12) yielded a peculiarly small numerical coefficient insufficient to explain the rapid variation of the data. We believe this is due to the choice of what was calculated, namely the phase of the asymptotically largest amplitude. What is needed, however, is the amplitude relevant at lab energies which has the largest imaginary anomalous dimension, which would cause the most rapid oscillations. There should be a systematic way to isolate this among the many diagrams, but it is a highly non–trivial calculation. Until this is done, one again has to continue with the data and see what can be learned.

High Energy Hadronic Helicity Flip

Recall that the hadron helicity conservation rule represents an exact symmetry of the quark - counting factorization, because the hard scattering is "small" and "cylindrical". We believe this will be a good description due to filtering in the nuclear medium for $A >> 1$. But it should not be true in free space, if independent scattering is really contributing there. The crucial question is whether (or not) the symmetry of the short–distance model is a property of the entire perturbative theory. For this we follow the treatment of Ref. (35).

The non-perturbative Hamiltonian of QCD does not conserve spin and orbital angular momentum separately, but instead generates mixing between them. This is why a list was given earlier of all the orbital and spin projections in a meson wave function. Thus *if* a non-zero orbital angular momentum component somehow enters the hard scattering - and this is a crucial point - *then* the long-time evolution before or after the scattering can convert this angular momentum into the observed hadron spin. It is not necessary to flip a quark spin in the hard interaction, because the asymptotic hadron spin generally fails to equal the sum of the quark spins. Such a mechanism is totally consistent with the impulse approximation.

The challenge is therefore to find those large Q^2 processes in which non-zero orbital angular momentum enters, or in other words, to finds those which are not "round". It turns out that the independent scattering processes are not "round" but instead are "flat" (Fig (15)).

The origin of the asymmetry is kinematic.* As before let $\hat{x}$ be the direction perpendicular to the scattering plane and $\hat{y}$ be a vector in the scattering plane. In the independent scattering mechanism, the two (or more) uncorrelated scattering planes

---

* It is embarrassing that independent scattering has been studied for almost 20 years before this was realized. One problem was the practice of theorists (including this author) to do covariant calculations which did not have simple interpretation.

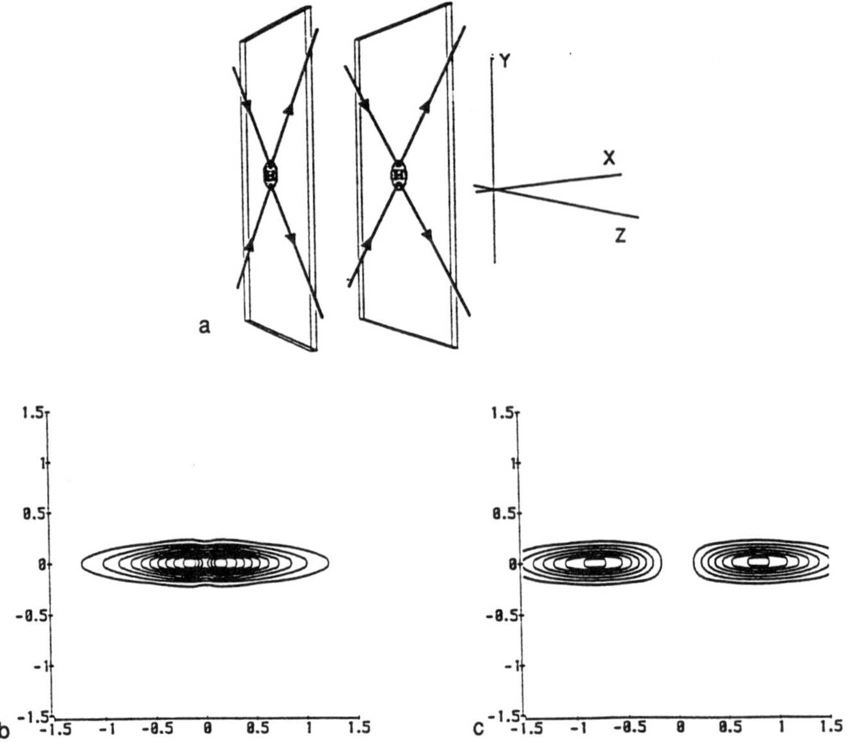

Fig (15). a) Coordinate space pictures of meson-meson independent quark scattering. (b) Contour map of the real part of leading Sudakov "wave functions". (c) Same wave function multiplied by a polynomial representing m = 2. From Ref (35).

are separated at the collision point by a transverse out-of-plane distance b. The in-plane transverse position separations Δb_y that contribute are as small as possible, namely of order the uncertainty principle estimate $1/Q$, because this is the direction of large momentum flow carried by the gluons. The kinematic "short-distance" in the problem is, however, only the in-plane distance. The relevant out-of plane transverse distance is set by the hadronic wave functions.

Recalling (2.15), the amplitude for independent scattering is proportional to the integration over b_x. Note the role of the scattering plane, namely the breaking of rotational symmetry with the out-of-plane direction x. Immediately one is struck by the absence of any selection rule favoring $m = 0$; instead, all orbital angular momenta in the wave functions are allowed to participate.[35] It is as if hadrons "flatten" under impact in the in-plane direction y, forming a cigar-shaped hard scattering region.

We therefore have a new rule of "hadron helicity non-conservation" when independent scattering occurs:

$$\lambda_A + \lambda_B \neq \lambda_C + \lambda_D \tag{2.18}$$

There are a number of tests of this idea. Basically, we want to have spin information on every possible reaction in free space, and every similar reaction in the nuclear target. *When one observes large Q^2 hadronic helicity violation one is observing the orbital* *angular momentum makeup of the hadron!* This will be discussed further below. At this point, the fact that the free space $pp \rightarrow pp$ scattering shows large spin effects, apparently correlated with the oscillations, is more evidence for consistency of the picture that the independent scattering regions are responsible.

Hadronic Basis Approach: The Expansion Time Scale Problem

The calculation of the soft re–scattering of quarks traveling through a nuclear medium is not claimed by anyone to be very reliable. With plenty of room for different models of the distribution amplitude, claims are controversial. The theory so far has been presented in the quark Fock space basis. If a different quantum mechanical basis is used, the physical interpretation will be different.

Many authors have used the basis of asymptotic hadron states to study color transparency. This basis is certainly the most effective one for studying problems in nuclear physics. One weakness of the hadronic basis approach is that a unified picture of the hard scattering and nuclear effects is not attempted; the hard scattering rate is assumed to be "given" arbitrarily. That is not a strong criticism, because any non-perturbative input to the problem would be helpful.

One approach in calculating ϕ_A is to saturate the intermediate states with a complete set of hadrons creating an s–channel dispersion relation. However, calculations in the literature are not yet formalized as statements about the distribution amplitude, because in the hadronic basis one does not have a clear picture of factorization. Jennings and Miller have made a number of studies of their own model in successively more detailed approximations[16]. One way or the other, one must put in physics for the propagation of a number of hadrons through a nuclear medium. There is no "model–independent" calculation.

The "expansion time scale" controversy arose due to the following argument offered by several authors[36]. It is adapted from ancient theoretical technology for diffractive processes occurring in free space. Consider the initially struck state as a hadronic wave-packet, a superposition of on-shell energy eigenstates of invariant mass m_n. Assume that the energies of the particles are given by the free space formula, $E = \sqrt{(p^2 + m^2)} \cong p_z + (m^2 + p_T^2)/2p_z$. For a wave packet with one value of $p_z >> GeV$, then the time evolution of the system is:

$$\psi(t) = e^{-ip_z t} \sum_n c_n e^{-it[m^2 + p_T^2 - (m^2 + p_{T,n}^2)]/2p_z} \tag{2.19}$$

This is a formal argument; the coefficients c_n are unknown, since one generally does not know how to express quark-picture concepts in a hadronic basis. However, somehow the wave packet is to correspond to the small transverse spatial separation region identified in the perturbative picture. It should have reduced strong interactions due to destructive interference among the hadronic components. When the time t is large enough so that the relative phases in (2.19) change by an amount of order unity, or maybe π, the time evolution of the wave packet has become a big effect. It is assumed to have "expanded" to a size that is no longer small, and, (the argument goes on to conclude), color transparency must be lost.

This allows one to calculate a characteristic time from the initial hard interaction in which to observe color transparency, in order to avoid this expansion problem. The longest time is set by the lowest values of $m^2 - m_n^2$. (Differences in p_T^2 are ignored.) That time is the same as the target length since all relativistic systems move at the speed of light. Thus one wants

$$L(m^{*2} - m^2)/2p_z < 1$$

or, in numbers, one has an expansion length L for wave packets given by

$$L \approx 0.4 fm(p_z/GeV) \tag{2.20}$$

using $m^* = 1.4\ GeV$ and $m = 1\ GeV$. For large nuclei, and for reasonably large Q^2 in the range of a few GeV^2 to 20 GeV^2, then from this argument one might not expect to see color transparency due to rapid time evolution.

The argument is valuable because it uses hadronic degrees of freedom. However, this is an order of magnitude argument, uncertain by factors of π, etc. To promote it to a real bound is just wishful thinking. For an estimate of the errors in the estimate one can examine numerical calculations using similar assumptions[37]. The "time scale" seems to be uncertain by about a factor of five. Although we should be able to use our information about hadrons to study the time evolution of a hard struck system, there are several other problems with the estimate:

- Nobody said we should neglect interactions, and what has been produced is a "nothing happened, no interactions" criterion. Interactions are crucial to the phenomenon of nuclear filtering; we actually want them to happen. When they happen, the argument ceases to apply.
- There is a basic misconception, or vagueness, about the initial state. As emphasized earlier, the integration region of small quark separation enters the perturbative argument, without making the assertion that the quarks or proton were prepared into this region.*
- The expansion of a wave packet is a picture of the entire event, which is mainly inelastic. In fact, the argument is well applied to describe highly inelastic circumstances such as deeply inelastic shadowing. But coherence is very important in an elastic event, and the argument has just not addressed the contribution of this kind of channel. Basically, it is the channel which by definition does not "die" after the first time click, so the "decoherence" rule is inapplicable.
- The system is interacting, and the time evolution of interest depends on the interaction energies, not the free space spectrum. If one looked at an atom in an external field, then two properties - the atom's structure and the external field - combine for the interaction energy.† This information just cannot be obtained from the unperturbed spectrum. Taking small differences between the big energies assuming the free space spectrum is just not a safe step. Even if we used the free space spectrum of energies as an approximation for the interacting system, the multi-pion and the transverse momentum continuum contributions to the spectrum have been omitted. But these are contributions which begin right at the proton mass, and are therefore the smallest energy differences or longest time scales that we actually wanted to retain.

Despite these criticisms, the picture in the hadronic basis is useful, and the idea that the system is changing its makeup as we measure it in the nucleus is interesting and important. FLFS introduced an effective attenuation cross section which is time dependent in Ref. (15). Can this idea be tested?

On the "pro's" side, the argument gives an energy dependence to cross sections which can be used again and again: from hard scattering processes, to J/ψ formation[38], to diffractive processes. Its emphasis on energy rather that Q^2 is a feature that is intuitively appealing. Note however that Q^2 and the system energy in the lab are proportional in many standard processes. In the "cons" category, one may have to give up the factorization if one is going to study the k^- (light cone time) dependence, and then much of the perturbative formalism becomes less clear. Experimentally, the data analysis presented in Section 1 should *not* work if there is a rapid expansion of the cross section: instead, one would see a rather large cross section due to the low energies involved in the data. But the fits are quite good, providing evidence against the rapid expansion idea. Next, transparency scaling should *not* work if the system is expanding rapidly, because the interaction size sampled over the target would not be

---

* This is the answer to the mistaken argument using the $\Delta x \Delta p$ uncertainty principle that color transparency could not exist.

† The Stark effect, which occurs at second order perturbation theory, seems to be the non-relativistic phenomenon most like the two-gluon exchange responsible for geometric cross sections in the gauge theory.

set by Q^2. (If, for example, the cross section grew proportionately to the time interval after formation, maybe the system would scale in the variable $Q^2/A^{2/3}$.) *But there is evidence that transparency scaling does work.*

The scaling law analysis and extraction of a cross section from the BNL data are rather new. Taken together with the evidence for filtering of the oscillations, this author believes they provide conclusive evidence that we are on the right track. The two points of view, in the quark basis and hadron basis, are compatible if we realize that *the system that the exclusive measurement quantum mechanically selects* settles into a steady state in some (so far) unobservable time(s). Nevertheless, at this point proponents of the rapid expansion idea tend to look to the electroproduction data as evidence in favor of their approach. Perhaps more data will settle the question.

The following experimental trick can be used to investigate the question. Consider events with the same Q^2 but different lab energies. This can be arranged using the Fermi motion of the target - one can sample different center of mass motions in the lab for $2 \rightarrow 2$ events with the same s, t, u. By doing this one can investigate events that should be identical except for spending more or less time inside the nuclear target. The expanding cross section will have more time to expand (if this is an issue). Perhaps what would actually be observable would be a different σ_{eff}. We look forward to data on this question from the new BNL experiment soon.

III. TESTS, PREDICTIONS, SPECULATIONS

Color transparency needs be tested in several ways. Assuming these tests will be passed, we can then be confident in the picture and turn to other applications using color transparency as a tool for exploring hadron structure.

We believe that focusing on small details would not be the right road to establishing a general picture. A particular value of an observable cannot be considered a test at all so long as one can fit it with some unknown parameter. So, for the first level of tests, we want to look at qualitative trends and to make sure that we have all of the big effects under control.

A: Tests of the Overall Picture
Hadron Helicity Conservation

We have seen that hadron helicity conservation (hhc) is a test of short-distance dominance. If nuclear filtering is correct, then for $A >> 1$ we should be approaching a short distance limit, even in processes where Q^2 is not large enough to select short distance.

One therefore looks for processes which can test for restoration of hadron helicity conservation in a nuclear target:

The Analyzing Power. Experiments on free space $pp \rightarrow pp$ scattering show that an unpolarized initial state produces a polarized final state. The final state spin direction is set by the scattering plane, consistent with parity conservation. This effect violates hhc, although it has been measured at the large momentum transfer of 13 GeV^2. It should disappear when measured for $A >> 1$. The same goes for transitions to more exotic final states: for example, $p\bar{p} \rightarrow pn^*, p\bar{p}, \Lambda, \bar{\Lambda}$. The helicity conserving amplitudes should be bigger than the helicity violating ones for $A >> 1$.

$\pi\mathrm{p} \rightarrow \rho\mathrm{p}$. In this experiment, Heppelmann *et al* observed the final state ρ to have a spin density matrix containing off diagonal elements forbidden by hhc. The same experiment can be done on a large nuclear target, where we should see hhc restored. Reversing the reaction, in photoproduction of ρ's, an asymmetry violating hhc due to vector meson dominance will occur on a proton target and disappear for $A >> 1$.

Inclusive Production of Mesons and Hyperons at Large x. Several experiments show an inexplicable polarization of final state Σ and Λ hyperons produced inclusively on proton targets. There is also experimental information on the asymmetry of inclusive pions produced as a function of x_F. These experiments indicate that orbital angular momentum flows from the initial state to the final state in a leading particle effect. Perturbative QCD cannot explain these effects, which appear to actually increase as a function of momentum transfer, but it is not clear how leading particle effects are to be put in to a perturbative description anyway. *If* the production is associated with a unit of orbital angular momentum, *then* they should be preferentially filtered away when going through a large nuclear target. This is more an application of nuclear filtering than color transparency.

We can estimate the rate of suppression of these processes due to nuclear filtering. It takes at least one unit of orbital angular momentum to violate hhc. When the b-space integrations are performed, we need to combine at least two terms going like b to make the amplitude. The ratio of short distance to helicity violating amplitudes thus goes like at least like b^2. Since the dominant integration regions that give survivors have $b^2 A^{1/3} <$ const, we expect the helicity violators to be suppressed by about $\text{const}/A^{1/3}$ relative to the helicity conservers.

The Limit s >> t

Elastic $pp \to pp$ scattering in the limit $s >> t$, $t \gtrsim \text{GeV}^2$ is a very interesting subject. The arguments of factorization and the quark counting process can be applied, implying that $d\sigma/dt = t^{-10} f(t/s)$, modulo logarithms. This is a dangerous limit, however, since there are many ways to generate large $\log(s/t)$; if a sum of logarithms is exponentiated, then one can build a power of s/t. At the same time, the independent scattering diagrams are also present, and these predict $d\sigma/dt = t^{-8}$ modulo logarithms, as shown by Donnachie and Landshoff[40]. Their prediction of t^{-8} behavior is stunningly confirmed by data over many decades of cross section. It is rather striking evidence that independent scattering is at work.

But if the process is going through independent scattering then the systems are "large"and should be filtered away in a nuclear target. For $A >> 1$, we should see the short distance prediction confirmed.

Flavor Flow

With some modeling of the distribution amplitudes, theory at present can calculate the relative normalizations of different flavor channels in "quark counting" exclusive processes: for example, the ratio of $\pi p \to \pi p$ compared to $\pi p \to k\Lambda$. The calculations are complicated but can be automated by computer (as suggested by Farrar). However, the independent scattering region ruins this program. It is interesting, then, to extend the study to nuclear targets and see what relative normalizations change. Would the relative normalizations predicted by quark-counting be correct after nuclear filtering?

Another question is the normalization of the $p\bar{p}$ $90°$ differential cross section compared to the pp case. It is approximately 50 time smaller. Naively using crossing symmetry $-t \to 2s$, one obtains a ratio of $(1/2)^8 = 1/64$. This rough agreement may be an indication that independent scattering may be responsible.[41]. It would be interesting to test in a nuclear target. One would expect the ratio of $p\bar{p}/pp$ processes to decrease with A as the independent scattering proportion is depleted.

B: Probes of Hadron Structure

Angular Momentum. From nuclear filtering we have seen that states containing a large component of quark orbital angular momentum should be suppressed in traveling through large nuclear targets. Some interesting processes require a unit of orbital angular momentum. For example, a transition form factor such as $\gamma^* p \to \Delta_{3/2}$ with a helicity flip cannot go through $m = 0$ states. Recently we showed[42] that certain helicity-flip form factors are "calculable" in pQCD although they are zero in the quark counting factorization. By studying this sort of observable and comparing free space and nuclear target experiments, one might be able to determine the relative amplitudes of different angular momentum components.

Anomalous Color Transparency. In exclusive photoproduction of $\pi^{o'}$s off large nuclei by the "Primakoff effect", the anomalous $\gamma\gamma\pi^o$ vertex couples the photons to a two-quark component of the pion. Because of the anomaly, and because the process probes the two-quark, rather than multi-quark pion wave function, one can argue[43] that the experiments will see an anomalously small pion-nucleus cross section as a function of energy. This is an extremely interesting probe into the space-time characteristics of an anomalous interaction.

Diffractive ρ's. ρ mesons quasi-elastically produced in photoproduction can be studied as a function of the virtual photon Q^2, the momentum transfer t, the polarization, and the nuclear size A. Very recent data[44] from Fermilab shows a small rise with Q^2 of the ratio of production off large nuclear targets to the production on ^2D. This seems to be evidence for color transparency. The data's Q^2 dependence seems to agree well with the prediction of Kopeliovich, Nemchick, Nikolaev and Zakharov.[45] Our own prediction is that the polarization should be transverse (non-short distance) at low Q^2, evolving to longitudinal (short-distance) at high Q^2. Due to filtering the same evolution towards longitudinal polarization should occur at $Q^2 \simeq GeV^2$ fixed as A increases. The t-dependence is also fascinating, because it can be used to move from coherent to incoherent production on the target. If these expectations work out, then we will have a laboratory with a number of experimental knobs to use to look inside the quark structure of the ρ.

States in Nuclei. If a ρ meson is produced in a hard exclusive reaction and also decays inside a large nuclear target, will the invariant mass and width of the rho be the same? According to our ideas about filtering, the wave functions will be significantly disturbed, and we would expect the effective mass to shift and the width to change. Clearly the idea can be extended to a host of resonances. Calculations are in progress on this.

Epilogue

To complete one of many stories here, I offer an explanation of whatever happened to *Le Passe-muraille*. If you recall, he had visited a physician, a very wise man who gave him a powder to cure the problem. And one day upon having a headache, our hero took the medicine. He walked out of his house through the wall, just as the medicine worked, and found himself trapped forever in stone.

We do not yet know if the ideas and calculations studying color transparency are a correct description of Nature. Some scientists find it all too incredible to believe. It is also possible that certain theoretical medicines prescribed by participating *physicien* can cure the disease disasterously. If what they say is corrrect, then shortly after a hard scattering a proton will grow too fat to be able to penetrate the nuclear medium. We hope not: the incredible predictions of pQCD have a tendency to come true! We need experiments to tell us what happens, and there will be plenty of those.

Acknowledgments: This work was supported in part by the Department of Energy Grant No. FG02-85ER40214.A009 and the Kansas Institute for Theoretical and Computational Science. I thank my collaborators Bernard Pire and Pankaj Jain for their insights and many hours of patient discussions. I also thank NATO for sponsoring the school, and the organizing committee and the participants of the Dronten summer school for their friendly hospitality.

REFERENCES

1) M. Aymé, *Le Passe-muraille*, Editions Gallimard (1943).
2) F. Low, *Phys. Rev.* **D12**, 163 (1975); S. Nussinov, *Phys. Rev. Lett.* **34**, 1286 (1975); J. Gunion and D.E. Soper, *Phys Rev* **D15**, 2617 (1977).
3) A. H. Mueller, in *Proceedings of the Seventeenth Rencontres de Moriond* (Les Arcs, France 1982) edited by J. Tran Thanh Van (Editions Frontiers, Gif-sur Yvette 1982); S. J. Brodsky, in *Proceedings of the Thirteenth International Symposium on Multiparticle Dynamic*, (Vollendam, 1982) edited by W. Kittel *et al* (World Scientific 1982); S. J. Brodsky and A. H. Mueller, *Phys. Lett.* B **206**, 685 (1988).
4) P. V. Landshoff, *Phys. Rev.* **D10**, 1024 (1974).

5) A. S. Carroll *et al.*, *Phys. Rev. Lett.* **61**, 1698 (1988).

6) S. Heppelmann, in *Proceedings of the Workshop on Future Directions in Particle and Nuclear Physics at Multi-GeV Energies*, edited by D. Geesaman *et al* (Brookhaven National Laboratory 1993) (in press); *Nucl. Phys. B* (Proc. Suppl.) **12**, 159 (1990).

7) B. Pire and J.P. Ralston, *Phys Lett.* **117 B**, 233 (1982).

8) J.P. Ralston and B. Pire, *Phys. Rev. Lett.* **49**, 1605 (1982).

9) S. J. Brodsky and G. F. de Teramond, *Phys. Rev. Lett.* **60**, 1924 (1988).

10) J.P. Ralston and B. Pire, *Phys. Rev. Lett.* **61**, 1823 (1988).

11) A. Sen, *Phys. Rev. D* **28**, 860 (1983).

12) J. Botts and G. Sterman, *Nuc. Phys. B* **325**, 62 (1989).

13) G. Ramsay and D. Sivers, *Phys Rev.* **D45**, 79 (1992).

14) C. Carlson, F. Myrher and V. Chashkuhnashvilli, *Phys Rev.* **D46**, 2891 (1992).

15) G. R. Farrar, H. Liu, L. Frankfurt and M. Strikman, *Phys. Rev. Lett.* **61**, 686 (1988).

16) B. Jennings and G. Miller, *Phys. Lett. B* **236**, 209 (1990); *Phys. Rev. D* **44**, 692 (1991); *Phys. Lett. B* **274**, 442 (1992); TRIUMF preprint 1993.

17) B. Z. Kopeliovich, *Sov J Part Nuc* **21**, 117 (1990); B. Z. Kopeliovich and B. G. Zakharov, *Phys. Lett. B* **264**, 434 (1991).

18) N.N. Nikolaev, Landau Institute preprint 9–93, to be published in *Surveys in High Energy Physics* and references therein; V. V. Anisovich *et al. Phys. Lett. B* (1992); Genoa preprint 1993.

19) T.-S. H. Lee and G. A. Miller, *Phys. Rev. C* **45**, 1863 (1992).

20) O. Benhar, A. Fabrocini, S. Fantoni, G.A. Miller, V. R . Panharipande and I. Sick, *Phys Rev. C* **44**, 2328 (1991).

21) P. Jain and J. P Ralston, *Phys. Rev.* **D48**, 1104 (1993).

22) S. J. Brodsky and G. P. Lepage, *Phys. Rev.* **D22**, 2157 (1980); *Phys. Rev.* **D24**, 2848 (1981).

23) See, e.g. B. Fillipone and SLAC NE-18 collaboration, in *Proceedings of PANIC 93* (Perugia, Italy) (in press).

24) N. Isgur and C. Lleyellyn-Smith, *Phys Rev. Lett.* **52**, 1080 (1984); *Phys. Lett. B* **217**, 535 (1989).

25) H. Li and G. Sterman, *Nuc. Phys.* **B381**, 129 (1992); H. Li, Stony Brook preprint ITP–SB–92–25.

26) J. P. Ralston, *Phys. Rev. A* **40**, 4872 (1989).

27) S. Brodsky and G. R. Farrar, *Phys. Rev. Lett* **31**, 1153 (1973); *Phys. Rev.* **D11**, 1309 (1975); V. Matveev *et al, Lett Nuovo Cim* **7**, 719 (1972).

28) See. e.g. J. D. Bjorken and S. D. Drell, Relativistic Quantum Fields, (McGraw-Hill, 1965) Ch 18.4 -18.6.

29) See A. H. Mueller, *Phys. Rep.* **73**, 237 (1981). This review is one of the classics on Sudakov effects.

30) J. P. Ralston and B. Pire, *Phys. Rev. Lett.* **65**, 2343 (1990).

31) B. Pire and J. P. Ralston, *Phys. Lett. B* **256**, 523 (1991).

32) P. Jain and J. P. Ralston, in *Proceeding of the XXVII International Rencontre de Moriond* (Les Arcs, France, 1993) edited by J. Tran Thanh Van (Editions Frontiers, Gif sur Yvette, France) in press.

33) V. Sudakov, *Soviet JETP* **3**, 65 (1956).

34) See, e.g. Landshoff Ref (4); G.R. Farrar and C.–C.Wu, *Nuc. Phys.* **B85**, 50 (1975).

35) J. P. Ralston and B. Pire, Kansas and Ecole Polytechnique preprint (1992).

36) See, e.g. G. R. Farrar, L. Frankfurt and M. Strikman, *Phys. Rev. Lett.* (Comments) **67**, 2111(1991); J. P. Ralston and B. Pire *ibid* (response) 2112 (1991); L. Frankfurt and M. Strikman, in *Progress in Nuclear and Particle Physics* (1991) v. 27.

37) Compare the argument to B. Jennings and G. A. Miller, TRIUMF preprint 1993.

38) G. R. Farrar, H. Liu, L. Frankfurt and M. Strikman, *Phys. Rev. Lett.* **64**, 2296(1990).

39) J. Botts, *Nuc Phys B* **353**, 20 (1991).

40) A. Donnachi and P. Landshoff, *Zeit. Phys* **C2**, 55 (1979); erratum **C2**, 372 (1979).
41) G. Sterman (private communication) has made the same observation.
42) P. Jain and J. P. Ralston, in *DPF 92* (Fermilab 1992) edited by C. Albright et al. (World Scientific, 1993).
43) J. P. Ralston, *Phys Lett. B* **269**, 439 (1991).
44) G. Fang, in *Proceedings of PANIC 93* (Perugia, Italy) (in press).
45) B.Z. Kopeliovich, J. Nemchick, N.N. Nikolaev and B.G. Zakharov, *Phys. Lett.* **B309**, 179 (1993).

PHOTON AND MESON PRODUCTION IN ULTRA-RELATIVISTIC NUCLEUS–NUCLEUS COLLISIONS

Herbert Löhner

Kernfysisch Versneller Instituut
Rijksuniversiteit Groningen
The Netherlands

INTRODUCTION

In high-energy collisions of heavy nuclei the interaction of baryons and their parton constituents leads to abundant production of mesons, so that a hadronic system is created with a particle density much larger than the mere baryonic superposition density of $2\gamma\rho_0$. Here ρ_0 is the normal nuclear density of $0.17/\text{fm}^3$ and γ the center-of-mass Lorentz factor with a value $\gamma \approx 10$ at the SPS energy of 200 GeV/nucleon. Collisions among the constituents of the dense hadronic system lead to thermal excitation. Eventually, a physical situation similar to the big-bang scenario might thus be established in the laboratory. This allows to investigate the existence and the nature of the phase transition to quark matter[1] which is predicted by QCD lattice calculations[2, 3, 4, 5] in hadronic-matter systems at high energy density. The study of the thermodynamic behaviour of a strongly-interacting-matter system will provide an interesting test of the confinement property of QCD. The interaction of elementary quark and gluon constituents is weakend by colour screening at very short distances. Sufficiently hot and dense matter should therefore become a gas of noninteracting quarks and gluons which move quasifreely in a deconfined but overall colour neutral environment given by the dimensions of the dense-matter volume. Numerical simulations of statistical QCD in the non-perturbative sector of QCD are very time-consuming calculations and still suffer from severe approximations. These refer to the quark-masses and the limited lattice size. The order of the phase transition is still not firmly predicted. Calculations with infinitly large or zero quark masses yield a first-order phase transition, while finite quark masses ($m_{u,d} = 10$ MeV, $m_s = 200$ MeV) seem to indicate a second-order phase change[5]. The required energy density is about $1 - 3$ GeV/fm^3 which is well above the energy density inside a single nucleon (≈ 0.5 GeV/fm^3) and about 10 times the energy density of normal nuclear matter. The change in energy density across the phase boundary reflects the latent heat residing in the about 10 times larger number of degrees of freedom in the quark-gluon system as compared to a π gas. A transition temperature $T_c \approx 160$ MeV appears to be a reasonable value[6] at zero baryonic chem-

Perspectives in the Structure of Hadronic Systems
Edited by M.N. Harakeh *et al.*, Plenum Press, New York, 1994

ical potential or zero baryonic density. This situation of purely mesonic matter might be accomplished in central collisions of heavy nuclei and observed in the phase-space region around midrapidity ($y_{cm} = 0$). The center-of-mass rapidity is defined as $y_{cm} = \tanh^{-1}(\beta_{\parallel}^{cm})$ with $\beta_{\parallel}^{cm}$ the longitudinal particle velocity in the center-of-mass of the colliding nuclei.

The experimental study of particles produced along the dynamical path of the colliding system in the plane of density and temperature allows to study the medium-modified properties of the hadronic interaction. These studies are performed at SIS (GSI) where 2 times normal nuclear density and temperatures of about 100 MeV are reached. Interactions among nucleons, produced mesons and nucleon resonances (Δ, N^*) seem to play a significant role at 1 GeV/nucleon incident energy[7]. At the SPS energy of 200 GeV/nucleon the conditions of a hadronic resonance gas[8] are met and the transition to the quark-gluon plasma might be reached. Special attention has to be paid to carefully investigate and understand the nature of hot hadronic matter before any conclusions on the observation of an abnormal phase can be drawn. The main experimental objective is to achieve a sufficiently large and long lived volume of matter close to thermal equilibrium at the required energy density. Further, multiple scattering of baryons and rescattering of produced particles has to be demonstrated to ensure thermalization in hadronic matter. Finally, observation of real or virtual photons from elementary interactions of the matter constituents should allow to deduce the temperature evolution of the hot matter from the thermal emission rates. Photons are particularly suited, since their long mean free path leaves the photon signal free from distortions due to hadronic final-state interactions. If quark matter has been produced in the course of the collision, a signal of colour deconfinement is expected from the spectroscopy of heavy quark bound states which would be suppressed due to colour screening[6].

Besides interactions with hadron beams a systematic study has been carried out since 1986 at the CERN SPS with ^{16}O and ^{32}S ions at beam energies of 200 GeV/nucleon and at BNL with Si ions of 14.6 GeV/nucleon. Experiments could demonstrate that indeed dense-matter systems in a transient state can be created this way[9].

In this lecture we discuss the experimental and theoretical investigations of real photon production in heavy-ion reactions at SPS energies. Photons and neutral mesons have been measured in the WA80[10, 11] experiment and with an upgraded setup in WA93. Lepton pairs and thus photons through lepton pair conversion have also been studied in the HELIOS[12] and CERES[13] experiments.

First we present the WA80 experiment. For the description of the experimental apparatus of other experiments we refer to the literature[9, 14]. Results from the recent heavy-ion experiments giving indications for hot and dense hadronic matter will be summarized. Then we discuss the theoretical investigations of the various sources of photons in nuclear collisions at high energies. Next we introduce the experimental work on the single photon analysis and present results on the photon signal compared to theoretical expectations. We close with an outlook to the forthcoming Pb+Pb experiments.

THE WA80 EXPERIMENTAL SETUP

The WA80 experiment served the purpose to study global characteristics of heavy-ion reaction events, and to investigate the momentum distributions of photons and neutral mesons in the high particle-multiplicity environment at midrapidity. The setup shown in fig. 1 combines an almost full coverage for charged particles with a calorimetric measurement of the energy flux at midrapidity, in the target-fragmentation region and at zero degrees. Details of the detection equipment are described in ref.[15, 16, 17].

After passing the beam-detector and start-timing system the beam hits the target in the center of the Plastic-Ball sphere[18]. This charged-particle spectrometer covers the backward angles and measures the distribution and the energy of target fragments in order to determine the excitation of target matter[19].

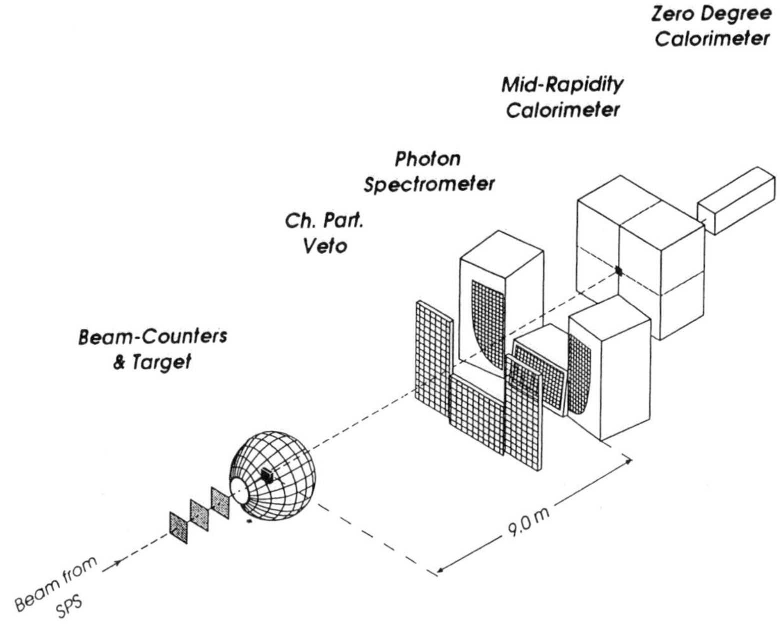

Figure 1. The setup of the WA80 heavy-ion experiment at the CERN SPS.

For the spectroscopy of photons a lead glass array of 3800 detector elements is employed and arranged in 3 sections. The detector design follows closely the original photon detector of WA80[17]. Photons are measured in a rapidity range $2.1 \leq y_{lab} \leq 2.9$ for 60% of the full azimuthal ring around the beam direction. Charged-particle rejection is achieved by two layers of streamer tube arrays[16] with an overall efficiency of 98%. .

The global and geometrical characteristics of the reaction are finally measured in sampling calorimeters at intermediate and forward rapidity. On basis of these calorimeters the online selection and offline definition of central reactions has occurred.

Neutral pions and η mesons are identified by their decay photons $(\pi^0, \eta \rightarrow 2\gamma)$ which develop an electromagnetic shower in the photon detector and deposit energy in a number of adjacent detector cells. Thus the position of an incident photon can be determined to better than the detector cell size. Transverse-momentum (p_T) distributions are obtained from the meson yield as extracted from the invariant-mass spectra. These are accumulated for different event classes in various bins of $p_T(\gamma, \gamma)$ from pairs of neutral hits. The meson efficiency depends sensitively on the signal/background ratio and the accurate subtraction of the combinatorial background which is due to unavoidable false pair combinations. However, hadron contaminations and merging showers in a high-multiplicity environment would lead to an increased background. An improved and fast photon and hadron identification method based on extensive test-beam studies of the shower profile has been developed recently[20]. A shower unfolding procedure is applied with good results down to relative shower distances of 1.5 detector cell units.

The precise shape of the combinatorial background is determined from polynomial fits to the invariant-mass spectra. For unfavorably small signal/background ratios, dominantly at low transverse momenta, the mixed events method has proven to be a very powerful tool. Compared to previous analyses the accessible lower transverse-momentum limit could be improved from about 500 MeV/c down to 200 MeV/c[21].

Finally, the obtained meson-momentum distributions are corrected for the photon reconstruction efficiency and the geometrical acceptance obtained by a Monte Carlo calculation. The details of this technique are described in ref.[17, 10]. The efficiency estimation is based on the actual experimental data and is performed by superimposing single hadronic and electromagnetic showers on the raw data level to the measured showers from single heavy-ion reaction events. The artificial events are then analysed with the same chain of shower reconstruction routines as is used for real data events. The quality of recognizing the superimposed showers is a measure of the efficiency which can be determined up to an accuracy (systematic + statistical) of about 3% .

INDICATIONS FOR HOT AND DENSE MATTER

Before any specific signal like e.g. thermal photons from hadronic interactions in hot hadronic matter or parton interactions in quark-gluon matter can be discussed and judged in comparison to theoretical studies, the general conditions prevailing in the excited-matter volume need to be analysed.

The degree of excitation is characterized by the amount of energy deposited in the volume of interacting target and projectile nucleons, the energy density ϵ_0. The energy density can be derived from final-state observables under assumptions on the reaction geometry and the kinematical evolution. In order to estimate the uncertainty, extreme cases have been considered. In the partial-stopping regime[22] at SPS energies the longitudinal growth of the reaction volume dominates the transverse expansion. Therefore the transverse size of the reaction volume is given by the projectile area πR_A^2 while the longitudinal extension of a unit volume for the calculation of energy concentration is determined by the formation time τ_0, the time needed before excited quark-objects appear as hadrons on the mass shell and interact with hadronic cross section. The energy concentration is given by the mean transverse energy per unit rapidity at midrapidity. These data are shown in fig. 2 for different selections of impact parameters as defined experimentally by different amounts of missing forward energy which thus must appear at midrapidity. The marked maximum of $\approx$ 100 GeV per unit rapidity corresponds to $\epsilon_0 = 2$ GeV/fm^3 if free hydrodynamic expansion[23] is assumed as derived by Bjorken:

$$\epsilon_0 = (dE_T/d\eta)_{max} / (\pi R_A^2 \cdot \tau_0).$$ (1)

The formation time is assumed to be 1 fm/c and will be discussed later in connection with the space-time evolution. If entropy conservation is considered, then for an isentropically expanding quark-gluon system an upper limit of the energy density of 3 GeV/fm^3 is estimated[14] for the SPS energy of 200 GeV/nucleon. In any case, the amount of energy stopping appears to be sufficient to create the conditions for a

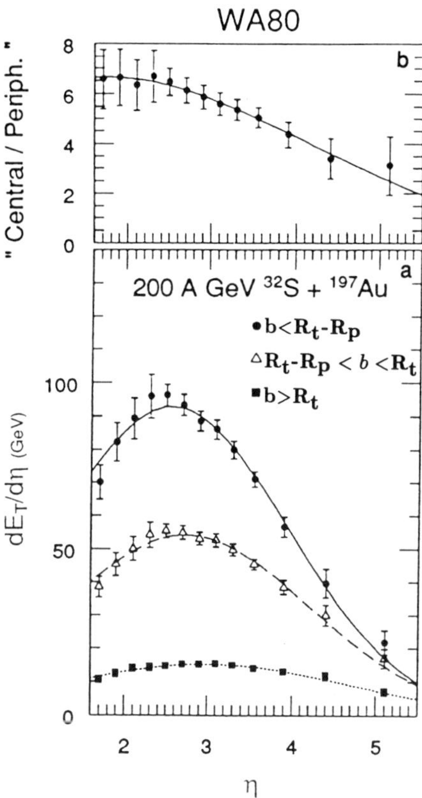

Figure 2. (a) Measured transverse energy distribution for 200 GeV/nucleon S+Au reactions as function of the pseudorapidity η = -ln tan($\theta/2$) for different estimated impact parameter ranges. The curves are gaussian fits to the data points. (b) The ratio of $dE_T/d\eta$ distributions from central and peripheral reaction events.

quark-gluon plasma transition if thermalization is taken for granted. The momentum distributions of identified mesons are studied to obtain further insight in the interaction dynamics of produced particles. The observed transverse-momentum spectra in p and heavy-ion induced reactions clearly differ from a pure exponential shape as expected for thermal behaviour but rather show a concave shape. These spectra are influenced by resonance decays, emission from different stages in the thermal evolution of the interaction and possibly by collective expansion in heavy-ion reaction systems. Also hard scattering of partons will contribute at large p_T. A thermal interpretation of transverse-momentum slope parameters determined at p_T near 1 GeV/c reveals the remarkable systematics shown in fig. 3. The "Boltzmann temperature" approaches $\approx$160 MeV rather independent of beam energy and for different hadron species at midrapidity.

The comparison between O+nucleus or S+nucleus and p+nucleus spectra shows a systematic dependence on A similar to the nuclear enhancement observed earlier[24] in p + nucleus as compared to p+p cross sections (Cronin effect). The target dependence has been parametrized by

$$d\sigma/dp_T(p + A) = A^{\alpha(p_T)} d\sigma/dp_T(p + p) \qquad (2)$$

with an empirical p_T dependent exponent $1 \leq \alpha(p_T) \leq 1.2$ where $\alpha = 1$ is the limit of mere superposition of independent nucleon nucleon collisions. An explanation for the

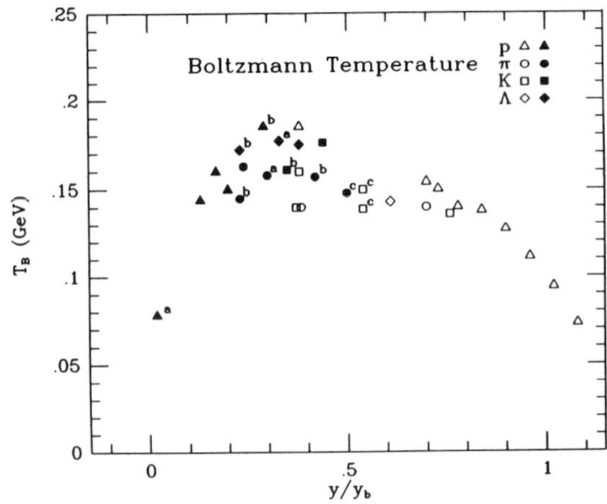

Figure 3. Boltzmann temperatures T_B derived from slope parameters of hadron-transverse-momentum distributions in the p_T range $0.5 - 1$ GeV/c as function of the rapidity normalized to the beam rapidity, from[14]. Data are from 200 GeV/nucleon S+(Au,W) (solid symbols) and 14.6 GeV/nucleon central Si+Pb reactions (open symbols). For comparison data for O-projectiles (a), S+S reactions (b) and minimum-bias collisions (c) are included.

power-law dependence on A consistent with the experimental systematics is multiple low-momentum scattering of partons in the hadronic medium. Quantitative calculations have been carried out for p+A reactions[25]. The projectile dependence was found to follow a similar parameterization[26]

$$d\sigma/dp_T(B + A) = B^{\alpha(p_T)} d\sigma/dp_T(p + A). \qquad (3)$$

The cross sections of central and peripheral S + Au reactions have been compared to p+p data in a similar rapidity range near midrapidity. The π^0 cross sections divided by the charged π cross sections from p+p reactions[27] are shown in fig. 4. The ratio shows a marked increase at large p_T which is more pronounced for the central reactions. Here the larger effective interaction volume leads to a stronger influence of multiple interactions.

Enhanced strangeness production has been predicted as another signal for the quark-gluon plasma. Gluon fusion would lead to increased $s\bar{s}$ production which is energetically more favoured than the associated production p+n $\rightarrow$ Λ^0 + K$^+$ + n. At high baryon density, as expected in the fragmentation region, the light quark pair production is hindered due to the large baryon chemical potential in contrast to $s\bar{s}$ production. The dominating amount of light quarks over light anti-quarks favours the production of K$^+(u\bar{s})$ over K$^-(\bar{u}s)$ and would lead to a significant change in the K$^+/\pi^+$ and K$^-/\pi^-$ ratios.

These ratios have been measured at BNL and SPS energies. Fig. 5 shows recent heavy-ion results from NA34[28] compared to the corresponding ratios from p+p reactions. A marked increase above the p+p level is observed for S+W reactions in the range $1\leq y_{lab} \leq 1.5$ near the baryon rich target fragmentation region. The same increase in K$^+/\pi^+$ from the p+p level to p+nucleus and and nucleus+nucleus data is observed at the BNL energy. This suggests that rescattering is the responsible mechanism which

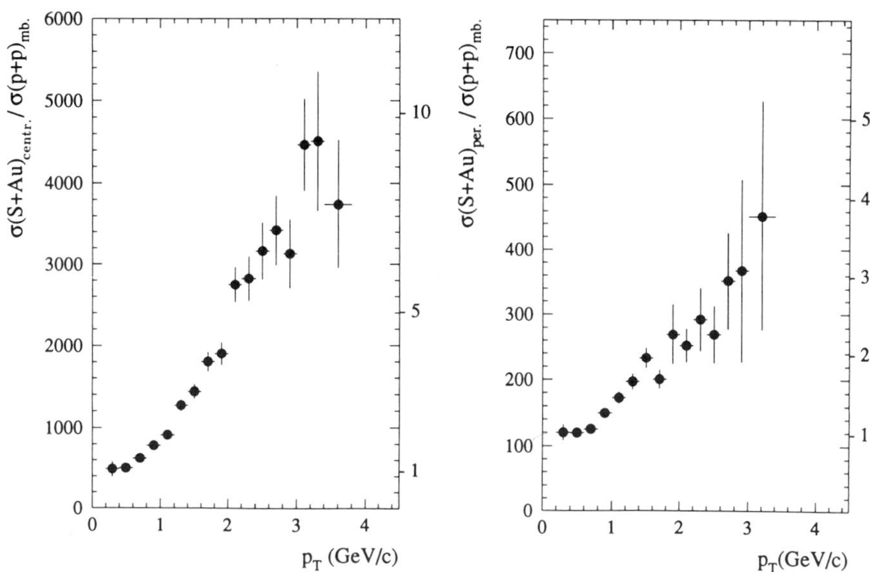

Figure 4. Ratios of π^0 p_T distributions from central and peripheral S+Au reactions at 200 GeV/nucleon to charged π cross sections from p+p reactions.

may lead the heavy reaction system close to equilibrium. The parameterization of K and π spectra with the α dependence as discussed in connection with the Cronin effect reveals the interesting result that the above K/π ratios are satisfactorily described by the nuclear enhancement[29]. Non-equilibrium-model calculations[30] including multi-hadron dynamics have been applied to the BNL energy regime and explain successfully the $\pi^\pm$ and $K^\pm$ transverse-momentum distributions. On the other hand, the independent fragmentation mechanism as contained in the FRITIOF model[31] fails. Thus hadron interactions in dense matter appear to be of crucial importance to explain the observed meson ratios.

An impressive set of data has been collected by the NA38 collaboration on the production of J/ψ mesons in p+nucleus and nucleus+nucleus collisions as compared to p+p reactions[32]. A suppression of J/ψ mesons as a consequence of the Debye-screening of the $c\bar{c}$ binding potential in the presence of freely moving colour charges in a quark-gluon plasma has been predicted[33]. Experimental data on invariant-mass spectra of the $\mu^+\mu^-$ decay channel show indeed a marked decrease of the J/ψ production rate above the Drell-Yan continuum by a factor 3 for estimates of the energy density increasing from 1 to 3 GeV/fm^3. This observation is consistent with the expected plasma signature. However, scattering and absorption in the initial and final state have been investigated and reveal significant influence on the production and propagation of J/ψ mesons in dense matter, such that the experimentally studied transverse-momentum and energy-density dependence can be explained. A spectral analysis of heavy quark bound states is proposed[6] to differentiate the deconfinement scenario from scattering and absorption in dense hadronic matter.

In summary, sufficient indications have been gathered in recent high-energy heavy-ion experiments for the existence of dense and highly excited hadronic matter. Interactions of hadrons are of significant influence to explain observed data so that approach to thermal equilibrium can be concluded. Direct observations of initial-state features,

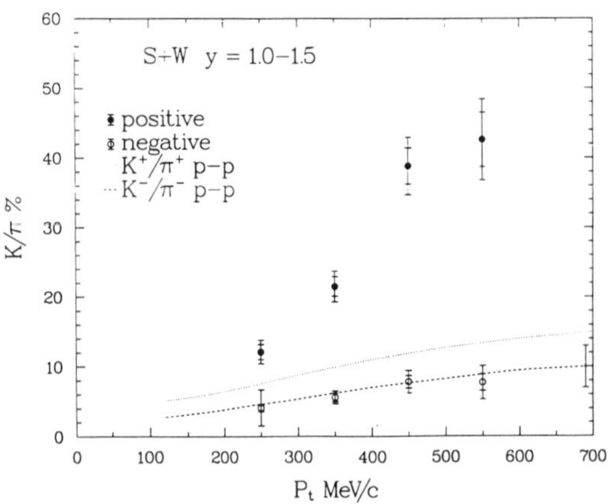

Figure 5. K/π ratios measured at $1.0 \leq y \leq 1.5$ as function of p_T for 200 GeV/nucleon S+W reactions. The dotted an dashed lines represent the K^+/π^+ and the K^-/π^- ratios from p+p data, respectively. From [28].

unmodified by hadronic interactions, are thus desirable and expected from electromagnetic probes. Real photon production has been investigated. The photon yield should allow to draw a conclusion on the temperature of the emitting source.

SOURCES OF SINGLE PHOTONS

The emission rates of real photons and lepton pairs from virtual photon decay are a useful tool to study the interaction rate among constituents of matter investigated in high-energy hadronic or nuclear collisions. Since the mean free path of electromagnetically interacting probes is much larger than the transverse size of colliding nuclei, the signal may be observed undisturbed by final-state interactions. Lepton pairs are being studied in a large-solid-angle pair spectrometer[13] which requires careful investigation of background from like sign pairs. Here we concentrate on real photons, called single photons in contrast to decay photons from meson decays which constitute a severe but manageable background for the single photon signal.

Photons from hard parton scattering

The production of high-energy direct photons from large momentum transfer hadron-hadron collisions has been studied extensively[34, 35]. In the first stage of a hadronic collision photons are produced in elementary interactions of valence and sea quarks. Their distribution is described by the structure functions $G(A, a, x)$ giving the probability of finding a parton a in a hadron A with a fraction of the hadron momentum between x and $x+dx$. The structure functions are obtained by e.g. deep inelastic lepton scattering experiments and must be extrapolated, if necessary, to the proper momentum transfer Q^2 at which they are applied. The photon cross section σ_γ for a hadron collision $A + B$ is obtained by integration over the elementary subprocess cross section $\hat{\sigma}(a + b \rightarrow \gamma + c)$ for hard scattering of parton a and b into parton c and γ weighted by the structure

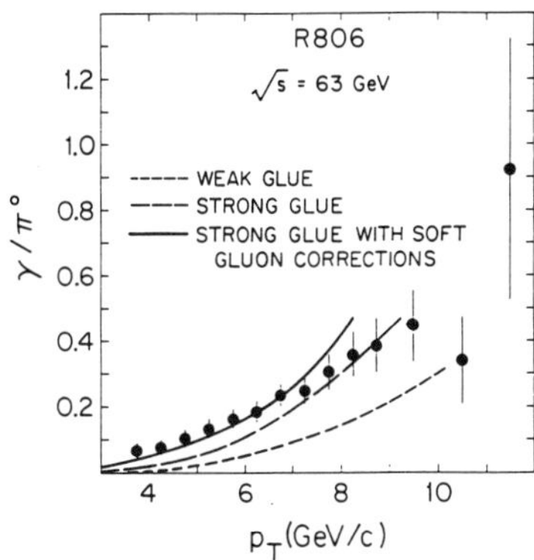

Figure 6. The γ/π^0 ratio for p+p reactions at 63 GeV center-of-mass energy. The curves are fits for different assumptions concerning the gluon structure functions, from [36].

function:

$$E_\gamma \frac{d^3\sigma_\gamma}{d^3p_\gamma}(A+B \to \gamma + X) \; = \; \sum_{abc} \int dx_a \int dx_b \, G(A,a,x_a,Q^2) \, G(B,b,x_b,Q^2) \, E_\gamma \frac{d^3\hat{\sigma}}{d^3p_\gamma} \tag{4}$$

The elementary parton subprocesses for high-p_T γ production correspond to 2-body scattering and are calculated in lowest-order perturbation theory. These are the 'QCD-Compton' process $qg \to q\gamma$ and the annihilation $q\bar{q} \to g\gamma$ which are of order $\alpha\alpha_S$ in electromagnetic and strong coupling. Higher-order terms like bremsstrahlung and soft gluon corrections are found to be important. The photon cross section is usually compared to hadron production represented by the π^0 cross section. Thus conveniently the ratio of cross sections for single γ and π^0 production, in short γ/π^0, is presented as a function of transverse momentum p_T. An instructive example[36] is shown in fig. 6 for ISR results from p+p reactions at center-of-mass energy $\sqrt{s} = 63$ GeV. Information on the importance of gluon contributions to the structure functions could thus be obtained. The ratio γ/π^0 increases with p_T since hadrons originate from jet fragmentation and receive a momentum fraction while photons carry essentially all the final-state momentum of the partonic subprocess. On the other hand, extrapolating the hard scattering γ production to small Q^2 into the thermal regime at low $p_T \approx 1$ GeV/c, where no rigorous calculations are available yet, we expect a contribution of $\gamma/\pi^0 \approx 0.01$.

Photons from the quark-gluon plasma

If high-energy nuclear collisions lead to formation of a quark-gluon plasma the emission rate of photons from parton interactions in a thermal environment may provide information on the temperature of the system[37]. The calculations proceed in analogy to those introduced in the previous section. Instead of the structure functions the thermal distribution functions of partons describe the momentum spectrum of elementary interacting constituents. However, low Q^2 processes have to be considered which are complicated by infrared divergences. In a first attempt this problem was

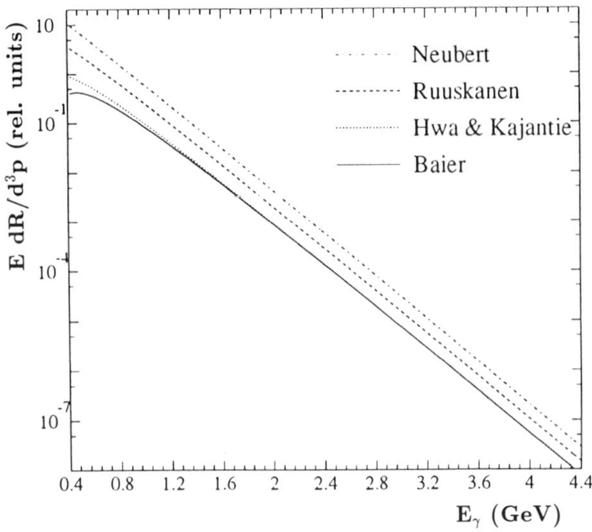

Figure 7. Calculated thermal photon emission rates for different theoretical approaches at a temperature of 200 MeV.

bypassed by introducing an infrared cutoff at the current light-quark mass $m_q \approx 5$ MeV[38].

Recently, the resummation method for finite-temperature perturbation theory[39] was applied to real-photon production[40, 41]. The quark-mass singularity is dynamically screened by many-body effects: One-loop corrections to the photon self-energy are calculated in the high-temperature limit with effective (dressed) quark propagators. An effective cutoff $k_c^2 = 2m_q^2(T)$ on the 4-momentum transfer is thus introduced which depends on the thermal quark mass $m_q^2(T) = \frac{2}{3}\pi\alpha_s(T)T^2$. For $T = 200$ MeV one obtains $\alpha_S \approx 0.3$ [42] which leads to $m_q(T{=}200\text{MeV}) \approx 280$ MeV which is considerably higher than the current light-quark mass. An analytic result for the photon rate at temperature T is obtained for the sum of annihilation and Compton contributions:

$$E_\gamma \frac{d^3 R_\gamma}{d^3 p_\gamma} = C \left(\sum_{n_f} e_q^2\right) \frac{\alpha\alpha_S}{2\pi^2} T^2 e^{-E_\gamma/T} \left(\kappa_1 + ln\frac{\kappa_2\,E_\gamma}{\alpha_S\,T}\right) \tag{5}$$

with $C = 1$, $\kappa_1 = 0$ and $\kappa_2 = 0.23$. Various approximations have been used which lead to a similar analytic dependence on E_γ and T but differ in the constants κ_1 and κ_2 and the absolute normalization C. Hwa and Kajantie[43] introduced an infrared cutoff at the thermal quark mass $k_c^2 = m_q^2(T)$. Ruuskanen[44] applied the full momentum dependence of the dressed quark propagator for the soft contributions to the photon rate like Baier[41], but used thermal distribution functions of partons in the Maxwell-Boltzmann approximation. Results of different evaluations at a temperature $T = 200$ MeV are compiled in[45] and compared in fig. 7 in the photon energy range where the high-temperature approximation gives reliable results. The thermal propagator approach of Baier gives about a factor 5 lower photon rates than the evaluation of Neubert, where many-body screening effects were not taken into account.

Photons from hot hadronic matter

If nuclear collisions lead to the formation of hot hadronic matter, then photon emission from the electromagnetic coupling of interacting mesons (dominantly $\pi^{\pm}$) needs to be taken into account. The relevant processes have been studied recently by Kapusta[40]. Due to the large effective $\pi - \rho$ coupling strength the ρ meson interaction gives much stronger contributions than η or ω interactions. Annihilation and Compton processes were considered among which $\pi\pi \to \rho\gamma$ dominates at small photon energies and $\pi\rho \to \pi\gamma$, with higher available energy due to the ρ-mass in the initial state, contributes most in the tail of the photon spectrum. With thermal momentum-distribution functions at temperature T the hadronic collisions give rise to similar photon rates as expected from parton interactions at the same temperature. Recently also the importance of the $A_1(1260)$ resonance formation in the process $\pi\rho \to A_1 \to \pi\gamma$ has been pointed out[46] with a photon rate competing with hadron scattering.

Thus a photon signal is expected even if the plasma phase has not been reached. Since the hadronic-matter phase is expected to dominate at a lower temperature than the quark-gluon plasma, the spectral shape of the photon signal needs to be analysed and should allow to separate the different contributions.

SPACE-TIME EVOLUTION

In order to evaluate the photon yield that can be compared to experimental data the space-time evolution of the interacting system needs to be considered. The reaction volume develops from proper time $\tau=0$ when projectile and target nuclei overlap to maximum energy density in a highly excited and thermalized stage at τ_i with initial temperature $T_i(\tau_i)$. Subsequently the system cools down to the freeze-out temperature $T_f(\tau_f)$ where interactions cease. A first-order phase transition at proper time τ_c with $\tau_i < \tau_c < \tau_h < \tau_f$ and critical temperature T_c might lead to an extended mixed phase at constant temperature T_c until at time τ_h the plasma is completely converted to the hadronic phase. The expected temperature profile will influence considerably the photon yield so that estimates for the space-time evolution are highly desirable. Since microscopic calculations for the complete dynamical evolution of hadronic and quark-gluon matter are unavailable or only being studied recently[47], various attempts based on hydrodynamical expansion and an ideal gas equation-of-state have been made[38, 48, 49]

The initial time or thermalization time τ_i is of the order 1 fm/c[23] and is related to the matter formation time τ_0 deduced from hadron production in high-energy hadronic collisions[50]. Since τ_i is related to the colour neutralization rate in strong colour fields, which in turn depends on the number of colour charges and binary collisions, a dependence on nuclear mass A like $\tau_i = \tau_0\,A^{-\delta}$ has been derived[51] with δ in the range of 1/6 to 1/3 [52]. Thus faster thermalization in larger volumes of matter is expected and values of $0.1 \leq \tau_i \leq 1$ fm/c indicate the range of uncertainty.

Hydrodynamical model calculations have been applied to describe the expansion scenario which proceeds isentropically and is dominated by longitudinal expansion. Hadron distributions have been analysed to derive the additional contribution from transverse expansion[53] which is neglected in the following. Conservation of the total entropy S thus yields

$$dS/dy = s(\tau)\,dx_T\,\tau = const. \qquad (6)$$

with entropy density $s(\tau)$ and the transverse size dx_T given by the area πR_A^2 of the projectile nucleus A. In a given rapidity interval dy we thus obtain $s(\tau) \cdot \tau = const.$

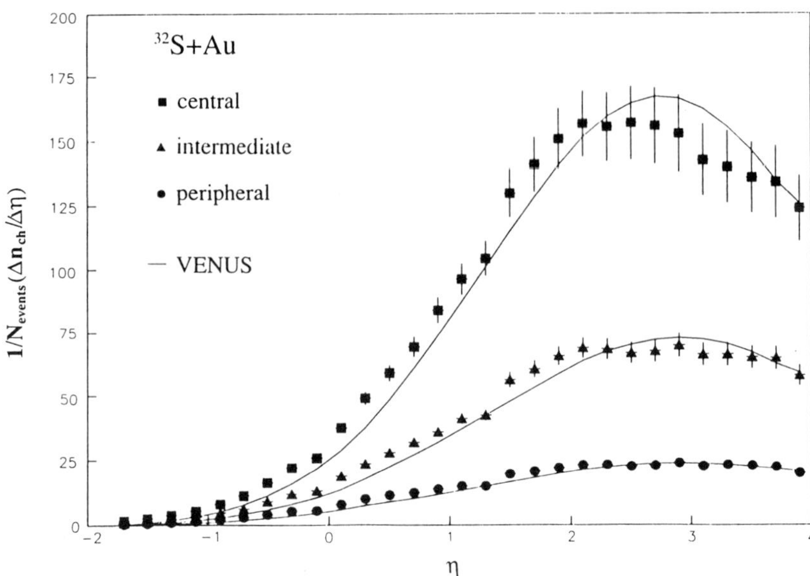

Figure 8. Measured charged-particle density for 200 GeV/nucleon S+Au reactions as function of the pseudorapidity η. Curves are calculations from the VENUS string fragmentation model [55].

and with the ideal gas equation-of-state $s \sim T^3$ also $T^3 \cdot \tau = const.$. The initial entropy density can be related to the final-state particle (i.e. dominantly π) density dN/dy assuming adiabatic decoupling of pions[43, 49, 54]:

$$s(\tau_i) = 3.6 \frac{dN}{dy} / (\pi R_A^2 \tau_i) = \frac{4\pi^2}{90} a_p T_i^3 \tag{7}$$

with the number of degrees-of-freedom (d.o.f.) for an ideal quark-gluon plasma $a_p = 47.5$ for 3 flavours. Thus with reasonable assumptions on τ_i the initial temperature can be derived from the measured particle density. The charged-particle-density distribution from S+Au reactions is shown in fig. 8 as function of the pseudorapidity $\eta = -\ln \tan(\theta/2) \approx y$. Correcting the charged-particle distribution with a factor 1.5 for π^0 a $dN/dy \approx 240$ is obtained at midrapidity ($\eta \approx 2.8$). The correspondingly derived initial temperature for $\tau_i = 0.5$ fm/c is $T_i = 260$ MeV. The uncertainty in τ_i mentioned earlier translates into $210 \leq T_i \leq 440$ MeV in S+Au reactions at 200 GeV/nucleon incident energy. The cooling of the hot plasma proceeds isentropically with $T \sim \tau^{-1/3}$ until the transition temperature $T_c = 160$ MeV is reached at $\tau_c = (T_i/T_c)^3 \cdot \tau_i = 2.1$ fm/c. Again, an uncertainty in T_c between 160 and 200 MeV leaves us with $1 \leq \tau_c \leq 2$ fm/c. Assuming a first-order phase transition, the system enters the mixed phase where plasma and hadronic matter coexist during a time $\Delta\tau_{mix}$ dependent on the latent heat of the phase transition and thus on the difference in d.o.f. in both phases. Entropy conservation demands for hadronic and plasma entropy density s_h and s_p, respectively:

$$s_h(T = T_c) \cdot \tau_h = s_p(T = T_c) \cdot \tau_c \tag{8}$$

and this leads to:

$$\Delta\tau_{mix} = \tau_h - \tau_c = \frac{s_p - s_h}{s_h}\tau_c = \frac{a_p - a_h}{a_h}\tau_c. \tag{9}$$

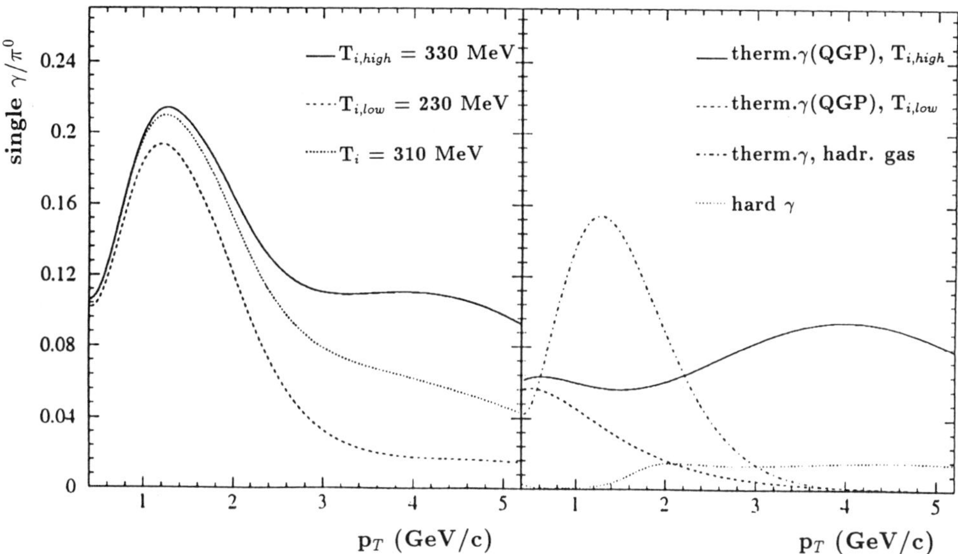

Figure 9. The calculated γ/π^0 ratio based on a dynamical situation as expected for 200 GeV/nucleon S+Au reactions for different initial temperatures and a critical temperature of 160 MeV.

Here a_h and a_p are the numbers of d.o.f. for the hadronic and the plasma phase, respectively, with $a_h = 3$ for the ideal π gas. A value $a_h = 6.6$ has been proposed[56] as an average value for an ideal gas of π, ρ and ω. Thus an extension of the mixed phase $\Delta\tau_{mix} = 30$ fm/c is estimated for $a_h = 3$ and $T_c = 160$ MeV. In spite of the large uncertainty range of 6 – 30 fm/c due to a_h and $\tau_c(T_c)$ the mixed phase is found to extend much longer than the plasma cooling time $\tau_c - \tau_i$. A considerable contribution to the spectral shape of the photon yield can thus be expected from the mixed phase corresponding to the transition temperature T_c.

Subsequently, the hadron gas cools down emitting photons at a rate determined by $T \sim \tau^{-1/3}$ for $\tau_h \leq \tau \leq \tau_f$ until the hadrons stream freely beyond the freeze-out time τ_f. A freeze-out temperature $T_f = 100$ MeV was assumed and the space-time integration was carried out[45] based on the photon rate calculated in[44] in order to obtain the photon yield

$$E_\gamma \frac{d^3\sigma_\gamma}{d^3p_\gamma} = \int_{\tau_i}^{\tau_f} d\tau\, dy\, \tau\, \pi R_A^2\, E_\gamma \frac{d^3 R_\gamma(T(\tau))}{d^3 p_\gamma} \tag{10}$$

in the rapidity interval dy at midrapidity. Results from these calculations for different initial temperatures T_i, $T_c=160$ MeV and $a_h=6.6$ are shown in fig. 9, where photon cross sections have been normalized to the measured π^0 production cross section at the corresponding p_T.

The summed production yield of all photon processes discussed earlier peaks slightly above $p_T = 1$ GeV/c with significant differences at large p_T for different initial temperatures. A considerable yield above the currently feasible detection limit of $\gamma/\pi^0 \approx 0.05$ may then be expected. Looking at the individual contributions to the photon yield, the yield from the hadron phase clearly dominates at low p_T and the relatively low initial temperatures expected for the S+Au reactions at SPS energies. Photons from hard parton scattering stay below $\gamma/\pi^0 \approx 0.02$ in the photon energy range currently accessible by experiment and thus below the detection limit.

SINGLE PHOTON ANALYSIS

The measured momentum distributions of π^0 mesons and the determination of the η/π^0 production ratio provide the basis for extracting the rate of single photons from the measured inclusive photon yield. Severe problems arise from the high particle multiplicity and the tremendous hadronic background of photons from neutral meson decays. The single photon to π^0 ratio (γ/π^0) is calculated from the total measured photon yield N_γ and the measured pion yield N_{π^0} according to the following expression:

$$\frac{\gamma}{\pi^0} = \frac{N_\gamma}{N_{\pi^0}} \cdot \frac{\epsilon_{\pi^0}}{\epsilon_\gamma} \cdot A_{geo} - (R_{\pi^0} + R_\eta + R_x). \tag{11}$$

The ϵ_{π^0} and ϵ_γ are the π^0 and photon reconstruction efficiencies, respectively, A_{geo} is the geometrical acceptance of the detector for π^0 and the R_i are the Monte Carlo calculated ratios of observed background photon yield from meson decays to the detected π^0 yield for each of the sources i. Besides photons originating from π^0 and η decays we also take into account the decays of η', ω, ϕ, K^0 and Σ^0 with contributions well below that from the η mesons. Measured p_T distributions for π^0 [10] and η mesons are the basis for the Monte-Carlo background calculation. The non-measured resonance decays contribute only $\approx 2\%$ to the photon background and are based on p+p data and scaling with the transverse mass $m_T = \sqrt{p_T^2 + m^2}$ [57] as

$$f(m_T; h) = C \cdot f(m_T; \pi^0); \quad h = \eta, \eta', \omega. \tag{12}$$

The η production is measured with the photon detector array exploiting the 38.9% branching into the $\gamma\gamma$ decay channel. For 200 GeV/nucleon O+Au reactions a value $\eta/\pi^0 = 0.61 \pm 0.2$ was measured for minimum-bias reactions in the p_T region 2 GeV/c $\leq p_T(\gamma\gamma) \leq 2.4$ GeV/c. This ratio is consistent with the η/π^0 ratio from p+p reactions. For the analysis of recent S+Au data the mixed events method to subtract the combinatorial background was developed and tuned to simulated data[21, 58, 59]. The accessible transverse-momentum range could be considerably extended into the region with a signal/background ratio as small as 0.2%. The η cross sections for S+Au reactions as measured in the range 0.5 GeV/c $\leq p_T(\gamma\gamma) \leq 2.5$ GeV/c are shown in fig. 10 together with the π^0 results. Following a suggestion by Hagedorn[8], the momentum distributions are parametrized by the following expression:

$$E \cdot \frac{d^3\sigma}{d^3p} = C \cdot \left(\frac{p_0}{p_T + p_0}\right)^n \tag{13}$$

with free parameters C, p_0 and n. The power-law ansatz is deduced in a model of hard collisions of quarks based on QCD. Transformed into the m_T scale, the resulting fit is included in fig. 10. Except for the absolute normalization, the same parameters also fit the η distribution, so that a ratio $\eta/\pi^0 = 0.61 \pm 0.08$ is obtained for the minimum-bias S+Au data.

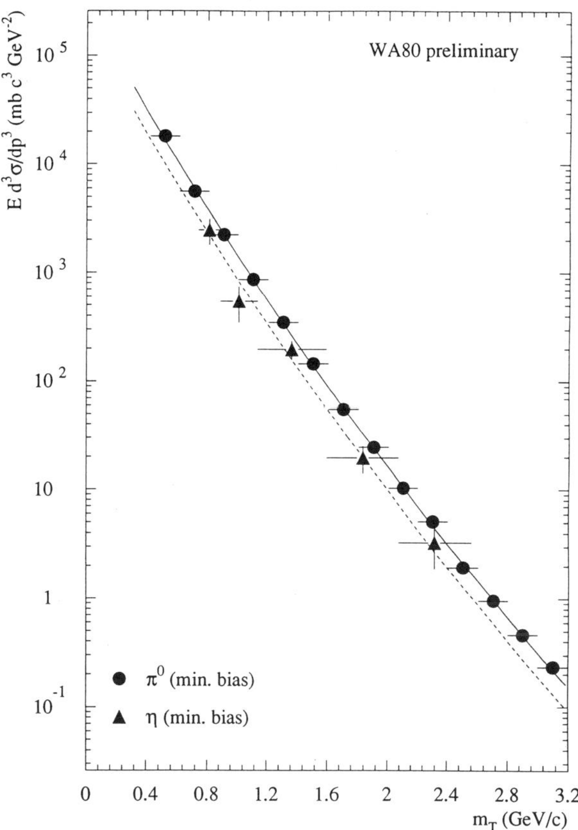

Figure 10. The invariant cross sections for π^0 and η mesons as function of the transverse mass m_T for minimum-bias reactions of S+Au at 200 GeV/nucleon.

RESULTS ON SINGLE PHOTONS

Since the transition to a quark-gluon-plasma state and hence an excess of single photons is expected to occur more likely in central than in peripheral reactions, the 200 GeV/nucleon S+Au data sample has been divided into peripheral and central events. This selection was made according to the transverse energy measured in the midrapidity calorimeter and results are shown in fig. 11. The circles indicate the experimental inclusive photon yield and the histograms are Monte Carlo results based on the measured π^0 and η yield and represent the hadronic background. A difference between experimental inclusive photon yield and the histograms must be attributed to the contribution of single photons. The general p_T dependence of the data is well reproduced by the Monte Carlo simulations. For central O+Au reactions investigated earlier[11] an upper limit for single photons of $\gamma/\pi^0 < 0.15$ could be determined with a 90% confidence level. Due to better statistical accuracy, improved η reconstruction capability and more reliable efficiency calculations the total error has appreciably been reduced for the new data. The central S+Au data show a slight overall enhancement over the hadronic background. Collecting all sources of statistical and systematic errors and assuming them to follow a gaussian distribution, we arrive at an uncertainty of 6–8% and 8–15% for p_T bins below and above 2 GeV/c, respectively. The excess of photons in central data over the expected yield from the hadronic background observed for $p_T < 2$ GeV/c reaches a level of 2 standard deviations[58, 60].

The remaining single photon yield, obtained by subtracting the calculated decay contribution from the inclusive photon yield, is shown in fig. 12 for central S+Au reactions. Other than in the peripheral data the central sample shows a positive signal and a slight increase towards low p_T. This feature resembles the spectral shape expected from hadronic interactions at a temperature below 200 MeV/c. Inclusive photon production has also been studied by the NA34 collaboration[12] for O+W data in the range 0.1 GeV/c $\leq p_T \leq$ 1.5 GeV/c using the conversion method where no π^0's were reconstructed. The hadronic background photons were instead calculated by assuming relations between measured charged pion cross section and the π^0 yield. In the region of mutual overlap (0.4 GeV/c $\leq p_T \leq$ 1.5 GeV/c) the O+W and O+Au inclusive photon results agree and are compatible with the expectations from hadronic decays. Recent data from the NA45-CERES dilepton spectrometer[13] could provide a single photon estimate of $\gamma/\pi^0 = 0.08 \pm 0.11$ averaged in the range 0.4 GeV/c $\leq p_T \leq$ 2.4 GeV/c for S+W reactions.

At this stage of the data analysis, however, no quantitative comparison to calculated photon spectra will be made. This is because large uncertainties in the estimation of the initial conditions only allow rough predictions. The experimental analysis is extremely tedious due to critical Monte-Carlo procedures in the determination of the photon and π^0 efficiency. Therefore an independent re-analysis is planned in order to gain extra confidence in the estimates of systematic errors. The measured photon yield could then be used to estimate the hadronic temperature from the low p_T yield and to set limits on the contribution from a high-temperature quark-gluon plasma phase at high p_T.

OUTLOOK TO FORTHCOMING EXPERIMENTS

New experiments are being planned at the CERN SPS for the near future. For the first time Pb ions will be accelerated to 160 GeV/nucleon. In Pb+Pb collisions

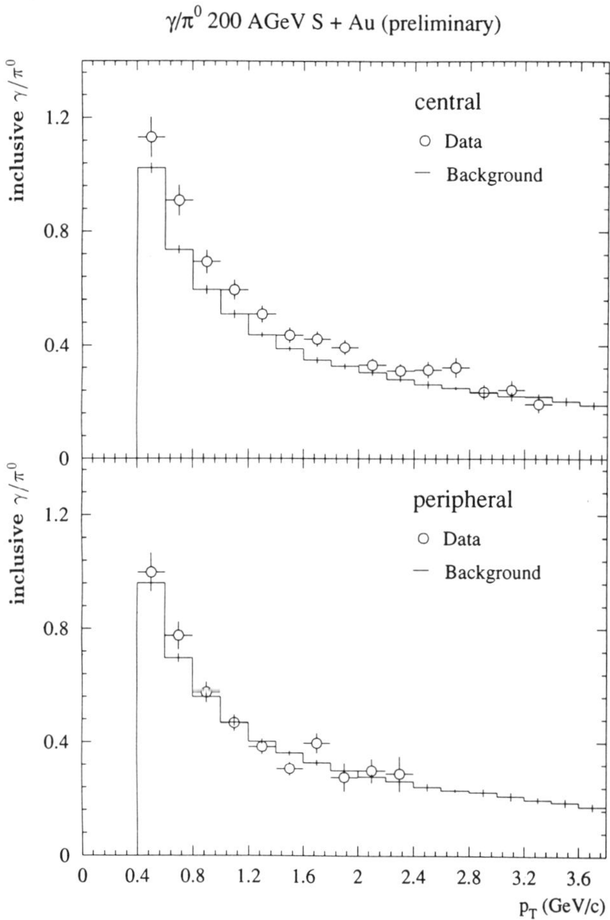

Figure 11. Ratio of the inclusive photon cross section to π^0 cross section as a function of transverse momentum for central and peripheral S+Au reactions at 200 GeV/nucleon. The circles indicate the data, and the histograms are the Monte Carlo results representing the hadronic decay background.

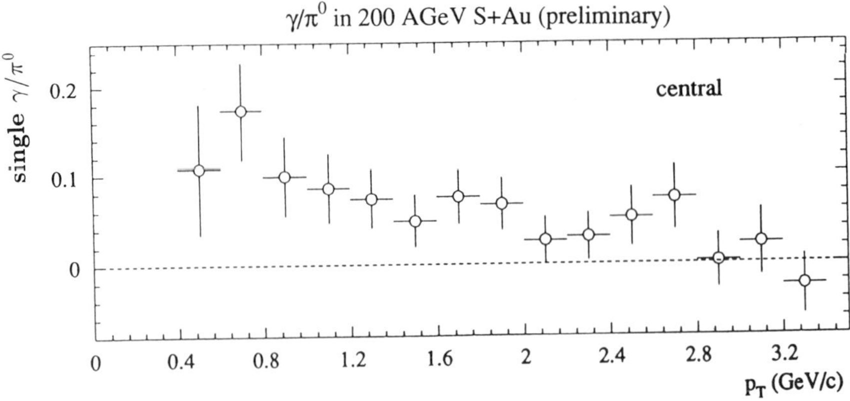

Figure 12. Single photon yield for central S+Au reactions at 200 GeV/nucleon. The data are obtained by subtracting the Monte Carlo calculated hadronic decay background from the inclusive γ/π^0 spectrum as shown in fig. 11.

at this energy interactions volumes above 7000 fm^3 are likely[6] so that truely thermal behaviour of hot hadronic matter may be expected. The above discussed A dependence of the thermalization time could lead to an initial temperature as high as 400 MeV. The photon yield expected in this case and in more pessimistic situations is shown in fig. 13[45]. At sufficiently high initial temperature above 300 MeV the contribution from the quark-gluon plasma at large p_T is obvious. The spectral shape of the photon yield could thus not only provide the temperature of the emitting system but also might indicate a possible phase change and reveal the critical temperature from the low p_T photon yield.

Recently, photon intensity interferometry has been studied theoretically[61]. The correlation functions reflect the different time scales of the system in the plasma, hadronic or mixed phase. Since photons from the plasma phase dominate at high p_T, the correlation functions for high-p_T photons are found to be sensitive to the space-time evolution of the quark-gluon plasma.

The largely extended WA80 photon detector will be implemented in the WA98 setup for Pb+Pb collisions. About 10000 Pb-glass elements might provide sufficient spatial resolution to investigate photon correlations experimentally. A photon multiplicity detector is included to allow selection of photon-rich events. Charged hadrons will be measured in a dipole spectrometer, so that electromagnetic and hadronic signals can be correlated. In this way a sensitive selection of reaction events is expected that may reveal the quark-gluon plasma signature.

SUMMARY

The thermal photon signal to reveal the quark-gluon plasma in high-energy nuclear collisions has been investigated theoretically and experimentally. A limit in the ratio of single photons divided by the π^0 yield of about 0.05 has been achieved with advanced

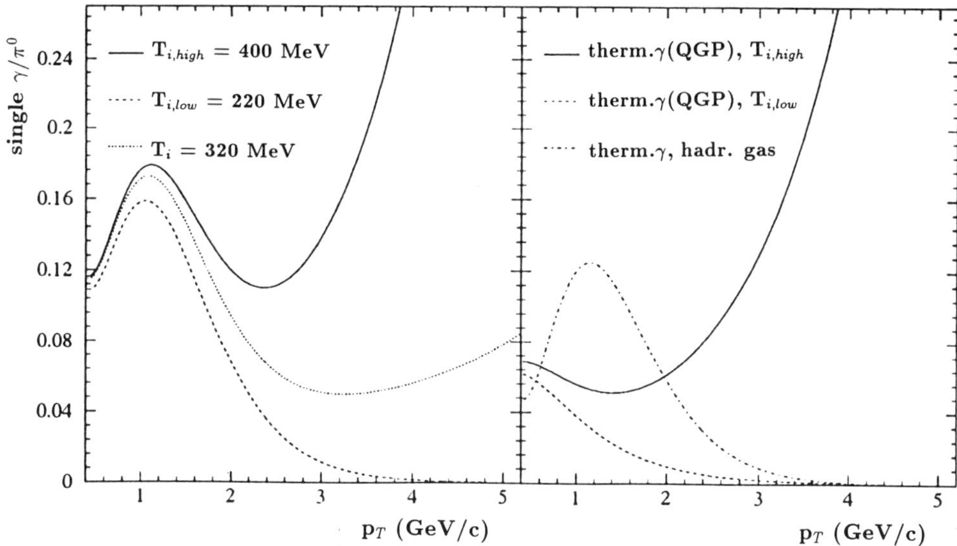

Figure 13. The calculated γ/π^0 ratio based on a dynamical situation as expected for 160 GeV/nucleon Pb+Pb reactions for different initial temperatures and a critical temperature of 160 MeV.

photon analysis techniques in the WA80 heavy-ion experiment at the SPS Central reaction events exhibit an energy density of about 4 times the energy density inside a single nucleon. A large degree of rescattering of primary and secondary hadrons can explain the measured K^+/π^+ ratio and the suppression of the J/ψ production in central collisions. Rescattering is a prerequisite to achieve thermalization.

Calculations of single photon production from various sources have been reviewed. Photons from hot hadronic matter occur at the same rate and exhibit a similar spectral shape as photons from the quark-gluon plasma at the same temperature. Within a hydrodynamical scenario the space-time evolution of hot matter is discussed and the photon yield is calculated by integration over the temperature evolution.

Experimental data reveal for the first time an excess of single photons above the yield expected from hadronic decay sources. The spectral shape suggests photon emission from hot hadronic matter.

ACKNOWLEDGMENTS

The author wishes to thank all members of the WA80 Collaboration for their support and fruitful cooperation. Results presented in this lecture are based to a large extent on thorough experimental investigations conducted by the photon teams of WA80 at Univ. Münster and the Kurchatov Institute Moscow. Inspiring discussions with Prof. R. Santo, Dr. K.H. Kampert, Dr. G. Clewing, G. Hölker and D. Bucher are gratefully acknowledged. This work was supported in part by the German BMFT and DFG, the United States DOE, the Dutch Foundation FOM, the Swedish NFR, and the CERN PPE division.

References

[1] H. Satz, Phys. Rep. **88** (1982) 349

[2] H. Satz, Ann. Rev. Nucl. Part. Sci. **35** (1985) 245

[3] F. Karsch and R. Petronzio, Z. Physik **C 37** (1988) 627

[4] J. Engels et al., Phys. Lett. **B 252** (1990) 625

[5] F.R. Brown et al., Phys. Rev. Lett. **65** (1990) 2491

[6] H. Satz, CERN-TH.6666/92 preprint and BI-TP 92/37 preprint

[7] U. Mosel and V. Metag, Phys. Bl. **49** (1993) 426;
S.A. Bass, C. Hartnack, H. Stöcker, and W. Greiner, Phys. Rev. Lett. **71** (1993) 1144

[8] R. Hagedorn, Rivista del Nuovo Cimento **6** (1983) 1

[9] H.R. Schmidt and J. Schukraft, GSI preprint GSI-92-19, to be published in Journ. of Modern Physics G

[10] R. Albrecht et al., WA80 Collaboration, Z. Physik **C 47** (1990) 367

[11] R. Albrecht at al., WA80 Collaboration, Z. Physik **C 51** (1991) 1

[12] T. Åkesson et al., NA34/HELIOS Collaboration, Z. Physik **C 46** (1990) 369

[13] A. Drees, NA45/CERES Collaboration, invited talk presented at the Tenth International Conference on Ultra-Relativistic Nucleus-Nucleus Collisions, Borlänge, Sweden 1993, to be published in Nucl. Phys. A

[14] J. Stachel and G.R. Young, Ann. Rev. Nucl. Part. Sc. **42** (1992) 537

[15] G. R. Young et al., Nucl. Instr. Meth. **A 279** (1989) 503;
T. C. Awes et al., Nucl. Instr. Meth. **A 279** (1989) 479

[16] R. Albrecht et al., Nucl. Instr. Meth. **A 276** (1989) 131

[17] H. Baumeister et al., Nucl. Instr. Meth. **A 292** (1990) 81

[18] A. Baden et al., Nucl. Instr. Meth. **203** (1982) 189

[19] K.H. Kampert et al., Prog. Part. Nucl. Phys. **30** (1993) 171

[20] F. Berger et al., Nucl. Instr. Meth. **A 321** (1992) 152

[21] S. Lebedev, WA80 Collaboration, invited talk presented at the Tenth International Conference on Ultra-Relativistic Nucleus-Nucleus Collisions, Borlänge, Sweden 1993, to be published in Nucl. Phys. A

[22] R. Albrecht et al., WA80 Collaboration, Phys. Rev. **C 44** (1991) 2736

[23] J.D Bjorken, Phys. Rev. **D 27** (1983) 140

[24] J. W. Cronin et al., Phys. Rev. **D 11** (1975) 3105;
D. Antreasyan et al., Phys. Rev. **D 19** (1979) 764

[25] M. Lev and B. Petersson, Z. Physik **C 21** (1983) 155

[26] T. Åkesson et al., NA34/HELIOS Collaboration, Z. Physik **C 46** (1990) 361

[27] B. Alper et al., BS Collaboration, Nucl. Phys. **B 100** (1975) 237

[28] H. van Hecke at al., NA34 Collaboration, Nucl. Phys. **A 525** (1991) 227c;
T. Åkesson et al., NA34/HELIOS Collaboration, Phys. Lett. **B 296** (1992) 273

[29] R.A. Salmeron, invited talk presented at the Tenth International Conference on Ultra-Relativistic Nucleus-Nucleus Collisions, Borlänge, Sweden 1993, to be published in Nucl. Phys. A

[30] R. Matiello, H. Sorge, H. Stöcker and W. Greiner, Phys. Rev. Lett. **63** (1989) 1459

[31] B. Andersson, G. Gustafson and B. Nilsson-Almquist, Nucl. Phys. **B 281** (1987) 289

[32] C. Baglin et al., NA38 Collaboration, Phys. Lett. **B 251** (1990) 465; Phys. Lett. **B 255** (1991) 459

[33] T. Matsui and H. Satz, Phys. Lett. **B 178** (1986) 416

[34] T. Ferbel and W. R. Molzon, Rev. Mod. Phys. **56** (1984) 181

[35] J.F. Owens, Rev. Mod. Phys. **59** (1987) 465

[36] E. Anassontzis et al., Z. Physik **C 13** (1982) 277;
A.P. Contogouris, S. Papadopoulos and J. Ralston, Phys. Rev. **D 25** (1982) 1280

[37] E. L. Feinberg, Nuovo Cimento **34 A** (1976) 391

[38] M. Neubert, Z. Physik **C 42** (1989) 231

[39] N.P. Landsman and Ch.G. van Weert, Phys. Rep. **145** (1987) 141

[40] J. Kapusta, P. Lichard and D. Seibert, Phys. Rev. **D 44** 1991

[41] R. Baier, H. Nakkagawa, A. Niegawa and K. Redlich, Z. Physik **C 53** (1992) 433

[42] F. Karsch, Z. Physik **C 38** (1988) 147

[43] R. C. Hwa and K. Kajantie, Phys. Rev. **D 32** (1985) 1109

[44] P.V. Ruuskanen, Nucl. Phys. **A 544** (1992) 169c;
K. Kajantie and P.V. Ruuskanen, Phys. Lett. **B 121** (1983) 352

[45] D. Bucher, Diploma thesis, Univ. Münster 1993

[46] L. Xiong, E. Shuryak and G.E. Brown, Phys. Rev. **D 46** (1992) 3798

[47] K. Geiger and J.I. Kapusta, Phys. Rev. **D 37** (1993) 4905

[48] S. Chakrabarty et al., Phys. Rev. **D 46** (1992) 3802

[49] D. Seibert, Z. Physik **C 58** (1993) 307

[50] W. Busza and A. Goldhaber, Phys. Lett. **B 139** (1984) 235

[51] H. von Gersdorff et al., Phys. Rev. **D 34** (1986) 794

[52] L. McLerran and T. Toimela, Phys. Rev. **D 31** (1985) 545

[53] K. S. Lee and U. Heinz, Z. Physik **C 43** (1989) 425;
E. Schnedermann and U. Heinz, Phys. Rev. Lett. **69** (1992) 2908

[54] M. Gyulassy and T. Matsui, Phys. Rev. **D 29** (1984) 419

[55] K. Werner, Phys. Rev. Lett. **62** (1989) 2460

[56] D.K. Srivastava et al., Phys. Lett. **B 276** (1992) 285

[57] J. Bartke et al., Nucl. Phys. **B 120** (1977) 14;
E. V. Shuryak, Phys. Rep. **61** (1980) 71

[58] G. Clewing, PhD thesis, Univ. Münster 1993

[59] G. Hölker, PhD thesis, Univ. Münster 1993

[60] R. Santo, WA80 Collaboration, invited talk presented at the Tenth International Conference on Ultra-Relativistic Nucleus-Nucleus Collisions, Borlänge, Sweden 1993, to be published in Nucl. Phys. A; IKP-MS-93/0701 preprint 1993

[61] D.K. Srivastava and J.I. Kapusta, Phys. Lett. **B 307** (1993) 1

NEAR-THRESHOLD PARTICLE PRODUCTION:
A PROBE OF RESONANCE-MATTER FORMATION
IN HEAVY-ION COLLISIONS

Volker Metag

II. Physikalisches Institut
Universität Gießen, Germany

INTRODUCTION

The observation of the lowest excited state of the nucleon, the Δ-resonance, by Fermi and collaborators[1] in 1952-55 was the first experimental evidence for an internal structure of the nucleon. It took another 15 years until Friedman, Kendall and Taylor[2] discovered quarks and gluons as constituents of the nucleon in their pioneering experiments on deeply inelastic electron scattering. As any composite system the nucleon can be excited to a rich excitation energy spectrum. The excited states of the nucleon can be characterized by their mass, spin J, parity P, and isospin I. Fig. 1 shows the excitation spectrum up to a total mass of 1.7 GeV/c^2. The excited states of the nucleon decay by the strong interaction into a nucleon in its ground state and one or several mesons with lifetimes of about $5 \cdot 10^{-24}$ s, corresponding to a typical width of roughly 100 MeV. Because of their large widths the excited states are often called resonances. While all excited states have a strong decay branch into the pion + nucleon channel, only the $S_{11}(1535)$ resonance (J=1/2, I=1/2) with a mass of 1535 MeV is strongly coupled to the η meson. The observation of an η meson is thus a selective signal for the excitation of the $S_{11}(1535)$ resonance.

Nucleon resonances and their decay into mesons and nucleons have been extensively studied in electromagnetic as well as in hadronic interactions. If pion production does not occur on a free nucleon but within nuclei as, e.g., in photon-nucleus or proton-nucleus interactions, two qualitatively new processes occur:

(i) the nucleon from the Δ-decay may find the final states occupied by other nucleons of the nucleus and, as a consequence, this decay is forbidden by the Pauli principle;

(ii) a Δ surrounded by nucleons within a nucleus has an additional decay option via the process $\Delta N \rightarrow NN$. Both are typical *in-medium* processes, the first one leading to a reduction of the decay width, the second one to an increase. Experiments on photon absorption by nuclei[3] show that at normal nuclear matter density the width of the Δ is only moderately increased by 10-20% while the resonance energy is nearly unchanged.

Perspectives in the Structure of Hadronic Systems
Edited by M.N. Harakeh *et al.*, Plenum Press, New York, 1994

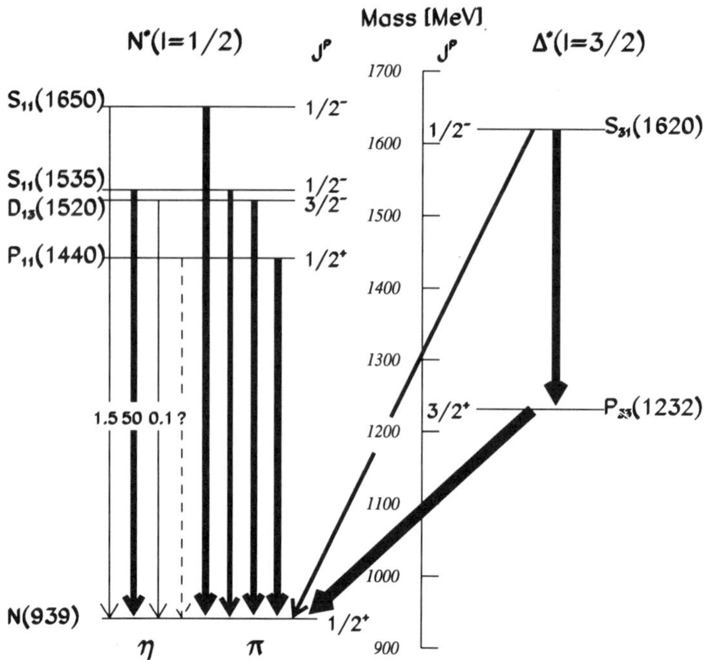

Figure 1. Excited states of the nucleon characterized by their mass, spin and isospin. Arrows indicate predominant decay branches.

Both effects mentioned above thus seem to almost compensate each other at normal nuclear-matter density. Experiments with hadronic probes, however, indicate a lowering of the Δ-mass by about 10 %. It is not yet clear whether this observation may be attributed to different form factors in the excitation[4].

While the above-mentioned studies focus on the properties of one single excited nucleon in a nucleus a new perspective is opened by relativistic heavy-ion collisions. In collisions of heavy nuclei at relativistic energies $(T_{lab}/A \approx m_p c^2)$ a large number of simultaneous nucleon-nucleon interactions occur in which a sizable fraction of nucleons is excited to resonance states. Pion multiplicities measured in central heavy-ion collisions at 1-2 A·GeV indicate that in this energy regime 20-30 % of all nucleons are excited. Taking the compression of nuclear matter to 2-3 times the normal density ρ_0 into account a resonance density comparable to ρ_0 is deduced for baryon resonances in the center of the collision zone, as discussed below. Nuclear matter initially under normal conditions in the projectile and target nuclei thus undergoes a transition to *resonance matter*, a system of strongly interacting baryon resonances and nucleons. The properties and consequences of this new form of hadronic matter will be discussed.

The study of high-density nuclear matter becomes even more interesting in view of theoretical predictions[5, 6, 7] that for high baryon densities chiral symmetry may be restored, a fundamental symmetry of Quantum Chromo Dynamics (QCD) which is spontaneously broken in our world. As a consequence, the properties of nucleons and their resonances are expected to change: a lowering of masses and widths is predicted. It is the aim of many ongoing studies of high energy heavy-ion collisions to search for corresponding indications or precursor phenomena even at densities of 2-3 times the normal nuclear-matter density.

EXCITATION OF BARYON RESONANCES IN HEAVY-ION COLLISIONS

Based on hydrodynamical calculations Scheid, Müller and Greiner[8] pointed out that nuclear matter can be compressed in the collision of two heavy nuclei. The properties of high-density nuclear matter play an important role also in the understanding of astrophysical phenomena such as supernova explosions or the formation and characteristics of neutron stars. With the availability of relativistic heavy-ion beams a study of high-density nuclear matter under controlled conditions in the laboratory has become feasible.

The main complication is that in these reactions the compression of nuclear matter can only be maintained over volumes of ≈ 50 fm^3 and times of the order of $\leq 5 \cdot 10^{-23}$s $\approx$ 15 fm/c as shown in transport-model calculations which have become a widely accepted theoretical description of relativistic heavy-ion collisions (for reviews see e.g.,[9, 10]). It is the main task of the experimentalists to devise methods which provide information as directly as possible on the high-density phase of the heavy-ion collision. Two approaches have mainly been pursued:

- the study of emission patterns of nucleons and complex fragments formed in the reaction.

- the measurement of particles not present in the entrance channel like mesons or antiprotons.

The first approach has the advantage of large yields and a straightforward analysis in terms of a global event characterization. The disadvantage is that in some cases it is difficult to disentangle participants from spectator nucleons not directly involved in the reaction. In the second approach the observed particles are always true participants since they must have been produced in the reaction. On the other hand, production cross sections are usually quite low, at least at energies below the production threshold in free nucleon-nucleon interactions. Consequently, many experiments suffer from a lack of statistics.

Both types of experiments have been systematically pursued at the BEVALAC (Berkeley, USA), at SATURNE (Saclay, France), Dubna (Russia) and more recently at the heavy-ion synchrotron SIS at GSI (Darmstadt, Germany). Phenomena like *sideward flow*[11], *squeeze out*[12], and *expansion of the collision zone*[13] have revealed collective features of relativistic heavy-ion collisions and the role of compression phenomena. Particle-production experiments[14] have provided direct information on the excitation of nucleon resonances in the collision zone. The experimental techniques applied at GSI and recent results on $\pi^{0,\pm}, \eta$, K$^+$, and $\bar{p}$ production are outlined in the subsequent sections.

EXPERIMENTAL TECHNIQUES

Although the detection of π^0, η, K$^+$- mesons and anti-protons requires rather different experimental techniques all experiments have certain features in common. The detectors have to provide highly redundant information for a unique identification of the produced particles in view of the low production cross sections and the high number of nucleons and fragments emitted per event. Mesons and anti-protons are identified via their rest mass. While the mass of charged particles is usually determined from a momentum and time-of-flight measurement the mass of neutral mesons is reconstructed from their decay photons in an invariant-mass analysis. Results obtained for the different particle species are presented in the following subsections.

Neutral-Meson (π^0, η) Production

π^0 and η mesons decay into two photons with a branching ratio of 99% and 39%, respectively. The 4-momentum vectors $p_i = (E_i, \vec{p}_i)$ of the decay photons have to be measured with sufficient energy and position resolution for a reconstruction of the mesons in an invariant-mass analysis

$$m_{\gamma\gamma}^2 = (p_1 + p_2)^2 = (E_1 + E_2)^2 - (\vec{p}_1 + \vec{p}_2)^2 = 2E_1 E_2 (1 - cos\theta_{12}) \tag{1}$$

For the experiments at SIS the Two-Arm Photon Spectrometer **TAPS**[15] has been used in a configuration of 256 BaF$_2$ modules with individual veto counters arranged in 2 towers with 2 blocks each (fig. 2). Event characterization was obtained from the

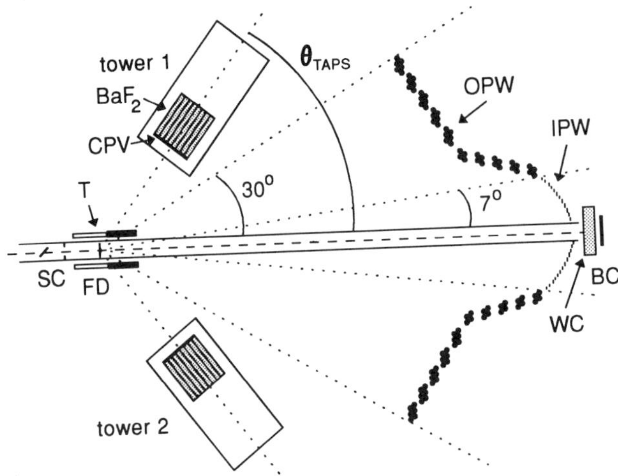

Figure 2. The experimental setup of the TAPS detector system in front of the Forward Wall at SIS. (SC: beam counter, FD: plastic-scintillator array around target, CPV: charged-particle veto counters, WC, BC: wire chamber and beam monitor). The angles θ and ϕ are used to define the BaF$_2$-block positions. The figure is taken from[16].

charged-particle multiplicity measured in the outer forward wall of the FOPI detector[17], comprising 512 plastic strips at laboratory angles between 8^0 and 30^0.

A complication of the invariant-mass analysis is the high multiplicity of π^0 mesons and decay photons which cause a coincident registration of uncorrelated photon pairs in the detector. While the invariant mass of a photon pair originating from one meson gives the rest mass of this meson within the resolution of $\sigma_m/m \approx 5\%$, pairs of photons from different mesons lead to m$_{\gamma,\gamma}$ values representing the combinatorial background. This background can be corrected for as it can be generated from the data with high statistical accuracy by pairing photons from different events (event mixing). For illustration fig. 3 shows the invariant-mass spectrum for a subset of data from the

Au+Au 1.0 A·GeV run. Apart from the π^0 signal at 135 MeV and the η signal at 547 MeV the $m_{\gamma,\gamma}$ distribution is reproduced to a level of few percent by the combinatorial background obtained from event mixing.

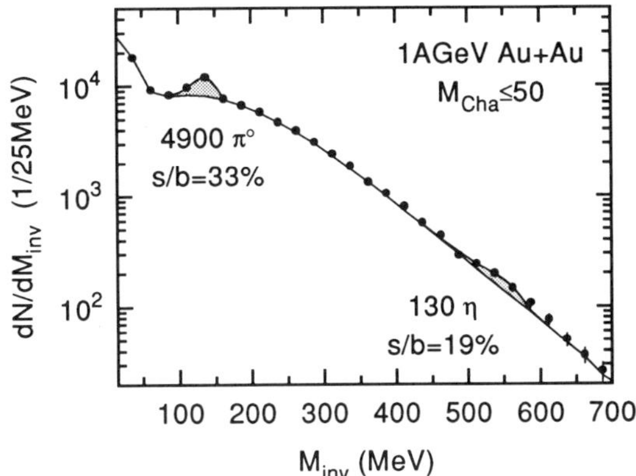

Figure 3. Invariant-mass spectrum of photon pairs measured with TAPS in the reaction Au+Au at 1.0 A·GeV for a subset of data with the charged-particle multiplicity in the Forward Wall $\leq$ 50. The π^0 peak and the structure assigned to $\eta \longrightarrow 2\gamma$ decay are indicated. The meson intensities and signal to background ratios are also given. The solid curve represents the combinatorial background determined by event mixing. The figure is taken from[18].

Transverse-momentum spectra for π^0 and η mesons are obtained by subtracting the transverse-momentum distributions of photon pairs for real and mixed events with cuts on the meson masses in the invariant-mass spectra. In contrast to charged-pion measurements p_t spectra of π^0 mesons can easily be measured over the full dynamic range down to $p_t = 0$. Even for a π^0 meson emerging at 0^0 the decay photons are emitted at finite angles with respect to the beam axis and can thus be registered without placing detectors near or in the beam. Furthermore, distortions of transverse-momentum spectra due to the Coulomb interaction with spectator matter are absent.

Subthreshold Production of K$^+$ Mesons

The magnetic spectrometer **KaoS**[19] has been used for systematic studies of K$^+$ meson production from heavy-ion collisions over a broad mass range from Ne+NaF to Au+Au. A double-focussing quadrupole/dipole configuration combines a compact geometry to minimize meson decay in flight with a large acceptance in solid angle ($\Omega = $ 30 msr) and momentum ($p_{max}/p_{min} \approx 2$ up to 1.8 GeV/c). It is thus particularly optimized for the detection of low-yield kaons. Time-of-flight scintillators, wire chambers and Cherenkov detectors are used for particle tracking and identification. The quality of the mass separation is demonstrated in fig. 4.

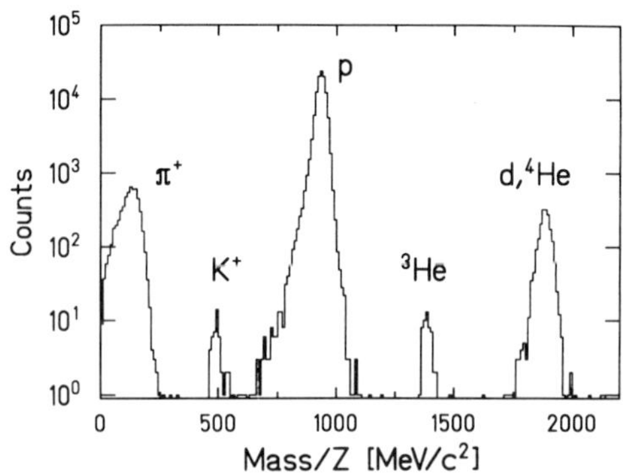

Figure 4. mass spectrum of positively charged particles derived from time-of-flight and momentum measurements for the ^{197}Au+^{197}Au reaction at 1 A·GeV within a momentum band of 650 MeV/c $< p_{lab} <$ 1150 MeV/c. The particle yields are not efficiency corrected. The figure is taken from[19].

Subthreshold Production of Antiprotons

Encouraged by pioneering experiments at the BEVALC[20] the production of antiprotons has been studied at GSI in the bombarding energy range of 1.6 - 2.0 A·GeV using Ne and Ni beams[21]. The fragment separator (**FRS**)[22] was used in a high transmision mode ($\Delta\Omega = 3$ msr, Δ p/p ≈ 6 %). Negatively charged light particles emitted at 0^0 with momenta of 1.0 and 1.5 GeV/c, respectively, were identified as $\bar{p}$, K$^-$ and π^- using time-of-flight techniques and additional Cherenkov detectors. The high selectivity and rejection power of the set-up is demonstrated in fig. 5 which shows a clean and well separated $\bar{p}$ peak despite the unfavourable production ratio of $\bar{p}/K^-/\pi^- = 1/3{\cdot}10^3/10^7$ in the target.

EXPERIMENTAL RESULTS: THE BARYONIC COMPOSITION OF HADRONIC MATTER IN THE COLLISION ZONE

In this section, results from the SIS experiments described above are presented. The interpretation of the data provides information on the excitation of baryon resonances and their mutual interactions in the hadronic matter of the collision zone. This information is extracted in 3 steps:

- The abundance of Δ-resonances is deduced from the inclusive π^0 yield.

- The contribution of higher lying resonances N(1440), N(1535) is extracted from the π^0 yield at high transverse momenta and the inclusive η multiplicity, respectively.

- Evidence for ΔN and $\Delta\Delta$ interactions emerges from the enhanced production cross sections for K$^+$ mesons and antiprotons.

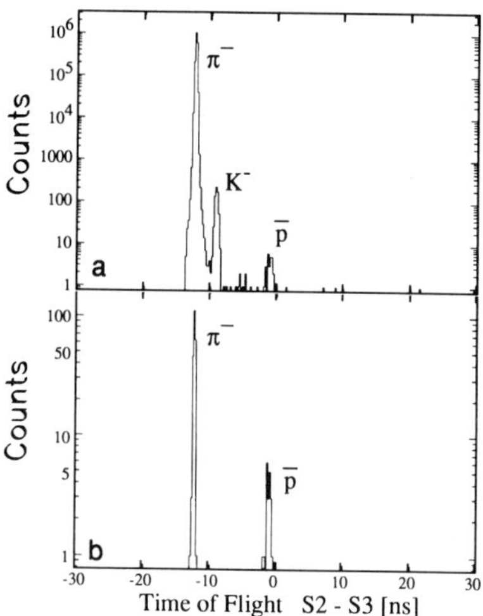

Figure 5. Time-of-flight spectrum for negatively charged particles emitted at 0^0 with a momentum of 1.5 GeV/c in the reaction ^{20}Ne + Sn at 2.0 A·GeV. Threshold Cherenkov detectors are used to suppress π^- and K^- mesons. The figure is taken from[21].

Abundance of Δ Resonances

Since the Δ resonance is the lowest excited state of the nucleon it is predominantly excited in the large number of nucleon-nucleon collisions which occur throughout the high-density phase of the heavy-ion collision. Consequently, the pion decay of this resonance will dominate the observed π^0 spectra. Fig. 6 shows transverse-momentum spectra measured at mid-rapidity for the 3 systems Ne+Al at 350 A·MeV, Ar+Ca at 1.0 A·GeV and Ar+Ca at 1.5 A·GeV. The solid curve represents fits to the spectra with the expression:

$$\frac{1}{p_t} \cdot \frac{d\sigma}{dp_t} \sim m_t \cdot e^{-\frac{m_t}{T}} \quad with \quad m_t = \sqrt{m_\pi^2 + p_t^2} \tag{2}$$

Combining these results with an earlier TAPS-measurement of the Xe+Au system at 44 A·MeV at GANIL[24] the average π^0 transverse momentum is plotted as a function of the bombarding energy in fig. 7. A steep rise at bombarding energies below 1 A·GeV flattens out above this energy. The plateau-like behaviour is supported by an additional data point from the C+C reaction at 3.4 A·GeV[25] where π^- mesons have been measured at mid-rapidity down to a threshold of 70 MeV/c.

While the average p_t stays almost constant the pion-production probability per participant nucleon (fig. 8) derived from the data of[23] and existing systematics[26] increases by a factor 4 when increasing the bombarding energy from 0.8 to 2 A·GeV. The additional translational energy is not used to increase the kinetic energy of few emitted pions but rather to increase the number of pions with almost no change in their average kinetic energy. The observation is easily explained in the picture of Δ resonance

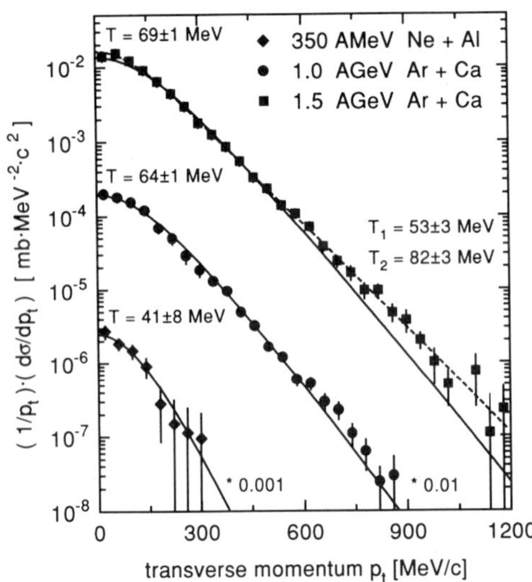

Figure 6. Inclusive invariant π^0 -cross section as a function of the transverse momentum for the target-projectile combinations Ne+Al 350 A·MeV, Ar+Ca 1.0 A·GeV, and Ar+Ca at 1.5 A·GeV. The solid and dashed lines are thermal one- and two temperature fits to the spectra, respectively (see text). The error bars represent statistical errors. The figure is taken from[23].

excitation. The additional energy available in individual nucleon-nucleon collisions is not converted into 'heat' but rather into mass by exciting a large fraction of nucleons to the Δ resonance which subsequently decays by pion emission. It follows from relativistic decay kinematics that due to the large mass difference between the pion and the nucleon the pion momentum distribution is largely determined by the mass and very little by the kinetic motion of the decaying baryon resonance. The slight rise in $< p_t >$ seen in fig. 7 thus results from a weak but increasing population of higher-lying resonances to be discussed in the next section.

Fig. 8 shows that the π-production probability per participant nucleon and pion species reaches a value of $\approx 10\%$ at 2 A·GeV. Taking all 3 charge states of the pion into account and averaging over isospin a pion-emission probability of 30% is derived. Since at these energies pions almost exclusively originate from Δ decays this result shows that at least every third participating nucleon has gone through the Δ resonance. The true fraction is certainly even larger because of π and Δ absorption.

While the experiment provides a lower limit on the number of excited nucleon resonances, transport-model calculations as e.g. BUU simulations[27] allow an estimate of the number of Δ resonances simultaneously present in the reaction volume. As shown in fig. 9 the density of Δ resonances is maintained over a period of up to 10 fm/c $\approx 6 \cdot \tau_\Delta$ in heavy collision systems by regeneration of Δ resonances after decay or absorption via the processes $NN \leftrightarrow \Delta N$ and $\pi N \leftrightarrow \Delta$. Since the baryon density ρ_B is about 2-3 times higher than at equilibrium, the Δ density ρ_Δ becomes comparable to the density ρ_0 of nucleons in normal nuclear matter:

$$\rho_\Delta \approx \frac{1}{3}\rho_B = \frac{1}{3}((2-3)\rho_0) \approx (0.6 - 1.0)\rho_0 \qquad (3)$$

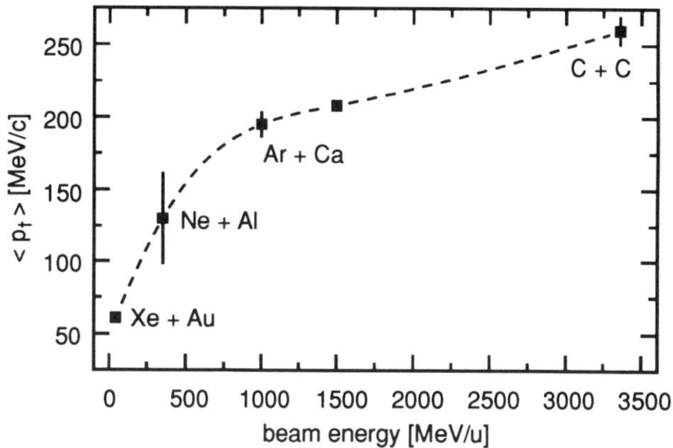

Figure 7. Average transverse momentum of π^0 mesons as a function of the bombarding energy. The data points for Xe+Au and C+C (π^-) are from[24] and[25], respectively. The other data are taken from[23].

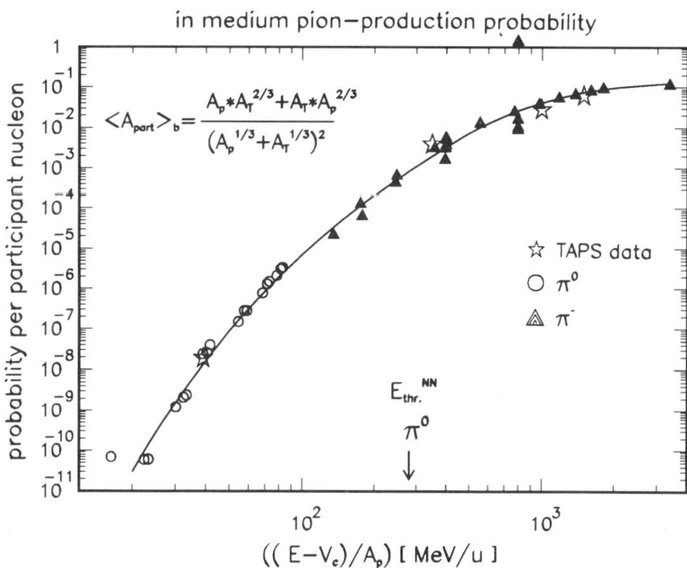

Figure 8. Pion-production probabilities per participant nucleon in heavy-ion collisions as a function of the Coulomb-corrected bombarding energy.

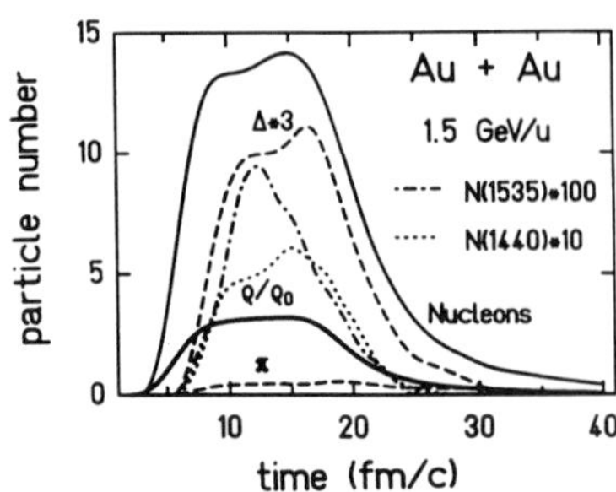

Figure 9. Calculated time evolution of baryon composition and density in the innermost 35 fm^3 of the collision zone. The figure is taken from[27].

corresponding to an average distance between the Δ's of about 1.6-2.6 fm. We are obviously dealing with a strongly interacting system of relatively dense *resonance matter*, a new form of hadronic matter.

It should be noted that the pion-production probability P_{π^0} per participant nucleon shown in fig. 8 is independent of the collision system within the experimental uncertainties. This justifies the parameterization of the inclusive cross section as

$$\sigma_{incl} = \sigma_R \cdot < A_{part} >_b \cdot P_{part}. \tag{4}$$

Here, σ_R is the heavy-ion reaction cross section[28] and $< A_{part} >_b$ represents the number of participant nucleons obtained in a geometrical model[29] by averaging over the impact parameter b. As a consequence, the pion multiplicity M_{π^0} is found to be proportional to the number of participant nucleons. This indicates - as discussed in[30] - that Δ excitations occur over the full collision volume. All Δ resonances not absorbed by the $\Delta N \Rightarrow NN$ process will eventually decay by pion emission. Consequently, the observed final-state π yield is determined by the $\pi + \Delta$ abundance at freeze-out.

While there is evidence for an equilibration of the $\Delta N \Leftrightarrow NN$ process within the collision zone the probability for this reaction to occur appears to drop rapidly as the nucleon density falls off during the expansion phase of the heavy-ion reaction. As a result, the pion multiplicities observed in the final state remains proportional to the number of participating nucleons.

Abundance of Higher Baryon Resonances

The neutral-meson data have also been analyzed to extract the population of higher-lying baryon resonances. The most direct information on the S_{11} (1535) resonance is obtained from the observation of η mesons (see fig. 3) which almost exclusively come from the decay of this resonance. The η/π^0 ratio of (2.2 ± 0.4) % at 1.5 A·GeV can be converted into an S_{11}-excitation probability of 0.4% using the observed π^0-production probability per nucleon of (6.5 ± 0.5) %, assuming comparable absorption for π^0 and η mesons. The corresponding S_{11}-excitation probability at 1 A·GeV is 0.1%.

As mentioned above, the decay of heavier-mass resonances will contribute to the pion spectra at higher transverse momenta. Pion spectra measured in the Ar+Ca system at 1.0 and 1.5 A·GeV have been analyzed within a simple thermal model to

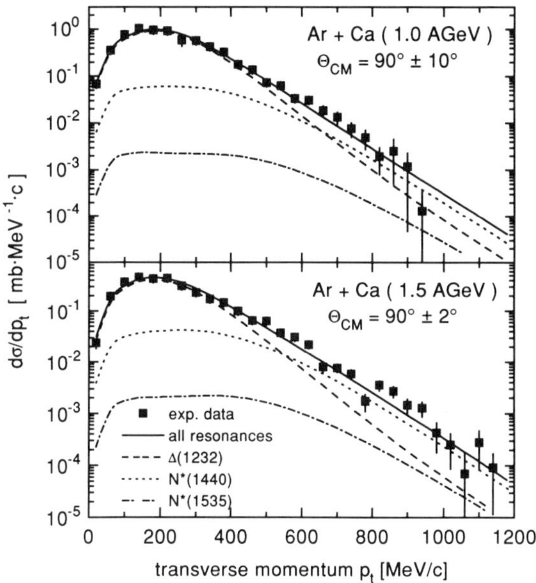

Figure 10. π^0 transverse-momentum spectra for 1.0 and 1.5 A·GeV Ar+Ca collisions. The curves indicate the decomposition into the contributions from Δ, N(1440) and N(1535) resonance decays as determined in a thermal-model analysis[23].

extract the contribution of higher-lying resonances as illustrated in fig. 10. At 1.5 A·GeV the following baryon abundancies have been derived: nucleons 80%, $\Delta(1232)$ 15.2%, N(1440) 4.4%, N(1535) 0.4 %. The resonance abundancies are close to the expectations for matter in chemical equilibrium at temperatures of 100 and 120 MeV for 1 A·GeV and 1.5 A·GeV, respectively. This indicates that the high-density phase of the heavy-ion collision lasts long enough to approximately achieve chemical equilibrium, as discussed in[31].

Evidence for Baryon Resonance - Nucleon Interactions

A criterion to justify the term *resonance matter* would be evidence that baryon resonances excited in the reaction zone interact among themselves and with nucleons. In the following subsections it will be shown that the production rates of K^+ mesons and antiprotons can only be understood within the current theoretical descriptions if ΔN and $\Delta\Delta$ collisions are taken into account.

K^+ Production. As strangeness is conserved in strong interactions the production of K^+ mesons ($u\bar{s}$) is associated with Λ (uds) production in the reaction $NN \rightarrow K^+ \Lambda N$ which raises the energy threshold for this process in free nucleon-nucleon collisions to 1.56 GeV. Furthermore, due to their open strangeness K^+ mesons have a mean free path exceeding the dimensions of the collision zone, in contrast to the other mesons discussed here. Consequently, K^+ momentum spectra are very little affected by final-state interactions and provide almost direct and undistorted information on the collision dynamics. In addition, theoretical calculations indicate an appreciable sensitivity of the K^+ yield on the compressibility of nuclear matter[32] which makes K^+ mesons a very valuable probe for the properties of the high-density phase in heavy-ion reactions.

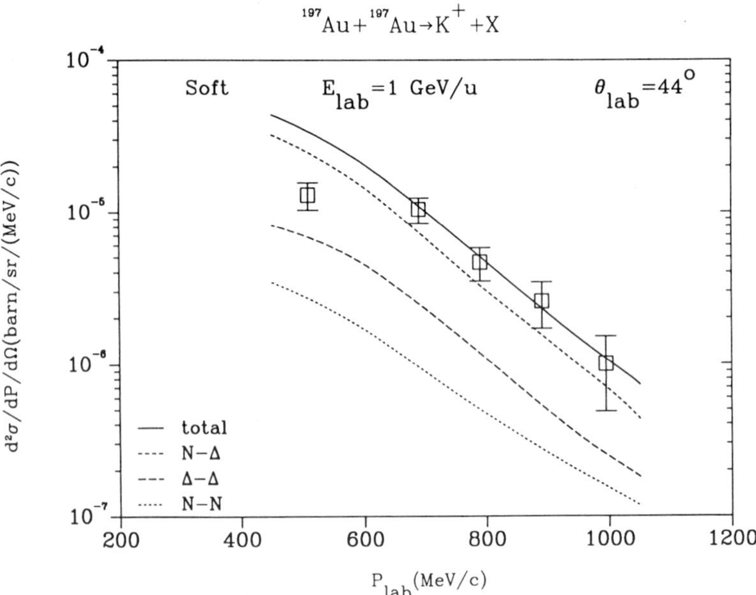

Figure 11. K^+ momentum spectrum measured with the KaoS spectrometer[19] at a laboratory angle of $(44\pm4)^0$ for the Au+Au reaction at 1 A·GeV[34]. The curves represent the result of QMD calculations taken from[36] which show a decomposition of the total K^+-production cross section into contributions from single- and multi-step processes involving baryon resonances as energy storage. The figure is taken from[36].

The K^+ spectrum of fig. 11 measured for 1 A·GeV Au+Au collisions[33, 34] corresponds to 64% of the free production threshold. Several groups[35, 36] have calculated K^+ production within various transport model approaches. As an example, the results of Quantum Molecular Dynamics (QMD) calculations by Huang el al.[36] are shown in comparison to the data. For nucleon-nucleon collisions without resonance excitation a cross section one order of magnitude below the measured K^+ yield is expected. The dominant contribution to K^+ production comes from ΔN or $\Delta\Delta$ collisions, i.e. K^+ mesons are produced in secondary or even higher-order collisions. In a first nucleon-nucleon collision a Δ resonance is excited. In a subsequent collision of this Δ with a third nucleon the excitation energy stored in the Δ resonance is exploited for meson production. The latter process can only occur with a substantial probability if the time between two baryon collisions in the eigensystem of the resonance is shorter than or comparable to the lifetime ($\tau_\Delta \approx 1.6$ fm/c) of the Δ resonance. For cross sections of typically $\sigma \approx 30$ mb and relative velocities of the collision partners of v/c $\approx 1/2$ this condition can only be met for baryon densities ρ of the order of 2 - 3 ρ_0:

$$< \tau_{coll} >= 1/(< \sigma \cdot v > \cdot \rho) \approx \tau_\Delta = 1.6 fm/c$$

$$\rho \approx 2/(\sigma \cdot \tau_\Delta \cdot c) = 0.42 fm^{-3} = 2.6\rho_0 \tag{5}$$

Although this is only a crude estimate which considers the fate of an individual resonance it nevertheless shows that an appreciable compression is indeed reached in relativistic heavy-ion collisions in the 1-2 A·GeV energy regime.

$\bar{p}$ **production.** Of all particle-production processes discussed in this article the creation of p$\bar{p}$ pairs has the highest threshold (5.6 GeV/u). At bombarding energies of 1.5–2 A·GeV, corresponding to $\approx$ 30 % of the production threshold in free nucleon-nucleon collisions this process is thus most sensitive to multi-step processes and possible cooperative phenomena in the production mechanism.

The $\bar{p}$-momentum spectrum measured by Schröter et al.[21] is shown in fig. 12 in comparison to BUU calculations by Teis et al.[37]. The most difficult task in these

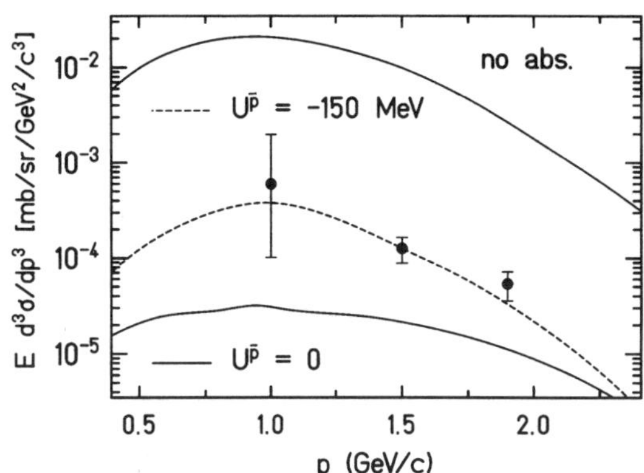

Figure 12. $\bar{p}$-momentum spectrum measured with the FRS spectrometer at 0 ^{0} for the Ni+Ni reaction at 1.85 A·GeV[21]. The curves represent the result of BUU calculations[37].

calculations is to properly treat the re-absorption of antiprotons as well as their self-energy which is related to the antiproton potential in the medium. The main result for the present discussion is that again multi-step processes involving $\Delta\Delta$ and ΔN collisions are found to contribute predominantly to $\bar{p}$ production, as illustrated in fig. 13.

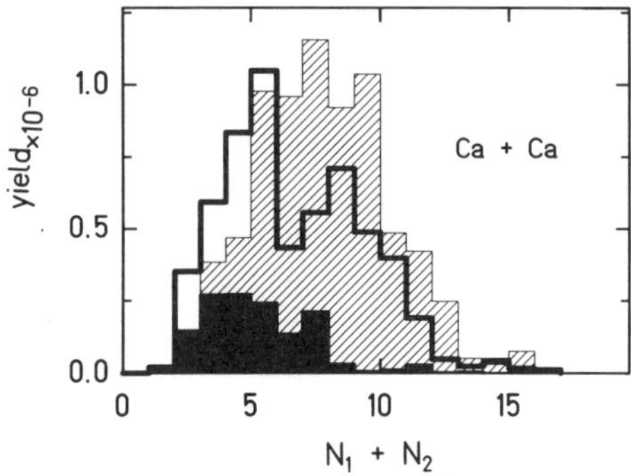

Figure 13. $\bar{p}$-production rate as a function of the number of baryon collisions calculated in BUU simulations for the Ca+Ca system at 2 A·GeV. The histograms indicate a decomposition into the reaction channels NN (full), NΔ (open) and $\Delta\Delta$ (hatched). The figure is taken from[38].

Despite differences in quantitative details all model calculations of K$^+$ an $\bar{p}$ production agree on the qualitative conclusion that ΔN and $\Delta\Delta$ interactions with intermediate resonances as energy storage have to be included to reproduce the high production cross sections at energies far below the production threshold in free nucleon-nucleon collisions. This conclusion supports the scenario of strongly interacting, resonance-enriched hadronic matter.

DILEPTON SPECTROSCOPY IN HEAVY-ION REACTIONS

All particle species discussed so far - except for K$^+$ mesons - undergo at least some final-state interactions which may lead to appreciable distortions of the measured momentum spectra. Dilepton pairs (i.e. e$^+$e$^-$), on the other hand, experience negligible final-state interactions and thus emerge from the reaction zone without rescattering. Various sources contributing to the dilepton yield in heavy-ion collisions have been considered in recent transport-model calculations.

1. Dalitz decays of neutral mesons, in particular $\pi^0 \to \gamma$e$^+$e$^-$ and $\eta \to \gamma$e$^+$e$^-$
2. nucleon-nucleon bremsstrahlung NN $\to$ NNe$^+$e$^-$
3. Dalitz decay of the Δ resonance: $\Delta \to$ Ne$^+$e$^-$
4. vector-meson decays: $\rho, \omega, \Phi \to$ e$^+$e$^-$

The last channel is considered to have the highest potential for obtaining information on possible in-medium modifications of hadrons.

Despite small vector-meson production cross sections at a few GeV and an unfavourable branching ratio for the dilepton channel of the order of $5 \cdot 10^{-5}$ the DLS collaboration has observed a clear ρ, ω signature in the the e$^+$e$^-$ invariant-mass spectrum in elementary p+p and p+d reactions at 4.9 GeV[39]. The same group has also studied dilepton production in heavy-ion collisions[40]. Because of the high pion density in these reactions $\pi^+\pi^-$ annihilation is believed to be an additional channel for dilepton production in central heavy-ion collisions. According to the Feynman diagram shown in fig. 14 the coupling to the dilepton channel is mediated by the ρ meson through

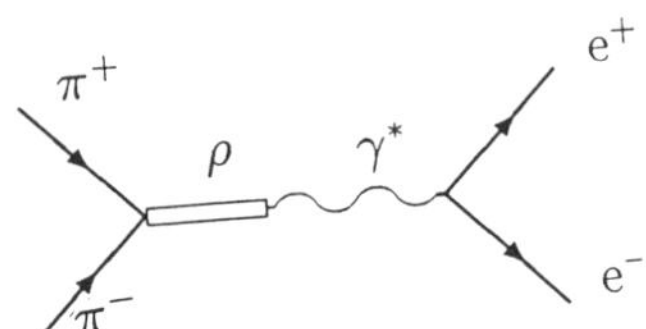

Figure 14. Diagram illustrating dilepton production via $\pi^+\pi^-$ annihilation and ρ-meson decay

vector-meson dominance. Here, the vector-meson pole in the pion electromagnetic form factor leads to an appreciable cross section enhancement. A comparison of the DLS data in fig. 15 to BUU simulations[41] shows that at higher masses the strongest process expected theoretically is indeed $\rho \longrightarrow$ e$^+$e$^-$ while at M$_{e^+e^-}$<500 MeV the spectrum is dominated by the $\eta \longrightarrow \gamma$ e$^+$e$^-$ Dalitz decay. This opens up the possibility to study properties of the ρ meson in the high-density collision zone by measuring the high-mass part of the dilepton spectrum.

Several theoretical papers (e.g. ref.[5, 6, 7]) have addressed the question whether at high densities and/or temperatures vector-meson properties may be significantly changed, indicating a transition towards chiral symmetry restoration. It is one of the in-

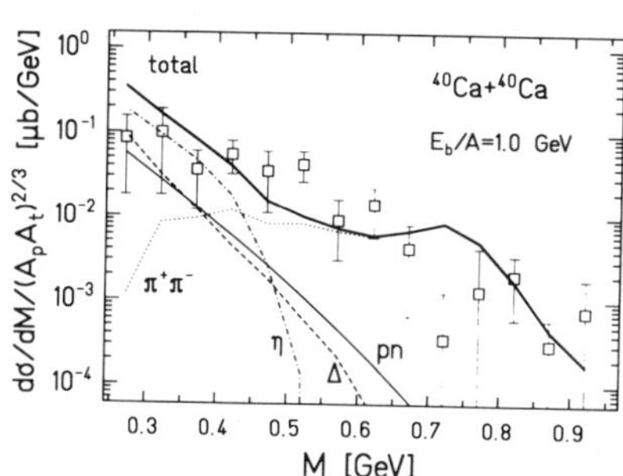

Figure 15. Dilepton mass spectrum for the Ca+Ca reaction at 1.0 A·GeV. The contribution from η Dalitz decay and $\pi^+\pi^-$ annihilation according to BUU simulations are indicated. The figure is taken from[41].

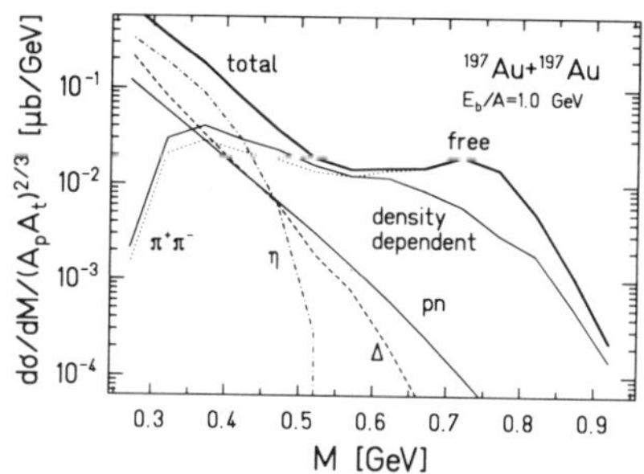

Figure 16. BUU predictions[41] for the dilepton mass spectrum in Au+Au at 1 A·GeV using the free and the density-dependent pion form factor of[47], respectively.

triguing questions whether signatures of this kind may already be observable in hadronic matter formed in central heavy-ion collisions in the several A·GeV bombarding-energy regime.

Details on the density and temperature dependence of the ρ-meson mass and width appear to be still an open theoretical problem in view of several conflicting predictions based on QCD descriptions in the non-perturbative regime[42, 43] and on models on the meson-baryon level[44, 45, 46, 47]. As one example for the many ongoing calculations Herrmann et al.[47] consider the coupling of the ρ meson to two pions as well as the strong mixing of pions and delta-nucleon-hole states in nuclear matter. They find the ρ-meson width to increase drastically with the nuclear density while the ρ mass remains almost unchanged. Using the results of Herrmann et al.[47] as input Wolf et al.[41] have performed extensive transport calculations and have investigated the observable consequences of in-medium modifications of ρ-meson properties. They predict changes in the dilepton invariant-mass spectrum which amount to factors of 3–4 in the yield near the ρ mass as illustrated in fig. 16 for the Au+Au system at 1 A·GeV. The expectation for the free pion form factor is indicated for comparison.

In order to measure differences of this magnitude high statistics data are needed, calling for a high resolution dilepton spectrometer with large acceptance. To address this physics the High Acceptance Di Electron Spectrometer (HADES) is presently under consideration as a second-generation detector system at SIS.

SUMMARY

The role of baryon resonances in the 1-2 GeV energy regime has been discussed. While photo-nuclear and hadronic interactions probe the properties of individual baryon resonances an ensemble of mutually interacting resonances can be excited in the collision zone of relativistic heavy-ion reactions. The excitation of resonances has been studied experimentally by observing their π^-, π^0 and η decays. Measured pion multiplicities indicate that at bombarding energies of 2 A·GeV at least 30% of all nucleons are excited to the Δ or heavier resonances. Combined with information on the baryon density in the collision zone ($\rho_B \approx 2 - 3\rho_0$) a density of Δ resonances comparable to that of nucleons in nuclear matter at saturation is deduced. According to model descriptions the resonance density is maintained over about 10-15 fm/c by a sequence of Δ-decay, absorption and regeneration processes until the high-density zone expands. During the high-density phase interactions among baryon resonances and nucleons lead to the observed enhancement in K$^+$ and $\bar{p}$ production via multi-step processes, exploiting baryon resonances as energy storage. All experimental observations can be consistently described by assuming that a strongly interacting resonance-enriched hadronic matter (*resonance matter*) is formed in the course of the relativistic heavy-ion collision. The future perspectives of the field will focus on dilepton production with the hope to find changes of the ρ-meson mass and width in dense hadronic matter as a first indication for a transition to chiral symmetry restoration. Such studies require the construction of a second-generation dilepton spectrometer with high mass resolution and acceptance such as the HADES detector to be built at GSI.

ACKNOWLEDGEMENT

I would like to thank the colleagues in the TAPS collaboration who have carried the major load of taking and analyzing data included in this contribution, in particular

F. D. Berg, W. Kühn, H. Löhner, R. Novotny, M. Pfeiffer, O. Schwalb, R. S. Simon, and L. Venema. I thank E. Grosse, P. Kienle, W. König, G. Roche, and P. Senger for discussions on their very recent results and for their permission to include them in this overview. Valuable comments by W. Kühn and illuminating discussions with W. Cassing, U. Mosel, and Gy. Wolf on the theoretical aspects of the field are gratefully acknowledged. This work was supported in part by Gesellschaft für Schwerionenforschung under contract GI Met K and by Bundesministerium für Forschung und Technologie under contract 06 GI 174 I.

References

[1] H. L. Anderson et al., Total cross section of negative pions in hydrogen, Phys. Rev. 85 (1952) 934, and later work

[2] J. I. Friedman and R. W. Kendall, Deep inelastic electron scattering, Ann. Rev. Nucl. Sci. 22 (1972) 203

[3] Th. Frommhold et al., Total photofission cross section for ^{238}U as a substitute for the photon absorption cross section in the energy range of the first baryon resonances, Phys. Lett. B 295 (1992) 28

[4] T. Hennino et al., Study of decay and absorption of the Δ resonance in nuclei with a 4π detector, Phys. Lett. B 283 (1992) 42

[5] G. E. Brown and M. Rho, Scaling effective lagrangians in a dense medium, Phys. Rev. Lett. 66 (1991) 2720

[6] W. Weise, Nuclear aspects of chiral symmetry, Nucl. Phys. A 553 (1993) 59c

[7] G. E. Brown, M. Rho, and M. Soyeur, Vector-meson masses in the nuclear medium, Nucl. Phys. A 553 (1993) 705c

[8] W. Scheid, H. Müller, and W.Greiner, Nuclear shock waves in heavy-ion collisions, Phys. Rev. Lett. 32 (1974) 741

[9] W. Cassing, V. Metag, U. Mosel and K. Niita, Production of energetic particles in heavy-ion collisions, Phys. Rep. 188 (1990) 363

[10] U. Mosel, Subthreshold particle production in heavy-ion collisions, Ann. Rev. Nucl. Sci. Part. 41 (1991) 29

[11] H. A. Gustafson et al., Collective flow observed in relativistic nuclear collisions, Phys. Rev. Lett. 52 (1984) 1590

[12] H. H. Gutbrod et al., Squeeze-out of nuclear matter as a function of projectile energy and mass, Phys. Rev. C 42 (1990) 640

[13] S. C. Jeong et al., submitted to Phys. Rev. Lett. (1993)

[14] R. Stock, Particle production in high energy nucleus-nucleus collisions, Phys. Rep. 135 (1986) 259

[15] R. Novotny, The BaF$_2$ photon spectrometer TAPS, IEEE, Trans. Nucl. Sci. 38 (1991) 379

[16] F. D. Berg et al., Neutral meson production in relativistic heavy-ion collisions,
Z. Phys. A 340 (1991) 297

[17] A. Gobbi et al., A highly-segmented ΔE-time-of-flight wall as forward detector of
the 4π-system for charged particles at the SIS/ESR accelerator,
Nucl. Instr. Meth. Phys. Res. A 324 (1993) 156.

[18] F. D. Berg et al., Transverse-momentum distributions of η mesons in near-
threshold relativistic heavy-ion reactions, subm. to Phys. Rev. Lett. (1993)

[19] P. Senger et al., The KaoS spectrometer at SIS,
Nucl. Instr. Meth. Phys. Res. A 327 (1993) 393

[20] J. B. Carrol et al., Subthreshold antiproton production in ^{28}Si+^{28}Si collisions at
2.1 GeV/nucleon, Phys. Rev. Lett. 62 (1989) 1829

[21] A. Schröter et al., Subthreshold antiproton production in heavy-ion collisions at
SIS-energies, Nucl. Phys. A 553 (1993) 775c

[22] H. Geissel et al., The GSI projectile fragment separator (FRS): a versatile magnetic
system for relativistic heavy ions, Nucl. Instr. Meth. B 70 (1992) 286

[23] M.Pfeiffer et al., Energy dependence of π^0-production in heavy-ion reactions, submitted to Phys. Lett. B

[24] R. S. Mayer et al., Investigation of pion absorption in heavy-ion induced subtresh-
old π^0 production, Phys. Rev. Lett. 70 (1993) 904

[25] Lj. Simic et al., Dependence of average characteristics of π^- mesons on number of
interacting protons in nucleus-nucleus collisions at 4.2 GeV/c per nucleon,
Phys. Rev. D 34 (1986) 692

[26] V. Metag, Near-threshold particle production: a probe for resonance matter formation in relativistic heavy-ion collisions,
Prog. Part. Nucl. Phys. 30 (1993) 75

[27] W. Ehehalt et al., Resonance properties in nuclear matter,
Phys. Rev. C 47 (1993) 2467

[28] W. Shen et al., Total reaction cross section for heavy-ion collisions and its relation
to the neutron excess degree of freedom, Nucl. Phys. A 491 (1989) 130

[29] J. B. Cugnon et al., Participant intimacy. A cluster analysis of the intranuclear
cascade, Nucl. Phys. A 360 (1981) 444

[30] J. W. Harris et al., Pion production in high-energy nucleus-nucleus collisions, Phys.
Rev. Lett. 58 (1987) 463

[31] J. W. Harris et al., Pion Production as a Probe of the nuclear-matter equation of
state, Phys. Lett. 153 B (1985) 377

[32] J. Aichelin, "Quantum" molecular dynamics – a dynamical microscopic n-body
approach to investigate fragment formation and the nuclear equation of state in
heavy-ion collisions, Phys. Rep. 202 (1991) 233

[33] P. Senger et al., Kaon production in heavy-ion collisions and the nuclear equation of state, Nucl. Phys. A 553 (1993) 757c

[34] E. Grosse, Meson production in nuclear collisions and the equation of state, Prog. Part. Nucl. Phys. 30 (1993) 89

[35] A. Lang et al., Covariant calculation of K^+ production in nucleus-nucleus collisions at SIS-energies, Nucl. Phys. A 541 (1992) 507

[36] S. W. Huang et al., Subthreshold K^+ production in 1 GeV/u ^{197}Au+^{197}Au collisions, Phys. Lett. B 298 (1993) 41

[37] S. Teis et al., Antiproton production in p-nucleus and nucleus-nucleus collisions within a relaytivistic transport approach, submitted to Phys. Lett. B

[38] S. Teis, Diplom thesis, Univ. Gießen (1993)

[39] H.Z. Huang et al., Dielectron yields in p+d and p+p collisions at 4.9 GeV, Phys. Lett. B 293 (1992) 233

[40] G. Roche et al., Dielectron production in Ca+Ca collisions at 1.0 and 2.0 A·GeV, Phys. Lett. B 226 (1989) 228

[41] Gy. Wolf et al., Eta and dilepton production in heavy-ion reactions, Nucl. Phys. A 545 (1992) 139

[42] S. Klimt et al., Chiral phase transition in the SU(3) Nambu and Jona-Lasinio model, Phys. Lett. B 249 (1990) 386

[43] T. Hatsuda and S. H. Lee, QCD-sum rules for vector mesons in the nuclear medium, Phys. Rev. C 46 (1992) 34

[44] Y. Asakawa et al., Rho meson in dense hadronic matter, Phys. Rev. C 46 (1992) 1159

[45] V. Mull et al., Modifications of scalar and vector mesons in nuclear matter, Phys. Lett. B 286 (1992) 13

[46] G. Chanfray et al., σ- and ρ-meson strength distributions from in-medium corrected $\pi - \pi$ correlations, Phys. Lett. B 256 (1991) 325

[47] M. Herrmann, B. L. Friman, W. Nörenberg, Properties of the ρ meson in dense nuclear matter, Z. Phys. A 343 (1992) 119

QUARK MATTER AND NUCLEAR COLLISIONS

Helmut Satz

Theory Division, CERN
CH-1211 Geneva 23, Switzerland and
Fakultät für Physik, Universität Bielefeld
D-33501 Bielefeld, Germany

I summarize briefly and qualitatively the content of my lectures. They treated on one hand quark deconfinement and its theoretical basis in finite temperature QCD, and on the other hand high energy nuclear collisions as a way to produce a deconfined state of matter in the laboratory. The material presented will shortly appear in greater detail as a book under the same title as these lectures.

What happens to matter when its density becomes so high that the relevant constituents are no longer molecules, atoms or nuclei, but instead the elementary particles of strong interaction physics? This question has in the past twenty years led to a new field of physics research, the thermodynamics of strongly interacting matter, and this field will be the subject of my lectures. It is a rather interdisciplinary topic, in which elementary particle and nuclear physics overlap with statistical mechanics; its implications reach from quark interactions to the astrophysics of neutron stars and the cosmology of the early universe. The study itself is pursued both in theory and in experiment. Modern supercomputers allow us to evaluate the statistical mechanics based on quantum chromodynamics (QCD), and high energy nuclear collisions are expected to produce strongly interacting matter in the laboratory. Today many theorists are engaged in large scale computer simulations of finite temperature QCD, and experimental teams from all over the world are carrying out nuclear collision experiments at CERN and at Brookhaven National Laboratory. Let us see why strong interaction thermodynamics is of such great interest.

If we could compress a gas of nucleons to ever higher densities, then eventually the nucleons would become densely packed. This would be the high density limit of nuclear matter, if nucleons were indeed elementary particles, i.e., primordial hard spheres with a radius of about 1 fm. But we know today that this is not the case, that nucleons are bound states of quarks. Hence we can continue compressing our densely packed nucleonic system, and the nucleons will now interpenetrate each other. At sufficiently high density, this means that any given quark will find in its immediate neighbourhood, say in a sphere of 1 fm radius, very many other quarks. It has no way to tell which two of these many quarks were in a more dilute past its partners in some nucleon. The concept of a nucleon, or more generally that of any hadron, thus becomes meaningless at high enough density: the relevant constituents of matter are now quarks. We thus expect in strong interaction thermodynamics that with increasing density there will be a transition from hadronic matter (with mesons and nucleons as constituents) to a plasma of deconfined quarks.

On the other hand, we know that the forces binding quarks into hadrons are confining: the potential between quarks grows with increasing distance of separation, and an infinite amount of energy would be needed to tear a quark out of its parent hadron. How can we then arrive at a state of deconfined quarks? The answer is found in a phenomenon well-known in condensed matter physics: charge screening. The Coulomb potential of an electric charge in vacuum,

$$V_o(r) \ = \ \left(\frac{e^2}{r} \right) \, , \tag{1}$$

becomes Debye-screened in a medium of many other (positive and negative) electric charges:

$$V_D(r) \ = \ \left(\frac{e^2}{r} \right) e^{-r/r_D} \, , \tag{2}$$

with r_D denoting the Debye screening radius. The origin of this effect is clear: positive charges will tend to cluster around a negative test charge and thus effectively reduce the range of the Coulomb force in a medium of charges. The Debye radius decreases with increasing density, so that the force becomes more and more short-ranged. For matter consisting of bound states of electric charges, e.g., for a system of hydrogen atoms, this leads to a transition point: when $r_D \leq r_B$, where r_B is the binding radius of the atom, the electron can no longer "see" its proton and hence will become liberated. As a result, the insulating hydrogen matter turns into a conductor, in which electrons can move around freely. Screening is thus a short-range effect; the presence of many other charges suppresses long-range phenomena. This can thus also occur in QCD; the long-range confining aspects of strong (colour) interactions simply do not come into play in a medium of very high quark density. Quark deconfinement thus corresponds to a transition from a colour insulator (hadronic matter) to a colour conductor (quark matter).

The qualitative arguments we have just considered can be made much more quantitative. We have today with QCD a basic theory of strong interaction dynamics, and we can use it to formulate strong interaction thermodynamics. The real problem is the evaluation of this thermodynamics, i.e., the calculation of the relevant thermodynamic observables, such as pressure, energy density, or specific heat. QCD is an interacting relativistic field theory, and expressions derived from such a theory, as we know from QED, generally have to be renormalised. Moreover, the perturbative methods used in QED break down when we cannot assume the interaction to be small - and around the transition point from colour insulator to colour conductor that is not the case. We thus need a non-perturbative, renormalising method to evaluate our thermodynamic observables. Up to now, the only viable approach for this is the computer simulation of finite temperature QCD formulated on the lattice. The QCD partition function is formulated as a path integral with discrete space-time coordinates; it then has a similar structure as the partition functions of spin systems, and for the evaluation of these we have simulation methods. The past dozen years have seen intensive activity along such lines of research, and we know today directly from QCD that the hadron-quark transition will occur for vanishing baryon number density at a temperature of about 150 MeV. This means that an energy density $\epsilon_o \simeq 1$ GeV/fm^3 is necessary to turn hadronic matter into quark matter.

If we want to produce the predicted new form of matter in high energy nuclear collisions, we have to reach at least this density. But there are other problems to be addressed in such collisions. When two heavy nuclei collide "head on" at high energies, they pass through each other (nuclear transparency), leaving behind a "vapour trail" of deposited energy. A given bubble of this trail consists of a considerable amount of energy within a certain volume; it is the system which we want to use for our study of QCD thermodynamics. Before we can do this, we have to ask if this system is large enough to be considered as (macroscopic) matter, and we have to check that it is indeed thermalised and not just a superposition of independent nucleon-nucleon collisions. If both these conditions are fulfilled, there is the question of the initial energy density of the system: was it high enough for quark deconfinement? And if it was, how can we establish experimentally that there was quark matter? Our bubble is not contained in any way; it expands, cools off, and eventually breaks up into hadrons. This hadronic final state will in general have lost the memory of its earlier phases, and so we have to find a way to probe the primordial state of the matter produced in nuclear collisions.

Before we turn to these questions, let us briefly look at the present state and the future prospects of quark matter studies in high energy nuclear collision experiments. High energy nuclear beams are presently provided at Brookhaven National Laboratory (about 15 GeV/nucleon) and at CERN (about 200 GeV/nucleon). In a first phase, both labs used existing facilities (injectors, detectors, and of course their existing accelerators) to check the feasibility of such experiments. The existing injectors allowed only nuclei with equal numbers of protons and neutrons, so that with $A \simeq 30$ rather light beam nuclei could be collided with heavy targets. This phase has been successfully completed; it has led to a number of quite interesting observations. Both labs therefore initiated the construction of new injectors, to be able to provide really heavy nuclear beams. BNL is already now conducting $Au - Au$ collisions; at CERN, $Pb - Pb$ experiments are expected to start towards the end of 1994. We are thus at the start of the study of true heavy-ion collisions, at the mentioned energies determined by the existing accelerators (BNL-AGS and CERN-SPS). Following this second phase, the end of this decade should bring new accelerators of much higher collision energy. BNL is constructing a dedicated relativistic heavy-ion collider (RHIC), which will permit arbitrarily heavy nuclei to collide at a centre-of-mass (com) energy of 200 GeV/nucleon. CERN is planning to construct a large hadron collider (LHC), which will from its start run part-time as a heavy-ion collider, at com energies of 8 TeV/nucleon, and also for arbitrarily heavy nuclei. Around the year 2000, we should thus be able to reach very much higher energy densities than can be attained with the present accelerators.

We now return to the questions noted above. We first look at the range of energy densities attainable, then at the size of the bubbles produced and at ways to check thermalisation. Finally we consider primordial state probe.

To estimate the initial energy density present in the bubble of matter produced in a high energy nucleus-nucleus collision, we let the evolution of the collision "run backwards". The hadrons eventually emerging from the bubble can be detected and their energy measured. The initial bubble size V_o can be estimated from the transverse nuclear dimensions and from isotropy requirements to fix the longitudinal size. We now just have to multiply the number of emerging hadrons by their energies and divide this total energy by V_o to obtain ϵ_o. The resulting values lie around 1 GeV/fm^3 for the BNL-AGS and around 2.5 GeV/fm^3 for the CERN-SPS; the BNL-RHIC will bring ϵ_o up to around 5 GeV/fm^3, the CERN-LHC to about 10 GeV/fm^3. Although these values still contain some theoretical uncertainties, they indicate that it should be possible to reach the energy densities needed to produce quark deconfinement.

The size of the system at the time it breaks up into hadrons (at "freeze-out") can be measured experimentally. If this size is sufficiently large, also earlier stages are presumably macroscopic enough. At freeze-out the strong interactions stop, but the emitted hadrons are still subject to effective interactions due to quantum statistics. This means that the wave functions of identical bosons have to be symmetrised, those of identical fermions anti-symmetrised. For the bosons, that gives rise to a size-dependent attraction, for the fermions to a size-dependent repulsion. The correlation functions of identical hadrons can thus be used to establish the freeze-out size of the system. One finds that this size grows with increasing beam energy, indicating that the initial system was hotter at higher incident energy and thus expands to a larger volume before freezing out. The actual freeze-out volumes go up to about a thousand fm^3 already at present CERN energies; at RHIC and LHC they are expected to increase by another one to two orders of magnitude. Thus there is hope that the bubbles can be considered as reasonably macroscopic.

To check that the constituents in these bubbles are indeed thermalised, we can compare the ratios of different produced hadron species. Thus the ratio of K^+ to π^+ production is found to be about 0.05 for proton-proton collisions. In contrast, it is expected to be about 0.25 for an equilibrium hadron gas of the temperature and baryon number density in question. Such an increase can only occur through multiple rescattering among primaries and secondaries, and among the secondaries themselves, in the course of the collision process. The final stage of such rescattering is evidently thermalisation. Measurements of particle ratios (the noted K^+/π^+ as well as a number of others) indeed show a striking increase from the values found in $p - p$ collisions towards the values expected from a thermal system. There thus seems to be evidence for the onset of thermalisation.

That brings us to our final problem, probing the primordial state. The observed hadrons will in general carry direct information only about the state of the system at the time they were formed – and that was after the quark-hadron transition. To get back to earlier times, at which the system hopefully was quark matter, we have to resort to probes which were then already present. So far, the most promising such probe is the study of charmonium production. Heavy charm quark pairs are formed only very early in the collision and then bind to form $c\bar{c}$ vector mesons (J/ψ, ψ'), which decay into dileptons and are thus observable. In a deconfined medium, the binding into resonances is inhibited, so that the c and the $\bar{c}$ will fly apart. At hadronisation, they will then have to combine with light quarks present in the medium, since the thermal production of further c or $\bar{c}$ quarks is excluded because of their high mass. The existence of a deconfining medium will thus lead to the suppression of J/ψ or ψ' production in nuclear collisions. Experiments at CERN do indeed show a strong suppression of J/ψ production. Before these results can be interpreted as evidence for deconfinement, alternative explanations based on hadronic interactions (absorption) have to be ruled out. A comparison of J/ψ and ψ' suppression may provide a suitable tool for this. In any case, the suppression provides already now strong evidence for the existence of extremely dense matter in the early primordial stages of the bubble.

In summary, we note that high energy nucleus-nucleus collisions have led to three rather striking "nuclear" effects: large freeze-out volumes, a strong enhancement of hadron production ratios towards thermalisation, and a strong suppression of charmonium production. All three are in accord with the expectation that such collisions will produce dense strongly interacting matter, dense enough to provide quark deconfinement.

THE STRING MODEL
OF NUCLEAR SCATTERING:
THEORETICAL CONCEPTS

K. Werner[1]

Institut für Theoretische Physik, Universität Heidelberg
Philosophenweg 19
6900 Heidelberg, Germany
[1] Heisenberg fellow

INTRODUCTION

The early universe was probably a hot and dense "fireball" of quarks and gluons, before, due to expansion and cooling, hadrons emerged. Presently, there are considerable efforts to create such a "quark-gluon plasma" (QGP) in nucleus–nucleus scattering at ultrarelativistic energies ($\gg 1$ GeV per nucleon)[1].

In ultrarelativistiv collisions there is certainly enough energy available to produce high enough energy densities for a QGP, provided all the energy is used to heat up the system. This is, however, not the case. We know that the nuclei are to some extent tranparent, they go through each other by keeping a large fraction of their original momentum. But still, the nucleons do loose energy, which shows up as baryon–poor matter in the central region. To be more precise: the system has roughly the form of a cylinder which expands essentially longitudinally, the forward and backward front moving almost with the velocity of light. The forward and backward region of the cylinder are baryon–rich, the central region is baryon–poor. On the other hand, the energy density is largest in the central region, and the big question is whether the energy density in this region is large enough to form a plasma.

Theoretically, such questions related to the space–time structure can be investigated by using string models[2-6]. These models provide a full description of ultrarelativistic nucleus–nucleus collisions, starting really with two incident nuclei and not at some vaguely known intermediate stage as many hydrodynamical models. Basic features of high-energy hadronic collisions as tranparency, longitudinal structure ... can be easily understood in terms of string models, even more, these are basic properties of string models and not just a specific choice of parameters. String models are also very useful to study signals of the QGP, since practically all signals are strongly affected by the space–time evolution of the system.

Perspectives in the Structure of Hadronic Systems
Edited by M.N. Harakeh *et al.*, Plenum Press, New York, 1994

There is quite a number of microscopic dynamical models, a fact which is at first sight somewhat disturbing. Looking closer, one finds, however, similarities. There is a whole class of models (VENUS[2], the dual parton model (DPM)[3, 4], the quark-gluon string model (QGSM)[5, 6]) with all the models being strictly based on Gribov-Regge theory and Veneziano's cylinder hypothesis (these models are referred to as Gribov–Regge models, GRM's). This general framework is referred to as "the string model" in this article. The different models of this class (VENUS, DPM, QGSM) differ in details concerning the precise formulation of string formation and decay. A nice feature of the Gribov–Regge approach is the consistent formulation within relativistic quantum theory. The classical string is only used as a phenomenological parametrization of particle production, this is not a classical model.

There is no real alternative to the string model discussed in this article. There is a "classical string model"[7], which seems to be quite different but nevertheless successful. However, also here the hadronic interaction amounts to string formation and decay. Lacking theoretical guidance one simply assumes something about string formation, which is actually very similar to the "colour exchange mechanism" of the GRM's. Another successful approach is the RQMD model[8], which uses classical trajectories and measured hadron–hadron cross sections. But also here, for energetic hadron–hadron interactions, a string approach is applied. Furthermore, at high energies, the Gribov cross sections are used to introduce multiple scattering. So all successful models use at least elements of GRT, even when they are not formulated within this framework.

In this article we treat only the class of models based on Gribov–Regge theory (VENUS, DPM, QGSM),in the general framework referred to as "the string model". We discuss the theoretical concepts as relativistic strings and Gribov–Regge theory, and demonstrate how both are linked to provide "the string model of hadronic interactions". We also discuss the generalization to nucleus–nucleus scattering.Many details and references missing in this article can be found in a recent review article[2].

STRING DYNAMICS

The dynamics and fragmentation of relativistic strings are crucial ingredients of the model for hadronic interactions to be introduced later. We shall discuss strings in a general fashion in order to demonstrate that dynamics and fragmentation are almost fixed from symmetry requirements. This is the reason that string-fragmentation models are much less arbitrary than one might think.

We review classical string theory[9, 10, 11]. We discuss how to obtain a string action, we derive equations of motion for the space-time evolution of strings, and we discuss the general solution as well as conservation laws. We then treat the simplest-possible string solution, the so called "yo-yo" string.

A Gauge-Invariant String Action

We discuss in this section how to obtain a string action from invariance requirements.

A classical string is a two-dimensional surface in the four-dimensional Minkowski space,

$$x = x(\tau, \sigma) \,, \tag{1}$$

with a spacelike parameter σ and a timelike one τ. Of course this is only one of infinitely many parametrizations of this surface. A transformation

$$\begin{pmatrix} \tau \\ \sigma \end{pmatrix} \longrightarrow \begin{pmatrix} \tilde{\tau}(\tau,\sigma) \\ \tilde{\sigma}(\tau,\sigma) \end{pmatrix} \tag{2}$$

from one parameter space to another is called a *gauge transformation*, and the group of such transformations is called a *gauge group*. One assumes that the string action should not depend on the parametrization, so gauge invariance is a necessary requirement. Further restrictions should be locality and covariance. Concerning the question of gauge invariance, it is useful to relate a metric g to a certain string parametrization via

$$g_{\alpha\beta} = \partial_\alpha x^\mu \partial_\beta x_\mu \, , \tag{3}$$

where α and β assume the values 1 and 2, and where we used $\partial_1 \equiv \frac{\partial}{\partial\tau}$ and $\partial_2 \equiv \frac{\partial}{\partial\sigma}$. By taking "dot" and "prime" as abbreviations for $\frac{\partial}{\partial\tau}$ and $\frac{\partial}{\partial\sigma}$, the metric can be written as

$$g = \begin{pmatrix} \dot{x}\dot{x} & \dot{x}x' \\ x'\dot{x} & x'x' \end{pmatrix} \, . \tag{4}$$

How does the metric g transform under the gauge transformations given in eq. (2)? Defining the two component variable ξ via

$$\xi_1 \equiv \tau \, , \quad \xi_2 \equiv \sigma \, , \tag{5}$$

and using

$$\tilde{\partial}_\alpha \equiv \frac{\partial}{\partial\tilde{\xi}_\alpha} \, , \tag{6}$$

we get

$$\begin{aligned}
g_{\alpha\beta} &= \partial_\alpha x^\mu \partial_\beta x_\mu \tag{7} \\
&= \tilde{\partial}_i x^\mu \partial_\alpha \tilde{\xi}_i \tilde{\partial}_j x_\mu \partial_\beta \tilde{\xi}_j \tag{8} \\
&= \partial_\alpha \tilde{\xi}_i \tilde{g}_{ij} \partial_\beta \tilde{\xi}_j \, . \tag{9}
\end{aligned}$$

Since the components of the Jacobi matrix M of the gauge transformation eq. (2) are given as

$$M_{ab} = \partial_b \tilde{\xi}_a \, , \tag{10}$$

we can write eq. (9) in matrix notation as

$$g = M^T \tilde{g} M \, . \tag{11}$$

This leads to the identity

$$\sqrt{|\det g|} = \sqrt{|\det \tilde{g}|} \, |\det M| \, . \tag{12}$$

On the other hand we have

$$d^2\tilde{\xi} = |\det M| \, d^2\xi \, , \tag{13}$$

which together with eq. (12) immediately suggests that a ξ integration over $\sqrt{|\det g|}$ is invariant under gauge transformations,

$$\tilde{I} \equiv \int \sqrt{|\det \tilde{g}|} \, d^2\tilde{\xi} = \int \sqrt{|\det g|} \, d^2\xi \equiv I \, . \tag{14}$$

Writing the integral I explicitly as

$$I = \int \sqrt{(x'\dot{x})^2 - x'^2 \dot{x}^2}\, d\tau d\sigma \qquad (15)$$

shows that I is also local and covariant. In fact, I is the simplest local, covariant, and gauge-invariant expression, and so I is a very attractive candidate for a string action. Therefore we define the action of a relativistic string to be[9]

$$S = \int L\, d\tau d\sigma\ , \qquad (16)$$

with

$$L = -\kappa \sqrt{-\det g} = -\kappa \sqrt{(x'\dot{x})^2 - x'^2 \dot{x}^2}, \qquad (17)$$

where we used $|\det g| = -\det g$. We will see later that the proportionality constant κ can be identified with the "string tension", the energy per unit length of the string. This action can be written as

$$S = -\kappa \int d^2A\ , \qquad (18)$$

with d^2A being a string-surface element, and therefore S measures the area of the string surface. This becomes obvious when we choose τ to be the time t and σ to be the length of the string. By defining

$$\gamma \equiv (1 - v_\perp^2)^{-1}\ , \qquad \vec{v}_\perp \equiv \frac{\partial \vec{x}}{\partial t} - \frac{\partial \vec{x}}{\partial l}\Big(\frac{\partial \vec{x}}{\partial t}\frac{\partial \vec{x}}{\partial l}\Big)\ , \qquad (19)$$

and using $\left|\frac{\partial \vec{x}}{\partial l}\right| = 1$, we find

$$S = -\kappa \int \frac{1}{\gamma}\, dl dt\ , \qquad (20)$$

which is a surface integral as stated in eq. (18).

Equations of Motion, Conservation Laws

We rewrite the action defined in the last section more explicitly as

$$S = \int_{\tau_1}^{\tau_2} d\tau \int_0^\pi d\sigma\, L\ , \qquad (21)$$

with

$$L = -\kappa \sqrt{(x'\dot{x})^2 - x'^2 \dot{x}^2}\ . \qquad (22)$$

We use the convention $\sigma_{\min} = 0$ and $\sigma_{\max} = \pi$, where π is an arbitrary number at the moment. The symbols τ_1 and τ_2 represent initial and final times. To obtain the equations of motion, we require

$$\delta S = 0 \qquad (23)$$

under infinitesimal variations $\delta x(\sigma, \tau)$ of the string surface. We find the equations of motion

$$\frac{\partial}{\partial \tau}\frac{\partial L}{\partial \dot{x}_\mu} + \frac{\partial}{\partial \sigma}\frac{\partial L}{\partial x'_\mu} = 0\ , \qquad (24)$$

and the boundary conditions

$$\frac{\partial L}{\partial x'_\mu} = 0 \qquad \text{at } \sigma = 0, \pi\ . \qquad (25)$$

From the invariance of the action under translations one obtains conservation laws for energy and momentum. Defining the energy–momentum currents as

$$P_\tau^\mu := -\frac{\partial L}{\partial \dot{x}_\mu} \, , \qquad P_\sigma^\mu := -\frac{\partial L}{\partial x'_\mu} \, , \tag{26}$$

we may introduce the string momentum in various ways, for example as

$$P^\mu(\text{string}) := \int_{C_\tau} d\sigma \, P_\tau^\mu \, , \tag{27}$$

where integration at constant τ is implied. From $\delta S = 0$, one concludes that the string momentum is conserved.

Using the currents defined in eq. (26), we may rewrite the equations of motion eq. (24) as

$$\frac{\partial}{\partial \tau} P_\tau^\mu + \frac{\partial}{\partial \sigma} P_\sigma^\mu = 0 \, , \tag{28}$$

and the boundary condition eq. (25) reads

$$P_\sigma^\mu = 0 \, , \qquad \text{at } \sigma = 0, \pi \, . \tag{29}$$

Our next aim will be to solve these equations of motion.

Solutions of the String Equations

To solve the equations of motion, we choose a gauge which simplifies the equations of motion. The orthonormal gauge

$$x' \dot{x} = 0 \, , \qquad \dot{x}^2 + x'^2 = 0 \tag{30}$$

does so. The currents eq. (26) are now

$$P_\tau = \kappa \dot{x} \, , \qquad P_\sigma = -\kappa x' \, , \tag{31}$$

the equations of motion eq. (28) are simply wave equations

$$\ddot{x}_\mu - x''_\mu = 0 \, , \tag{32}$$

and the boundary conditions eq. (29) are

$$x'(t, 0) = x'(t, \pi) = 0 \, . \tag{33}$$

The gauge conditions eq. (30) do not yet specify the choice of coordinates completely, because there are infinitely many orthonormal systems on a surface. All gauge transformations $\tilde{\tau} = \tilde{\tau}(\tau, \sigma)$, $\tilde{\sigma} = \tilde{\sigma}(\tau, \sigma)$ which conserve orthonormality satisfy

$$\ddot{\tilde{\tau}} - \tilde{\tau}'' = 0 \, , \quad \ddot{\tilde{\sigma}} - \tilde{\sigma}'' = 0 \, . \tag{34}$$

Since the x_μ satisfy the same equation, we may fix the parametrization by requiring

$$nx = n^\mu x_\mu = \lambda \tau \, , \tag{35}$$

with some arbitrary timelike vector n $(n^2 \geq 0)$. Because of $P_\tau^\mu = \kappa \dot{x}^\mu$, we have $\lambda = n P_\tau / \kappa$, or

$$\lambda = n P(\text{string}) / \pi \kappa \, . \tag{36}$$

A possible choice of n is $n = (1, -1, 0, 0)$, which defines the transverse gauge. The simplest choice is the *lab-frame parametrization* obtained from $n = (1, 0, 0, 0)$, which leads to

$$x_0 \equiv t = \lambda\tau \, , \quad \lambda = E/\pi\kappa \, , \tag{37}$$

with E being the string energy $P^0(\text{string})$. With π being the number $\pi = 3.14\ldots$, the parameter τ is dimensionless and represents the time in units of the length scale $\lambda = E/\pi\kappa$. We use a different option,

$$\pi := \frac{E}{\kappa} \, , \tag{38}$$

which means $\lambda = 1$ and therefore

$$x_0 \equiv t = \tau \, . \tag{39}$$

This implies that τ and σ have length dimensions. For the following we use this *lab-frame parametrization*, and consider only the space components of x. Eq. (30) now reads

$$\vec{x}'\dot{\vec{x}} = 0, \qquad (\dot{\vec{x}})^2 + (\vec{x}')^2 = 1 \, . \tag{40}$$

The solution of eqs. (32, 33) is

$$\vec{x}(t, \sigma) = \frac{1}{2}[\vec{y}(t + \sigma) + \vec{y}(t - \sigma)] \, , \tag{41}$$

where $\vec{y}(t)$ is obviously the trajectory of one endpoint, $\vec{x}(t, 0)$, called the directrix. The directrix has to be periodic,

$$\vec{y}(t + 2\pi) - \vec{y}(t) = \frac{2\vec{P}}{\kappa} \, , \tag{42}$$

where $\vec{P}$ is the string momentum (to be shown later). Eq. (41) has the following meaning: each point on the string may be obtained by a simple geometrical construction once the directrix $\vec{y}(t)$ is known. From eqs. (27, 31, 41), we also see that the momentum of a piece of string is generated by the momenta of the two corresponding directrix pieces,

$$dP(t, \sigma) = \frac{\kappa}{2}[\dot{y}(t + \sigma)d\sigma + \dot{y}(t - \sigma)d\sigma] \, , \tag{43}$$

so that the momentum for the string piece corresponding to $[0, \sigma]$ (from A to B in fig. 1) is

$$P[0, \sigma] = \int_0^\sigma dP = \frac{\kappa}{2} \int_{t-\sigma}^{t+\sigma} dt' \, \dot{y}(t') = \frac{\kappa}{2}[y(t + \sigma) - y(t - \sigma)] \, , \tag{44}$$

which is proportional to the distance vector between two directrix points (DE in fig. 1). In particular, we obtain $\vec{P}[0, \pi] = \vec{P}$, so $\vec{P}$ introduced in eq. (42) is indeed the string momentum. All this shows that the directrix piece from $D = \vec{y}(t - \sigma)$ to $E = \vec{y}(t + \sigma)$ determines the string piece from $A = \vec{x}(t, 0)$ to $B = \vec{x}(t, \sigma)$. We also find a section of the directrix related to the other string part from $B = \vec{x}(t, \sigma)$ to $C = \vec{x}(t, \pi)$. However, we can equally well relate the "antidirectrix" (trajectory of the other end $\vec{x}(t, \pi)$) from F over C to G to this string piece. It is obvious from fig. 1 that putting together the directrix piece D to E and the antidirectrix piece F to G, we recover the full directrix (after a shift of the antidirectrix by $\frac{1}{\kappa}\vec{P}$, which is the constant vector by which directrix and antidirectrix differ).

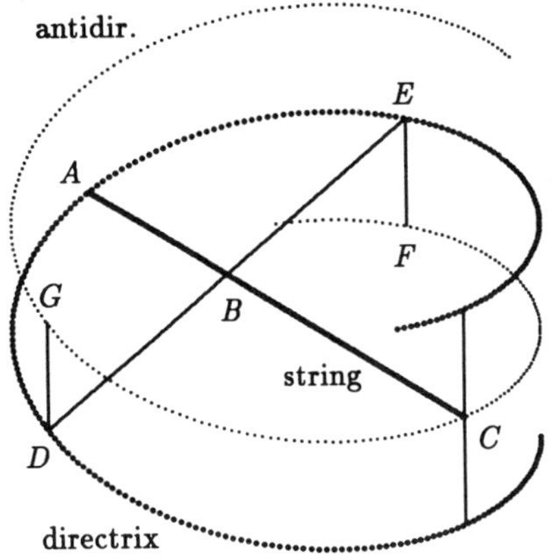

Figure 1. A string with its directrix and antidirectrix. The directrix segment DAE defines the string piece AB and the antidirectrix FCG the string piece CB.

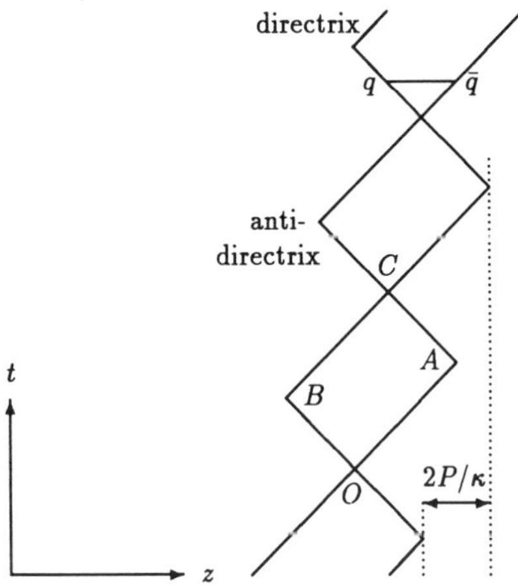

Figure 2. Space-time picture of a "yo-yo" string with period $2P/\kappa$.

The Yo-Yo String

We are now going to discuss a simple but important example: a one-dimensional directrix with one period consisting of two linear segments ("yo-yo string"). In case of a one-dimensional directrix, straight lines with a tilt of 45° against vertical (in space-time) are mandatory, because the string end (represented by $\vec{y}(t)$) moves with the velocity of light (because of eqs. (40, 33)). From eq. (41) it is clear that the corresponding string is a simple straight line ($q\bar{q}$ in fig. 2) stretched between directrix and antidirectrix.

It is very instructive to investigate energy and momentum distribution along a one-dimensional yo-yo string. For the following, we use $E \equiv p_0$ and $P \equiv p_3$ for energy and longitudinal momentum, the space-time coordinates are t and $z \equiv x_3$. By employing eqs. (27, 31), we obtain for an arbitrary string element

$$dE = \kappa \, d\sigma \, , \qquad dP = \kappa \dot{z} \, d\sigma \, . \tag{45}$$

The gauge fixing condition $z'\dot{z} = 0$, which is a consequence of eq. (40), requires either z' or $\dot{z}$ to be zero. An ordinary yo-yo has exactly two points with $z' = 0$: the two endpoints of the string (because of the boundary condition eq. (29), every string has to fulfil $\vec{x}' = 0$ at the endpoints). Due to the gauge fixing conditions $z'\dot{z} = 0$ and $\dot{z}^2 + z'^2 = 1$, we obtain

$$\dot{z} \equiv \frac{\partial z}{\partial t} = 0 \ , \qquad |z'| \equiv \left|\frac{\partial z}{\partial \sigma}\right| = 1 \ , \qquad \text{for } \sigma \neq 0, \pi \tag{46}$$

inside the string, and

$$z' = \frac{\partial z}{\partial \sigma} = 0 \ , \qquad |\dot{z}| \equiv \left|\frac{\partial z}{\partial t}\right| = 1 \ , \qquad \text{for } \sigma = 0, \pi \tag{47}$$

at the endpoints. Since these two domains (characterized by $z' = 0$ and $z' \neq 0$) behave so differently, we discuss their contributions to energy and momentum separately. We use an index g (for glue) for the interior and an index q (for quark) for the endpoints. By using eqs. (45, 46, 47), the energy and momentum of an inner piece of string of length dl can be shown to be

$$dE_g = \kappa \, dl \ , \qquad dP_g = 0 \ , \tag{48}$$

whereas

$$dE_q = \kappa s \, dt \ , \qquad dP_q = \kappa s \, dz \tag{49}$$

represent the change of energy and momentum of an endpoint during a time step dt with a corresponding movement dz of the endpoint. The symbol $s = +1(-1)$ indicates that the endpoint has absorbed (emitted) a piece of string, or a piece of parameter space, to be more precise. Eqs. (48, 49) demonstrate, among other things, energy conservation: the energy dE_q gained by an endpoint by absorbing a piece of string is equal to the energy loss $-dE_g$ of the string due to its contraction. It is also easy to see from eq. (49) that the two endpoints change momentum in an opposite way: $dP_{\bar{q}} = -dP_q$, which guarantees momentum conservation. We are now going to integrate eq. (49). Let us consider one "basic cell", $OACB$ in fig. 2. The polygon OBC is half a period of the directrix. Instead of the other half, we consider the corresponding half period OAC of the antidirectrix. So $OACB$ defines the string completely. We use for the left end (OBC) the index $\bar{q}$, for the right end the index q. At the turning points the momenta vanish,

$$P_q(A) = P_{\bar{q}}(B) = 0 \ , \tag{50}$$

and since this implies that at these points the parameter space specifying the endpoints consists of just one point (0 and π respectively), eq. (50) also requires the energy to be zero,

$$E_q(A) = E_{\bar{q}}(B) = 0 \ . \tag{51}$$

Now we can easily integrate eq. (49) backwards to point O to obtain the initial energy and momentum,

$$E_q(O) = \kappa \, t_A \ , \qquad P_q(O) = \kappa \, z_A \ , \tag{52}$$

$$E_{\bar{q}}(O) = \kappa \, t_B \ , \qquad P_{\bar{q}}(O) = \kappa \, z_B \ , \tag{53}$$

where we use $t_O = z_O = 0$. Employing light–cone coordinates $x^{\pm} = t \pm z$ and light–cone momenta $p^{\pm} = E \pm P$, we get

$$p_q^+(O) = \kappa \, x^+(A) \ , \qquad p_{\bar{q}}^-(O) = \kappa \, x^-(B) \ , \tag{54}$$

together with $p_q^-(O) = 0$ and $p_{\bar{q}}^+(O) = 0$. So eq. (54) provides a simple relation between initial momenta and the length of directrix pieces. We have a mapping of "momentum space" to "real space" via

$$\Delta p = \kappa \, \Delta x \ . \tag{55}$$

STRING FRAGMENTATION

The classical string might be considered as a model for a QCD fluxtube between two colour charges. The colour field induces the production of q–$\bar{q}$ pairs, which screen the field and thus produce two substrings, which (may) continue to break up. We do not want to mix up different models here, so we discuss string breaking completely within the framework of classical strings, and propose a breaking law from symmetry requirements. The corresponding fragmentation procedure is considered a model for particle production in e^+e^- annihilation and also more complicated reactions.

We discuss the rules for string breaking in the framework of classical relativistic string theory in general and in particular for yo-yo strings. Although, in classical string theory, the time evolution is fixed once a string breakpoint is known, the determination of locations of breakpoints requires further input. For this purpose we introduce the AMOR model (the fragmentation model used in VENUS), where the same symmetry arguments are used which led earlier to the string action. This procedure leads to an "area law" for string breaking (Artru–Mennessier approach). We finally discuss relations of this fragmentation model to others.

String Breaking

We do not know within our classical treatment where a string breaks, but once the breakpoint is fixed, we know how to proceed. As for the action, we assume locality. If a break occurs at $x(t,\sigma)$, we have to make sure that for the future as well as the past we have periodic (anti-) directrices, and that the directrices for future and past match properly in the present. The only way to ensure this is to periodically continue the directrix corresponding to one string piece and the antidirectrix corresponding to the other string piece into the future. This fully determines the time evolution of either string piece also for all the future, at least untill the next break.

Let us now discuss these "cutting rules" for a yo-yo string (see fig. 3). Without interaction, the string stretches between directrix $(t, y(t))$ and antidirectrix $(t, \bar{y}(t)) = (t, x(t, \pi))$. Let the point $B = (t, x(t, \sigma))$ be a breakpoint on the string at time t, dividing the string into two segments AB and BC with $A = (t, x(t,0))$ and $C = (t, x(t, \pi))$. The directrix and antidirectrix corresponding to these segments are DEF with

$$
\begin{aligned}
D &= (t - \sigma, y(t - \sigma))\,, & (56)\\
F &= (t + \sigma, y(t + \sigma)) & (57)
\end{aligned}
$$

and GHI with

$$
\begin{aligned}
G &= (t - (\pi - \sigma), \bar{y}(t - (\pi - \sigma)))\,, & (58)\\
I &= (t + (\pi - \sigma), \bar{y}(t + (\pi - \sigma)))\,. & (59)
\end{aligned}
$$

Using

$$
\bar{y}(t) = y(t - \pi) + \frac{P}{\kappa}\,, \tag{60}
$$

we verify easily that, after the appropriate shift, the segments DEF and GHI provide a full period of the unperturbed string. As discussed earlier, we obtain the directrices of the two segments after the break by continuation of DEF ($\to DEFS\cdots$) and of GHI ($\to GHIN\cdots$). The corresponding antidirectrices can be easily constructed from the relation

$$
\bar{y}(t) = \frac{1}{2}(y(t + \pi) + y(t - \pi)) \tag{61}
$$

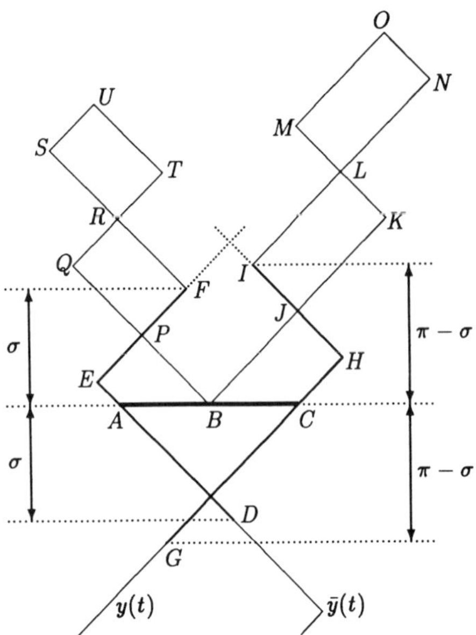

Figure 3. Breaking of a "yo-yo" string: we first determine the segments of the (anti)directrix corresponding to the string pieces AB and CB; these segments are then periodically continued into the future.

between directrix y and antidirectrix $\bar{y}$, and we get $BKM\cdots$ and $BQT\cdots$. We realize the identities

$$\|BJ\| = \|JK\| , \quad \|HJ\| = \|JI\| \tag{62}$$

and

$$\|BP\| = \|PQ\| , \quad \|EP\| = \|PF\| , \tag{63}$$

which provide a very simple procedure for actually constructing the new directrices in numerical applications.

The string-fragmentation Model AMOR

In the following we introduce the fragmentation procedure AMOR, which is the fragmentation model used in VENUS. Our appraoch is close to the Artru-Mennessier model. AMOR is the abbreviation for *Artru–Mennessier Off–shell Resonance* model.

We discussed in the last chapters string dynamics, including the case of a breaking string. For a simple one-dimensional yo-yo string, this is illustrated in fig. 3. We have not yet specified any law determining where the string breaks. This clearly goes beyond any classical treatment. However, we can restrict the variety of possible breaking laws by requiring certain properties. In the same way as for the string action S (see previous section), it can be shown that the simplest local, covariant, and gauge-invariant breaking law can be written as

$$dP(\tau,\sigma) \sim \sqrt{-\det g}\, d\tau d\sigma , \tag{64}$$

with dP being the probability for a break at $x(\tau,\sigma)$ and g being the metric (eq. (3)).

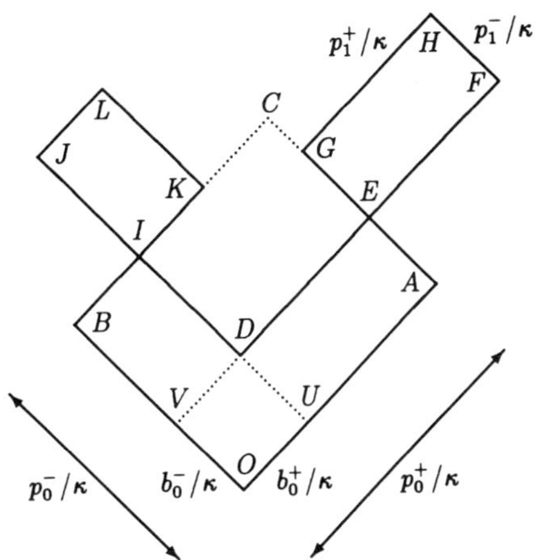

Figure 4. Relation between (anti)directrix segments and parton momenta. Energy discontinuities (at the endpoints in the case of a yo-yo) can be identified with partons.

This means that the breaking probability is proportional to the corresponding area on the string surface, $dP \sim d^2A$, or

$$dP = (1 - P)\,\alpha\,d^2A\,, \tag{65}$$

with the "break probability" α as a parameter. This is the fragmentation law first suggested by Artru and Mennessier[12, 13] and later also used by other authors[14]. It is so appealing, because it is not just a good guess but rather a strict consequence of requiring very plausible properties: locality, covariance, and gauge invariance. Another nice feature is that there is only one parameter (α) which should be the same for processes as different as for example diquark fragmentation into baryons or heavy-quark fragmentation into heavy mesons.

The AMOR model, to be described in the following, has been introduced in[15]. We restrict ourselves to yo-yo strings, which form a closed group among all possible strings in the sense that a yo-yo breaks into two yo-yo's again. In fig. 4 we show an "elementary cell" of a half period directrix OAC ("quark" q_0) and the corresponding antidirectrix OBC ("antiquark" $\bar{q}_0$). Because of eq. (55), the coordinates $x_0^+ = t_A + z_A$ and $x_0^- = t_B - z_B$ are related to the initial momenta of the quark (p_0^+) and the antiquark (p_0^-) via

$$\|OA\| = x_0^+ = \frac{p_0^+}{\kappa}\,, \qquad \|OB\| = x_0^- = \frac{p_0^-}{\kappa}\,. \tag{66}$$

Let us consider a break-up at D into a "quark" ($DIJ\cdots$) and an "antiquark" ($DEF\cdots$). We may define "break-up momenta" $b_0^\pm$ via

$$\|OU\| = \frac{b_0^+}{\kappa}\,, \qquad \|OV\| = \frac{b_0^-}{\kappa}\,. \tag{67}$$

With $p_1^\pm$ being the parton momenta of the right substring at E, we find

$$\|EF\| = \frac{p_1^+}{\kappa} = \frac{p_0^+ - b_0^+}{\kappa}\,, \tag{68}$$

$$\|EG\| = \frac{p_1^-}{\kappa} = \frac{b_0^-}{\kappa} \ . \tag{69}$$

A corresponding formula holds for the other substring. The area of absolute past with respect to the break-up point D is given as

$$A \equiv \frac{1}{\kappa^2}\mathcal{A} = \frac{1}{\kappa^2} b_0^+ b_0^- \ . \tag{70}$$

All points D having the same value of A lie on a hyperbola in space-time, given by

$$(t+z)(t-z) = A \ . \tag{71}$$

As the other variable required to fix D completely, we choose the space-time rapidity

$$\eta \equiv \frac{1}{2}\ln\frac{t+z}{t-z} = \frac{1}{2}\ln\frac{b_0^+}{b_0^-} \ . \tag{72}$$

Using these variables $\mathcal{A}$ and η, eq. (65) becomes

$$dP = (1-P)\,\frac{\alpha}{\kappa^2}\,d\mathcal{A}\,d\eta \ , \tag{73}$$

leading to

$$dP(\mathcal{A}) = \alpha_0\,e^{-\alpha_0\mathcal{A}}\,d\mathcal{A} \ . \tag{74}$$

We are now in a position to define exactly, step by step, how we proceed to fragment a yo-yo string into two substrings. In order to fix the break point D, we first determine $\mathcal{A} = \kappa^2 A$ via integrating and inverting eq. (74), and obtain

$$\mathcal{A} = -\frac{1}{\alpha_0}\ln r \ , \tag{75}$$

with $r \in [0,1]$ being a random number. Before fixing D completely by determining η, we have to be more specific about the break-up. We create a qq-$\bar{q}\bar{q}$ pair with probability p_{diq} (fit parameter) and a q-$\bar{q}$ with $(1-p_{\mathrm{diq}})$. Concerning flavour, we create a strange quark with probability p_{s} (fit parameter) and u as well d quarks with $(1-p_{\mathrm{s}})/2$. We then look into a resonance table (see VENUS writeup) to determine for each substring the minimum mass m_{min} for the corresponding parton content. So the minimum mass for a $u\bar{d}$ system is the π^+ mass and so on. Suggested by the uncertainty principle, we generate transverse momenta $\vec{p}_t$ and $-\vec{p}_t$ for the two partons at D according to an exponential distribution

$$f(p_t) \sim p_t \exp\left[-\frac{p_t}{2\bar{p}_t^f}\right] \ , \tag{76}$$

with a fit parameter $\bar{p}_t^f$ to be chosen in the order of the inverse proton size. Taking this value of p_t together with the minimum mass m_{min} we get a minimum transverse mass

$$\mu_{\mathrm{min}} = \sqrt{m_{\mathrm{min}}^2 + p_t^2} \tag{77}$$

for each substring. Using

$$\mu_+^2 = p_0^+ b_0^- - \mathcal{A} \ , \tag{78}$$
$$\mu_-^2 = p_0^- b_0^+ - \mathcal{A} \tag{79}$$

for the transverse masses $\mu_\pm$ of the two substrings, and using

$$b_0^+ = \sqrt{\mathcal{A}}e^\eta \ , \tag{80}$$
$$b_0^- = \sqrt{\mathcal{A}}e^{-\eta} \ , \tag{81}$$

we see that the requirement of minimum transverse masses restricts the rapidity η to be

$$\eta_- < \eta < \eta_+ \,, \tag{82}$$

with

$$\eta_+ = \ln \frac{\sqrt{\mathcal{A}}\, p_0^+}{\mu_{+\,\mathrm{min}}^2 + \mathcal{A}} \,, \qquad \eta_- = \ln \frac{\mu_{-\,\mathrm{min}}^2 + \mathcal{A}}{\sqrt{\mathcal{A}}\, p_0^-} \,. \tag{83}$$

For $\eta_+ < \eta_-$ there is no solution, the string cannot be broken. Otherwise, we determine the rapidity according to a constant distribution between η_- and η_+, so η is given as

$$\eta = \eta_- + r(\eta_+ - \eta_-) \,, \tag{84}$$

with a random number $r \in [0, 1]$. From eqs. (80, 81), we see that the breakpoint is now fully determined. In case of low-mass strings a correction procedure applies in order to have on–shell hadrons.

We want to stress that the break-up of a string into two substrings occurs completely arbitrarily in the sense that each of the substrings may be a stable hadron, a resonance, or a high-mass string. This is the major difference compared to the Lund model, where one fragment has to be a hadron with a discrete mass. So in our model we have a "tree structure": a string decays into two substrings, each substring may then decay into two subsubstrings, and so on. The Lund model has a "salami structure": a hadron is chopped off at the end, then another one from the remaining string, and so on.

The last step of the fragmentation procedure is resonance decay. All the primary hadrons from string break-up decay, if they are unstable, according to standard branching ratios. The off-shellness of resonances poses no difficulties. One only has to consider that some of the partial decays cannot occur, because the energy is not available. We simply discard such decay modes.

Finally, we comment on the relation of AMOR to other fragmentation models. AMOR, as well as other models[14] based on the Artru-Mennessier model[12, 13], take the string picture seriously and provide a covariant, gauge-invariant (= reparametriza tion invariant), energy and momentum-conserving string-breaking procedure. We refer to models based on the Artru–Mennessier model, including AMOR, as *area law models*. A very different approach is the Field-Feynman model[16]. Instead of strings, here one considers two independent partons moving into opposite directions and radiating hadrons. This approach has been inspired by the experimentally observed jet-structure of produced particles. The model is not covariant and of course not gauge invariant. And since the two jets are independent, energy and momentum are not conserved. An advantage of the model is its simplicity. The Lund model JETSET[17] is similar to the Field-Feynman model in the sense that again two partons are considered, and that one fragmentation step amounts to forming a hadron which contains one of the partons. However, the two partons are linked by a colour field, which makes it possible to achieve energy and momentum conservation as well as covariance. Therefore also the Lund Model is within the framework of classical relativistic strings.

GRIBOV–REGGE THEORY (GRT)

Gribov's multiple-scattering theory of ultrarelativistic hadronic interactions, referred to as Gribov–Regge theory (GRT), is the theoretical basis of the string model of hadronic/nuclear scattering to be introduced later.

We briefly discuss the Pomeron. We then introduce an expression for the amplitude of elastic hadron–hadron scattering due to multiple Pomeron exchange. We calculate the elastic and, via the optical theorem, the total cross section. We discuss the AGK cutting rules, which provide a technique to calculate the imaginary part of elastic amplitudes. Finally we apply the AGK rules to expand the total cross section as $\sigma_{\text{tot}} = \sum \sigma_m$, with "topological cross sections" σ_m referring to m elementary inelastic processes (m cut Pomerons).

The Pomeron

In the next section we introduce a multiple-scattering theory of hadron–hadron interactions, implying multiple exchange of elementary objects called Pomerons. From general considerations of the high-energy limit of elastic amplitudes, one parametrizes the amplitude associated with Pomeron exchange as

$$A(s,t) \sim s^{\alpha(t)} \approx s^{\alpha(0)+\alpha't},$$ (85)

with s and t being the Mandelstam variables $s = (p_1 + p_2)^2$ and $t = (p_1 - p_3)^2$, with p_1, p_2 (p_3, p_4) being the momenta of the incoming (outgoing) hadrons.

The nature of the Pomeron in terms of quarks and gluons is still not known. Initially, Pomerons were thought to be ladder diagrams (gluon ladders), or gluon networks of cylindrical topology, but QCD calculations are not conclusive. We adopt Venezianos picture of a Pomeron being a cylinder of gluons and quark loops (being a generalized gluon ladder).

The Multi–Pomeron Amplitude

Gribov[18] derived his results on the basis of ϕ^3 theory, being used as a theoretical laboratory for introducing and testing concepts which cannot be handled (yet) in the true theory. Many general features should survive in the true theory.

Crucial for the derivations is the fact that, in the limit of large s, integrations over longitudinal momenta can be carried out, with the consequence that in the final results only two-dimensional integrations $\int d^2 p_\perp$ occur. This is due to the identity

$$q = q_\perp$$ (86)

for $s \to \infty$ and limited q^2. Consequently, whenever a function of q^2 appears in an Feynman integral, it may be replaced by $q_\perp^2$, and it is, therefore, a constant regarding the longitudinal integration.

We do not intend to give the details of Gribov's derivation here, which are quite technical[18]. We rather provide the results[19, 20]. Using the normalization

$$A = \frac{T}{8\pi\, w(s, m^2, m^2)} ,$$ (87)

with T being the T-matrix, or the asymptotic form

$$A = \frac{T}{8\pi s}$$ (88)

for $s \to \infty$, one gets

$$A_{2\to 2}(s,t) = \sum_n A_n(s,t) ,$$ (89)

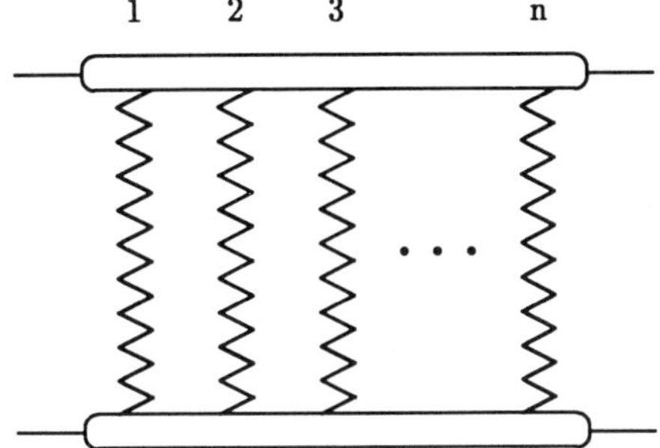

Figure 5.
n Pomeron exchange.

with

$$A_n(s,t) = \frac{i^{n-1}\pi^{1-n}}{n!} \int \prod_{i=1}^{n} d^2k_i \, \delta^{(2)}(k - \sum k_i) \, N_n(k_1 \cdots k_n) \, D(s, k_1^2) \cdots D(s, k_n^2) \, , \quad (90)$$

corresponding to the so called *nonenhanced diagrams* shown in fig. 5, representing n Pomerons exchange. The variables k and k_i represent transverse momenta, and we use $t = -k^2$. So we have only two-dimensional momentum integrations, as discussed above. The Pomeron Green's function is, as discussed earlier,

$$D(s, k^2) = \eta \left(\frac{s}{s_0}\right)^{\alpha(-k^2)-1} . \quad (91)$$

Introducing a *rapidity gap* y via $y = \ln s/s_0$ and defining

$$\Delta := \alpha(0) - 1 \, , \quad (92)$$

we find

$$D(s, k^2) \approx \eta \, \exp(\Delta y) \, \exp\left[-\alpha'(0) \, y \, k^2\right] \, , \quad (93)$$

with $\exp(\Delta y)$ being 1, if we have an ideal Pomeron with $\alpha(0) = 1$. As we are going to argue later, the data, however, suggest that $\alpha(0)$ is slightly larger than 1. The simple form of the Pomeron Green's function is a pure assumption (of Regge-pole dominance), only justified by the success of the theory, and it is, therefore, most important to justify this assumption within QCD.

There are of course other diagrams possible with triple- (or more) Pomeron vertices. We do not consider them, assuming the multiple Pomeron coupling to be small.

Eqs. (90, 93) are the basis for applications to be discussed in the following.

Elastic Scattering and Total Cross Section

As a first application of the Gribov–Regge theory, we consider elastic scattering. The amplitude for elastic scattering is given as

$$A(s,t) = \sum_{n=1}^{\infty} A_n(s,t) \, , \quad (94)$$

where A_n represents n Pomeron exchanges, and is given by eqs. (90, 93). Assuming factorization of the vertex function,

$$N_n(k_1, \cdots, k_n) = C_n \prod_{i=1}^{\infty} N(k_i^2) \, , \quad (95)$$

we have a simple convolution,

$$A_n(s,t) = \frac{i^{n-1}\pi^{1-n}C_n}{n!} \int \prod_{i=1}^{n} d^2k_i \, \delta^{(2)}(k - \sum k_i) \prod_{j=1}^{n} N(k_j^2)\, D(s,k_j^2) \tag{96}$$

(see eq. (90)). Using

$$\delta^{(2)}(k - \sum k_i) = \int \frac{d^2b}{(2\pi)^2} \exp\left[i(\vec{k} - \sum \vec{k_i})\vec{b}\right], \tag{97}$$

and introducing the Fourier transform of the Pomeron propagator,

$$\omega(s,b) := \frac{1}{i\pi} \int d^2k \, N(k^2)\, D(s,k^2) \exp(-i\vec{k}\vec{b}) , \tag{98}$$

we get

$$A_n(s,t) = \frac{C_n}{4\pi i} \int d^2b \, \exp(i\vec{k}\vec{b}) \frac{\left[-\omega(s,b)\right]^n}{n!} . \tag{99}$$

Summing over n, we get

$$A(s,t) = \frac{i}{4\pi} \int d^2b \, \exp(i\vec{k}\vec{b}) \, \gamma(s,b) , \tag{100}$$

with

$$\gamma(s,b) := -\sum_{n=1}^{\infty} C_n \frac{\left[-\omega(s,b)\right]^n}{n!} . \tag{101}$$

Setting

$$C_n = C^{n-1} , \tag{102}$$

we get

$$\gamma(s,b) = \frac{1}{C}\left\{1 - \exp\left[-C\,\omega(s,b)\right]\right\} \tag{103}$$

from eq. (101). What remains to be done is the calculation of ω according to eq. (98). Assuming

$$N(k^2) = N_0 \, \exp(-R^2 k^2) , \tag{104}$$

and approximating

$$\eta = \eta_0 \, \exp(i\pi\alpha' k^2/2) , \tag{105}$$

we get

$$\omega(s,b) = \frac{N_0 \, \exp(\Delta y)}{\pi} \int d^2k \, \exp\left[-(R^2 + \alpha'y)k^2\right] \exp(-i\vec{k}\vec{b}) \tag{106}$$

from eqs. (93, 98), neglecting $i\pi/2$ compared to y and using $\eta_0 = i$. Performing the integration, we get

$$\omega(s,b) = \frac{N_0 \, \exp(\Delta y)}{R^2 + \alpha'y} \exp\left[-\frac{b^2/4}{R^2 + \alpha'y}\right] . \tag{107}$$

Eqs. (100, 103, 107) are the final results, which may be used to calculate cross sections as

$$\sigma_{\text{tot}} = 8\pi\,\text{Im}A(s,0) = \int d^2b\, 2\,\text{Re}\gamma \tag{108}$$

and

$$\sigma_{\text{el}} = \int dk^2\, 4\pi\, |A|^2 = \int d^2b\, |\gamma|^2 . \tag{109}$$

There are a couple of free parameters: N_0, Δ, R^2, α', C. By comparing with data, one finds

$$
\begin{aligned}
N_0 &= 3.64 \text{ GeV}^{-2} \ , \\
R^2 &= 3.56 \text{ GeV}^{-2} \ , \\
\Delta &= 0.07 \ , \\
\alpha' &= 0.25 \text{ GeV}^{-2} \ , \\
C &= 1.5 \ .
\end{aligned}
\tag{110}
$$

In fact, the increase of $\sigma_{\text{tot}}(s)$ and $\sigma_{\text{in}}(s)$ with s as well as many elastic scattering data can be nicely reproduced, not so for $\Delta = 0$.

By comparing $\omega(s, b)$ and $\gamma(s, b) = 1 - \exp[-\omega(s, b)]$ (in the eikonal approximation, $C = 1$) as a function of b, one can easily see that multiple Pomeron exchange cures unitarity, which would be violated for single Pomeron exchange. In this case, γ has to be replaced by ω in the formulas to calculate cross sections. The function $\omega(s, b)$ increases for decreasing b, and at high energies it increases well beyond 1, with the maximum at $b = 0$ increasing linearly with s. This implies, after b-integration, a linear increase of the total cross section with s. This violates the Froissart bound and therefore unitarity. Due to multiple Pomeron exchange, we have to consider $\gamma = 1 - e^{-\omega}$ rather than ω. The quantity γ behaves properly. For large b, where ω is small, γ and ω are similar. For small b, however, where ω grows beyond 1, they differ: γ is always smaller than 1, approaching 1 from below for decreasing b. The b dependence of γ is similar to a Fermi function. So it is as it should be: a disc cannot be blacker than black!

The Abramovskiǐ–Gribov–Kancheli Cutting Rules

Due to the optical theorem, the discontinuity (or imaginary part) of the elastic amplitude is related to inelastic processes. One therefore investigates disc$A$ in order to study inelastic scattering. The Abramovskiǐ–Gribov–Kancheli (AGK) Cutting Rules[20] provide a technique to express disc$A$ (or ImA) in terms of disc$G$, representing elementary inelastic processes associated with the exchange of a single Pomeron.

We start with the elastic amplitude for multiple Pomeron exchange, given in eq. (90), which we write as

$$
i A_n = \int d\Omega \prod_{\gamma=1}^{n} i\, G_\gamma \ ,
\tag{111}
$$

with $G_i \equiv G(s, k_i^2) = N(k_i^2)D(s, k_i^2)$. The Cutkoski cutting rules state for a Feynman diagram (or a sum of graphs) having the structure eq. (111):

$$
\frac{1}{i} \text{disc} A_n = \sum_{\text{cuts}} \int d\Omega \prod_{\substack{\text{left} \\ \text{of cut}}} i\, G_\alpha \prod_{\substack{\text{right} \\ \text{of cut}}} \frac{1}{i} G_\beta^* \prod_{\substack{\text{cut} \\ \text{Pomerons}}} \frac{1}{i} \text{disc} G_\mu = \sum_m A_{nm} \ ,
\tag{112}
$$

where m represents the number of elementary inelastic interactions ("cut Pomerons"). One finds analytic expressions for $A_{nm}{}^2$.

An Expansion of the Total Cross Section

We are now going to use the expansion of $\frac{1}{i}\text{disc} A_n$ in terms of the number of cut Pomerons to obtain an expansion of the total cross section in terms of *topological cross sections*: $\sigma_{\text{tot}} = \sum \sigma_m$. Here σ_m corresponds to the cross section of m elementary inelastic processes (m cut Pomerons). We have

$$
\sigma_{\text{tot}}(s) = \frac{1}{2is} \text{disc} T(s, 0) = \frac{4\pi}{i} \text{disc} A(s, 0)
\tag{113}
$$

$$= 4\pi \sum_{n=1}^{\infty} \frac{1}{i}\, \mathrm{disc}\, A_n(s,0) \,, \tag{114}$$

where n reflects n Pomeron exchanges. We obtain

$$\sigma_{\mathrm{tot}}(s) = \sum_{m=0}^{\infty} \sigma_m(s) \,, \tag{115}$$

with

$$\sigma_m(s) = \frac{8\,\pi\,N_0\,\exp(\Delta y)}{m\,z}\Big[1 - e^{-z}\sum_{k=0}^{m-1}\frac{z^k}{k!}\Big] \quad ; m > 0 \,, \tag{116}$$

which may be evaluated easily numerically. For $m = 0$, we get

$$\sigma_0(s) = 8\,\pi\,N_0\,\exp(\Delta y)\sum_{n=1}^{\infty}\frac{1-2^{n-1}}{n\,n!}\left(-\frac{z}{2}\right)^{n-1} . \tag{117}$$

The σ_m are strictly positive!

THE VENUS STRING MODEL

The Gribov–Regge theory provides a framework to calculate amplitudes $A_{2\to2}$ for elastic hadron–hadron scattering, and therefore elastic and total cross sections can be calculated. For more detailed investigations, one needs inelastic amplitudes $A_{2\to n}$, describing particle production. These amplitudes are, however, not calculable within GRT. On the other hand, we know that discontinuities (or imaginary parts) of elastic amplitudes are related to inelastic scattering, so one may take the expansion of $\mathrm{disc}\, A_{2\to2}$ (or of σ_{tot}) from the last chapter as a guideline to construct a model. This is exactly what is done in the VENUS model:

- In VENUS, elastic amplitudes are calculated strictly according to Gribov–Regge Theory (GRT), and so are the elastic and total cross sections.

- VENUS provides a model for calculating inelastic amplitudes, guided by the expansion of σ_{tot} in terms of topological cross sections.

The VENUS model to be introduced in the following is closely related to the dual parton model (DPM)[3, 4], introduced by Capella et al., and the quark-gluon string model (QGSM) by Kaidalov et al.[5, 6].

In order to formulate a model for inelastic scattering, one needs to know something about the nature of the Pomeron. According to Veneziano[21], a Pomeron is a cylinder, see fig. 6. Correspondingly we identify the discontinuity of the Pomeron propagator (or more precisely $-i\,\mathrm{disc}\,G = 2\,\mathrm{Im}\,G$) with a squared cut cylinder, with a cut cylinder shown in fig. 7. The two cutting edges of the cut cylinder are identified with relativistic strings, as shown in fig. 8, where the thin lines represent (anti)quarks and the thick lines socalled remnants, each one representing an incident hadron minus the corresponding (anti)quark. So, before the interaction, (anti)quark and remnant of the same hadron are connected, after the interaction, (anti)quark and remnant from different hadrons. Therefore the term "colour exchange" is used, because it looks like colour being exchanged between the two (anti)quarks.

Considering two incoming hadrons $f_1(\tilde{p}_1)$, $f_2(\tilde{p}_2)$, with $f(p)$ refering to flavour (momentum), and a colour exchange between the (anti)quarks $i(k)$ and $j(l)$, one obtains two strings S_+ and S_-. The string

$$S_+ := \big\{\tilde{f}_1(\tilde{p}_1) - i(k)\,\big|\,j(l)\big\} \tag{118}$$

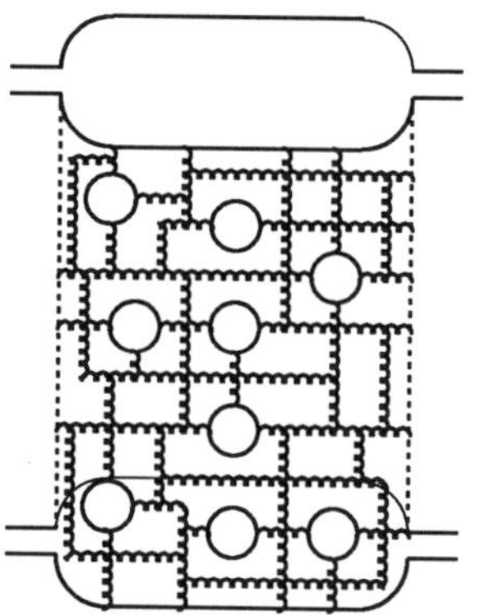

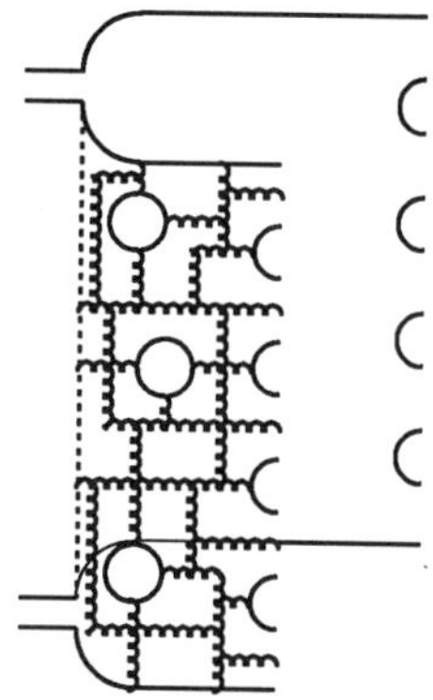

Figure 6. Cylinder diagram (gluons and quark loops on the back sheet are not drawn).

Figure 7. Cut cylinder diagram (gluons and quark loops on the back sheet are not drawn).

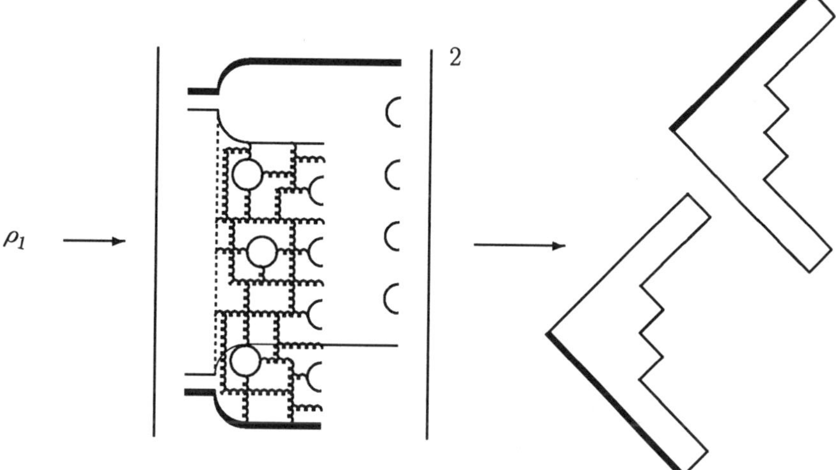

Figure 8. Cut cylinder corresponding to two strings.

contains at one end the projectile remnant $\tilde{f}_1(\tilde{p}_1)-i(k)$, being the projectile $\tilde{f}_1$ reduced by the parton (quark or antiquark) i. This string end carries the projectile momentum $\tilde{p}_1$ reduced by the parton momentum k. The other end consists of the parton j with momentum l. Since this string contains the projectile remnant, which has a large forward momentum, we refer to this string as forward string or forward baryonic string (it carries baryon number 1). The other string

$$S_- := \left\{ \tilde{f}_2(\tilde{p}_2)-j(l) \,\middle|\, i(k) \right\} \tag{119}$$

contains at one end the target remnant $\tilde{f}_2(\tilde{p}_2)-j(l)$, which is the target $\tilde{f}_2$ reduced by the parton j, with the target momentum $\tilde{p}_2$ reduced by the parton momentum l. On the other end, we have the parton i with momentum k. This string is referred to as backward string or backward baryonic string (it carries as well baryon number 1). So if, for example, $\tilde{f}_1$ and $\tilde{f}_2$ are protons and i and j are u quarks, the two strings are ud–u

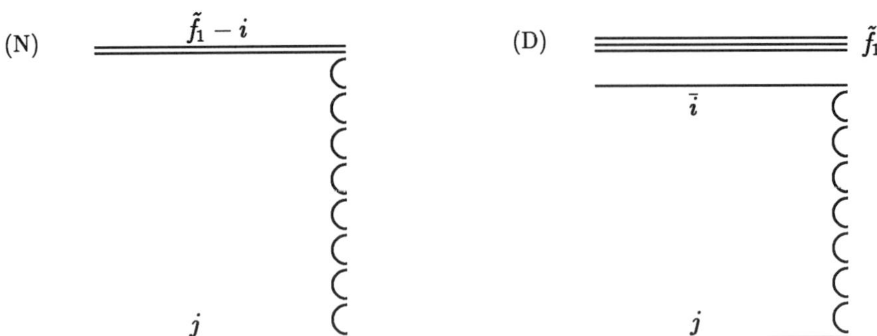

Figure 9. Nondiffractive (N) and diffractive (D) type.

strings, with the diquark ud being in one case the forward end and in the other case the backward end.

The forward string S_+ consists of a backward parton j and a forward hadron $\tilde{f}$ minus a parton i. It might be by chance that $\tilde{f}_1 - i$ is equal to $\tilde{f}_1 + \bar{i}$ with $\tilde{f}_1$ having survived as a singlet, or, in other words, the Color Exchange (CE) involved only a colour singlet $i - \bar{i}$ pair and left $\tilde{f}_1$ surviving as a spectator. The same arguments apply of course for the backward string S_-. We refer to such interactions as "diffractive" (D) and to others as "nondiffractive" (N) (see fig. 9). We say, the hadron suffers a D–type or N–type interaction (we use weights w and $1 - w$).

For the full 1–cylinder contribution (consisting of two strings), we have four combinations, NN, ND, DN, and DD, with weights $(1 - w)^2$, $(1 - w)w$, $w(1 - w)$, and w^2, which are referred to as

$$
\begin{array}{lll}
\text{nondiffractive scattering} & \text{(NN)}\,, & \\
\text{diffractive projectile excitation} & \text{(ND)}\,, & \\
\text{diffractive target excitation} & \text{(DN)}\,, & (120) \\
\text{Pomeron} - \text{Pomeron scattering} & \text{(DD)}\,. &
\end{array}
$$

Quark-Line Diagrams

We proceed in two steps to generate a particle sample for a given number of colour exchanges: we first generate a sample of strings and then decay the strings independently. The iterative procedure to decay strings has been discussed in detail earlier, so for the following we only consider the generation of string samples. Since the strings are specified by quark contents and momenta of the endpoints, it turns out to be very useful to introduce quark–line diagrams in order to construct the final strings by successive CE's. We use the following rules:

A1 An (anti)quark is represented by a horizontal line. If neccessary, an arrow may be used to distinguish between quarks and antiquarks.

A2 A colour exchange is represented by a vertical arrow between the two (anti)quark lines of the (anti)quarks participating in the CE.

A3 A string is represented by a vertical line between (anti)quarks, with the corresponding (anti)quarks being highlighted by dots. An intersection of quark and string lines without dots has no meaning, it is like a bridge being involved.

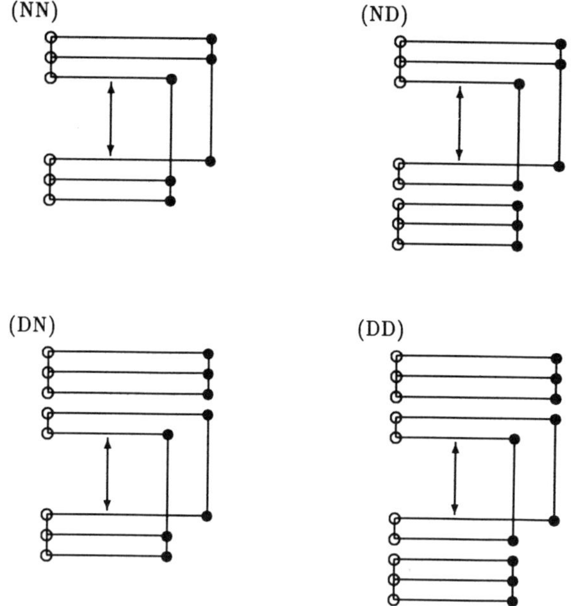

Figure 10. Quark line diagrams representing the four contributions to single colour exchange between quarks.

A4 Each initial hadron is represented as a main string (containing the baryon number) and an arbitrary number of q–$\bar{q}$ strings.

A CE amounts to providing an additional link of each remnant to a quark of the other remnant. With a probability w each of the (anti)quarks involved in the CE may be part of a colourless q–$\bar{q}$ pair. The momentum distributions of all participating (anti)quarks are determined from measured distribution functions. So the following rules hold:

A5 If, before the CE, i_1 and i_2 (in the projectile) are linked, and j_1 and j_2 (in the target) are linked, and if the CE involves i_1 and j_1, then, after the CE, i_2 and j_1 are linked as well as i_1 and j_2.

A6 With the probability w, the colour exchange involves a q–$\bar{q}$ pair (D–type interaction). Otherwise (with probability $1 - w$), the colour exchange involves a the main string (N–type interaction).

A7 We assign momenta and flavours to each participating (anti)quark (and also to the q–$\bar{q}$ partners) according to measured distribution functions. Whether a quark or an antiquark is involved in the CE, is determined according to the integrals of the quark and antiquark distribution functions.

In order to easily identify the final strings, we use the convention:

A8 Only the quarks involved in the final strings are plotted as full dots, in initial or intermediate configurations, they are represented as open dots.

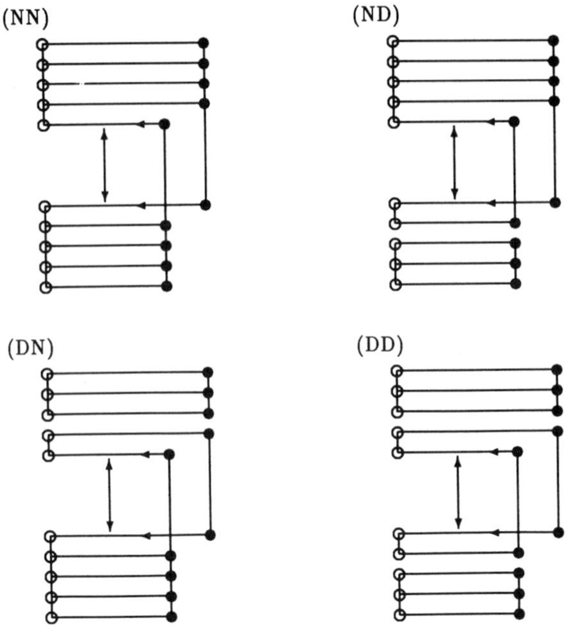

Figure 11. Quark line diagrams representing the four contributions to single colour exchange between antiquarks. Antiquarks are indicated by backward arrows.

Since both, projectile and target, may undergo a D–type or N–type interaction, we have four types of colour exchanges. They are referred to as nondiffractive scattering (NN), diffractive projectile excitation (ND), diffractive target excitation (DN), and Pomeron-Pomeron scattering (DD). For the first CE, for example, we obtain the four quark line diagrams shown in fig. 10. We have to keep in mind that for antiquark colour exchange, the diquarks in fig. 10 have to be replaced by four quarks (four lines), see fig. 11.

NUCLEUS-NUCLEUS INTERACTIONS

We are going to generalize the results of the preceeding chapters to nucleus-nucleus $(A-B)$ scattering in a straightforward way. We present a derivation of cross sections, using Gribov–Regge theory, introduce the Monte Carlo method, and discuss quark–line diagrams to handle multiple colour exchange.

GRT for Nuclear Scattering

For elastic scattering, we consider diagrams of the type shown in fig. 12, where the dashed line with index n represents a block of n Pomeron exchanges. The same nucleon may (or may not) be involved in more than one interaction. We write the

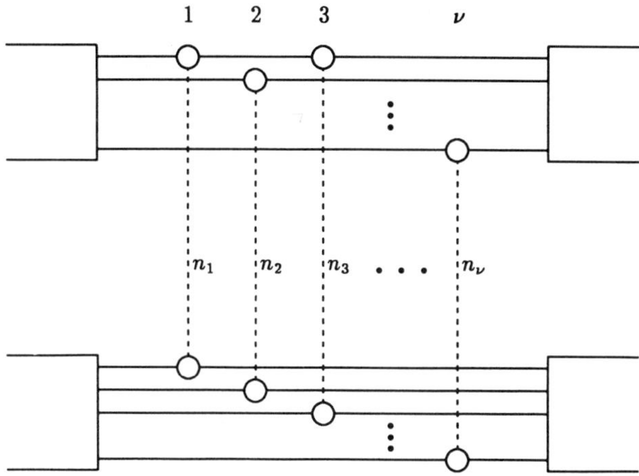

Figure 12. Elastic nucleus-nucleus scattering: nucleons of the projectile nucleus interacts with nucleons from the target nucleus via a "block" of several Pomeron exchanges (dashed line) for each N–N interaction.

elastic amplitude for A–B scattering as

$$A = \sum_\nu \sum_{\alpha_1\beta_1...\alpha_\nu\beta_\nu} \sum_{n_1...n_\nu} A_{n_1...n_\nu}^{\alpha_1\beta_1...\alpha_\nu\beta_\nu} \, , \tag{121}$$

with

$$i\,A_{n_1...n_\nu}^{\alpha_1\beta_1...\alpha_\nu\beta_\nu} = \int d\Omega \; N_{n_1...n_\nu}^{\alpha_1\beta_1...\alpha_\nu\beta_\nu} \prod_{\mu=1}^{\nu} \frac{1}{n_\mu!} \prod_{j=1}^{n_\mu} i\,D \, , \tag{122}$$

where $\sum_{\alpha_1\beta_1...}$ represents a sum over all possible "collision sequences" $\alpha_1\beta_1 \ldots \alpha_\nu\beta_\nu$, with α_μ and β_μ being indices refering to the projectile and target nucleon involved in the μth interaction. Ordering in the sequence is irrelevant, and the double indices have to be pairwise different: $\alpha_i\beta_i \neq \alpha_j\beta_j$. Eq. (122) corresponds to a sequence of ν N–N collisions, each of them representing a "block" of several pomeron exchanges (see fig. 12). Now we expand the absorptive part of $i\,A_{n_1...n_\nu}^{\alpha_1\beta_1...\alpha_\nu\beta_\nu}$ as a sum of terms corresponding to m_μ out of the n_μ Pomerons being cut,

$$2\,\mathrm{Im}A_{n_1...n_\nu}^{\alpha_1\beta_1...\alpha_\nu\beta_\nu} = \frac{1}{i}\mathrm{disc}\,A_{n_1...n_\nu}^{\alpha_1\beta_1...\alpha_\nu\beta_\nu} = \sum_{m_1...m_\nu} A_{n_1...n_\nu,m_1...m_\nu}^{\alpha_1\beta_1...\alpha_\nu\beta_\nu} \tag{123}$$

(see fig. 13), with

$$A_{n_1...n_\nu,m_1...m_\nu}^{\alpha_1\beta_1...\alpha_\nu\beta_\nu} = \sum_{I_{\mathrm{el}}I_{\mathrm{ib}}} \int d\Omega \; N_{n_1...n_\nu}^{\alpha_1\beta_1...\alpha_\nu\beta_\nu}$$

$$\prod_{\mu\in I_{\mathrm{in}}} (-1)^{n_\mu - m_\mu} \binom{n_\mu}{m_\mu} \frac{1}{n_\mu!} \prod_{j=1}^{n_\mu} 2\,\mathrm{Im}D$$

$$\prod_{\mu\in I_{\mathrm{el}}} (-1)^{n_\mu} \frac{1}{n_\mu!} \left[\prod_{j=1}^{n_\mu} 2\,\mathrm{Im}D - 2\,\mathrm{Re}\prod_{j=1}^{n_\mu} \frac{1}{i}D \right]$$

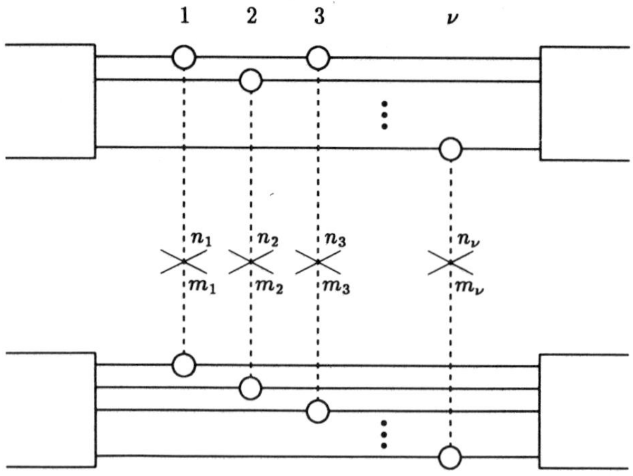

Figure 13. A contribution to the absorptive part of the elastic amplitude for nucleus-nucleus scattering, related to cutting m_μ out of the "block" of n_μ exchanged Pomerons in the μth nucleon-nucleon interaction.

$$\prod_{\mu \in I_{\text{ib}}} (-1)^{n_\mu} \frac{1}{n_\mu!} 2 \, \text{Re} \prod_{j=1}^{n_\mu} \frac{1}{i} \, D \; . \tag{124}$$

I_{in}, I_{el}, and I_{ib} are sets of indices, corresponding to inelastic, elastic, and inter–block cuts: I_{in} is the set of indices μ with $m_\mu > 0$, and I_{el}, I_{ib} are partitions of $I_\nu \setminus I_{\text{in}}$, with $I_\nu = \{1 \ldots \nu\}$, which means $I_{\text{el}} \cup I_{\text{ib}} = I_\nu \setminus I_{\text{in}}$. The sum $\sum_{I_{\text{el}} I_{\text{ib}}}$ is meant to sum over all partitions. Using the optical theorem, one obtains

$$\sigma_{\text{in}}^{AB} = \int d^2b \, \tilde{\sigma}_{\text{in}}^{AB}(b) \; , \tag{125}$$

with

$$\tilde{\sigma}_{\text{in}}^{AB}(b) = \sum_{\nu>0} \sum_{\alpha_1 \beta_1 \ldots \alpha_\nu \beta_\nu} \sum_{m_1 \ldots m_\nu} \tilde{\sigma}_{m_1 \ldots m_\nu}^{\alpha_1 \beta_1 \ldots \alpha_\nu \beta_\nu}(b) \; , \tag{126}$$

$$\tilde{\sigma}_{m_1 \ldots m_\nu}^{\alpha_1 \beta_1 \ldots \alpha_\nu \beta_\nu}(b) = \int dT_{AB} \prod_{\mu=1}^{\nu} \xi_{m_\mu}(b_\mu) \prod_{\mu=\nu+1}^{AB} [1 - \xi(b_\mu)] \; , \tag{127}$$

$$dT_{AB} := \prod_{\alpha=1}^{A} d^2b_\alpha^A \, T(b_\alpha^A) \prod_{\beta=1}^{B} d^2b_\beta^B \, T(b_\beta^B) \; , \tag{128}$$

$$b_\mu := b - b_{\alpha_\mu}^A + b_{\beta_\mu}^B \; , \tag{129}$$

$$\xi(b) := \frac{1}{C} \left\{ 1 - \exp\left[- 2C\omega(b) \right] \right\} \; , \tag{130}$$

$$\xi_m(b) := \frac{1}{C} \frac{\left[2C\omega(b) \right]^m}{m!} \exp\left[- 2C\omega(b) \right] \; . \tag{131}$$

Performing the sum in eq. (126), one obtains

$$\tilde{\sigma}_{\text{in}}^{AB}(b) = 1 - \int dT_{AB} \prod_{\mu=1}^{AB} [1 - \xi(b_\mu)] \; . \tag{132}$$

This formula, derived in the GRT framework without approximations (apart from assuming a simple form of the nucleus–nucleon vertex) is the starting point of many applications. Eq. (132) is often referred to as the Glauber formula, being a generalization of Glauber's result for h–A scattering, which has been derived, however, within nonrelativistic scattering theory. Eq. (132) is still very complicated, being a AB–fold integration, which cannot be reduced to a product of integrations as in the h–A case.

The Monte Carlo Method

The Monte Carlo method amounts to constructing particle ensembles. As usual, we proceed in two steps. We first construct strings and then decay the strings as discussed in the first chapter. We first determine a collision sequence $\alpha_1\beta_1\ldots\alpha_\nu\beta_\nu$, and a sequence $m_1\ldots m_\nu$ of numbers of CE's, according to $\sigma^{\alpha_1\beta_1\ldots\alpha_\nu\beta_\nu}_{m_1\ldots m_\nu}$ (to be dicussed later in this section) and then do the CE's explicitly with the help of quark line diagrams (next section).

We begin with considering the inelastic cross section. Although, as discussed above, it is impossible to calculate this quantity by standard methods, we demonstrate that it becomes a trivial problem by using Monte Carlo techniques. The inelastic cross section may be written as

$$
\tilde{\sigma}^{AB}_{\text{in}}(b) = \sum_\nu \sum_{\alpha_1\beta_1\ldots\alpha_\nu\beta_\nu} \int \prod_{\alpha=1}^{A} d^2b^A_\alpha\, T(b^A_\alpha) \prod_{\beta=1}^{B} d^2b^B_\beta\, T(b^B_\beta)
$$

$$
\prod_{\mu=1}^{\nu} \xi(b - b^A_{\alpha_\mu} + b^B_{\beta_\mu}) \prod_{\mu=\nu+1}^{AB} [1 - \xi(b - b^A_{\alpha_\mu} + b^B_{\beta_\mu})] , \tag{133}
$$

using the explicit expressions for dT_{AB} and b_μ. The quantity $\xi(b - b^A_{\alpha_\mu} + b^B_{\beta_\mu})$ may be regarded as the mean value of a random variable, which assumes only values 0 or 1, the latter one with probability ξ. Correspondingly, we may use the following procedure.

C1 Generate b^A_α ($\alpha = 1\ldots A$) and b^B_β ($\beta = 1\ldots B$) according to $T(b)$. This is equivalent to distributing nucleons according to nuclear densities in A and B, and then ignoring the longitudinal coordinates.

C2 Determine, for given b, the number $\xi_{\alpha\beta} := \xi(b - b^A_\alpha + b^B_\beta)$ for each α and β, as well as AB random numbers $r_{\alpha\beta}$. If $r_{\alpha\beta} \leq \xi_{\alpha\beta}$, consider the N–N interaction as being existent, otherwise not. The "existent" N–N interactions provide a sequence $\alpha_1\beta_1\ldots\alpha_\nu\beta_\nu$ of interactions.

C4 With $N_\nu(b)$ being the number of samples with at least ν N–N interactions, the inelastic cross section is $\tilde{\sigma}^{AB}_{\text{in}}(b) = N_1(b)/N_0(b)$.

C5 The b–integrated cross section σ^{AB}_{in} is obtained by (rather than taking fixed b) generating b according to a uniform two–dimensional distribution covering an area B, and finally taking $\sigma^{AB}_{\text{in}} = B\, N_1/N_0$ (with $N_\nu = \sum_\alpha N_\nu(b^{(\alpha)})$).

The procedure **C1** – **C4** is an exact numerical solution of eq. (133). No approximation is needed. The quantity ξ has been shown to be identical to the inelastic N–N cross section, $\xi(b) = \tilde{\sigma}^{NN}_{\text{in}}(b)$, and has qualitatively the shape of a Fermi function. If the latter one were taken as a step function $\Theta(R - |b|)$, which implies $R = \sqrt{\sigma^{NN}_{\text{in}}/\pi}$, then **C2** is replaced by

C2' (Step function approximation for ξ) If, for given b, the relation $|b - b_\alpha^A + b_\beta^B| \leq \sqrt{\sigma_{\rm in}^{NN}/\pi}$ holds, consider the N–N interaction involving nucleons α and β as being existent, otherwise not. ...

So in this approximation, the cross section is calculated as for the case of a classical collision of two bags containing hard spheres. But, of course, it is only one aspect that is equivalent to a classical system, other aspects are not, so this is not a justification for using classical models.

We now proceed to generate collision sequences $\alpha_1\beta_1 \ldots \alpha_\nu\beta_\nu$ as well as sequences $m_1 \ldots m_\nu$ of numbers of CE's according to eq. (127),

$$\tilde{\sigma}_{m_1 \ldots m_\nu}^{\alpha_1\beta_1 \ldots \alpha_\nu\beta_\nu}(b) \;=\; \int \prod_{\alpha=1}^{A} d^2b_\alpha^A \, T(b_\alpha^A) \prod_{\beta=1}^{B} d^2b_\beta^B \, T(b_\beta^B) \tag{134}$$

$$\prod_{\mu=1}^{\nu} \xi_{m_\mu}(b - b_{\alpha_\mu}^A + b_{\beta_\mu}^B) \prod_{\mu=\nu+1}^{AB} [1 - \xi(b - b_{\alpha_\mu}^A + b_{\beta_\mu}^B)] \,,$$

where we used the explicit expressions for dT_{AB} and b_μ. Eq. (134) is very similar to eq. (133). Since ξ_{m_μ} is stricly positive and $\sum_m \xi_m = \xi \leq 1$, we can apply a similar Monte Carlo procedure:

C3 [Generate b_α^A and b_β^B (**C1**) and determine a sequence $\alpha_1\beta_1 \ldots \alpha_\nu\beta_\nu$ of N–N interactions (**C2**)]. For each of the ν N–N interactions $\alpha_\mu\beta_\mu$ determine the number m_μ of CE's from $\sum_1^{m_\mu - 1} \xi_m < r_{\alpha\beta} \leq \sum_1^{m_\mu} \xi_m$, with $r_{\alpha\beta}$ being the random number used in **C2** to determine whether or not an interaction occurs.

This is an exact procedure to generate $\alpha_1\beta_1 \ldots \alpha_\nu\beta_\nu$ and $m_1 \ldots m_\nu$ according to eq. (134). Of course the $\alpha_1\beta_1 \ldots \alpha_\nu\beta_\nu$, $m_1 \ldots m_\nu$ and the $\sigma_{\rm in}^{AB}$ are determined in *one* procedure (**C1, C2, C3, C4, C5**), it was just for illustration to first calculate $\sigma_{\rm in}^{AB}$ separately.

The determination of m_μ according to $\xi_{m_\mu}(b\mu)$ is different to the N–N case, where m is generated according to

$$\frac{\sigma_m}{\sigma_{\rm in}^{NN}} = \frac{\int d^2b \, \xi_m(b)}{\sigma_{\rm in}^{NN}} \,. \tag{135}$$

In the case of N–N, we have an average over impact parameters; in the case of A–B, we have a specific impact parameter. For peripheral interactions, small m's are preferred, for central collisions, large m's. If we ignore the difference between A–B and N–N, and replace $\xi_{m_\mu}(b_\mu)$ by the average $\sigma_{m_\mu}/\sigma_{\rm in}^{NN}$, then **C3** is replaced by

C3' (Averaging approximation for ξ_m) [Generate b_α^A and b_β^B (**C1**) and determine a sequence $\alpha_1\beta_1 \ldots \alpha_\nu\beta_\nu$ of N–N interactions (**C2**)]. For each of the ν N–N interactions $\alpha_\mu\beta_\mu$ determine the number of CE's m_μ according to $\sigma_{m_\mu}/\sigma_{\rm in}^{NN}$ (with σ_m given, for example, in eq. (116)).

String Formation

After having obtained collision sequences $\alpha_1\beta_1 \ldots \alpha_\nu\beta_\nu$ and the numbers $m_1 \ldots m_\nu$ of CE's per N–N interaction (last section), we are going to discuss how to get string samples for given $\alpha_1\beta_1 \ldots \alpha_\nu\beta_\nu$ and $m_1 \ldots m_\nu$ with the help of quark line diagrams.

Actually the rules **A1-A8** of the previous section still apply. For A–B interactions, we use:

A9a The projectile participants are represented as unconnected strings. A string is represented by a vertical line connection, with the (anti)quarks involved being highlighted by dots.

A9b The target participants are represented as unconnected strings.

Projectile and target participants are those α's and β's appearing in the sequence $\alpha_1\beta_1 \ldots \alpha_\nu\beta_\nu$. With $\alpha_\mu\beta_\mu$ refering to the projectile and target nucleon involved in the μth interaction, we use:

A10 The m_μ CE's of the μth N–N interaction are performed between projectile nucleon α_μ and target nucleon β_μ.

The order of the ν CE's is irrelevant. We have 8 basic CE's, namely NN, ND, DN, DD, NN', ND', DN', DD'. Quark line diagrams are specified by symbols $A_{k_1 l_1}$–$B_{k_2 l_2}$–$C_{k_3 l_3} \ldots$, with A, B, C being any of the eight symbols NN, ND, ..., and $k_i l_i$ referring to the projectile nucleon and the target nucleon. We are going to discuss an example in the following.

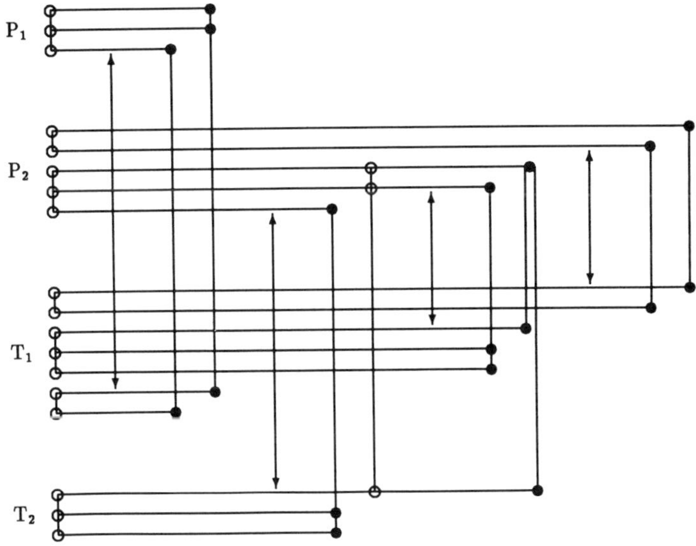

Figure 14. Quark line diagram representing an interaction of the type ND_{11}–NN_{22}–NN_{21}–DD_{21} of two projectile nucleons (P_1 and P_2) with two target nucleons (T_1 and T_2).

Let us consider the simplest nontrivial case of two projectile participants (1,2) and two target participants (1,2). We consider a collision sequence 11,22,21 and numbers of CE: $m_1 = 1$, $m_2 = 1$, $m_3 = 2$. One possible contribution would be ND_{11}–NN_{22}–NN_{21}–DD_{21}, as shown in fig. 14. It is very easy to construct these structures by using a computer code; it is only the graphical presentation on a sheet of paper that gets messy for complex interactions.

Since NN has by far the largest weight among the eight 1–CE contributions, also for A–B scattering it is most likely that diquarks "break up", whenever a nucleon is involved in at least two N–N interactions. The likelihood of this depends, of course, on the impact parameter.

References

[1] Proc. of "Quark Matter 1991", Gatlinburg, USA, Nov. 1991, Nucl. Phys. A544 (1992)

[2] K. Werner, Phys. Rep. 232 (1993) 87

[3] A. Capella, U. Sukhatme, Chung-I Tan and J. Tran Thanh Van, preprint LPTHE 92-38, to appear in Physics Reports

[4] H. J. Möhring, A. Capella, J. Ranft, J. Tran Thanh Van, C. Merino, Nucl. Phys. A525 (1991) 493c

[5] A. Kaidalov, Nucl. Phys. A525 (1991)39c

[6] V. D. Toneev, A. S. Amelin and K. K. Gudima preprint GSI-89-52, 1989

[7] B. Andersen, G. Gustafson and B. Nielsson-Almqvist, Nucl. Phys. B281 (1987) 289

[8] H. Sorge, H. Stöcker and W. Greiner, Nucl. Phys. A498 (1989) 567c

[9] Y. Nambu, Proc. Intl. Conf. on Symmetries and Quark Models, Wayne State Univ., 1969

[10] J. Scherk, Rev. Mod. Phys. 47 (1975) 123

[11] C. Rebbi, Phys. Rep. 12 (1974) 1

[12] X. Artru and G. Mennessier, Nucl. Phys. B70 (1974) 93

[13] X. Artru, Phys. Rep. 97 (1983) 147

[14] D. Gottschalk and D.A. Morris, Nucl. Phys. B288 (1987) 729

[15] K. Werner and P. Koch, Z. Phys. C47 (1990) 215, preprint CERN-TH-5575/89

[16] R.D. Field and R.P. Feynman, Nucl. Phys. B136 (1978) 1

[17] B. Andersson, G. Gustafson, G. Ingelman and T. Sjöstrand, Phys. Rep. 97 (1983) 31

[18] V. N. Gribov, Sov. Phys. JETP 26 (1968) 414

[19] K. A. Ter-Martirosyan, Nucl. Phys. B36 (1972) 566

[20] V. A. Abramovskiǐ, V. N. Gribov, O. V. Kancheli, Sov. J. Nucl. Phys. 18 (1974) 308

[21] M. Ciafaloni, G. Marchesini, G. Veneziano, Nucl. Phys. B98 (1975) 472, 493

INTRODUCTION TO THE DUAL PARTON MODEL

A. Capella

Laboratoire de Physique Théorique et Hautes Energies[1]
Université de Paris XI, bâtiment 211
91405 Orsay Cedex, France

INTRODUCTION

The Dual parton model is a phenomenological model of multiparticle production in hadronic and nuclear collisions. It is based on the large-N expansion of non-perturbative QCD and the Reggeon field theory. In these lectures I will concentrate on the theoretical basis of the model and to its generalization to hadron-nucleus and nucleus-nucleus collisions. I will also discuss two recent applications of the model : heavy flavor production off nuclei and strangeness enhancement in central nucleus-nucleus collisions.

THE MODEL

The aim of the dual parton model (DPM) is to determine the mechanism of multiparticle production in soft hadronic and nuclear collisions[1]. For soft (low $p_\perp$) collisions the strong coupling constant is large and perturbative QCD cannot be applied. For this reason one has to use a different approach. This is provided by the large-N expansion of QCD. Consider an $SU(N_c)$ color gauge theory coupled to N_f flavors of quarks in the fundamental representation. We are interested in an expansion where N_c is large and $\lambda = g^2 N_c$ held fixed, where g is the coupling constant. The conventional large-N_c expansion was first considered by 't Hooft[2], where the flavor number N_f was also held fixed. For high energy collisions where inelastic particle production dominates, Veneziano[3] proposed a modified expansion in which $N_c \to \infty$, $N_f \to \infty$, with both λ and N_f/N_c held fixed. In the following we shall be interested in the Veneziano limit. From a simple diagrammatic analysis it can be proven that an n-meson amplitude can be expressed as a sum of terms over topologies of a two-dimensional orientable surface, each weighted by an appropriate power of N^{-1}. This power of N^{-1} is found to be larger as the topological complexity increases. Each surface is labelled by three indices, (b, w, h), for the number of boundaries, windows,

---

[1] Laboratoire associé au Centre National de la Recherche Scientifique

Perspectives in the Structure of Hadronic Systems
Edited by M.N. Harakeh *et al.*, Plenum Press, New York, 1994

and handles respectively ($1 \leq b \leq n$ and $0 \leq h$, w). The N_c and N_f dependence of each contribution is given by

$$A_n^{(b,w,h)} \sim f_n^{(b,w,h)}(\lambda)(N_f/N_c)^w N_c^{2-\frac{n}{2}-2h-b} \, , \qquad (1.1)$$

where $f_n^{(b,w,h)}(\lambda)$ is independent of N_c. Note that the number of internal quark loops (or "windows") does not change the order in N in the Veneziano limit. However, the order changes with the number of "boundaries" (i.e. quark loops with attached external particles). As a consequence of (1.1), when N_c is large, scattering amplitudes are dominated by planar graphs (which correspond to h = 0, b = 1). The order of these graphs for two-body amplitudes (n = 4) is 1/N where N denotes either the number of flavors or colors since we consider the Veneziano limit. The next to leading order $1/N^2$ corresponds to the topology of a cylinder (h = 0, b = 2), etc.

Let us now come back to the main aim of DPM which is to determine the mechanism of multiparticle production in hadronic collisions. In hard processes like e^+e^- annihilation or lepton-proton scattering it consists in a 3-$\bar{3}$ color separation (q-$\bar{q}$ for e^+e^-, q-qq for ℓp) with the production of one string of hadrons (see Fig. 1). Here the word string is equivalent to two jets emitted back to back. The "jettiness" or string-like shape is due to the collimation of the color field lines produced by confinement.

Turning next to hadron-hadron scattering, single string production is also possible (see Fig. 2). This is precisely the planar diagram which is of leading order (1/N) in the large-N expansion. However, in this case we have an exchange of flavor quantum numbers between the colliding hadrons, and this results in a cross-section which decreases as $1/\sqrt{s}$, i.e. the same energy behaviour as for a charge-exchange process such as $\pi^- p \to \pi^0 n$. The simplest way to avoid this exchange of flavour quantum numbers is not to annihilate the quark and antiquark and to produce instead a second string of type q-$\bar{q}$ (see Figs. 3, 4 and 5). These graphs when squared in the sense of unitarity have the topology of a cylinder which is the next to leading one in the large-N expansion and will give the dominant contribution at large energies (Fig. 6). Physically, the graph in Fig. 3 can be interpreted as follows. In the case of a pp collision, the interaction takes place in the form of a color exchange between the two protons producing a splitting of each proton into two constituents : a valence quark and the rest of the proton (diquark). This color separation results in the production of two strings stretched between a valence quark of one proton and the diquark of the other proton.

There are also more complicated graphs producing 4 (Fig. 7), 6, 8 ... strings while the contribution of graphs with an odd number of strings tends to zero as the energy increases. These graphs are non-dominant in the large-N expansion. For instance, a graph with 2 strings when squared has the topology of a cylinder, the one with 4 strings will correspond to the topology of a cylinder with a handle the one with 6 strings to that of a cylinder with two handles, etc. Therefore, according to Eq. (1.1) the order of these contributions will be $1/N^4$, $1/N^6$, etc. However, contrary to the graphs with an odd number of strings their contributions will not vanish at asymptotic energies.

It is clear from energy-momentum conservation that the sum of the momentum fractions x_i taken by each constituent adds up to unity. For instance in Fig. 4 one has $x_1 + x_2 + x_3 + x_4 = 1$, where x_1 denotes the momentum fraction of the valence quark, x_2 and x_3 those of the sea quark and antiquark and x_4 that of the diquark.

Although the order of a given graph in the large N-expansion is known, it is not possible at present to compute their weights from the QCD lagrangian. Interestingly (and fortunately) it has been shown[4] that there is a one-to-one correspondence between term in the large-N expansion of QCD and those in Reggeon Field Theory (RFT)[5]. Therefore perturbative RFT provides a phenomenological mean of computing the weights of the various graphs in the large N-expansion. More precisely, the graph with the topology of a cylinder corresponds to the Born term representing a single scattering in the pertubative RFT (or in the eikonal model) ; a cylinder with a handle corresponds to a term with two interactions, etc. Explicit formulae for the weights of the various graphs in the large-N expansion and also for those with a given number of strings are given in Appendix A.

The mechanism of particle production described above is not particularly simple and one may wonder why it is so predictive. The reason is that the strings are universal building blocks. More precisely the hadronic spectra of the individual strings are obtained from a convolution of the momentum distribution functions (giving the probability that a specific constituent or fragment of a hadron carries a given fraction x of the hadron momentum) of the two constituents at the string ends and their fragmentation functions (giving the probability that a constituent produces a specific hadron carrying a given fraction z of the constituent momentum). The momentum distribution and fragmentation functions are universal - the same in all hadronic and nuclear processes. Moreover the fragmentation functions are also approximately the same as in hard processes at moderate values of Q^2.

A main assumption of DPM is string independence, i.e. the various strings hadronize independently from each other.

At this stage, one can combine the above ingredients to obtain physically observable quantities. For instance, the inclusive single particle spectra of charged particles in pp collisions is

$$\frac{dN^{pp}}{dy} = \frac{1}{\sigma_{in}}\frac{d\sigma^{pp}}{dy} = \frac{1}{\sum \sigma_n} \sum \sigma_n \left\{ N_n^{qq-q}(s,y) + N_n^{q-qq}(s,y) + (2n-2)N_n^{q-\bar{q}}(s,y) \right\} \ .$$

$$(1.2)$$

Here σ_n are the eikonal weights given by Eq. (A.2) and $N_n^{qq-q}(s,y)$, etc., are inclusive single particle spectra of charged hadrons produced in each chain. They can be obtained by folding the momentum distribution functions of the constituents at the string ends with their fragmentation functions

$$N_n^{qq-q}(s,y) = \int_0^1 \int_0^1 dx_1 dx_2 \rho_n^{qq}(x_1)\rho_n^{q}(x_2) \frac{dN^{qq-q}(y-\Delta; s_{ch})}{dy} \ . \qquad (1.3)$$

Here $\rho_n^{qq}(x)$ and $\rho_n^{q}(x)$ are respectively the momentum distribution functions of a diquark and a quark (the index n denotes a configuration in which 2n constituents of each incoming proton hadronize, i.e. a configuration with 2n strings). The last factor in the r.h.s. of Eq. (13) is the inclusive spectrum of charged particles produced in the qq-q string, for fixed positions x_1 and x_2 of the string ends. It is given by

$$\frac{dN^{qq-q}(y-\Delta; s_{ch})}{dy} = \begin{cases} \bar{x}_h\, D_{qq \to h}(x_h) & y \geq \Delta, \\ \bar{x}_h\, D_{q \to h}(x_h) & y < \Delta. \end{cases} \qquad (1.4)$$

where

$$x_h = \left| \frac{2\mu_h \sinh(y-\Delta)}{\sqrt{s_{ch}}} \right|, \qquad \bar{x}_h = \left(x_h^2 + \frac{4\mu_h^2}{s_{ch}} \right)^{\frac{1}{2}} . \qquad (1.5)$$

μ_h is the transverse mass of the detected particle h, and $D_{q \to h}$ and $D_{qq \to h}$ are the quark and diquark fragmentation functions. $s_{ch} = s x_1 x_2$, is the squared invariant mass of the chain. For massless quarks the rapidity shift Δ necessary to go from the overall pp CM frame to the CM of a chain is given by $\Delta = (1/2)\log \frac{x_1}{x_2}$.

The explicit forms of the momentum distribution functions (which are obtained using Regge arguments) and of the fragmentation functions can be found in Ref. 1.

Let me just recall here that in a two string graph the momentum distribution function of a valence quark in a meson is given by (see Fig. 8)

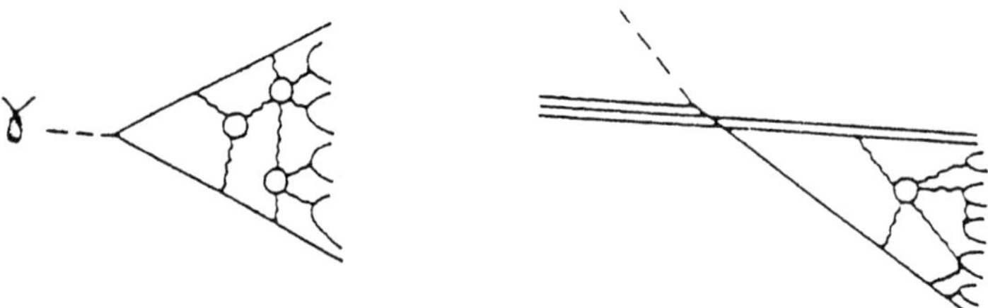

Figure 1. The mechanism of particle production in hard processes (e^+e^- annihilation and ℓp scattering).

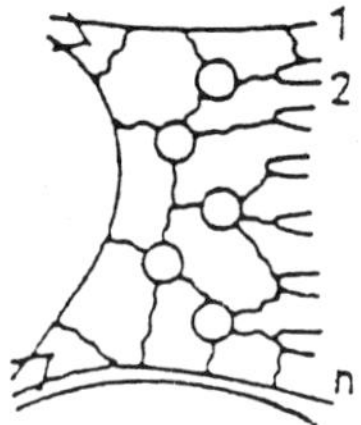

Figure 2. s-channel discontinuity of a one-chain planar contribution to high energy π^+-proton inelastic cross-section.

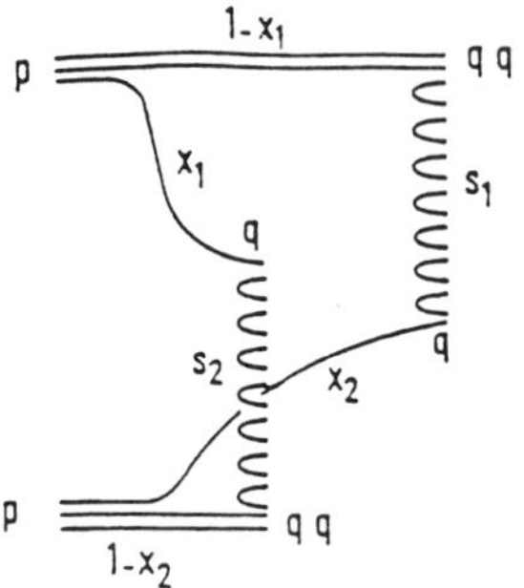

Figure 3. Dominant two-chain diagram describing multiparticle production in high energy proton-proton collisions. The two quark-diquark chain structure results from an s-channel unitarity cut of the cylindrical Pomeron shown in Fig. 6.

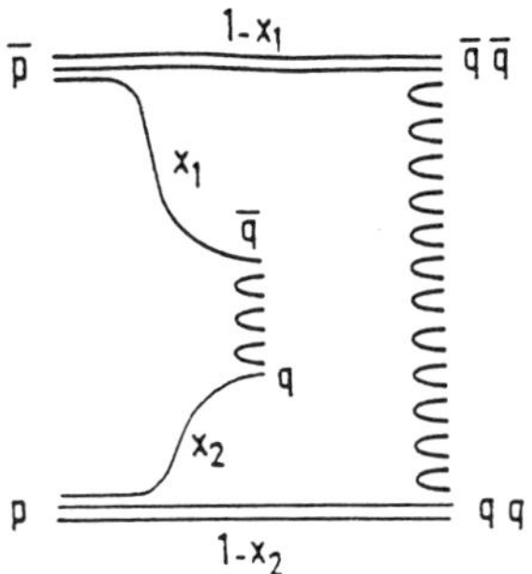

Figure 4. Dominant two-chain (single cut Pomeron) contribution to proton-antiproton collisions at high energies. Due to the held-back effect, one gets on average a long diquark-antidiquark chain and a short quark-antiquark chain, (see main text).

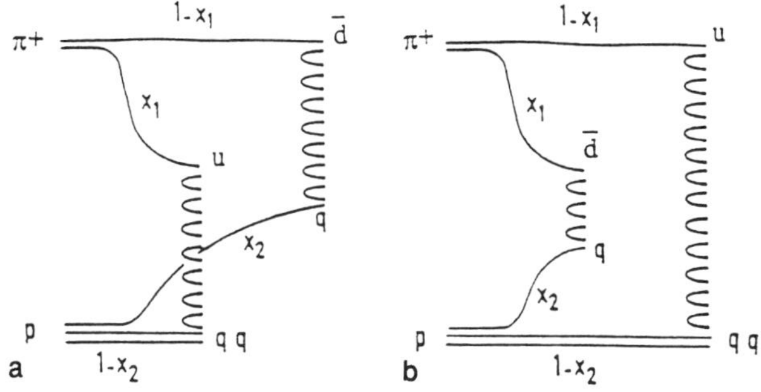

Figure 5. Dominant two-chain (single cut Pomeron) contributions to high energy π^+-proton collisions showing (a) a held-back u-quark, and (b) a held-back $\bar{d}$ antiquark. Similar diagrams can be drawn for π^- p and π^0 p collisions.

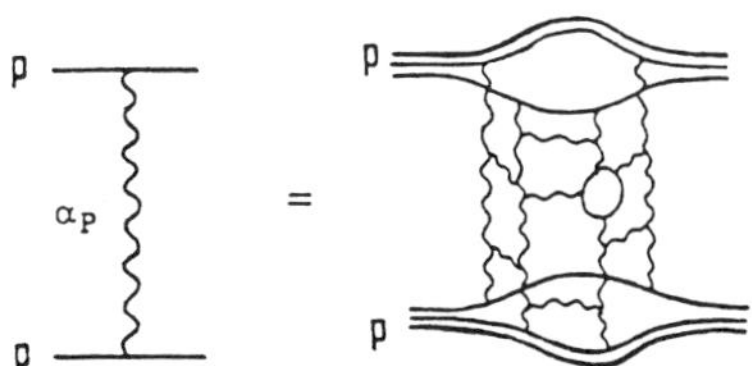

Figure 6. Single Pomeron exchange and its underlying cylindrical topology. This is the dominant contribution to proton-proton elastic scattering at high energies. The net of soft gluons and quark loops on the cylinder surface shown in this Figure have been omitted for simplicity in most other Figures.

$$\rho_1^m(x) = c_1^m x^{-1/2}(1-x)^{-1/2} \qquad (1.6)$$

and that of a valence quark in a proton by

$$\rho_1^p(x) = c_1^p x^{-1/2}(1-x)^{1.5} \quad . \qquad (1.7)$$

The coefficients c_1 are determined from the normalization to unity of the momentum distribution functions. An schematic illustration of Eq. (1.2) in the 2-string case (n = 1) is given in Fig. 9.

Although the mechanism of particle production, as discussed above, depends on both the momentum distribution functions and the fragmentation functions, the main features of particle production data in the central rapidity region can be understood in terms of two basic properties, which are consequences of the momentum distribution functions only and are very general. These properties, which determine the energy content of chains, will now be described in detail.

Held-Back Valence Quarks

From Eqs. (1.6) and (1.7) the valence quark momentum distribution function behaves as $x^{-1/2}$ at small x. As a consequence the valence quarks of each colliding hadron are separated into a slow held-back system and a fast colored system by the interaction. The dominant single cut Pomeron exchange, (two chain contribution) to several hadronic reactions is shown schematically in Figs. 3 to 5.

In the case of a meson, the held-back quark can be either the valence quark (with the antiquark being fast) or the antiquark (with the quark being fast). From Eq. (1.6) we see that both configurations have equal probabilities, while the configuration with both quark and antiquark sharing equal momenta is strongly suppressed. In the case of a proton, Eq. (1.7) leads to the preferred configuration where a valence quark is slowed down with the diquark remaining fast. Consequently, in nondiffractive events, the valence quark and its companion antiquark from an incoming meson prefer not to recombine into the same final hadron. Similarly, a quark tends to break away from an incoming proton and the probability of recombining with its original diquark is small. These important consequences of DPM have received experimental spectacular confirmation[6].

The $x^{-\frac{1}{2}}$ behavior of the valence quark momentum distribution functions corresponds to a distribution in the rapidity gap δy between the valence quark rapidity y and the maximal available rapidity $y_{\max}$, of the form $\exp(-\frac{\delta y}{2})$. Thus the average length of this gap is independent of the energy, and a simple calculation shows that this length is about 2.5 units. Therefore, up to ISR energies, the average rapidity of the held-back valence quark is very close to $y^* = 0$, and will influence the central rapidity region.

Short, Central q_s-$\bar{q}_s$ Multi-Chains

Corrections to the leading two-chain diagram come from multiple cut Pomeron exchanges, corresponding to multiple inelastic rescattering. The two new chains in a two-cut Pomeron exchange diagram terminate on sea quarks and antiquarks. Hence these chains span a smaller range of rapidity than those involving a diquark. They are forward-backward symmetric on the average, i.e. centered around $y^* = 0$, and span the central rapidity region. It is clear that each chain must at least have a minimum threshold CM energy in order for physical hadrons to materialize from it. Therefore, on the average, up until the highest CERN-ISR energies ($\sqrt{s} < 63$ GeV), the two-chain contribution dominates. However, at higher energies (like those at the

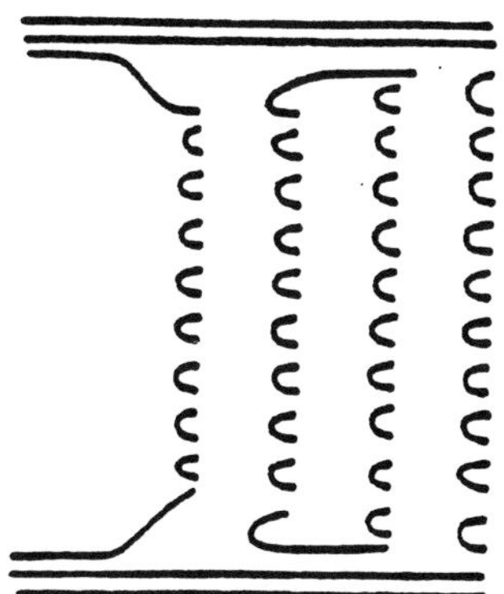

Figure 7. Two cut Pomeron (four-chain) diagram for proton-proton collisions.

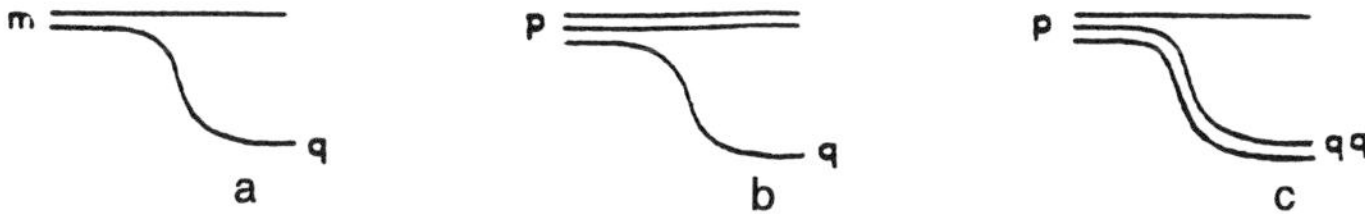

Figure 8. Quark line diagrams for momentum distribution functions of (a) quark (antiquark) in a meson, (b) quark in a proton, and (c) diquark in a proton.

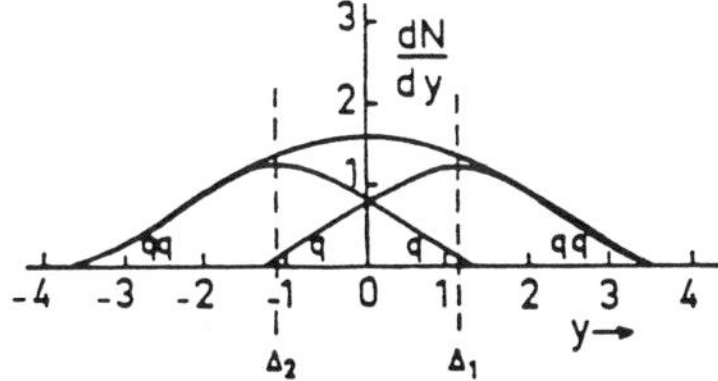

Figure 9. Diagram illustrating how the contributions from the two quark-diquark chains of Fig. 3 add up to give the total inclusive cross-section in pp scattering at CM energy $\sqrt{s} = 20$ GeV. At higher energies, multichain contributions of the type shown in Fig. 7 are also significant, and produce a rise if the central plateau.

CERN $\bar{p}p$ collider or the Fermilab Tevatron) there is sufficient energy available for q_s-$\bar{q}_s$ chains to have an average invariant mass larger than the threshold energy. The multi-chain contributions become increasingly important and the average number of strings produced increases with s. The effects of additional chains are mainly found in the central rapidity region. Moreover as a result of the convolution between momentum distribution and fragmentation functions (and of the existence of physical thresholds in the chains), it turns out that the short chains also have a smaller value of $\frac{dN}{dy}|_{y*=0}$. These additional short chains are responsible for the rise of the central plateau height with energy and for violations of KNO scaling. They also produce long-range correlations in rapidity which increase with s. Detailed comparison with experimental data can be found in Ref. 1.

Before concluding this section, we would like to comment on the role played by the gluons. First, the colored fragments of projectile and target between which the chains are stretched are composite objects which share the whole momentum of projectile and target. Thus gluons are part of these colored constituents. Moreover soft gluon and quark loops are present in the DPM graphs (see Figs. 1, 2 and 6). With increasing energies the role played by semi-hard gluons (minijets) will increase. In particular an increasingly large fraction of q_s-$\bar{q}_s$ pairs at the string ends (see Fig. 7) will come from a semi-hard gluon decay - with a corresponding increase of the intrinsic $p_\perp$ at the string ends. Details on the inclusion of minijets in DPM can be found in Ref. 1. Semi-hard scattering has also been implemented in a Monte Carlo code of DPM for pp and $\bar{p}p$ collisions called DTUJET[7].

HADRON-NUCLEUS COLLISIONS

So far, we have developed and discussed the dual parton model in the context of hadron-hadron collisions. The generalization of DPM to hadron-nucleus and nucleus-nucleus collisions is rather straightforward. The multiple-scattering framework used here is the Gribov-Glauber theory[8]. In this approach, the formulae for the hadron-nucleus cross-sections approximately coincide with those of the probabilistic Glauber model (see Section 2 of Appendix A). Previously proposed models of nuclear interactions formulated in this framework can be found in Ref. 9. The DPM for hadron-nucleus collisions is the result of various refinements introduced in these models. The complete multi-string version of DPM was formulated for hadron-nucleus collisions before being developed for hadron-hadron collisions[10]. Here multiple scatterings (and hence multi-string contributions) are enhanced by combinatorial factors when different nucleons of the nucleus are involved. For this reason, and also to keep our presentation as simple as possible, we shall neglect multiple-scattering in each individual nucleon-nucleon collision – which will therefore be hadronized in a two-string form. This simplified formulation is appropriate to study hadron-nucleus and nucleus-nucleus collisions up to the highest energies of the present CERN facility, i. e., 200 GeV per nucleon. However, at the higher energies of future machines like RHIC or LHC, multiple-scattering in each individual nucleon-nucleon collision can no longer be neglected. In particular the corresponding multi-string configurations contribute a substantial fraction of the multiplicity plateau height[11].

As in hadron-hadron collisions the hadronization mechanism in hadron-nucleus collisions is such that two strings are produced in each inelastic collision. Thus a configuration with ν struck nucleons of the nucleus will result in the production of 2ν strings. On the nucleus side these strings will be attached to the ν valence quarks and ν diquarks of the struck nucleons. On the projectile side two of the strings will involve the valence constituents while the other ones will involve sea constituents (see Fig. 5.1). One can now write down the formula for the inclusive spectrum in a proton-nucleus (pA) collision.

$$\frac{dN^{pA}}{dy}$$

$$= \frac{1}{\sigma_{pA}} \sum_{\nu=1}^{A} \sigma_{\nu}^{pA} \left[N_{\nu}^{qq_{v}^{p}-q_{v}^{A}}(y) + N_{\nu}^{q_{v}^{p}-qq^{A}}(y) + (\nu-1)\left(N_{\nu}^{q_{s}^{p}-qq_{v}^{A}}(y) + N_{\nu}^{\bar{q}_{s}^{p}-q_{v}^{A}}(y) \right) \right].$$

$$(2.1)$$

(For the special case of $\nu = 2$, the four contributions in Eq. (10) correspond to chains 4, 1, 3, 2 in Fig. 10 respectively). Here σ_{ν}^{pA} is the cross-section for ν inelastic collisions and given in Appendix A (Eq. A.5).

Average Multiplicities

If all the chains in DPM are assumed to be equal (both in average multiplicity and plateau height) one gets from Eq. (2.1)

$$\frac{<n>_{pA}}{<n>_{pp}} = \bar{\nu} = \frac{A\sigma_{pp}}{\sigma_{pA}}.$$

$$(2.2)$$

In Eq. (2.2) and in similar equations given below we have neglected the ν-dependence in the inclusive distributions N(y) due to energy-momentum conservation. Here $\bar{\nu}$ is defined as the ratio between $\sum_{\nu=1}^{A} \nu \, \sigma_{\nu}^{pA}$ and $\sum_{\nu=1}^{A} \sigma_{\nu}^{pA}$, and the second equality in Eq. (2.2) can be easily obtained using the expression for σ_{ν}^{pA} in Eq. (A.5).

More realistically, in DPM the chains in Eq. (2.1) are not equal. Due to the $x^{-\frac{1}{2}}$ behavior of the momentum distribution functions of the valence quarks, the latter are slowed down by the interaction (held-back effect). This has many interesting observable consequences, as discussed in Sec. 1. Likewise the q_{s} and $\bar{q}_{s}$ are also slow (on the average). Therefore, by momentum conservation, the diquarks will be fast in average, taking the remaining fraction of the incoming momentum. (It is interesting to note that in DPM the energy loss in each inelastic collision – nuclear stopping power – is entirely determined by momentum distribution functions and does not involve adjustable parameters). Thus, chains of the type q-qq have a larger invariant mass than ones of the type q-$\bar{q}$. Moreover, by performing the convolution between momentum distribution functions and fragmentation functions, it is easy to see that the density of particles dN/dy at y = 0 in the center of mass of each chain is much smaller for a q-$\bar{q}$ chain than for a q-qq one, especially at low energies. (With increasing s, the plateau height of the former increases very fast and both approach a common value).

Neglecting the contribution of the chains without diquarks (i. e. the q-$\bar{q}$ ones) as well as the difference between qq-q_v and qq -q_s, one would obtain instead

$$\frac{<n>_{pA}}{<n>_{pp}} = \frac{1}{2} + \frac{\bar{\nu}}{2}.$$

$$(2.3)$$

Numerical calculations[12] give a result which is close to the one in Eq. (2.3).

Inclusive Spectra

Let us now discuss in some detail the rapidity distribution resulting from Eq. (2.1). First, the rapidity distribution of the four strings involved is given in Fig. 11.

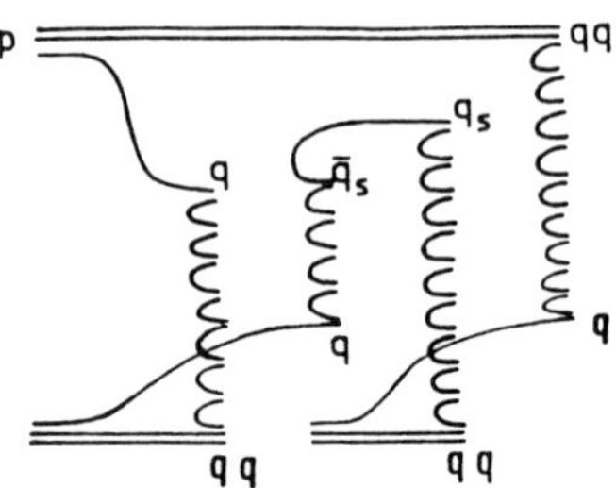

Figure 10. Diagram leading to a double inelastic scattering (two cut Pomerons, four chains) contribution to proton-nucleus scattering.

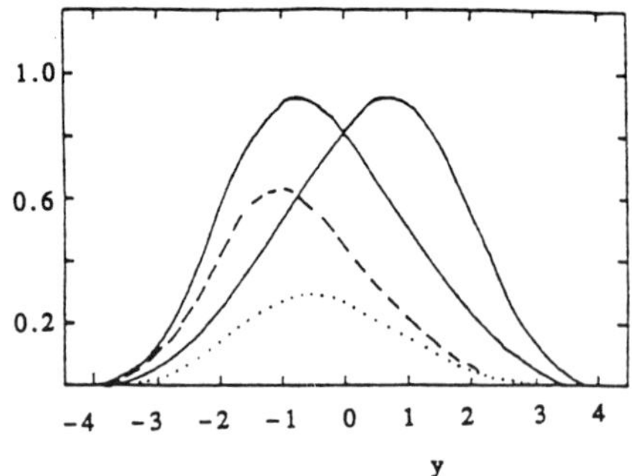

Figure 11. The rapidity distributions of the four chains in Eq. (1.1). The solid curves correspond to the valence chains $qq_A - q_p^v$ and $q_A^v - qq_p$. The dashed and dotted curves are the rescattering chains $qq_A - q_p^s$ and $q_A^v - \bar{q}_p^s$, respectively. The valence chains are computed for $\nu = 1$ and the rescattering chains for $\nu = 2$ (these chains are present only for $\nu \geq 2$).

For $\nu = 1$ one has two strings of type q_v-qq identical to the two main strings in a pp collision. Each new collision produces two strings q_s-qq and $\bar{q}_s$-q which are mainly located in the nucleus hemisphere. As in pp collisions, the detailed rapidity distribution of each string depends on the quark and diquark fragmentation functions. However, their rapidity position depends mainly on the momentum distribution functions. It is clear from Fig. 11 that the excess of produced secondaries in a proton-nucleus collision as compared to proton-proton is concentrated in the nucleus hemisphere and at mid rapidities. In the projectile fragmentation region, such excess is not present at all. In fact due to energy-momentum conservation, one produces less particles in this region in proton-nucleus than in proton-proton collisions. This phenomenon, called nuclear attenuation, has been observed experimentally and will be discussed below.

In order to be more quantitative, we discuss next the dependence of $(dN/dy)^{pA}$ on ν resulting[10] from Eq. (1.1). This result is more clearly illustrated by considering the ratio

$$R(y) = (dN/dy)^{pA} \,/\, (dN/dy)^{pp} \,. \tag{2.4}$$

It has been found experimentally that R(y) is practically constant and approximately equal to $\bar{\nu}$ in the nucleus fragmentation region ($y_{lab} = 1.5 \div 2$). It decreases with increasing y and becomes smaller than one in the projectile fragmentation region (nuclear attenuation). The above behavior of R(y) can be qualitatively understood from Eq. (2.1) and Fig. 11. Indeed in the nucleus fragmentation region, both $N^{qq_v^P - q_v^A}$ and $N^{\bar{q}_s^P - q_v^A}$ are small and $N^{q_v^P - qq_v^A}(y) \sim N^{q_s^P - qq_v^A}(y) \sim 1/2\, dN^{pp}/dy$. It then follows from Eq. (2.1) that

$$\frac{dN^{pA}}{dy} \sim \frac{1}{\sum \sigma_\nu^{pA}} \sum_\nu \nu\, \sigma_\nu^{pA} \frac{dN^{pp}}{dy} \equiv \bar{\nu}\, \frac{dN^{pp}}{dy} \tag{2.5}$$

which can also be written as

$$\frac{d\sigma^{pA}}{dy} \equiv \sigma_{pA} \frac{dN^{pA}}{dy} = A\, \frac{d\sigma^{pp}}{dy}\,. \tag{2.6}$$

Eq. (2.6) shows that inclusive cross-sections can behave linearly as A, (even for soft processes).

The same ratio $\bar{\nu}$ was obtained in Eq. (2.2) for the average multiplicity (assuming that all strings were equal). Under this assumption Eqs. (2.5) and (2.6) would also be valid at mid-rapidities. As argued in Sec. 2(a) this is not the case at present energies. If one makes the other extreme assumption of neglecting the contribution of the quark antiquark strings one obtains for the central plateau the same behavior as in Eq. (2.3) which, using $\sigma_{pA} \sim A^{2/3}\, \sigma_{pp}$, can be rewritten as

$$\frac{d\sigma^{pA}}{dy} = \frac{1}{2}\left(A + A^{2/3}\right) \frac{d\sigma^{pp}}{dy}\ (y \sim 0)\,. \tag{2.7}$$

This gives

$$\frac{d\sigma^{pA}}{dy} \sim A^\alpha\, \frac{d\sigma^{pp}}{dy}\ \text{with}\ \alpha \sim 0.9\,. \tag{2.8}$$

Finally it is easy to see that the ratio R(y) cannot exceed unity in the projectile fragmentation region. The "'cross-over" value, i.e., the value of y such that R(y) = 1, is approximately given by the smallest value of y for which only one string

$(qq_v^P - q_v^A)$ contributes, (see Fig. (11)). At this value of y one gets from Eq. (2.1) : $\frac{dN^{pA}}{dy} = N^{qq_v^P - q_v^A} = \frac{dN^{pp}}{dy}$ This cross-over value is thus approximately the same for all produced particles, in agreement with experiment. For larger value of y the ratio R is smaller than one due to energy-momentum conservation (with increasing ν the average momentum of the diquark and hence the value of $N^{qq_v^P - q_v^A}$ decreases). More quantitatively, in the limit $y \to Y_{MAX}$, Eq. (5.1) reduces to

$$\left(\frac{dN}{dy}\right)^{pA} \sim \frac{1}{\sum \sigma_n^{pA}} \sigma_1^{pA} \frac{dN^{pp}}{dy} \sim A^{-\alpha} \frac{dN^{pp}}{dy} \tag{2.9}$$

with $\alpha \sim -0.15$ to -0.25.

The agreement between the DPM results discussed above and experiment is satisfactory except for $y_{lab} < 1.5 \div 2$. It is well known that this region contains hadrons resulting from collisions of slow secondaries inside the nucleus. This intranuclear cascade is not included in the simple version of DPM described here.

NUCLEUS-NUCLEUS COLLISIONS

The generalization of DPM to nucleus-nucleus collisons is obtained as follows[13]. Consider a collision of nucleus A with nucleus B in a configuration with n_A participant nucleons of A, n_B participant nucleons of B and a total number n of inelastic collisions. Let us assume that $n_A < n_B$. In this configuration hadrons are produced in 2n chains (two chains for each inelastic collision). $2n_A$ of these chains are stretched between valence quarks and diquarks $(q_v^A - qq_v^B$ and $qq_v^A - q_v^B)$. The remaining $n_B - n_A$ valence quarks and diquarks of B have no valence partner in A and thus have to form $2n_B - 2n_A$ chains with sea constituents of A, $(q_s^A - qq_v^B$ and $\bar{q}_s^A - q_v^B)$. Finally the remaining 2n - $2n_B$ chains are stretched between sea constituents of A and B, $(q_s - \bar{q}_s)$.

In this way, one obtains the following formula :

$$\frac{dN^{AB}}{dy} = \frac{1}{\sigma_{AB}} \left[\sum_{n_A, n_B, n} \sigma_{n_A, n_B, n}^{AB} \, \theta \, (n_B - n_A) \right.$$
$$n_A \left(N_{\mu_A, \mu_B}^{qq_v^A - q_v^B}(y) + N_{\mu_A, \mu_B}^{q_v^A - qq_v^B}(y) \right)$$
$$+ (n_B - n_A) \left(N_{\mu_A, \mu_B}^{\bar{q}_s^A - q_v^B}(y) + N_{\mu_A, \mu_B}^{q_s^A - qq_v^B}(y) \right) \tag{3.1}$$
$$\left. + (n - n_B) \left(N_{\mu_A, \mu_B}^{q_s^A - \bar{q}_s^B}(y) + N_{\mu_A, \mu_B}^{\bar{q}_s^A - q_s^B}(y) \right) + \text{sym} \, (n_A \leftrightarrow n_B) \right].$$

Here $\sigma_{n_A, n_B, n}^{AB}$ is the cross-section for n inelastic nucleon-nucleon cross-sections, involving n_A nucleons of A and n_B of B and $\sum_{n_A, n_B, n} \sigma_{n_A, n_B, n}^{AB} = \sigma_{AB}^{abs.}$. The inclusive spectra N(y) of the individual strings are given as in pp and pA collisions by a convolution of momentum distribution and fragmentation functions. Here N(y) depends on two indices $\mu_A = n/n_A$ and $\mu_B = n/n_B$, via the product of momentum distribution functions $\rho_{\mu_A}\rho_{\mu_B}$ (cf. Eq. (1.3)). μ_A (resp. μ_B) is the number of times a nucleon of A (resp. B) has been hit in a configuration characterized by the cross-section $\sigma_{n_A, n_B, n}^{AB}$. In writing Eq. (3.1), we have neglected multiple-scattering in the individual nucleon-nucleon collisions (see the discussion of the beginning of section 2).

Average Multiplicities And Inclusive Spectra

If all chains would have the same average multiplicity one would get from (3.1)

$$\frac{\langle n \rangle_{AB}}{\langle n \rangle_{pp}} = \bar{n} = \frac{AB\,\sigma_{pp}}{\sigma_{AB}}\,.$$
(3.2)

The last equality is easily obtained[13] using the explicit formulae for $\sigma^{AB}_{n_A,n_B,n}$. As remarked in Sec. 2(a), Eq. (3.2) is only true if the dependence of the N's in μ_A and μ_B is neglected and if one assumes that all strings have the same average multiplicity. However, if one would neglect the short $(q - \bar{q}_s)$ chains, and retain only the chains containing a diquark (assumed to be all equal), one would get instead

$$\frac{\langle n \rangle_{AB}}{\langle n \rangle_{pp}} = \frac{\bar{n}_A + \bar{n}_B}{2},$$
(3.3)

where $\bar{n}_A$ and $\bar{n}_B$ are the average number of participating nucleons. Eq. (3.3) is rather obvious since the number of chains involving a diquark is equal to the number of participating nucleons.

The average number of collisions, $\bar{n}$, and the average number of participants, $\bar{n}_A$ and $\bar{n}_B$, are calculated from $\sigma_{n_A,n_B,n}$. Thus

$$\bar{n}_A = \sum_{n_A,n_B,n} \frac{n_A\,\sigma_{n_A,n_B,n}}{\sigma_{AB}},$$
(3.4)

where $\sigma_{AB} = \sum_{n_A,n_B,n} \sigma_{n_A,n_B,n}$. Using the expressions of $\sigma_{n_A,n_B,n}$ one gets[13]

$$\bar{n}_A = \frac{A\,\sigma_{pB}}{\sigma_{AB}}, \quad \bar{n}_B = \frac{B\,\sigma_{pA}}{\sigma_{AD}}\,.$$
(3.5)

The value of $\bar{n}$ is given by the last equality in Eq. (6.2).

Under the above assumptions Eqs. (3.2) and (3.3) also apply to the plateau heights dN^{AB}/dy at $y_{CM} \sim 0$. One has respectively

$$\frac{dN^{AB}}{dy} = \bar{n}\,\frac{dN^{pp}}{dy},$$
(3.6)

and

$$\frac{dN^{AB}}{dy} = \frac{\bar{n}_A + \bar{n}_B}{2}\,\frac{dN^{pp}}{dy},$$
(3.7)

or

$$\frac{d\sigma^{AB}}{dy} = AB\,\frac{d\sigma^{pp}}{dy},$$
(3.8)

and

$$\frac{d\sigma^{AB}}{dy} \sim \frac{1}{2}\left(AB^{2/3} + BA^{2/3}\right)\frac{d\sigma^{pp}}{dy}\,.$$
(3.9)

As in pA collisions the actual behavior of the plateau height is close to the one in Eq. (3.7) and (3.9). In the fragmentation regions, one has : $\frac{dN^{AB}}{dy} \sim \bar{n}_B \frac{dN^{PP}}{dy}$ in the fragmentation region of B, and $\frac{dN^{AB}}{dy} \sim \bar{n}_A \frac{dN^{PP}}{dy}$ in the fragmentation region of A. This can also be written as

$$\frac{d\sigma^{AB}}{dy} \sim BA^{2/3}\, \frac{d\sigma^{pp}}{dy} \quad \text{in the fragmentation region of B}$$

$$\frac{d\sigma^{AB}}{dy} \sim AB^{2/3}\, \frac{d\sigma^{pp}}{dy} \quad \text{in the fragmentation region of A .}$$

(3.10)

The above behavior in the nuclear fragmentation regions neglects the nuclear attenuation effect (see Sec. 2(b)) which results from energy-momentum conservation. When this is taken into account the power of 2/3 in Eq. (3.10) decreases and gets closer to $0.4 \div 0.5$ near the edges of phase space as in hA collisions (of Eq. (2.11)). One gets in this way the simple behavior[14]

$$\frac{d\sigma^{AB}}{dy}(y) \sim A^{\alpha(y)} B^{\alpha(-y)} \frac{d\sigma^{pp}}{dy},$$

where $\alpha(y)$ is the same function as in proton-nucleus collisions and the center of mass rapidity y is taken to be positive (negative) in the hemisphere of nucleus B (A).

HEAVY FLAVOR PRODUCTION

The production of J/Ψ and dilepton pairs (DY) in collisions off nuclei is important, not only to study J/Ψ suppression as a possible signal of QGP formation, but also in order to study the propagation of a $c\bar{c}$ or $\ell\bar{\ell}$ pair in nuclear matter.

We have seen in Section 2 that, if the inclusive cross-section off nuclei is parametrized as $A^{\alpha(x)}$ it is found experimentally that, for ordinary hadrons composed of light quarks, $\alpha(x \sim 0) \sim 0.9$ and for $\alpha(x \sim 1) \sim 0.4 \div 0.5$. We have also seen that such an A-dependence is obtained in DPM when energy conservation is taken into account. For the production of a J/ψ or a lepton pair of similar mass, the experimental situation is very different since one finds respectively at large x $\alpha_{J/\Psi} = 0.7 \div 0.8$ and $\alpha_{DY} = 1$ (except at the highest available energy of 800 GeV/c in the laboratory system where at large x $\alpha_{DY} = 0.95$).

In order to understand the situation let us first rederive the DPM results for ordinary hadrons using directly the diagrams of reggeon field theory (Fig. A.1), or, more concretely, the eikonal model (Appendix A).

It is useful to consider two different types of contributions to the inclusive cross-section I_A^a, for the reaction $h + A \rightarrow a + X$. When the final hadron a is produced in a cut Pomeron (the so-called direct contribution), one has from Eq. (A.5) :

$$I_A^{a(d)} \sim \sum_{n=1}^{A} n\, \sigma_n = A\, \sigma_{hN} \ . \tag{4.1}$$

The factor n is due to the fact that hadron a can be produced in any of the n cut Pomerons. However, when hadron a is present inside the "blob" of Fig. 15 (i.e.

when it is either present in the projectile wave function or produced inside the blob) the factor n in Eq. (4.1) is no longer present. We then have for this "intrinsic" contribution

$$I_A^{a(i)} \sim \sum_{n=1}^{A} \sigma_n = \sigma_{hA}^{in} \sim A^{2/3} \ .$$

(4.2)

In the limit $x \to 1$, only the term σ_1 in Eq. (4.2) can contribute - due to energy conservation.

As discussed in Section 2 one has $\sigma_1 \sim A^{1/3}$ for $A >> 1$ (edge effect). For actual values of A, $\sigma_1 \sim A^\alpha$ with $\alpha \sim 0.4 \div 0.5$. In this way we obtain the same results as in DPM.

When a heavy system is present in the blob, the situation is unchanged at asymptotic energies. However, at finite energies, there is a new energy threshold $E_M \sim M^2 R_A / x_+$, where M is the mass of the heavy state, x_+ its light cone momentum fraction and R_A the nuclear radius. Such a threshold is the result of a simple t_{min} effect. When $E > E_M$, some discontinuities of Fig. 1 involve a $t_{min} \neq 0$ and their contribution is suppressed by the nuclear form factor (15)(16). This leads to a change in the AGK cutting rules described in Appendix A (and thus in the expressions of σ_n) in such a way that the r.h.s. of Eq. (4.2) is now proportional to A.

In order to illustrate the situation let us consider the case of two interactions (k = 2 in Fig. 15). We have shown in Appendix A that the corresponding eikonal contributions $\sigma_1^{(2)}$ and $\sigma_2^{(2)}$ are related to each other by $\sigma_1^{(2)} = -2\sigma_2^{(2)}$. It can be shown that for $E > E_M$ the contribution $\sigma_2^{(2)}$ is unchanged, whereas $\sigma_1^{(2)}$ is reduced by a factor of 2 (i.e. one half of this contribution has $t_{min} \neq 0$ and is suppressed by the nuclear form factor). Then

$$\sigma_1^{(2)} = -\sigma_2^{(2)}$$

(4.3)

and these two contributions add up to zero in Eq. (4.2). The same cancellation occurs for any value of k. To summarize, the inclusive cross-section of a lepton pair of mass M, in the energy region $E < E_M$, is proportional to A. When the energy increases the behaviour will be A^α with $\alpha < 1$. This absorption will be first observed at large x : increasing further the energy it will extend to lower values of x.

The above situation applies to the production of a heavy system (lepton pair) which does not participate in the (strong) interaction. For a system that does interact (like a $c\bar{c}$ pair) the situation is different. Indeed for k = 2, the two contributions $\sigma_1^{(2)}$ and $\sigma_2^{(2)}$ will, in general, have a different x-dependence and thus their contributions will not cancel at a given x :

$$\sigma_1^{(2)} f_1(x) + \sigma_2^{(2)} f_2(x) \neq 0 \ .$$

(4.4)

Due to energy conservation $f_1(x) > f_2(x)$ at $x \sim 1$. Since $\sigma_2^{(2)} > 0$, the r.h.s. of (4.4) will be negative at $x \sim 1$ and absorption will be present - even at $E < E_M$. Again the generalization to any k is straightforward.

Eq. (4.4) leads to the following interesting conclusions :

1) The absorption of the $c\bar{c}$ system at a given x is not only given by its cross-section but it also depends on the difference $f_1(x) - f_2(x)$, which varies with x. This leads to an absorption with an "effective" cross-section that increases with increasing x.

2) If the system $c\bar{c}$ is not destroyed in its interactions with the nucleons one has : $\int f_1(x)dx = \int f_2(x)dx$, and the integrated inclusive cross-section will be proportional

to A at $E < E_M$. Since in some cases the system will convert into open charm a net absorption will remain after integration in x[15].

3) The A-dependence of the inclusive cross-section as a function of x will depend strongly on the x-distribution. In the case of hadroproduction of J/Ψ a slow decrease of the absorption with decreasing x is expected. However, for J/Ψ photoproduction, where the inclusive spectrum is peaked at large x, the fact that the difference $f_1(x) - f_2(x) > 0$ at $x \sim 1$ (where the spectrum is large) will lead to $f_1(x) - f_2(x) < 0$ (i.e. anti-shadowing) at some lower value of x[17]. This behaviour is in agreement with recent data from the NMC collaboration.

In the case of AB collisions all existing data are at mid-rapidities and the energies are such that $E < E_M$. Thus we expect that the dilepton continuum is proportional to AB whereas the J/Ψ behaves like $(AB)^{\alpha_{J/\Psi}}$ with $\alpha < 1$. It follows from the above considerations that the difference $\alpha_{J/\Psi} - 1$ is controlled at low x by the "absorptive" $\sigma^{abs.}_{J/\Psi-N}$ cross-section (i.e. the rate of destruction of the J/Ψ yielding open charm mesons). As we have said above $\alpha_{J/\Psi}(x \sim 0) \sim 0.9$ in pA collisions. Such a value of α can be obtained using $\sigma_{J/\Psi-N} \sim 6$ mb[15]. In AB collisions the experiments involve a limited number of projectiles and targets. Therefore, one measures the ratio $R(E_T)$ of J/Ψ over lepton pair continuum as a function of the total neutral energy E_T of the produced particles (mostly pions). Increasing E_T corresponds to increasing the average number of collisions and therefore is a good substitute to an increase of A and/or B. Numerical results for the E_T dependence of the ratio R in S-U and O-U collisions are given in Figs. 12 together with the DPM results[18] using the value of $\sigma^{abs}_{J/\Psi-N} \sim 6$ mb obtained in pA collisions. The agreement between theory[18,19] and experiment is good but the statistical errors are large. It is a feature of DPM (as well as of most absorptive models) that $R(E_T)$ decreases at low E_T and flattens at large E_T. An opposite situation is expected in a QGP formation scenario (where the suppression of the J/Ψ is large at large E_T). Forthcoming data with better statistics should allow to distinguish between these two behaviours.

STRANGENESS ENHANCEMENT MECHANISM[20]

Since the individual strings are universal building blocks of the model, the ratio of produced strange particles over non-strange ones will be approximately the same in all reactions. However, since some strings contain sea quarks at one or both ends and since strange quarks are present in the proton sea, it is clear, that, by increasing the number of those strings, the ratio of strange over non-strange particles will increase. This will be the case for instance, when increasing the centrality in a nucleus-nucleus collision. It is obvious, that the numerical importance of the effect will depend on the assumed fraction of strange over non-strange quarks in the proton sea. The rather extreme case leading to a maximum increase of strangeness is to assume a SU(3) symmetric sea (equal numbers of u, d and s flavors). A more conservative assumption is to take in the proton sea the same ratio of strange over non-strange quarks, which is produced in the chain fragmentation. We express the amount of SU(3) symmetry of the sea chain ends by our parameter s^{sea} defined as

$$ s^{sea} = \frac{2 < s_s >}{< u_s > + < d_s >} \tag{5.1} $$

where the $< q_s >$ give the average numbers of sea quarks at the sea chain ends. Full SU(3) symmetry of the sea is given by $s^{sea} = 1$; strangeness suppression at the sea chain ends equal to the one inside the chain fragmentation is obtained with $s^{sea} \approx 0.33$.

However, the above scenario has an important drawback. Since an antiquark from the sea is always attached to a valence- or sea-quark on the opposite hemisphere, and since the only important strings at CERN-energies are those containing at least a diquark at one end, it will be impossible to obtain an enhancement of antibaryons. In fact, due to energy-momentum conservation, the ratio $\bar{\Lambda}/h^-$ will in fact decrease with increasing centrality. In an attempt to solve this problem, we allow the creation of qq-$\bar{q}\bar{q}$ pairs from the proton sea, leading to the production of strings of type $\bar{q}\bar{q}$-$\bar{q}$ or $\bar{q}\bar{q}$-qq in which the production of strange antibaryons will be easier. The rate of diquark pairs to q-$\bar{q}$ pairs in the proton sea is assumed to be the same as the ratio of $q \rightarrow (qq)$ to $q \rightarrow q$ branching in the chain fragmentation (approximately $\alpha = 0.07$). The expression of the single particle density, Eq. (3.1), is now changed into

$$
\begin{aligned}
\frac{dN^{AB}}{dy} = \frac{1}{\sigma_{AB}} \sum \sigma_{n_A,n_b,n} &\left\{ \Theta(n_b - n_A) \left[n_A \left(N^{qq^A - q_v^B} + N^{q_v^A - qq^B} \right) \right.\right. \\
+(n_B - n_A) &\left((1 - \alpha)(N^{q_s^A - qq^B} + N^{\bar{q}_s^A - qq_v^B}) + \alpha(N^{(\bar{q}\bar{q})_s^A - qq^B} + N^{(qq)_s^A - q_v^B}) \right) \\
+(n - n_B) &\left((1 - 2\alpha)(N^{q_s^A - \bar{q}_s^B} + N^{\bar{q}_s^A - q_s^B}) \right. \\
+\alpha\ (N^{q_s^A - (qq)_s^B} &+ N^{(qq)_s^A - q_s^B}) + \alpha(N^{\bar{q}_s^A - (\bar{q}\bar{q})_s^B} + N^{(\bar{q}\bar{q})_s^A - \bar{q}_s^B}) \left.\left.\left. \right) \right] \right. \\
&\left. +(A \rightarrow B) \right\} \quad .
\end{aligned}
\tag{5.2}
$$

Following Regge arguments the momentum distribution function of sea diquarks and antidiquarks is taken to be the same as for valence diquarks. (For simplicity we ignore the difference between strange and non-strange ones).

A second new feature of the model is in the fragmentation scheme. Together with the usual diquark fragmentation (Fig. 13.a) we allow it to fragment into a meson at the first break-up of the string (Fig. 13.b). This produces a shift of the baryon and antibaryon mean rapidities towards the central region. This effet is more important for strange baryons since for a non-strange diquark the probability to become a strange baryon is twice as large in the fragmentation according to Fig. 13.b than in the one according to Fig. 13.a. A further enhancement of strange baryon production is obtained in this way. In Fig. 14 we show the single particle densities of Λ and $\bar{\Lambda}$ in SS, SAu central collisions and compare them with recent NA35 data[21]. The agreement between theory and experiment is quite good.

The WA85 collaboration has measured the relative yields of Λ-Hyperons and cascade particles in central S-W collisons[22] at $1 < p_\perp < 2$ GeV in the central rapidity window $2.3 < y < 2.8$. An important result of this experiment is the rise of the ratio $\bar{\Xi}/\bar{\Lambda}$ from p-p collisions to central S-W collisions. (Preliminary data from the NA36 collaboration are also available[23]). In Table 1 we give the computed ratios in the same window. They are in reasonable agreement with WA85 data[22]. Preliminary results for these ratios from the NA36 collaboration[23] in somewhat different phase-space windows are in rough agreement with those of WA85 - except for the ratio $\bar{\Xi}/\bar{\Lambda}$ which is 50 % lower in NA36, in better agreement with our results.

Table 1. Hyperon ratios in central S-W collisions computed in the phase space window $1 < p_T < 2$ GeV/c and $2.3 < y < 2.8$ are compared to the WA85 collaboration data[22] in the same window.

Ratio	$s^{sea} = 1$	$s^{sea} = 0.66$	WA85
$\bar{\Lambda}/\Lambda$	0.28	0.26	0.20 ± 0.01
$\bar{\Xi}/\Xi$	0.58	0.55	0.41 ± 0.05
Ξ/Λ	0.07	0.06	0.09 ± 0.01
$\bar{\Xi}/\bar{\Lambda}$	0.14	0.13	0.20 ± 0.03

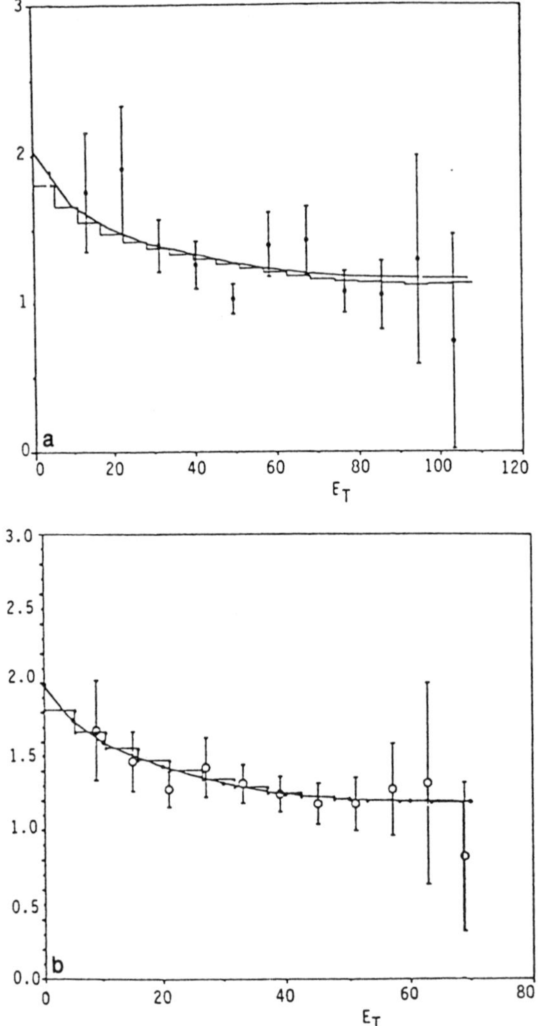

Figure 12. Ratios of E_T distributions of J/ψ over DY in (a) S-U and (b) O-U collisions obtained in DPM.

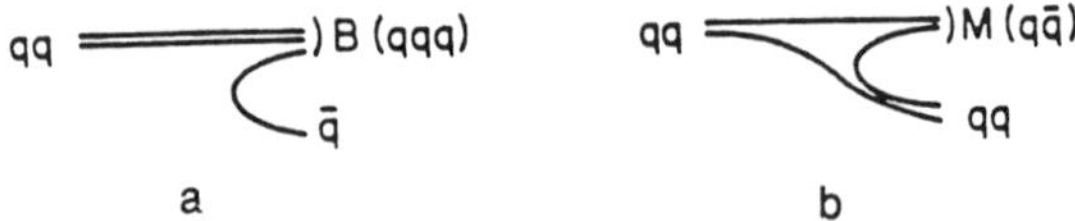

Figure 13. The two basic breakups of a recursive cascade model for diquark fragmentation : (a) diquark $\rightarrow$ baryon + leftover antiquark, and (b) diquark $\rightarrow$ meson + leftover diquark.

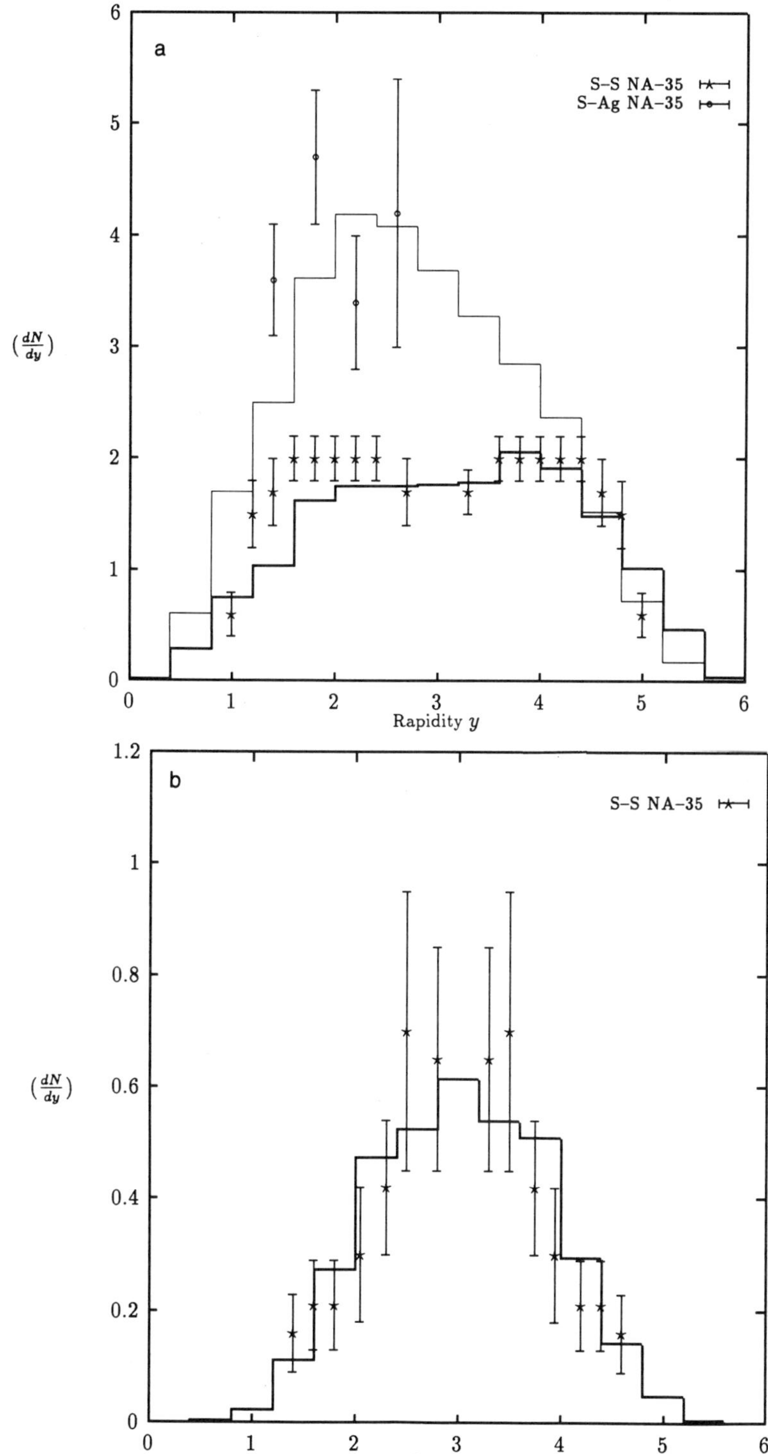

Figure 14. Rapidity distributions of Λ and $\bar{\Lambda}$ in central nucleus-nucleus collisions at 200 GeV. (a) Λ's in SS and SAg ; (b) $\bar{\Lambda}$ in SS. The data are from the NA35 colalboration[21]. The produced Σ^0's are included in the histograms. The histograms are calculated with $s^{sea} = 1$.

APPENDIX A

The Weights σ_n In Perturbative Reggeon Field Theory

The reggeon field theory (RFT) or reggeon calculus is a field theoretical approach to high energy hadronic collisions involving the Pomeron (as an effective field in two transverse plus one longitudinal dimensions) and its interactions. The simplest of these interactions (triple Pomeron interaction) is known experimentally to be small. If one neglects Pomeron interactions the elastic scattering amplitude for an elastic two-body process $a + b \rightarrow a + b$ is given by the sum of contributions of reggeon graphs of the type shown in Fig. 15, where the wavy lines denote the leading reggeon, called Pomeron. In order to compute reggeon graphs, one needs a model for the couplings of Pomerons to external hadrons. The simplest coupling is between two particles and two Pomerons, (Fig. 16). Note that the absorptive part of this vertex contains contributions from the elastic pole and from low mass diffractive intermediate states. The elastic pole contribution is the most important one - at present energies. Keeping only this contribution (the so-called elastic rescattering approximation) the reggeon calculus reduces to an eikonal model. Therefore we begin by briefly reviewing the standard eikonal model (without diffraction dissociation).

The Eikonal Model

An eikonal model is most easily described in impact parameter space. The elastic amplitude is given by $f(s, t) = i \int_0^\infty b\, db\, J_0\left(b\sqrt{-t}\right)\left[1 - e^{-\chi(s,b)}\right]$, where s, t are the usual Mandelstam variables. In general, the eikonal $\chi(s, b)$ is complex,

$$\chi(s, b) \equiv \chi^R(s, b) + i\, \chi^I(s, b). \tag{A.1}$$

The choice for $\chi(s,b)$ allows one to describe a specific inelastic production mechanism and hence a model for absorption. The total, elastic and inelastic cross sections are given by $\sigma_{\mathrm{TOT}}(s) = 4\pi\, \mathrm{Im}\, f(s, 0) = 4\pi\, \mathrm{Re} \int_0^\infty b\, db\left[1 - e^{-\chi(s,b)}\right]$, $\sigma_{\mathrm{EL}}(s) = \pi \int dt\, |f(s, t)|^2 = 2\pi \int_0^\infty b\, db\, |1 - e^{-\chi(s,b)}|^2$, and $\sigma_{\mathrm{INEL}}(s) \equiv \sigma_{\mathrm{TOT}} - \sigma_{\mathrm{EL}} = 2\pi \int_0^\infty b\, db\left[1 - e^{-2\chi^R(s,b)}\right]$. Unitarity of the elastic amplitude requires that $\chi^R(s, b)$ be positive.

The inelastic cross section can be written as a series of positive contributions

$$\sigma_{\mathrm{INEL}}(s) = \sum_{n=1}^\infty \sigma_n(s), \qquad \sigma_n(s) \equiv 2\pi \int_0^\infty b\, db\, e^{-2\chi^R}\, \frac{(2\chi^R)^n}{n!}. \tag{A.2}$$

The quantity $\sigma_n(s)$ can be identified as the cross section corresponding to n inelastic scatterings, summed over an arbitrary number of elastic ones. [Note that one has a Poisson distribution in n at a fixed impact parameter]. Eq. (A.2) has a simple probabilitistic interpretation : the factor $(2\chi^R)^n$ corresponds to the probability of n inelastic encounters at impact parameter b, and the exponential one to the probability that no other inelastic encounter occurs. It follows from the expression of $f(s, t)$ given above that the eikonal $\chi(s, b)$ is given by the inverse Bessel transform of the Born amplitude $f_B(s, t)$ (defined as the contribution to $f(s, t)$ of the term linear in $\chi(s,b)$). In the reggeon calculus f_B is given by the Pomeron exchange contribution

$$f_B(s, t) = i g_a\, g_b\, e^{(A_a + A_b)t/2}\left(se^{-i\pi/2}\right)^{\alpha_P(t)-1} \tag{A.3}$$

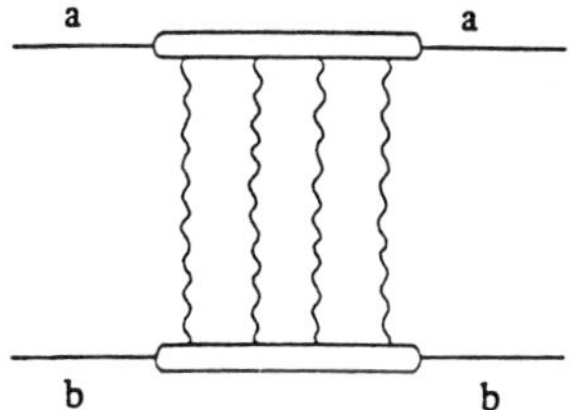

Figure 15. Typical t-channel Pomeron exchange graphs contributing to the elastic scattering process a + b → a + b.

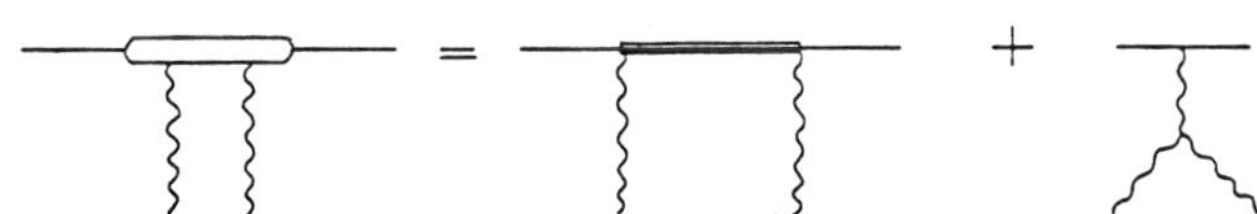

Figure 16. Contributions to the two-particle two-Pomeron coupling.

where $\alpha_P(t) = 1 + \Delta + \alpha' t$ is the Pomeron trajectory. The corresponding eikonal is

$$\chi(s,b) = -i \int_0^\infty d\sqrt{-t}\, \sqrt{-t}\, J_0\left(b\sqrt{-t}\right) f_B(s,t) = C(s) \exp\left(-\frac{b^2}{4B(s)}\right), \quad (A.4)$$

where $C(s) = \frac{g_a g_b}{4B(s)} \left(s e^{-i\pi/2}\right)^{\alpha_P(0)-1}$ and $B(s) = \frac{A_a + A_b}{2} + \alpha'\left(\ln s - i\frac{\pi}{2}\right)$. The parameters in (A.3), Δ, α', $g_a g_b$ and $A_a + A_b$ are obtained from a fit to the experimental data for the total and differential elastic cross-sections at various energies. The weights σ_n can then be computed from Eq. (A.2) without extra free parameters. Rising total cross-sections arise from choosing $\Delta > 0$, i.e. a Pomeron intercept $\alpha_P(0) > 1$. In this case the Born amplitude f_B violates the Froissart bound. The eikonal model is an s-channel unitarization procedure. It leads to an "eikonalized" amplitude f(s, t) which saturates the Froissard bound.

Generalization to Nuclear Collisions

Let us consider a hadron-nucleus collision. In this case Eq. (A.2) is replaced by (A >> 1)

$$\sigma_n^{hA}(s) = \int d^2 b \left(A\sigma_{in}^{hN}(s) T_A(\vec{b})\right)^n \exp\left(-A\sigma_{in}^{hN}(s) T_A(\vec{b})\right)/n! \ . \quad (A.5)$$

Here $\sigma_{in}^{hN}(s) = g_a g_b s^\Delta$ is the hadron-nucleon inelastic cross-section and $T_A(\vec{b})$ is the nuclear profile function (obtained from the standard Saxon-Wood nuclear density distribution by integrating over the longitudinal variable), normalized to unity. Note that in the case of a nuclear target, σ_n is determined from σ_{in}^{hN}. The parameters α' and $A_a + A_b$ which determine the t-dependence of the Born amplitude (A.3) are not present. This is due to the fact that this t-dependence (controlled by a hadronic scale) can be neglected as compared to that of the nuclear form factor (controlled by a nuclear scale). Again Eq. (A.5) has a simple probabilistic interpretation as the probability, at fixed impact parameter $\vec{b}$, to have n inelastic encounters with n nucleons in the target. From (A.5) we obtain the well known Glauber formula for the inelastic hA cross-section

$$\sigma_{in}^{hA}(s) = \int d^2 b \left(1 - exp\left(-A\, \sigma_{in}^{hN}(s)\, T_A(\vec{b})\right)\right) \ . \quad (A.6)$$

Note that in this case the Born term is given by

$$\sigma_{Born}^{hA}(s) = A\, \sigma_{in}^{hN}(s) \int d^2 b\, T_A(\vec{b}) = A\, \sigma_{in}^{hN}(s)$$

which is the so-called impulse approximation (no Glauber shadowing).

The generalization to nucleus-nucleus collisions is rather cumbersome and will not be given here. It is possible to obtain the cross-sections $\sigma_{n_A,n_B,n}$ for the configurations involving n inelastic collisions between n_A nucleons of the projectile and n_B nucleons of the target. These expressions are needed in the DPM expressions (Eq. (2.1)).

The AGK Cutting Rules

The total s-channel discontinuity of a reggeon graph, Fig. 15, (equal to twice its imaginary part) is obtained as a sum over all its possible s-channel discontinuities. The expression of σ_n given above, provides, in the elastic rescattering approximation, the contribution to the inelastic cross-section resulting from a cut of the (non-planar) diagram of Fig. A.1 through n-Pomerons (summed over the number k-n of uncut ones).

The Abramovsky, Gribov, Kancheli (AGK) cutting rules[24] relate to one another, via simple combinatorial factors, the various discontinuities of a reggeon graph with a fixed number k of exchanged Pomerons. They are valid not only in the eikonal model but for any form of the blobs in the diagram (provided they obey some general conditions : unitarity, analyticity and large p_T damping). They lead to very interesting properties of the (many particle) inclusive cross-sections. (See main text).

It is interesting that the AGK rules, although much more general than the eikonal model, are <u>entirely contained</u> in it. Let us denote by $\sigma_n^{(k)}$ the cross-section corresponding to the exchange of k Pomerons, n of which are cut and the remaining k-n uncut. The expression of σ_n^k can be obtained from Eqs. A.2 (or A.5) by developing the exponent in powers and keeping only the (k-n)-th one. The especific form of each $\sigma_n^{(k)}$ is, of course, valid only in the elastic rescattering approximation (i. e. eikonal model for hh and Glauber model without inelastic rescattering in hA). However, the ratios among them are exactly the ones given by the AGK cutting rules and thus are valid for a general blob. In particular we deduce from Eqs. (A.2) or (A.5) that

$$\sigma_n^{(k)} = (-)^{k-n} \binom{k}{n} \sigma_k^{(k)} \qquad (n > 0) \ . \qquad (A.7)$$

Perturbative RFT (or the eikonal model) is a multiple scattering model in which rescattering in the s-channel, imposed by unitarity, leads to a restoration of the Froissart bound when it is violated by single inelastic scattering (the Born term). Thus one may wonder why it is at all related with the large-N expansion of QCD - where the topology of the graphs plays a crucial role. The reason is that, due to the space time development of the strong interaction, the reggeon graphs in Fig. 15 vanish at asymptotic energies when they are planar. (Remember that the same is true for a planar graph in the large-N expansion). Due to the non-planarity of the vertex functions in Fig. 15, the complexity of the topology of the RFT graphs of Fig. 15 is increased with each extra exchange Pomeron (i.e. with each extra interaction). This simple argument makes plausible the one-to-one correspondence between graphs in the large-N expansion of QCD and graphs in the RFT demonstrated in Ref. 4. The precise formulation of this correspondence is the following : a one Pomeron graph (representing a single interaction) corresponds to the cylinder in the large-N expansion. A two Pomeron graph (corresponding to a double interaction in the sense of s-channel unitarity) corresponds to a cylinder with a handle, etc. Taking the s-channel discontinuity of a cylinder graph one obtains a contribution to the 2-string DPM graph of Fig. 2 (two two-strings of hadrons come from "cutting" the two "sheets" of the cylinder). The whole two-string contribution is obtained when cutting through a cylinder and suming over an arbitary number of uncut ones. In the eikonal model such a contribution has weight σ_1. Likewise a DPM graph with 2n strings represents the sum of all the topologies in the large-N expansion with n-cut cylinder (a summation is performed over an arbitrary number of uncut ones). In the eikonal model this summation has weight σ_n. <u>The above arguments provide the phenomenological justification for using the σ_n's given by the eikonal expression (A.2) as the weights of a 2n string configuration in DPM (Eq. (1.2)).</u>

REFERENCES

1. A recent review of DPM : A. Capella, U. Sukhatme, C. I. Tan and J. Tran Thanh Van, Orsay preprint LPTHE 92-38 to be published in Phys. Reports.
2. G. 't Hooft, *Nucl. Phys.* B72:461 (1974).
3. G. Veneziano, *Nucl. Phys.* B74:365 (1974), B117:519 (1976), and, "Color Symmetry and Quark Confinement", p. 113, edited by J. Tran Thanh Van, Editions Frontieres, (1977).
4. M. Ciafaloni, G. Marchesini and G. Veneziano, *Nucl. Phys.* B98:472 (1975).
5. See, e.g., V. N. Gribov, *Sov. Phys. JETP* 26:414 (1968) ; V.A. Abramovsky, V. N. Gribov and O. V. Kancheli, *Yad. Fiz.* 18:595 (1973) , *Sov. J. Nucl. Phys.* 18:308 (1974) ; M. Baker and K.A. Ter-Martirosyan, *Phys. Rep.* 28C:1 (1976) ; M. Baker and L. McLerran, Proc. of International Conference on High Energy Physics, EPS, Palermo, Italy (1975).
6. M. G. Ma et al., *Z. Phys.* C30:191 (1986) ; E. De Wolf, et al., *Z. Phys.* C31:13 (1986).
7. P. Aurenche, A. Capella, J. Kwiecinski, M. Maire, J. Ranft, and J. Tran T. V., Phys. Rev. D45:92 (1992). F. W. Bopp, A. Capella, J. Ranft and J. Tran Thanh Van, *Z. Phys.* C99:51 (1991).
8. R.J. Glauber, Lectures in Theor. Phys., edited by W.E. Butten, Inter. Science Pub., New York (1959), Vol. I, p.315 ; V.N. Gribov, *Zh ETF* 56:892 (1969), 57:1306 (1969) ; L. Bertocchi, *Nuovo Cimento* 11A:45 (1972).
9. A. Capella and A. Kaidalov, *Nucl. Phys.* B111:477 (1976) ; A. Capella and A. Krzywicki, *Phys. Lett.* B67:84 (1977) ; *Phys. Rev.* D18:3357 (1978). For later developments and/or related models, see: Yu M. Shabelski, *Nucl. Phys.* B132:491 (1978) ; K. Kinoshita, A. Minaka, and H. Sumiyosi, *Prog. Theor. Phys.* 61:165 (1979) ; 63:928 (1980) ; 63:1268 (1980) ; V. R. Zoller, *Z. Phys.* C44:207 (1989) ; V.V. Anisovich, Yu M. Shabelsky and V. M. Shekhter, *Nucl. Phys.* B133:477 (1978) ; R. Hwa and X. Wang, *Phys. Rev.* D39:2561 (1989) ; M. A. Braun, *Yad. Fiz.* 47:262 (1988), 52:257 (1990) ; V. V. Anisovich, L. G. Dakhno, and V. A. Nikonov, Phys. Rev. D44:1385 (1991).
10. A. Capella and J. Tran Thanh Van, *Z. Phys.* C10:249 (1981) and *Phys. Lett.* B93:146 (1980) .
11. A. Capella in : "Partons in Soft Hadronic Reactions, edited by R. Van de Walle, World Scientific Publishing Co., Singapore (1981).
12. J. Tran Thanh Van, Proc. XXIII Rencontre de Moriond (1988).
13. A. Capella, J. Kwiecinski, and J. Tran Thanh Van, *Phys. Lett.* B108:347 (1982). A. Capella, C. Pajares, and A.V. Ramallo, *Nucl. Phys.* B241:75 (1984).
14. K. G. Boreskov and A. B. Kaidalov, Yad. Fiz. 48:575 (1988). In this work, a summation of all graphs in the Glauber approach has been performed for the first time.
15. K. Boreskov, A. Capella, A. Kaidalov and J. Tran Thanh Van, *Phys. Rev.* D47:919 (1993).
16. M. Braun and A. Capella, Orsay preprint LPTHE Orsay 93/02.
17. K. Boreskov in Proceedings XXII International Symposium on Multiparticle Dynamics, Santiago de Compostela, Spain, June 1992.
18. A. Capella, et al., Proc. XXV Rencontre de Moriond (1990), Proc. Int. Conf. on High Energy Physics, Singapore (1990), and Proc. XX Int. Symposium on Multiparticle Dynamics, Dortmund RFA (Sept. 1990).
19. C. Gerschel and J. Hüfner, Proc. of Quark Matter (1991) , Gatlinburg, U.S.A, to be published in *Nucl. Phys. A*.
20. A. Capella, J. Ranft and J. Tran Thanh Van, Lund preprint LUTP 93-18.
21. NA35 collaboration, to be published in Proc. Quark Matter 93, Borlange, Sweden, (June 1993).
22. WA85 collaboration, Proc. Quark Matter 93, ibid.
23. NA36 collaboration, Proc. Quark Matter 93, ibid.
24. V. A. Abramovsky, V. N. Gribov and O. V. Kancheli, *Yad. Fiz.* 18:595 (1973).

NUCLEON-NUCLEON BREMSSTRAHLUNG

K. Nakayama

Institut für Kernphysik, Forschungszentrum Jülich
52425 Jülich, Germany
and
Dept. of Physics & Astronomy, University of Georgia
Athens, GA 30602, USA

INTRODUCTION

One of the most fundamental problems in Nuclear Physics is that of understanding
the nuclear force acting between nucleons as well as the limitations implied by eliminating other (subnucleonic) degrees-of-freedom. In principle, the nuclear force should
be derived from Quantum Chromodynamics(QCD) which nowadays is believed to be
the theory of strong interactions. This is a tremendous task because, in this theory, the
nucleon is a composite object formed from three quarks. Therefore, in order to obtain
the interaction between two nucleons, one has first to solve the nucleon problem, i.e., a
strongly interacting three-quark system, and then, the two-nucleon problem, i.e., two
interacting three-quark systems. Moreover, the low-energy nucleon-nucleon (NN) interaction is a prominent example of strong interactions which requires a non-perturbative
QCD treatment for its description. Although, considerable progress has been made
recently [1, 2], NN interactions derived from QCD are not yet of sufficient quality to be
used in realistic calculations. Therefore, in practice, nucleons are treated primarily as
point-like particles and the nuclear force acting between two such nucleons is described
by potential models which are either purely phenomenological or are based on meson-
exchange theory. The well known Hamada-Johnston [3] and Reid [4] potentials are some of
the examples of such phenomenologically constructed potentials, while those developed
more recently by the Bonn [5, 6] and Paris [7, 8] groups are examples of potentials based
on meson-exchange models. Nucleons and mesons are appropriate degrees-of-freedom
at large distances between nucleons, i.e., for describing low energy nuclear physics. In
a nucleon-meson picture the composite nature of nucleons is accounted for by form
factors which are introduced in a more or less ad hoc way into the model. These NN
potentials (V) are, then, used in an integral equation of the form

$$T = V + VGT \tag{1}$$

Perspectives in the Structure of Hadronic Systems
Edited by M.N. Harakeh *et al.*, Plenum Press, New York, 1994

in order to obtain the corresponding NN transition amplitudes, T, which are called NN T-matrix interactions or, simply, NN interactions. In a non-relativistic approach, the above integral equation represents the Lippmann-Schwinger (LS-) equation, while in a relativistic approach it may represent the Bethe-Salpeter (BS-) equation or some approximation to it. $\mathcal{G}$ denotes the two-nucleon propagator. The NN interactions contructed in this way are also called realistic interactions, whose parameters of the corresponding NN potentials are adjusted to fit the free NN scattering data and the ground state properties of the deuteron.

It should be stressed, however, that this procedure of obtaining the NN interaction is inadequate for determining it uniquely. Here, we note that the NN T-matrix interaction is, in general, a function of three independent variables, $T = T(\vec{p}\,', \vec{p}; z)$, where $\vec{p}$ and $\vec{p}\,'$ denote the initial and final relative momenta of two interacting nucleons and, z, the total energy of the NN system. The non-uniqueness of the interaction arises because the description of free NN scattering requires only on-energy-shell information about the interaction, i.e., the T-matrix with $|\vec{p}\,'| = |\vec{p}|$ and $z = z(\vec{p})$. Indeed, for a description of NN scattering it is sufficient to specify the incident energy (related to z) and the scattering angle (angle between the initial and final relative momenta). Therefore, fitting the free NN scattering data only, does not constrain the so-called off-energy-shell behavior of the interaction, i.e., the behavior of the interaction away from its on-energy-shell. Here, we define the full-off-energy-shell interaction as being the interaction when $|\vec{p}\,'| \neq |\vec{p}|$ and z independent on $\vec{p}\,'$ and $\vec{p}$. We also define the half-off-energy-shell interaction as being the interaction when $|\vec{p}\,'| \neq |\vec{p}|$ and $z = z(\vec{p})$. (As we shall discuss below, for a particular type of NN potentials, fitting the NN scattering data in the whole range of incident energy does constrain the off-energy-shell behavior of the corresponding T-matrices.) Hereafter, the terms on-shell and off-shell will mean on-energy-shell and off-energy-shell, respectively.

In general, the only constraint on the off-shell behavior of the interaction is given by the so-called off-shell unitarity relation which should be satisfied by any interaction that obeys the LS or BS-type integral equation with a Hermitian potential. For example, if an uncoupled partial-wave NN amplitude is constructed from a Hermitian potential by solving a LS-equation, it will satisfy the unitarity relation

$$Im[T_L(p', p; z)] = -mp T_L(p', p''; z) T_L^*(p'', p; z) \,, \tag{2}$$

where $p'' = \sqrt{mz}$ with m denoting the nucleon mass. For half-off-shell amplitudes $(z = z(\vec{p}) = p^2/m)$, the above relation leads to the constraint [9]

$$T_L(p', p; z) = e^{i\delta_L(p)} \tau_L(p', p; z) \,, \tag{3}$$

where δ_L denotes the NN on-shell phase-shift and, τ_L, a real function. However, the off-shell unitarity is not sufficient to determine the off-shell behavior of the interaction uniquely.

In the particular case of NN interactions which are on-shell equivalent for incident energies ranging from zero to infinity and which satisfy the LS-equation with NN potentials that are local in each partial-wave state and energy-independent, such as the Reid potential [4], the inverse scattering theory can be used to show that these interactions are also off-shell equivalent [10]. However, it should be noted that most modern NN potentials are non-local potentials and, therefore, off-shell informations are not accessible from on-shell phase-shifts. In an one-boson-exchange model, the dominant non-locality arises from the strong contributions of the scalar (sigma) and vector (omega) mesons. This can be easily seen by writting down the leading terms contributing to the NN

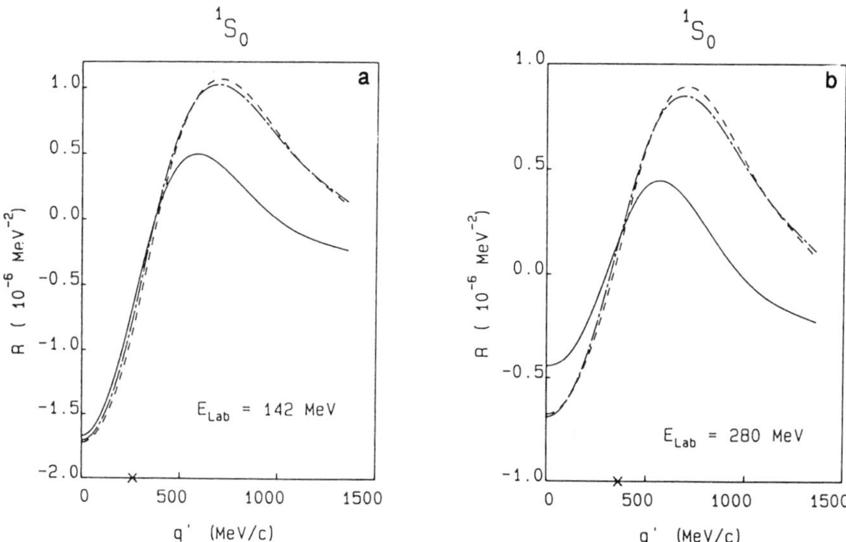

Figure 1. Half-off-shell R-matrices in the $^1S_0$ partial-wave state as a function of off-shell momentum at two fixed incident energies. The on-shell points are marked with crosses. Taken from [5].

potential from these mesons

$$V_\sigma(\vec{p}\,',\vec{p}) = -4\pi g_\sigma^2 \frac{1}{\vec{q}^2 + \mu_\sigma^2} \left\{ 1 + \frac{\vec{q}^2 - \vec{Q}^2}{8m^2} + \frac{i}{4m^2} \left(\frac{\vec{\sigma}_1 + \vec{\sigma}_2}{2} \right) \cdot (\vec{q} \wedge \vec{Q}) \right\} ,$$

$$V_\omega(\vec{p}\,',\vec{p}) = +4\pi g_\omega^2 \frac{1}{\vec{q}^2 + \mu_\omega^2} \left\{ 1 - \frac{\vec{q}^2 - 3\vec{Q}^2}{8m^2} - \frac{3i}{4m^2} \left(\frac{\vec{\sigma}_1 + \vec{\sigma}_2}{2} \right) \cdot (\vec{q} \wedge \vec{Q}) \right\} . \quad (4)$$

In the above equation, V_σ and V_ω denote the contributions from the σ- and ω-meson, respectively. The terms proportional to $(\vec{\sigma}_1 + \vec{\sigma}_2) \cdot (\vec{q} \wedge \vec{Q})$ are the spin-orbit components. $\vec{q} \equiv \vec{p} - \vec{p}\,'$ and $\vec{Q} \equiv \vec{p} + \vec{p}\,'$ denote the direct and exchange momentum transfer, respectively. A local potential means that it depends, in momentum space, only on the direct momentum transfer $\vec{q}$, i.e., $V = V(\vec{q})$. From eq. (4), it is immediately clear that these mesons not only generate a spin-orbit force which is non-local but they also introduce a considerable non-locality in the central component of the interaction. One should also bear in mind that most realistic NN interactions fit the on-shell phase-shifts only up to the pion-production threshold energy. Beyond that energy the inelasticity has to be taken into account if one wants to reproduce the phase-shifts but this is not the case for a large class of existing NN interactions. For these interactions the on-shell behavior beyond pion threshold is a prediction of the underlying potential model and it may not necessarily be the same for different interactions.

For the reasons discussed above, the procedure outlined here for obtaining realistic NN interactions yields on-shell equivalent interactions which may differ in their off-shell behavior. In general, apart from the constraint imposed by off-shell unitarity, there is only the underlying potential model constraint off-shell. Fig. (1) illustrates the different off-shell behavior of three such NN interactions. It is worth noticing that, in contrast to free NN scattering, the description of other processes (other than free NN scattering) requires, in principle, information about the NN interaction both on and off the energy-shell. The importance of taking off-shell effects of the NN interaction into

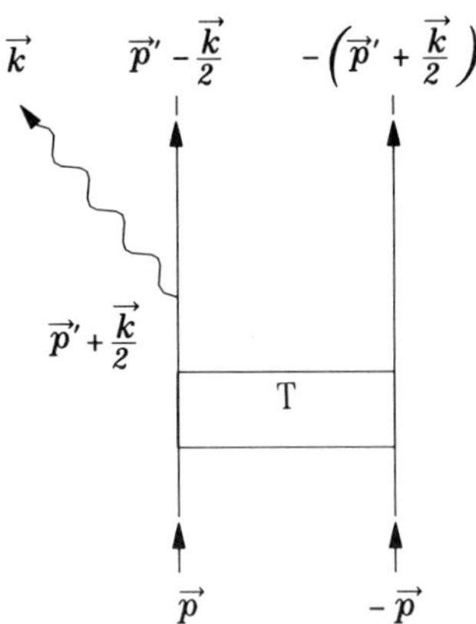

Figure 2. A diagrammatic representation of a process which contributes to the NNγ reaction. The wiggled line represents the photon while straight lines the nucleons. The block represents the NN T-matrix interaction.

account is, for example, manifested in recent investigations of the optical potential in describing N-nucleus elastic-scattering data at intermediate energies within the context of a full folding model [11]. Due to Fermi motion of the target nucleons, the full folding model requires the full-off-shell as well as the on-shell information of the effective NN interaction.

The simplest process for investigating the off-shell behavior of the NN interaction is the NN-bremsstrahlung (NNγ) reaction, $N_1 + N_2 \rightarrow N_1 + N_2 + \gamma$. One of the many diagrams which contribute to this reaction together with its kinematics is illustrated in Fig. (2). In this process the intermediate nucleon is off the mass-shell and the NN T-matrix interaction required is half off the energy-shell, i.e., $T = T(\vec{p}\,'+\vec{k}/2, \vec{p}; z(\vec{p}))$ with $|\vec{p}\,'+\vec{k}/2| \neq |\vec{p}|$. Therefore, by studing this reaction, we can learn about the half-off-shell behavior of the NN interaction. Of course, implicit in such a study is the assumption that we know the reaction mechanism and, in particular, the NN electromagnetic (e.m.) vertex.

The idea of using the NNγ reaction as a tool for investigating off-shell effects of the NN interaction has been first proposed by Ashkin and Marshak [12] in (as early as) 1949. One of the first ppγ experiment, however, was performed only in 1965 by Gottschalk, Shlaer, and Wang [13], motivated by the theoretical work of Sobel and Cromer [14]. Those authors [13] measured the exclusive proton-proton bremsstrahlung (ppγ) reaction cross sections in the coplanar geometry at an incident energy of $T_{lab} = 158 MeV$. Since then, the NNγ reaction has been studied intensively both theoretically and experimentally, especially from mid 60's throughout 70's [15]. However, in spite of the considerable effort devoted to this reaction very little has been learned about off-shell effects due, essentially, to relatively poor accuracy in the (then) existing experimental data and uncertainties involved in the theoretical predictions. In addition, there was even no firm indication for the need of any off-shell information of the NN interaction in order to describe those data. The bremsstrahlung amplitude obtained in the soft-photon approximation (SPA) via the low-energy theorem [16] and, therefore, containing no off-shell information, could account for an overall description of those data within the given uncertainties.

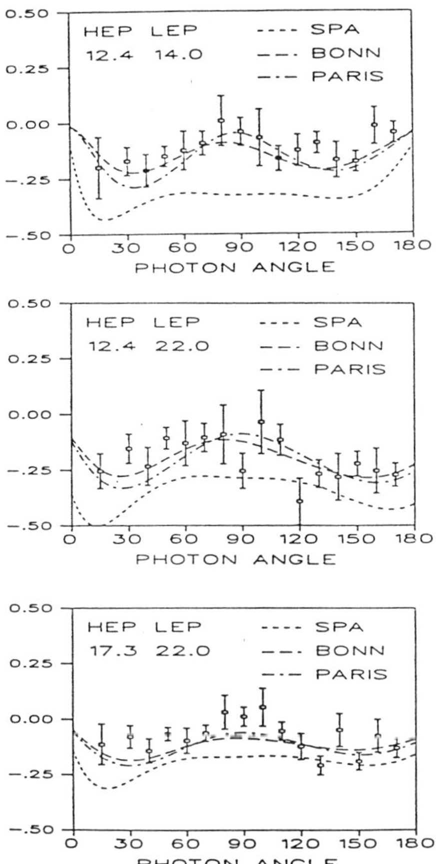

Figure 3. ppγ analyzing power in the coplanar geometry at an incident energy of $T_{lab} =$ 280MeV. Taken from [19].

This situation has changed since 1986 when the TRIUMF group [17, 18, 19, 20] investigated the analyzing power in ppγ reaction at beam energies near pion threshold, showing that this observable is quite sensitive to the off-shell behavior of the NN interaction. The obtained data clearly distinguishes the prediction of the SPA (which does not contain any off-shell information) from those of potential models as can be seen from Fig. (3). Due to the weakness of the e.m. interaction combined with the experimental techniques which were available, accurate measurements in bremsstrahlung processes, especially of spin-polarization observables, were extremely difficult to perform in the past. Indeed, most of those studies have been restricted to investigating NNγ cross sections and almost no attention has been given to the investigation of spin-polarization observables. Motivated by the TRIUMF group's finding a number of new high-precision ppγ experiments are presently scheduled at the major laboratories across the world [21, 22, 23, 24].

On the other hand, the observation of high energy photons produced in heavy-ion collisions at intermediate energies has also motivated a considerable effort to better understand the elementary NNγ reaction, in particular, the neutron-proton bremsstrahlung (npγ) process which is the main reaction mechanism for producing hard photons in these collisions [25, 26]. The npγ process is much more efficient in producing energetic photons than the ppγ process because of the presence of a large two-body exchange current [27] which is mostly absent in the latter case. Also, the recent investigations of dilepton productions in proton-nucleus collisions [28, 29] have shown that the npγ and Δ-decay processes are the dominant reaction mechanisms for producing these dileptons.

Therefore, the understanding of the elementary NNγ reaction is of crucial importance. With the development of modern facilities to perform the necessary detailed experiments in ppγ as well as in npγ reactions, combined with more sophisticated theoretical models than were available in the past, we are now in a much better position to address not only those questions about the off-shell behavior of the interaction, but also new questions related to the NNγ reaction mechanism itself, especially at energies beyond the pion-production threshold.

GAUGE INVARIANCE AND THE LOW-ENERGY THEOREM FOR NN BREMSSTRAHLUNG

In this chapter we shall review some of the basic conditions that any reasonable model for describing the NNγ process should satisfy. We start reviewing some of the basic aspects of the e.m. current operators emerging from the gauge-invariance condition. Then, the general structure of the off-shell NN interaction which is required in the description of the NNγ reaction is briefly discussed. Finally, the low-energy theorem for this reaction, which is a direct consequence of the gauge invariance requirement on the NN-bremsstrahlung amplitude, will be reviewed.

Electromagnetic current operators

In this section we discuss the e.m. current operators, in particular, the necessity of a two-body current as a consequence of the gauge-invariance requirement of the theory. Since the e.m. interaction is weak, it is reasonable to assume that the Hamiltonian of the system has a Taylor expansion with respect to the e.m. potential $A_\mu(x)$, with $x = (t, \vec{x})$. This is the basis for a perturbative treatment of the e.m. interaction. Up to first order in this expansion, we have

$$H(A) = H_o + \mathcal{V}_{em}(A) \,, \tag{5}$$

where the e.m. transition potential $\mathcal{V}_{em}(A)$ is given by

$$\mathcal{V}_{em}(A) = \int d^3x j_\mu(\vec{x}, t) A^\mu(\vec{x}, t) \,. \tag{6}$$

The above specific form is determined by the requirement of Lorentz invariance.

The expansion coefficient in eq. (6) defines the current density $j_\mu(x)$. It is obtained from the full Hamiltonian $H(A)$ as

$$j_\mu(x) \equiv \frac{\delta H(A)}{\delta A^\mu(x)}\bigg|_{A=0} \,. \tag{7}$$

The requirement of gauge invariance of the theory result in the continuity equation

$$\nabla \cdot \vec{j} + \frac{\partial \rho}{\partial t} = -i[H_o, \rho] \,, \tag{8}$$

where $j^\mu \equiv (\rho, \vec{j})$.

For simplicity, we resctrict our further discussions to a non-relativistic system of nucleons. Then, the part of the Hamiltonian which describes this system in the absence of an e.m. interaction has a simple form:

$$
\begin{aligned}
H_o &= \sum_j \frac{-\nabla_j^2}{2m} + \sum_{i>j} V_{ij} \\
&= K + V \,,
\end{aligned}
\tag{9}
$$

with m denoting the nucleon mass.

In order to derive the current density for this system, we need the Hamiltonian $H(A)$. The method of minimal substitution

$$
\begin{aligned}
\nabla_j &\rightarrow \nabla_j - i\hat{e}_j \vec{A}(x) \,, \\
H_o &\rightarrow H_o + \sum_j \hat{e}_j A_o(x) \,,
\end{aligned}
\tag{10}
$$

allows to obtain $H(A)$ in a gauge invariant way from the Hamiltonian H_o. In the above equation, $\hat{e}_j \equiv e(1 + \tau_{3j})/2$ stands for the nucleon charge operator. $e =$ proton charge. Then, from eq. (7) we have for the one-body-charge and convection-current densities

$$
\begin{aligned}
\rho_1(\vec{x}) &= \sum_j \hat{e}_j \delta(\vec{x} - \vec{r}_j) \,, \\
\vec{j}_1^{con}(\vec{x}) &= \frac{1}{2m} \sum_j \hat{e}_j \{\nabla_j, \delta(\vec{x} - \vec{r}_j)\} \,.
\end{aligned}
\tag{11}
$$

These operators arise from the minimal substitution in the kinetic energy (K) part of H_o and contain no explicit time dependence.

Obviously, the magnetic current is absent in the above equation because the kinetic energy operator, K, does not depend on the spin. We simply add to the convection-current density the magnetic contribution,

$$
\vec{j}_1^{mag}(\vec{x}) = \nabla_x \wedge \sum_j \frac{e\mu_j}{2m} \vec{\sigma}_j \delta(\vec{x} - \vec{r}_j) \,,
\tag{12}
$$

to yield the total one-body-current density

$$
\vec{j}_1(\vec{x}) = \vec{j}_1^{con}(\vec{x}) + \vec{j}_1^{mag}(\vec{x}) \,.
\tag{13}
$$

In eq. (12), $e\mu_j/2m$ is the magnetic moment of the j-th nucleon. If we wish, the magnetic current can be also generated from H_o by adding to this Hamiltonian the term $-i \sum_j \mu_j \vec{\sigma}_j \cdot (\nabla_j \wedge \nabla_j)/2m$ which is identically zero in the absence of an e.m. field.

It is clear that the one-body current given by eq. (13) obeys the relation

$$
\nabla_x \cdot \vec{j}_1(\vec{x}) = -i[K, \rho_1(\vec{x})] \,.
\tag{14}
$$

With the one-body current density given by eq. (13) and satisfying eq. (14), it is simple to verify that one needs an additional current in order to fulfill the continuity equation expressed by eq. (8). This additional current density which is called the two-body current density, $\vec{j}_2(\vec{x})$, should obey the following relation:

$$
\nabla_x \cdot \vec{j}_2(\vec{x}) = -i[V, \rho_1(\vec{x})] \,,
\tag{15}
$$

so that the total current density consistent with the continuity equation is

$$
\vec{j}(\vec{x}) = \vec{j}_1(\vec{x}) + \vec{j}_2(\vec{x}) \,.
\tag{16}
$$

In deriving eq. (15) we have assumed that the charge density remains unaltered by the interaction V. This is the Siegert's hypothesis [30]. The one-pion-exchange potential, for example, satisfies this assumption [31] (see, however, ref.[32], where possibilities of having two-body charge densities are discussed).

Eq. (15) shows that whenever the NN potential does not commute with the charge density, one needs a two-body current in order to satisfy the continuity equation. For a NN potential, V, that is local and isospin-independent the right-hand-side of eq. (15) vanishes identically. Consequently, this type of potentials do not require two-body currents in order to satisfy the continuity equation. However, for a potential that is non-local and/or isospin-dependent this is not the case and, therefore, this class of potentials do generate two-body currents. The most notable example of a two-body current is the meson-exchange current [33] whose clear signature was first observed in the radiative capture of thermal neutrons by protons.

One problem that arises here in the determination of the two-body current is that the gauge condition only is not sufficient to determine the two-body current uniquely. As can be seen from eq. (15), gauge invariance constrains only the divergence but not the curl. An excellent discussion of this problem may be found in ref.[32]. In order to determine the current uniquely, one needs an underlying dynamical model which allows to recover the structure of the operators in terms of the original degrees-of-freedom otherwise buried in these effective operators. This will cause an immediate problem if one wants to construct a two-body current from a NN potential which is constructed from purely phenomenological considerations, since for such a potential the underlying dynamics is not known. In principle, for NN potentials based on meson-exchange models, such a problem does not occur. For these potentials, where the underlying structure is known, it is possible to obtain the corresponding (two-body) meson-exchange currents, for example, from considerations of Feynman diagrams. In practice, even for this class of potentials the problem arises because they usually contain hadronic form factors which are introduced more or less in an ad hoc way into the model. This means that one has to consider the problem of how to couple e.m. fields to such form factors [34, 35].

The only way out of the problem mentioned above, which is inherent in any model whose underlying dynamics is unknown, is to find out prescriptions which yield "reasonable" results, since for such a model the precise form of the two-body current cannot be known. Presently, there are a variety of methods for constructing effective two-body currents [36, 37, 38].

We now turn to the consideration of the two-body current density in the SPA. In order to derive the two-body current density in the SPA, it is convenient to Fourier transform the charge and current densities:

$$\rho_i(\vec{k}) = \int d^3x \rho_i(\vec{x}) e^{i\vec{k}\cdot\vec{x}} ,$$

$$\vec{j}_i(\vec{k}) = \int d^3x \vec{j}_i(\vec{x}) e^{i\vec{k}\cdot\vec{x}} . \tag{17}$$

In terms of these Fourier transformed quantities, eq. (15) becomes

$$\vec{k} \cdot \vec{j}_2(\vec{k}) = [V, \rho_1(\vec{k})] . \tag{18}$$

Now, expanding both sides of the above equation in powers of $\vec{k}$, and equating the terms of the same order in the expansion, we have in zeroth order,

$$\vec{j}_2(\vec{k} = 0) = i[\vec{D}, V] , \tag{19}$$

where $\vec{D}$ denotes the charge dipole operator

$$\vec{D} \equiv \int d^3x \rho_1(\vec{x})\vec{x} \ . \tag{20}$$

Note that the result in eq. (19) is obtained under the assumption that both the charge and two-body current densities have Taylor expansions (see, however, discussions in ref.[32] where non-analytic two-body currents can, in principle, be constructed). We also note that in the SPA, arbitrary terms of the form $[\nabla_x \wedge \vec{\lambda}(\vec{x})]$ do not contribute to $\vec{j}_2(\vec{x})$. Eq. (19) is the so-called Siegert's Theorem which is one of the many low-energy theorems. Its virtue lies on the fact that it determines the current in the low frequency limit from the charge dipole operator, a quantity that is, in general, much better known in effective theories. Observe that in the SPA no explicit knowledge of the underlying structure of the NN potential is required in order to obtain the corresponding two-body current.

Inserting the explicit form of the charge density given in eq. (11) into eq. (20), we have for the current density

$$\vec{j}_2(\vec{k} = 0) = i[\sum_j \hat{e}_j \vec{r}_j, V] \ . \tag{21}$$

Although the above result has been obtained using a particular form of the charge density, it can also be derived directly via the minimal substitution in the NN potential [39].

For further convenience, we introduce the e.m. currents in momentum space. From eq. (6), the one-body e.m. transition potential corresponding to the one-body current density given by eq. (13) is

$$\begin{aligned}
\mathcal{V}_{em}^{(1)} &= \int d^3x \vec{j}_1(\vec{x}) \cdot \vec{A}(\vec{x}) \\
&\equiv \sum_j \mathcal{V}_{em}^{(1)}(\vec{r}_j{}', \vec{r}_j) \ ,
\end{aligned} \tag{22}$$

where the form given in the last line is due to the property of additivity of $\vec{j}_1(\vec{x})$. The Fourier transform of the above equation yields the corresponding one-body e.m. transition potential in momentum space,

$$\begin{aligned}
V_{em}^{(1)} &= \sum_j \int d^3r_j{}' d^3r_j e^{-i\vec{p}_j{}' \cdot \vec{r}_j{}'} \mathcal{V}_{em}^{(1)}(\vec{r}_j{}', \vec{r}_j) e^{i\vec{p}_j \cdot \vec{r}_j} \\
&\equiv \sum_j \vec{\epsilon} \cdot \vec{J}_1(\vec{p}_j{}', \vec{p}_j) \ ,
\end{aligned} \tag{23}$$

where the one-body current in momentum space, $\vec{J}_1(\vec{p}_j{}', \vec{p}_j)$, has been introduced in the last line of the above equation with $\vec{\epsilon}$ denoting the polarization vector of the photon.

Analogously, for the two-body current in momentum space, $\vec{J}_2(\vec{p}_j{}', \vec{p}_i{}'; \vec{p}_j, \vec{p}_i)$, we have

$$\begin{aligned}
V_{em}^{(2)} &= \sum_{i>j} \int d^3r_j{}' d^3r_i{}' d^3r_j d^3r_i e^{-i(\vec{p}_j{}' \cdot \vec{r}_j{}' + \vec{p}_i{}' \cdot \vec{r}_i{}')} \mathcal{V}_{em}^{(2)}(\vec{r}_j{}', \vec{r}_i{}'; \vec{r}_j, \vec{r}_i) e^{i(\vec{p}_j \cdot \vec{r}_j + \vec{p}_i \cdot \vec{r}_i)} \\
&\equiv \sum_{i>j} \vec{\epsilon} \cdot \vec{J}_2(\vec{p}_j{}', \vec{p}_i{}'; \vec{p}_j, \vec{p}_i) \ ,
\end{aligned} \tag{24}$$

where $\mathcal{V}_{em}^{(2)}(\vec{r}_j{}', \vec{r}_i{}'; \vec{r}_j, \vec{r}_i)$ is the analog of eq. (22) for the two-body e.m. transition potential due to the two-body current density $\vec{j}_2(\vec{x})$.

Relativistic NNγ electromagnetic vertex

In this section we discuss the one-body current operator from relativistic considerations, since the relativistic effects, especially, the so-called relativistic spin corrections
are known to be important in NNγ reactions [18, 40, 41, 42]. The most general form of
the NNγ vertex, consistent with invariance under the Lorentz group may be written as
[43]

$$\Gamma_\mu(p',p) = \sum_{j,l=0,1} (\gamma \cdot p')^j [\gamma_\mu A_1^{jl} + i\sigma_{\mu\nu}k^\nu A_2^{jl} + k_\mu A_3^{jl}](\gamma \cdot p)^l , \tag{25}$$

where, in this section, p and p' denote the four-momenta of the nucleon before and after
the coupling of a photon with four-momentum $k = p - p'$. The twelve coefficients A_i^{jl}
are functions of the three scalar variables at the vertex: k^2, $W^2 \equiv p^2$, and $W'^2 \equiv p'^2$.

In the description of NNγ processes, the nucleon in the intermediate state will be
off its mass-shell. For example, in the diagram illustrated in Fig. (2), the nucleon is
off the mass-shell before the emission of a photon, i.e., $W^2 \neq m^2$, whereas it is on the
mass-shell, $W'^2 = m^2$, after the photon emission. Therefore, the evaluation of that
diagram requires a half-off-shell vertex in which only the final nucleon is on the mass-
shell. In this case, the Dirac equation can be used for the final on-mass-shell nucleon
to reduce eq. (25) to a half-off-shell vertex of the form

$$\Gamma_\mu(p',p) = \Gamma_\mu^{(+)}(p',p) + \Gamma_\mu^{(-)}(p',p) , \tag{26}$$

with

$$\Gamma_\mu^{(\pm)}(p',p) = e[\gamma_\mu f_1^{(\pm)} - i\frac{\sigma_{\mu\nu}k^\nu}{2m} f_2^{(\pm)} - k_\mu f_3^{(\pm)}]\Lambda_\pm . \tag{27}$$

In the above equation, $\Lambda_\pm$ are the projection operators defined as $\Lambda_\pm \equiv (\pm\not{p}+W)/2W$.
We also define $W > 0$. For on-mass-shell nucleons, $W = m$, these operators are the
usual projection operators onto the positive- and negative-energy states. The coefficients $f_i^{(\pm)} = f_i^{(\pm)}(k^2, m, W)$ are linear combinations of A_i^{jl} appering in eq. (25).

The number of invariant functions $f_i^{(\pm)}$ in eq. (27) can be further reduced by using
the Ward-Takahashi identity

$$k^\mu\Gamma_\mu(p',p) = \hat{e}[S^{-1}(p) - S^{-1}(p')] , \tag{28}$$

where $S(p) \equiv 1/(\not{p} - m)$ denotes the nucleon propagator. Using the above equation in
eq. (27), we obtain the relation

$$f_1^{(\pm)} = e_N + \frac{k^2}{\pm W - m} f_3^{(\pm)} , \tag{29}$$

where e_N is the nucleon charge in units of e.

We now use the fact that in bremsstrahlung the emitted photon is a real photon,
i.e., $k^2 = 0$ and $\epsilon \cdot k = 0$ with ϵ_μ denoting the polarization of the photon. This fact
together with eq. (29), and noting that $f_3^{(\pm)}$ are analytic functions [43], reduces the
half-off-shell vertices given by eq. (27) to

$$\Gamma_\mu^{(\pm)}(p',p) = e[\gamma_\mu e_N - i\frac{\sigma_{\mu\nu}k^\nu}{2m} f_2^{(\pm)}]\Lambda_\pm . \tag{30}$$

Gauge invariance imposes no further constraints on the half-off-shell form factors $f_2^{(\pm)} = f_2^{(\pm)}(k^2 = 0, m, W)$ in the above equation, except that at on-mass-shell,
$f_2^{(+)}(k^2 = 0, m, m) = \kappa_N$, the anomalous magnetic moment of the nucleon in units of

nuclear magneton. Therefore, the determination of $f_2^{(-)}$ as well as $f_2^{(+)}$ off the mass-shell requires a dynamical model [44, 45, 46].

The NNγ vertices given by eq. (30) with the half-off-shell form factors $f_2^{(\pm)} = f_2^{(\pm)}(k^2 = 0, m, W)$ are required for describing the NNγ process of Fig. (2). There is another type of process, analogous to that of Fig. (2) (see Fig. (4)), in which a photon is emitted before the two nucleons interact with each other. For this type of process the NNγ vertices needed have also a form given by eq. (30) but with half-off-shell form factors $f_2^{(\pm)} = f_2^{(\pm)}(k^2 = 0, W', m)$ and the projection operators $\Lambda_\pm$ acting on the nucleon after the photon emission. It should be mentioned that there is also a more complicated diagram (the so-called rescattering diagram, see Fig. (4)) that contributes to the NNγ reaction which requires, in principle, full-off-shell e.m. vertices as for that diagram the nucleon is off the mass-shell before and after the photon emission. We stress that investigation of off-shell effects of the NN interaction via NNγ processes assumes that one knows these off-shell e.m. vertices.

The connection of the NNγ vertex discussed in this section to the non-relativistic one-body current, $\vec{J}_1(\vec{p}\,', \vec{p})$, introduced in the previous section is

$$\vec{\epsilon} \cdot \vec{J}_1(\vec{p}\,', \vec{p}) = \left\{ \sqrt{\frac{m}{\varepsilon_{p'}}} < \bar{u}(\vec{p}\,')|\vec{\Gamma}^{(+)}(p', p) \cdot \vec{A}(\vec{k})|u(\vec{p}) > \sqrt{\frac{m}{\varepsilon_p}} \right\}_{NR}, \qquad (31)$$

with $\vec{A}(\vec{k})$ denoting the Fourier transform of the e.m. vector potential $\vec{A}(\vec{x})$. $u(\vec{p})$ is the positive-energy Dirac spinor without the Pauli spin wave function, χ_s; it is normalized to $< \bar{u}|u > = 1$. We follow the notation of Bjorken and Drell [47]. $\varepsilon_p \equiv \sqrt{\vec{p}^2 + m^2}$. NR indicates that the right-hand-side of the above equation is taken in the non-relativistic limit. Note that in the above equation $\vec{\Gamma}^{(+)}$ is the on-shell e.m. vertex.

Half-off-shell NN interaction

The invariant four-component NN interaction can be written as

$$\hat{T} = \sum_{j=1,5} F_j \left[(\hat{\Omega}_j)_1 \cdot (\hat{\Omega}_j)_2\right], \qquad (32)$$

where F_j are scalar functions of the variables at the vertices 1 and 2 and, the operators $\hat{\Omega}_j$ are combinations of the Dirac γ-matrices as j runs from 1 to 5:

$$\hat{\Omega}_j = \{1, \gamma_\mu, \gamma_5, i\gamma_\mu\gamma_5, \sigma_{\mu\nu}\}. \qquad (33)$$

The subscripts 1 and 2 in the operators $\hat{\Omega}_j$ in eq. (32) indicate that they act on the interacting nucleons 1 and 2, respectively.

As we have stated previously, in order to describe free NN scattering processes one needs only the on-shell NN interaction. The structure of such an invariant interaction is of the form

$$\begin{aligned}
\tilde{T} &\equiv\ < \bar{u}_{1'}\bar{u}_{2'}|\hat{T}|u_1 u_2 > \\
&=\ \sum_{j=1,5} F_j(\nu, t)\Omega_j,
\end{aligned} \qquad (34)$$

where the five independent invariants Ω_j are defined as

$$\Omega_j = [\bar{u}_{1'}\hat{\Omega}_j u_1] \cdot [\bar{u}_{2'}\hat{\Omega}_j u_2], \qquad (35)$$

with u_i and $u_{i'}$ $(i = 1, 2)$ denoting the positive-energy Dirac spinors of the two interacting nucleons in the initial and final state, respectively. For on-shell interactions there are only two independent variables which we chose to be

$$
\begin{aligned}
\nu &= p_1 \cdot p_2 + p_1' \cdot p_2' \ , \\
t &= (p_1' - p_1)^2 \ ,
\end{aligned}
\tag{36}
$$

with p_i and p_i' denoting the initial and final four-momentum of the i-th nucleon.

The T-matrix interaction given by eq. (34) can be expressed in a representation used in a non-relativistic approach. In the NN center-of-mass (c.m.) system we have [48, 49]

$$
\begin{aligned}
T &\equiv \sqrt{\frac{m}{\varepsilon_{p'}}}\sqrt{\frac{m}{\varepsilon_{p'}}}\tilde{T}(\vec{p}\,',\vec{p})\sqrt{\frac{m}{\varepsilon_p}}\sqrt{\frac{m}{\varepsilon_p}} \\
&= \alpha P_{S=0} + \beta P_{S=1} + i\gamma(\vec{\sigma}_1 + \vec{\sigma}_2)\cdot\hat{n} + \delta S_{12}(\hat{q}) + \varepsilon S_{12}(\hat{Q}) \ ,
\end{aligned}
\tag{37}
$$

where $\vec{n} \equiv \vec{p} \wedge \vec{p}\,'$, $\vec{q} \equiv \vec{p} - \vec{p}\,'$, and $\vec{Q} \equiv \vec{p} + \vec{p}\,'$ with $\vec{p}$ and $\vec{p}\,'$ denoting the initial and final NN relative momentum, respectively. $\hat{n} \equiv \vec{n}/|\vec{n}|$ and so on. P_S stands for the total spin projection operator: $P_{S=0} = (1 - \vec{\sigma}_1 \cdot \vec{\sigma}_2)/4$ and $P_{S=1} = (3 + \vec{\sigma}_1 \cdot \vec{\sigma}_2)/4$. The tensor operator is defined as $S_{12}(\hat{q}) \equiv 3[\vec{\sigma}_1 \otimes \vec{\sigma}_2]^2 \cdot [\hat{q} \otimes \hat{q}]^2$. The coefficients α, β, etc, are scalar functions of $\vec{p}$ and $\vec{p}\,'$ with $|\vec{p}\,'| = |\vec{p}|$ for (on-shell) elastic scattering.

For a half-off-shell NN interaction, the five independent invariants as defined by eq. (35) are not sufficient to describe the interaction completely and one needs additional independent invariants. For example, consider the half-off-shell interaction required in the diagram of Fig. (2). In this case, such an additional invariant can be readily constructed (the other cases are treated analogously):

$$
\Omega_{j+5} = \left[\left(\frac{P_1' - m}{m}\right)\hat{\Omega}_j u_1\right] \cdot [\bar{u}_{2'}\hat{\Omega}_j u_2]
\tag{38}
$$

with P_1' denoting the four-momentum of the off-shell nucleon in that diagram and $\hat{\Omega}_j$ being any of the five matrices given in eq. (33). In fact, there is only one extra independent invariant as we know from non-relativistic considerations [48, 50], so that the value of j in the above equation can be fixed. For other values of j, eq. (38) is not independent from the invariants actually used.

We now write the invariant half-off-shell NN interaction as

$$
\begin{aligned}
\tilde{T} &= \sum_{j=1,5} F_j(\nu, t, \Delta_i)\Omega_j \\
&+ F_6(\nu, t, \Delta_{1'})\left[\left(\frac{p_1' - m}{m}\right)\hat{\Omega}_1 u_1\right] \cdot [\bar{u}_{2'}\hat{\Omega}_1 u_2] \\
&+ \text{(analogous terms for other cases)} \ .
\end{aligned}
\tag{39}
$$

Note that the extra invariants in the above equation vanish identically if the nucleons are on-shell. We also observe that half-off-shell interactions depend on an extra independent variable Δ_i which we define as

$$
\Delta_i \equiv l_i^2 - m^2 \ ,
\tag{40}
$$

where l_i denotes the four-momentum of the off-shell nucleon. For example, in the diagram of Fig. (2), $\Delta_i = \Delta_{1'} = P_1'^2 - m^2$. For an on-shell interaction, $\Delta_i \equiv 0$.

The most general form of the non-relativistic analog of eq. (39), consistent with symmetry principles is [48, 50, 51] (in the NN c. m. frame)

$$
T = \alpha P_{S=0} + \beta P_{S=1} + i\gamma(\vec{\sigma}_1 + \vec{\sigma}_2)\cdot\hat{n} + \delta S_{12}(\hat{q}) + \varepsilon S_{12}(\hat{Q}) + (\hat{q}\cdot\hat{Q})\omega S_{12}(\hat{q},\hat{Q}) \ , \tag{41}
$$

with the tensor operator $S_{12}(\hat{q}, \hat{Q}) \equiv 3[\vec{\sigma}_1 \otimes \vec{\sigma}_2]^2 \cdot [\hat{q} \otimes \hat{Q}]^2$. The last term in the above equation vanishes identically on-shell for $\hat{q} \cdot \hat{Q} = 0$ as a consequence of time reversal invariance.

The low-energy theorem

One of the basic conditions that any reasonable model for NNγ reactions should satisfy is the so-called low-energy theorem [16]. The essence of this theorem, which is a direct consequence of gauge invariance, is that the NNγ amplitude in the SPA $(\omega \equiv |\vec{k}| \to 0)$ is given uniquely in terms of quantities involving only on-shell information. Stated another way, the first two coefficients, A and B, of the expansion of the bremsstrahlung amplitude in powers of the photon energy ω,

$$\tilde{M} = \frac{A}{\omega} + B + C\omega + ... , \tag{42}$$

are determined uniquely in terms of on-shell NN phase-shifts and on-shell NN e.m. vertex.

The low-energy theorem, then, implies that any off-shell information is contained only in those coefficients appearing in higher order terms of the expansion, i.e., C, etc. This means that one has to go beyond the validity of the SPA if one wants to investigate off-shell properties, for example, of the NN interaction.

We now review briefly the major steps/assumptions in which the derivation of the low-energy theorem for the NNγ reaction is based upon. Its detailed derivation may be found elsewhere [52]. The low-energy theorem can be most easily derived if one uses a lemma due to Adler and Dothan [53]:

Let $\tilde{M}_\mu(x, k_\nu)$ and $\tilde{M}_\mu^{(a)}(x, k_\nu)$ be four-vector functions of a number of independent variables x and the four-vector k_ν, and let $\tilde{M}_\mu^{(b)}(x)$ be a function of the same variables x excluding k_ν, such that

$$\tilde{M}_\mu(x, k_\nu) = \tilde{M}_\mu^{(a)}(x, k_\nu) + \tilde{M}_\mu^{(b)}(x) + \mathcal{O}(k) . \tag{43}$$

Then, provided

$$k^\mu \tilde{M}_\mu(x, k_\nu) = k^\mu \tilde{M}_\mu^{(a)}(x, k_\nu) , \tag{44}$$

it follows that

$$\tilde{M}_\mu^{(b)}(x) = 0 . \tag{45}$$

The proof of the above lemma is immediate [53]. An important point to be emphasized here is that one has to distinguish between a quantity that is *independent of k* and a quantity that is *zeroth order in k*. For example, $k_\mu/(p \cdot k)$ is zeroth order in k but is not independent of k.

The above lemma may be applied to derive the NNγ amplitude in the SPA, with $k^\mu = (\omega, \vec{k})$ being the four-momentum of the emitted photon and each side of eq. (44) being zero. According to the lemma, in order to derive the NNγ amplitude $\tilde{M}_\mu$ in the SPA, it is sufficient to calculate only the amplitude $\tilde{M}_\mu^{(a)}$ and make sure that this amplitude fulfills the gauge-invariance requirement, $k^\mu \tilde{M}_\mu^{(a)} = 0$.

The total NNγ amplitude can be separated into various parts as shown in Fig. (4). The four diagrams under the column denoted as external emission, where the intermediate nucleons are in positive-energy states, are called external emission or one-body single-scattering contributions. The two diagrams in the upper part of the column denoted as internal emission, where the intermediate nucleons coupled to photons are also in positive-energy states, are called the one-body rescattering or double-scattering contributions. The one-body single- and double-scattering diagrams are due to the one-body current. The remaining diagrams represent the contributions from the effective two-body current. We mention that the one-body single- and double-scattering diagrams in which the intermediate nucleons are in negative-energy states are part of the effective two-body current contribution [31] and, as such, they are included in the diagrams due to the two-body current in Fig. (4).

It can be shown [52] that the one-body rescattering diagrams as well as those diagrams arising from the one-body current and involving the negative-energy intermediate states contribute only to the amplitude $\tilde{M}_\mu^{(b)}$ and to higher-order terms in eq. (43). The part of the two-body current which depends on the structure of the NN potential is *assumed* not to contribute to the amplitude $\tilde{M}_\mu^{(a)}$. Although the validity of this assumption can be strictly verified only if one knows the underlying structure of the NN potential, such as those based on meson-exchange models of NN force, it seems to be satisfied by most existing NN potentials. See, however, ref.[32] where possibilities of NN potentials that can violate this assumption are discussed.

Under the assumption discussed above, it is sufficient to consider only those diagrams corresponding to the external contribution in order to obtain the amplitude $\tilde{M}_\mu^{(a)}$. We have,

$$
\begin{aligned}
\tilde{M}_\mu^{ext} = & \; <\bar{u}(p_1')\bar{u}(p_2')| \left\{ [-i\Gamma_\mu^{(+)}(p_1', P_1')]iS(P_1')\hat{T}(P_1', p_2'; p_1, p_2) \right\} |u(p_1)u(p_2)> \\
& + \; <\bar{u}(p_1')\bar{u}(p_2')| \left\{ \hat{T}(p_1', p_2'; P_1, p_2)iS(P_1)[-i\Gamma_\mu^{(+)}(P_1, p_1)] \right\} |u(p_1)u(p_2)> \\
& + \; (1 \longleftrightarrow 2) ,
\end{aligned}
\tag{46}
$$

where $P_1 \equiv p_1 - k$ and $P_1' \equiv p_1' + k$, denote the four-momenta of off-shell nucleons in the intermediate states.

As has been discussed in section 2.3, the half-off-shell NN interactions depend actually on three independent variables which we have choosen to be the variables ν, t, and Δ_i as defined in eq. (36) and eq. (40). For example, $\hat{T}(P_1', p_2'; p_1, p_2) \to \hat{T}(\nu, t, \Delta_{1'})$. Similarly, the half-off-shell NNγ vertices can be parametrized in terms of the variables Δ_i and momentum of the emitted photon. So, for example, $\Gamma_\mu^{(+)}(p_1', P_1') \to \Gamma_\mu^{(+)}(k, \Delta_{1'})$. With these parametrizations, we now expand eq. (46) in powers of the photon energy ω, keeping terms through zeroth order. The result is

$$
\begin{aligned}
\tilde{M}_\mu^{ext} = & \; <\bar{u}(p_1')\bar{u}(p_2')| \left\{ \Gamma_\mu^{(+)}(k=0, \Delta_{1'}=0)\frac{1}{\not{P}_1' - m}\hat{T}(\nu, t, \Delta_{1'}=0) \right\} |u(p_1)u(p_2)> \\
& + \; <\bar{u}(p_1')\bar{u}(p_2')| \left\{ \hat{T}(\nu, t, \Delta_1=0)\frac{1}{\not{P}_1 - m}\Gamma_\mu^{(+)}(k=0, \Delta_1=0) \right\} |u(p_1)u(p_2)> \\
& + \; \left\{ e_1'\left(\frac{p_2'\cdot k}{p_1'\cdot k}\right)p_{1\mu}' + e_1\left(\frac{p_2\cdot k}{p_1\cdot k}\right)p_{1\mu} \right\}\left(\frac{\partial\tilde{T}}{\partial\nu}\right) \\
& + \; 2\left\{ e_1'\frac{p_{1\mu}'}{p_1'\cdot k} - e_1\frac{p_{1\mu}}{p_1\cdot k} \right\}(p_1' - p_1)\cdot k\left(\frac{\partial\tilde{T}}{\partial t}\right) \\
& + \; 2\left\{ e_1'p_{1\mu}'\left(\frac{\partial\tilde{T}}{\partial\Delta_{1'}}\right) + e_1 p_{1\mu}\left(\frac{\partial\tilde{T}}{\partial\Delta_1}\right) \right\} \\
& + \; <\bar{u}(p_1')\bar{u}(p_2')| \left\{ \left(\frac{\partial\Gamma_\mu^{(+)}}{\partial\Delta_{1'}}\right)[\not{p}_1' + m]\hat{T}(\nu, t, \Delta_{1'}=0) \right\} |u(p_1)u(p_2)> \\
& + \; <\bar{u}(p_1')\bar{u}(p_2')| \left\{ \hat{T}(\nu, t, \Delta_1=0)[\not{p}_1 + m]\left(\frac{\partial\Gamma_\mu^{(+)}}{\partial\Delta_1}\right) \right\} |u(p_1)u(p_2)> \\
& + \; (1 \longleftrightarrow 2) ,
\end{aligned}
\tag{47}
$$

where $\nu = p_1 \cdot p_2 + p_1' \cdot p_2'$, $t = (p_1' - p_1)^2$, and all the derivatives evaluated at the on-shell points. We note that in making the above expansion, all the external variables, p_i, p_i', and k, are kept as being independent from each other.

An important point to be emphasized, here, is that the above result is obtained under the assumption that the NN interaction as well as the e.m. vertex can be differentiated with respect to their arguments. This will not be the case if, for example, the system contains resonances. The NN interaction below pion threshold is certainly free of resonances, except at very low energies where it presents a quasideuteron resonance in the $^1S_0$ partial-wave state.

We now impose the gauge-invariance condition, $k^\mu \tilde{M}_\mu = 0$, on the amplitude $\tilde{M}_\mu^{ext}$ given by eq. (47) in order to obtain the amplitude $\tilde{M}_\mu^{(a)}$ and, consequently, the total amplitude $\tilde{M}_\mu$ through zeroth order in k (SPA). The way this is done is to add terms independent of k in eq. (47) such that we have

$$
\begin{aligned}
\tilde{M}_\mu &= \tilde{M}_\mu^{(a)} \\[2mm]
&= \ <\bar{u}(p_1')\bar{u}(p_2')| \left\{ \Gamma_\mu^{(+)}(k=0,\Delta_{1'}=0)\frac{1}{\not{p}_1' - m}\hat{T}(\nu,t,\Delta_{1'}=0) \right\} |u(p_1)u(p_2)> \\[2mm]
&+ \ <\bar{u}(p_1')\bar{u}(p_2')| \left\{ \hat{T}(\nu,t,\Delta_1=0)\frac{1}{\not{p}_1 - m}\Gamma_\mu^{(+)}(k=0,\Delta_1=0) \right\} |u(p_1)u(p_2)> \\[2mm]
&+ \ \left\{ e_1'\left[\left(\frac{p_2' \cdot k}{p_1' \cdot k}\right)p_{1\mu}' - p_{2\mu}'\right] + e_1\left[\left(\frac{p_2 \cdot k}{p_1 \cdot k}\right)p_{1\mu} - p_{2\mu}\right] \right\} \left(\frac{\partial \tilde{T}}{\partial \nu}\right) \\[2mm]
&+ \ 2\left\{ \left[e_1'\frac{p_{1\mu}'}{p_1' \cdot k} - e_1\frac{p_{1\mu}}{p_1 \cdot k}\right](p_1' - p_1)\cdot k - (e_1' - e_1)(p_{1\mu}' - p_{1\mu}) \right\} \left(\frac{\partial \tilde{T}}{\partial t}\right) \\[2mm]
&+ \ (1 \longleftrightarrow 2) \, .
\end{aligned}
$$

(48)

The above result is the low-energy theorem for NN bremsstrahlung which is a direct consequence of the gauge-invariance requirement. As we can see, the total NNγ amplitude in the SPA does not contain any term which requires off-shell information. Those terms requiring off-shell information (the last four terms in eq. (47)) drop out from the final result once gauge invariance is imposed. The uniqueness of the amplitude in eq. (48) is ensured by the lemma.

The amplitude in eq. (48) has been obtained by simply adding some terms to the amplitude $\tilde{M}_\mu^{ext}$ given by eq. (47) in order to yield the gauge-invariant amplitude. These additional terms should arise from some diagrams other than the external ones that have not been considered explicitly in the derivation of the low-energy theorem. In fact, it can be shown [54, 55] that these additional terms are generated precisely from the one-body rescattering and two-body current contributions. Therefore, in order to satisfy the gauge-invariance condition all the diagrams shown in Fig. (4) must be included in the NNγ amplitude. Leaving out any of these contributions will violate the gauge-invariance condition. This is the case not only in the SPA but also to all orders of the expansion in photon momentum.

POTENTIAL MODEL FOR NN BREMSSTRAHLUNG

In the following, we will work in momentum space and also in the Coulomb gauge, so that $\epsilon_0 = \vec{\epsilon} \cdot k = 0$. We write the total invariant amplitude, $\tilde{M}$, for producing a photon of momentum $\vec{k}$ $(\omega = |\vec{k}|)$ and polarization $\vec{\epsilon}$ in an NN collision as

$$
\tilde{M} = \sqrt{\epsilon_1'\epsilon_2'}\,\omega M \sqrt{\epsilon_1\epsilon_2} \, ,
$$

(49)

325

where $\varepsilon_1', \varepsilon_2'$ $(\varepsilon_1, \varepsilon_2)$ are the energies of the two interacting nucleons, 1 and 2, in the final (initial) state; they are defined as $\varepsilon_i \equiv \varepsilon_{p_i} \equiv \sqrt{\vec{p_i}^{\,2} + m^2}$ with $\vec{p_i}$ being the momentum of i–th nucleon. In the above equation M denotes the bremsstrahlung transition amplitude,

$$M \equiv\; < \vec{\epsilon}, \vec{k}; \Psi_f^{(-)} |\; V_{em}\; |0; \Psi_i^{(+)} > \, , \tag{50}$$

where V_{em} denotes the total e.m. transition potential and $\Psi_{i,f}^{(\pm)}$, the strongly interacting two-nucleon wave functions in the initial (i) and final (f) states; the superscripts $\pm$ refer to the outgoing $(+)$ and incoming $(-)$ waves.

The total e.m. transition potential in the above equation is the sum of the one- and two-body e.m. transition potentials as introduced in eq. (23) and eq. (24),

$$V_{em} = V_{em}^{(1)} + V_{em}^{(2)} \, . \tag{51}$$

For the one-body e.m. potential, $V_{em}^{(1)}$, we take the positive-energy part of the relativistic e.m. vertex given by eq. (30) in order to include relativistic corrections. We have,

$$
\begin{aligned}
V_{em}^{(1)}(\vec{p_1}', \vec{p_2}'; \vec{p_1}, \vec{p_2}) \;=\;& \sqrt{\frac{m}{\varepsilon_{p_1'}}} < \bar{u}(\vec{p_1}')|\; \vec{\Gamma}^{(+)} \cdot \vec{A}\; |u(\vec{p_1}) > \sqrt{\frac{m}{\varepsilon_{p_1}}} \\
+\;& \sqrt{\frac{m}{\varepsilon_{p_2'}}} < \bar{u}(\vec{p_2}')|\; \vec{\Gamma}^{(+)} \cdot \vec{A}\; |u(\vec{p_2}) > \sqrt{\frac{m}{\varepsilon_{p_2}}} \, .
\end{aligned}
\tag{52}
$$

Then, the effective one-body e.m. transition operator can be expressed as the sum of four terms

$$V_{em}^{(1)} \;=\; V_{con} + V_{mag} + V_{rsc} + V_{rem} \, , \tag{53}$$

where, in the NN c.m. frame,

$$
\begin{aligned}
V_{con} \;=\;& -\sqrt{\frac{2\pi}{\omega}}\frac{e}{2m}\vec{\epsilon}\cdot(\vec{p}+\vec{p}\,')\left\{ \tilde{e}_1^- \; \delta(\vec{p}-\vec{p}\,'-\frac{\vec{k}}{2}) - \tilde{e}_2^+ \; \delta(\vec{p}-\vec{p}\,'+\frac{\vec{k}}{2}) \right\} \, , \\
V_{mag} \;=\;& -i\sqrt{\frac{2\pi}{\omega}}\frac{e}{2m}\left\{ \tilde{\mu}_1^- \; \vec{\epsilon}\cdot(\vec{k}\wedge\vec{\sigma}_1)\delta(\vec{p}-\vec{p}\,'-\frac{\vec{k}}{2}) + \tilde{\mu}_2^+ \; \vec{\epsilon}\cdot(\vec{k}\wedge\vec{\sigma}_2)\delta(\vec{p}-\vec{p}\,'+\frac{\vec{k}}{2}) \right\} \, , \\
V_{rsc} \;=\;& i\sqrt{\frac{2\pi}{\omega}}\frac{e}{2m}\left\{ \tilde{\nu}_1^- \; \vec{\epsilon}\cdot(\vec{p}\,'\wedge\vec{\sigma}_1)\delta(\vec{p}-\vec{p}\,'-\frac{\vec{k}}{2}) - \tilde{\nu}_2^+ \; \vec{\epsilon}\cdot(\vec{p}\,'\wedge\vec{\sigma}_2)\delta(\vec{p}-\vec{p}\,'+\frac{\vec{k}}{2}) \right\} \, , \\
V_{rem} \;=\;& -i\sqrt{\frac{2\pi}{\omega}}\frac{e}{2m}\left\{ \tilde{\eta}_1^- \; \vec{\epsilon}\cdot(\vec{p}\,'\wedge\vec{k})\; \vec{\sigma}_1\cdot\vec{p}\; \delta(\vec{p}-\vec{p}\,'-\frac{\vec{k}}{2}) \right. \\
& \left. +\tilde{\eta}_2^+ \; \vec{\epsilon}\cdot(\vec{p}\,'\wedge\vec{k})\; \vec{\sigma}_2\cdot\vec{p}\; \delta(\vec{p}-\vec{p}\,'+\frac{\vec{k}}{2}) \right\} \, .
\end{aligned}
\tag{54}
$$

In the above equations, $\vec{p}$ and $\vec{p}\,'$ denote the relative momenta of the two interacting nucleons before and after the emission of a photon with momentum $\vec{k}$ and polarization $\vec{\epsilon}$. The factors $\tilde{e}_i^\pm$, $\tilde{\mu}_i^\pm$, $\tilde{\nu}_i^\pm$ and $\tilde{\eta}_i^\pm$ are functions of nucleon and photon momenta; they are given explicitly in ref.[42]. The leading terms in $\tilde{e}_i^\pm$ and $\tilde{\mu}_i^\pm$ are given by $\tilde{e}_i^\pm = \delta_{e,e_i}$ and $\tilde{\mu}_i^\pm = \delta_{e,e_i} + \kappa_i$ which give rise to the convection and magnetization current operators of the conventional non-relativistic approach. e_i stands for the nucleon charge ($e_i = e$ for proton and $e_i = 0$ for neutron) and κ_i the anomalous magnetic moment in units of nuclear magnetons ($\kappa_i = 1.793$ for protons and $\kappa_i = -1.913$ for neutrons). $\tilde{\nu}_i^\pm = \kappa_i k/m$ and $\tilde{\eta}_i^\pm = \kappa_i/(2m^2)$ to leading orders. V_{rsc} and V_{rem} are the relativistic spin corrections to the one-body e.m. transition potential. It should be observed that in eq. (54) we have set the half-off-shell form factor $f_2^{(+)}$, as discussed in section 2.2, to its on-shell value. We shall come back to this point in section 4.2.

The two-body e.m. potential, $V_{em}^{(2)}$, is taken only in the SPA. As discussed in section 2.2, with this approximation the model is gauge invariant only up to terms linear in ω. Beyond the SPA, the construction of a two-body current is subject to an ambiguity. As we shall discuss later, this violation of gauge invariance, however, is not as critical as it may appear at first. From eq. (21) and eq. (24), we have in the NN c.m. frame

$$V_{em}^{(2)} = \sqrt{\frac{2\pi}{\omega}\frac{\vec{\epsilon}}{2}} \cdot \left\{ (\hat{e}_2 - \hat{e}_1)\left[\nabla_{\vec{p}}V(\vec{p}\,',\vec{p})\right] + (\hat{e}_2' - \hat{e}_1')\left[\nabla_{\vec{p}\,'}V(\vec{p}\,',\vec{p})\right] \right\} . \tag{55}$$

It follows directly from the above equation that the two-body current contribution to $pp\gamma$ reactions vanishes identically in the SPA. Also, for a NN potential, $V(\vec{p}\,',\vec{p})$, based on a meson-exchange theory the one-pion-exchange should be the dominant source of the two-body current, since the pion is the lightest meson and, therefore, it generates the strongest momentum dependence in the NN potential.

The two-nucleon wave function, $\Psi^{(\pm)}$, in eq. (50) can be expressed in terms of the NN T-matrix interaction as

$$|\Psi^{(\pm)}> = |\phi> + \mathcal{G}^{(\pm)}T^{(\pm)}|\phi> , \tag{56}$$

where ϕ denotes the unperturbed two-nucleon wave function and $\mathcal{G}^{(\pm)}$ the two-nucleon propagators.

Inserting eq. (56) into eq. (50), we obtain

$$\begin{aligned}
M &= \; <\vec{\epsilon},\vec{k};\phi_f|\,V_{em}\,|0;\phi_i> \\
&+ \; <\vec{\epsilon},\vec{k};\phi_f|\,T(z_f)\mathcal{G}(z_f)V_{em}\,|0;\phi_i> + <\vec{\epsilon},\vec{k};\phi_f|\,V_{em}\mathcal{G}(z_i)T(z_i)\,|0;\phi_i> \\
&+ \; <\vec{\epsilon},\vec{k};\phi_f|\,T(z_f)\mathcal{G}(z_f)V_{em}\mathcal{G}(z_i)T(z_i)\,|0;\phi_i> ,
\end{aligned} \tag{57}$$

where $T(z_{i,f}) \equiv T^{(\pm)}(z_{i,f})$. z_i and z_f denote the total energy of the NN system before and after scattering, respectively. The first term in the above equation is called zeroth-scattering while the second and third terms are the single-scattering contributions. The last term is called rescattering or double-scattering contributuion. Fig. (4) shows the breakdown by type of current and number of scatterings into external and internal emission. Note that there is no contribution to the zeroth scattering term from the one-body current, because free particles cannot radiate.

For the two-nucleon propagators, $\mathcal{G}(z_{i,f})$, in eq. (57) most of the existing NNγ calculations assume an expression of the form (in the NN c.m. frame)

$$\mathcal{G}(z_{i,f}) = \frac{1}{z_{i,f} - 2\varepsilon_l''} \tag{58}$$

irrespective of the type of propagator entering the T-matrix integral equation. In the above equation, $z_i = 2\varepsilon_l$ and $z_f = 2\varepsilon_{l'}$ with $\vec{l}$ and $\vec{l}\,'$ denoting the NN relative momenta before and after scattering, respectively; the double prime denotes intermediate states. In a consistent approach the above propagator should be of the same type as that used for solving the T-matrix integral equation [42]. We should stress, however, that it is difficult to use consistently T-matrix interactions which are obtained by treating the two nucleons symmetrically, such as in the three-dimensional Blankenbecler-Sugar reduction of the BS-equation. The reason is that in the NNγ process one of the nucleons is necessarily off-shell while the other one is on-shell and, therefore, the nucleons cannot be treated symmetrically.

For the half-off-shell NN T-matrix interactions, $T(z_{i,f})$, entering in eq. (57) we take the form given by eq. (41) (including the appropriate isospin operator, $\tau_1 \cdot \tau_2$),

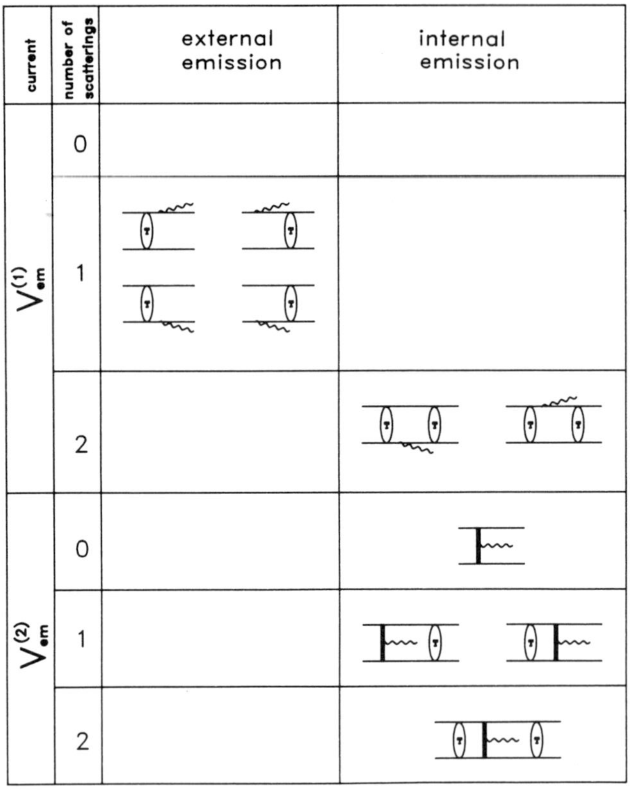

current	number of scatterings	external emission	internal emission
$V_{em}^{(1)}$	0		
	1		
	2		
$V_{em}^{(2)}$	0		
	1		
	2		

Figure 4. External and internal contributions to the NNγ amplitude from the one- and two-body e.m. transition potentials, $V_{em}^{(1)}$ and $V_{em}^{(2)}$. The latter potential is represented by a photon (wiggled line) attached to a region of NN potential (solid bar).

whose coefficients α, β, γ, δ, ε, and ω can be expressed as linear combinations of the partial-wave NN T-matrix elements. Their explicit expressions may be found in ref.[51]. The NN interactions in eq. (57) are related to the corresponding Lorentz invariant NN amplitudes, $\tilde{T}$, through

$$T(\vec{l}\,',\vec{l},z(\vec{l})) = \sqrt{\frac{m}{\varepsilon_{l'}}}\sqrt{\frac{m}{\varepsilon_{l'}}}\tilde{T}(\vec{l}\,',\vec{l},z(\vec{l}))\sqrt{\frac{m}{\varepsilon_{l}}}\sqrt{\frac{m}{\varepsilon_{l}}}\ . \tag{59}$$

With the T-matrix interaction given by eq. (41), the evaluation of the NNγ amplitude in eq. (57) reduces basically to the evaluation of the spin-isospin matrix elements involving the spin-isospin operators of the T-matrix interaction and those of the e.m. transition operator V_{em}.

The final remark before leaving this chapter concerns the different reference systems involved in the evaluation of the various terms in the NNγ amplitude. For the sake of discussion, let us restrict ourselves to the external emission diagrams (single-scattering terms from the one-body current contribution). The external contribution to the NNγ amplitude can be written as a sum of the two terms,

$$M^{ext} = M_a + M_b\ , \tag{60}$$

where

$$M_a = <\vec{\epsilon},\vec{k};\phi_f(\vec{p}\,')|T(\vec{p}_-,\vec{p}\,',z(p'))\mathcal{G}(\vec{p}_-,z(p'))V_{em;1}^{(1)}(\vec{p}_-,\vec{p}_+)|0;\phi_i(\vec{p}_+)>$$

$$+ \quad < \vec{\epsilon}, \vec{k}; \phi_f(\vec{p}\,')|T(\vec{p}_+, \vec{p}\,', z(p'))\mathcal{G}(-\vec{p}_+, z(p'))V_{em;2}^{(1)}(-\vec{p}_+, -\vec{p}_-)|0; \phi_i(\vec{p}_+) > ,(61)$$

corresponding to the single-scattering processes in which the nucleons interact after the photons are emitted, and

$$M_b \;=< \vec{\epsilon}, \vec{k}; \phi_f(\vec{p}\,'_-)|V_{em;1}^{(1)}(\vec{p}\,'_-, \vec{p}\,'_+)\mathcal{G}(\vec{p}\,'_+, z(p))T(\vec{p}\,'_+, \vec{p}, z(p))|0; \phi_i(\vec{p}) >$$
$$+ \quad < \vec{\epsilon}, \vec{k}; \phi_f(\vec{p}\,'_-)|V_{em;2}^{(1)}(-\vec{p}\,'_+, -\vec{p}\,'_-)\mathcal{G}(-\vec{p}\,'_-, z(p))T(\vec{p}\,'_-, \vec{p}, z(p))|0; \phi_i(\vec{p}) > ,(62)$$

corresponding to the single-scattering processes in which the nucleons interact before the photons are emitted. In the above equations, $\vec{p}_\pm \equiv \vec{p} \pm \vec{k}/2$ and $\vec{p}\,'_\pm \equiv \vec{p}\,' \pm \vec{k}/2$, with $\vec{p}$ and $\vec{p}\,'$ denoting the initial and final relative momenta of two interacting nucleons in the initial and final NN c.m. frame, respectively. $V_{em;j}^{(1)}$ stands for the one-body e.m. transition potential acting on the j-th nucleon.

We note that the amplitude M_a in eq. (61) is expressed in the final NN c.m. frame which is shifted by $-\vec{k}$ with respect to the initial NN c.m. frame. The amplitude M_b in eq. (62) is expressed in the initial NN c.m. frame. In order to obtain the total external emission amplitude the two amplitudes M_a and M_b should be added up. Of course, to do this both amplitudes must be first Lorentz boosted to a given single frame of reference. We accomplish this by forming the Lorentz invariant amplitudes, $\tilde{M}_a$ and $\tilde{M}_b$, as given by eq. (49) for M_a and M_b, separately.

For further details of the NN-bremsstrahlung formalism we refer to [27, 42, 56]. We mention that most of the currently available calculations of NN bremsstrahlung [18, 57, 58] use essentially the same formalism as we have presented here and, consequently, they lead to similar predictions. There are, however, differences in details such as the way relativistic corrections are taken into account. These are discussed in section 4.2. In ref.[59], the NNγ process is formulated in coordinate space.

PROTON-PROTON BREMSSTRAHLUNG

In this chapter we discuss the present status of the ppγ reaction for investigating the off-shell behavior of the modern realistic NN interactions. We start discussing some of the general characteristics of this reaction, especially at incident energies near pion threshold and small proton scattering angles ($< 20°$) where off-shell effects are expected to be most important. Also, the limitations and/or uncertainties present in the currently available theory of NN bremsstrahlung are discussed. Then, we compare the predictions of some selected potential models with the new available data.

General features of the ppγ reaction

We start examinig the inclusive cross sections where only photons are detected in the final state. In Fig. (5) the angular distributions in the initial pp c.m. frame at $T_{lab} = 300\ MeV$ and for various photon energies are shown. Here, the individual contributions from the convection and magnetic currents (V_{con} and V_{mag} in eq. (54)) are also displayed. These two contributions determine the basic features of the ppγ reaction [27]. Relativistic corrections (V_{rsc} and V_{rem}) are excluded in Fig. (5). The angular distributions are symmetric with respect to the photon angle $\theta = 90°$ because of the identity of the protons. We see a rather pronounced change in the angular distribution as the photon energy increases. For low ω the angular distribution is pronounced, showing a quadrupole shape which is due to the convection current contribution. As ω increases, the angular distribution changes to a $cos^2\theta$ dependence because of the

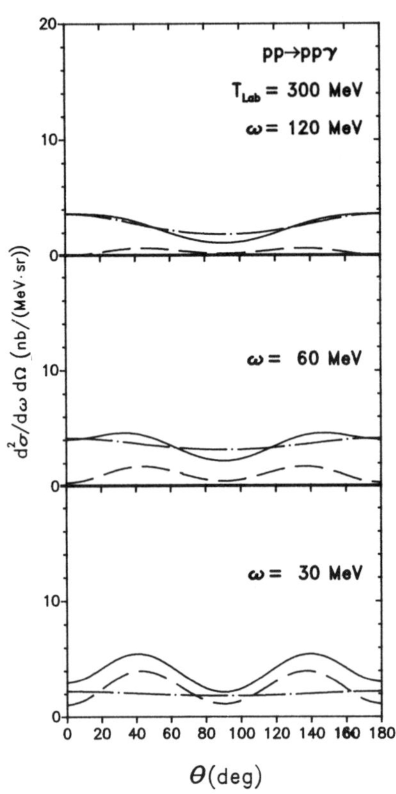

Figure 5. ppγ angular distribution in the initial proton-proton center–of–mass frame at $T_{lab} = 300 MeV$ and for various photon energies ω. The dashed and dash-dotted curves correspond to the convection and magnetization current contributions, respectively. The solid curves are the contributions from the sum of the two currents.

increasing contribution of the magnetic current. We also note in Fig. (5) that the convection contribution decreases while the magnetic contribution increases as ω increases. In particular, for high photon energies the ppγ cross section is almost entirely due to the magnetic current. These general features can be understood by examining the contribution arising from the external emission diagrams. For the convection current contribution we obtain in the leading order [27]

$$\frac{d^2\sigma_{con}}{d\omega d\Omega} = \frac{\alpha}{(2\pi)^2}\frac{1}{\omega}(\frac{p'}{p})\left[\frac{\varepsilon(p')\varepsilon(p)T^2_{pp}}{8\pi}\right]\left\{\frac{8}{15}v'^4 + v^4 sin^2 2\theta\right\} , \qquad (63)$$

and for the magnetic current contribution

$$\frac{d^2\sigma_{mag}}{d\omega d\Omega} = \frac{\alpha}{(2\pi)^2}\frac{\omega}{m^2}(\frac{p'}{p})\left[\frac{\varepsilon(p')\varepsilon(p)T^2_{pp}}{8\pi}\right]\mu_p^2\left\{[(g-d)^2 + \frac{g^2}{3}v'^2] + d^2v^2 cos^2\theta\right\} . \qquad (64)$$

Here, $\alpha = e^2$ is the fine-structure constant and T_{pp} stands for the pp T-matrix interaction. We have also omitted explicit reference to the arguments of T_{pp} and parameters d and g. v (v') is the velocity of the nucleon associated with the initial (final) relative momentum $\vec{p}$ ($\vec{p}\,'$) of the proton in units of c. The quantity in the square brackets becomes the pp total cross section (σ_{pp}) in the SPA limit. The classical bremsstrahlung formula for two equally charged particles may be recovered from eq. (63) if we drop there the phase space factor (p'/p) and replace the quantity in the square brackets by σ_{pp}.

Eq. (63) shows that the angular distribution from the convection current contribution has the well-known quadrupole shape. Moreover it becomes more pronounced as the incident energy increases because of the factor v^4, which is approximately proportional to T^2_{lab}. The photon cross section also goes basically as $1/\omega$. These features are what we observe qualitatively in Fig. (5). The magnetic contribution gives a $cos^2\theta$ dependence. However, the value of the parameters d and g are such that the term independent of the photon angle gives the dominant contribution. Eq. (64) also shows that the angular distribution becomes more pronounced as a function of incident energy. The cross section has a linear dependence on the photon energy. The deviations from eq. (63) and eq. (64) observed in Fig. (5) are due to the higher-order corrections which were neglected in the derivation of these equations. As a result of the behavior of the convection and magnetization current contributions, the total contribution exhibits the angular distribution shown in Fig. (5). We mention again that the dominant features of the ppγ reaction are determined by the external emission diagrams with convection and magnetization currents. For energetic photons the latter current dominates. All other contributions may, therefore, be regarded as higher-order corrections.

In Fig. (6) the half-off-shell behavior of the NN interaction is illustrated for fixed incident energies of T_{lab} = 20 and $280 MeV$. More specifically, Fig. (6) shows the angle averaged magnitude of the different components of the half-off-shell ($|\vec{p}\,'| \neq |\vec{p}|$) interaction, $T(\vec{p}\,',\vec{p},z(p))$, as a function of the magnitude of the off-shell momentum, $p' = |\vec{p}\,'|$. Since the interaction is angle averaged, $\overline{|\delta|} = \overline{|\varepsilon|}$. The vertical solid lines indicate the position of the on-shell momenta corresponding to the considered incident energy. Also, Fig. (6) shows three different interactions but for the discussion of the present section this is irrelevant. In the bottom parts of the figure we show the tensor $S_{12}(\hat{q},\hat{Q})$ component which is not present on the energy-shell. It is negligible compared to other components, unless one goes very far off-shell.

For a low incident energy of $T_{lab} = 20 MeV$, the spin-singlet ($P_{S=0}$) component exhibits much stronger off-shell dependence than the other components of the interaction. Observe also that it is, by far, the strongest component (note the difference in the scale) which is due to the dominat NN interaction in the $^1S_0$ partial-wave state. The spin-triplet components ($P_{S=1}$, spin-orbit, and tensor) of the interaction show a rather similar off-shell behavior. They are also of similar magnitude. We recall that off-shell T-matrices at such low incident energies are effectively required by NN-bremsstrahlung reactions in which energetic photons are produced even if the actual incident energy is high. At an higher incident energy of $T_{lab} = 280 MeV$, the off-shell behavior and also the strength of the $P_{S=0}$ component are much weaker than those exhibited at

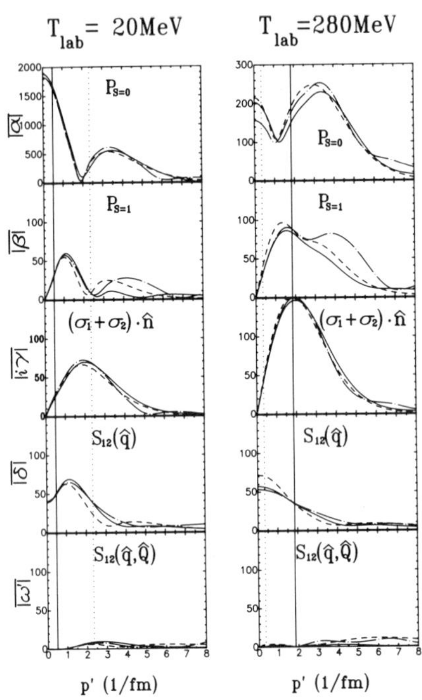

Figure 6. Absolute magnitude of the angle-averaged half-off-shell NN interaction, $T(\vec{p}\,',\vec{p},z(\vec{p}))$ (see eq. (41)), as a function of the off-shell momentum p' for incident energies of $T_{lab} = 20$ and $280 MeV$. The vertical solid lines indicate the position of the corresponding on-shell momentum. The solid, dashed and dash-dotted curves are the OBEPQ, OBEPT and PARIS potentials, respectively. $\omega' \equiv \omega(\hat{q} \cdot \hat{Q})$ of eq. (41).

$T_{lab} = 20 MeV$. This is mainly due to the strong energy and off-shell dependences of the $^1S_0$ state. The off-shell behavior of the spin-triplet components change only to a minor extend from that at lower energy. The magnitudes of the $P_{S=1}$ and spin-orbit components increase by a factor of ~ 2 and ~ 3, respectively, as compared to those at $T_{lab} = 20 MeV$.

The bremsstrahlung reaction requires half-off-shell T-matrix interactions as exhibited in eq. (61) and eq. (62). The magnitude of each component of the interaction, as it enters into the calculation of the ppγ amplitude at an incident energy of $T_{lab} = 280 MeV$ and for proton scattering angles of $\theta_1 = 12.4°$ and $\theta_2 = 12°$ in the coplanar geometry, is plotted in Fig. (7). Columns $a)$ and $b)$ show the interactions required in the two terms of the amplitude M_a given by eq. (61), where the nucleons interact after the emission of a photon. Columns $c)$ and $d)$ correspond to those required in the amplitude M_b of eq. (62), where the nucleons interact before the photon emission. The off-shell tensor channel $S_{12}(\hat{q}, \hat{Q})$ is not displayed here because it is negligible compared to the other

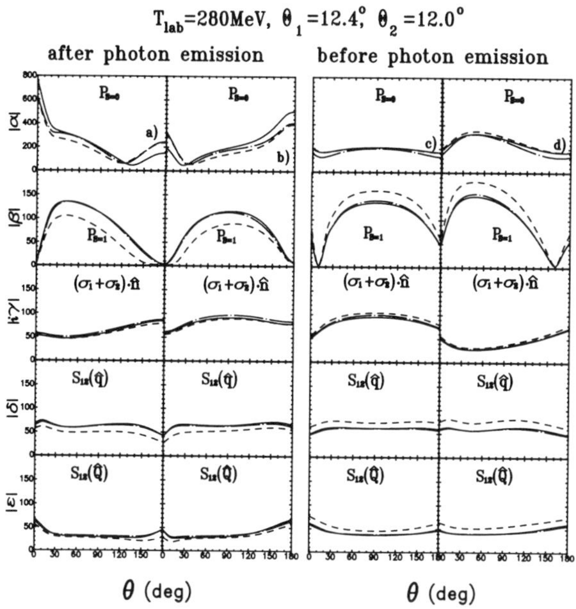

Figure 7. Half-off-shell T-matrix interactions (see eq. (41)) which enter in the calculation of the ppγ matrix element. Columns $a)$ and $b)$ correspond to the interactions entering in the external emission diagrams, where the strong interaction takes place after the photon emission. Columns $c)$ and $d)$ correspond to those with the strong interaction before photon emission. Labelling of the curves according to Fig. (6).

components of the interaction at the kinematics involved. Also, Fig. (7) shows three different interactions but this is irrelevant for the present discussion.

In columns $a)$ and $b)$, one observes that the spin-triplet components of the off-shell interaction are of the same order of magnitude to each other, while the $P_{S=0}$ component is much stronger than the corresponding spin-triplet components. This is due to the fact that the NNγ processes in which nucleons interact after emission of a photon samples effectively the NN interaction at small NN c.m. energies, where the off-shell interaction

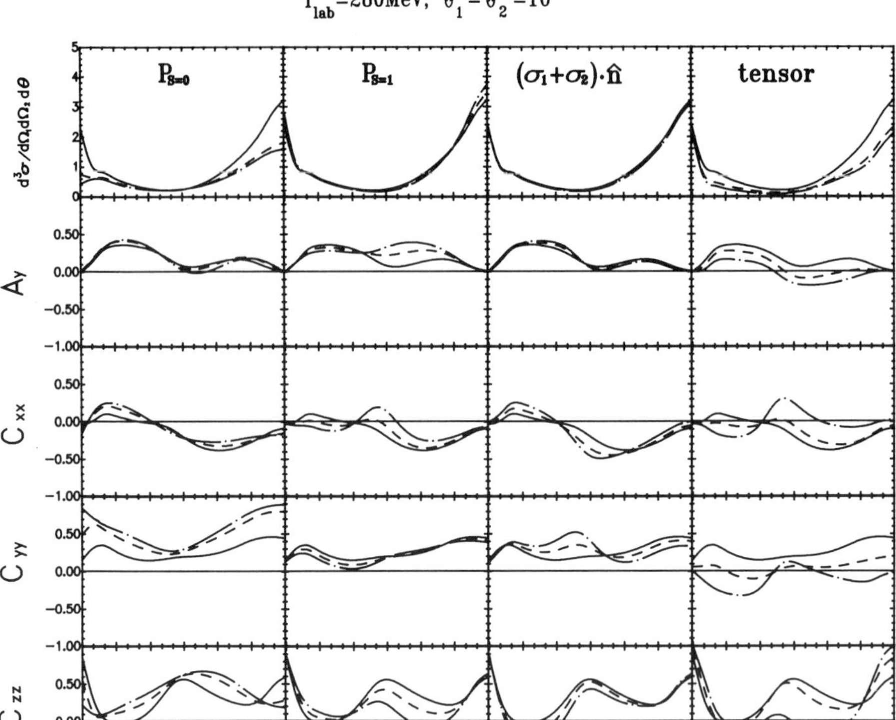

Figure 8. Effect of different components of the NN interaction on the ppγ cross-section, analyzing power (A_y), and spin correlation functions $(C_{xx}, C_{yy},$ and $C_{zz})$. The solid, dashed, and dash-dotted lines are the results when the T-matrix interaction in the considered components are multiplied by a factor of $r = 1$, 0.5, and 0, respectively. The incident energy is $T_{lab} = 280 MeV$ and proton scattering angles are 10°.

is dominated by a strong contribution from the $^1S_0$ partial-wave state similar to the on-shell interaction (see Fig. (6)). Also, the strong dependence on photon emission angle exhibited by the $P_{S=0}$ component is largely due to the strong energy dependence arising from the off-shell NN interaction in the $^1S_0$ state at low NN c.m. energies. Note that, here, both the NN c.m. energy in the final state and how far one goes off-shell depend on the photon emission angle. In columns c) and d), where the NN scattering takes place at the incident energy of $T_{lab} = 280 MeV$ independent on photon emission angle (strong interaction before the photon emission), the strength of the spin-singlet component is still much larger than those of the spin-triplet components. However, it is reduced by a factor of $3 - 5$ as compared to the situation of columns a) and b) which is mainly due to the strong energy dependence present in this component of the interaction as discussed before. The interaction strengths in the spin-triplet components remain nearly unchanged from the case of lower incident energies of columns a) and b) for the particular kinematics considered.

334

The sensitivity of the ppγ observables to the different components of the NN interaction is illustrated in Fig. (8). The columns indicated with the spin operators $P_{S=0}$, $P_{S=1}$, etc, correspond to the results when the T-matrix interaction in those indicated components (and only in those components) are modified by an arbitrary reduction factor r. The column indicated with "tensor" corresponds to the results when all the three tensor components are multiplied with the same reduction factor. The dashed curves correspond to the results with the reduction factor of $r = 0.5$ while the dash-dotted curves to those with $r = 0$. The solid curves correspond to $r = 1$, i.e., unmodified T-matrix interaction and are the same in all the columns. The cross section is seen to be sensitive only to the central spin-singlet and tensor components of the interaction. It remains practically unaffected by the changes of the $P_{S=1}$ or spin-orbit components. The behavior of the cross section is highly non-linear to the variation of the interaction in the $P_{S=0}$ or tensor components due to a non-trivial interference among various terms in the total bremsstrahlung amplitude. We note that, despite the fact that the T-matrix interaction in the $P_{S=0}$ channel is almost an order of magnitude larger than in the tensor channels (see Fig. (7)), the variations of these components by a same reduction factor lead to a similar amount of reductions in the cross section. The major reason for this is as follows: first of all, for energetic photons produced in the ppγ process, the dominant current is the magnetization current, V_{mag} (see Fig. (5)). Moreover, among the various ways in which a photon may be created in a NNγ reaction, the one in which the photon is emitted *before* the strong interaction takes place (corresponding to the amplitude M_a of eq. (61)) is favored due to the larger NN coupling at lower NN c.m. energy. The spin matrix element of V_{mag} is, then, proportional to [60]

$$< S'|V_{mag}|S >\propto (1 - \delta_{S',0}\delta_{S,0})[(-)^S \tilde{\mu}_1^- f(\vec{p}\,',\vec{p}-\vec{k}/2) + (-)^{S'} \tilde{\mu}_2^+ f(\vec{p}\,',\vec{p}+\vec{k}/2)] \,, \quad (65)$$

where $S(S')$ denotes the NN total spin in the initial(final) state and f is a function common to $\tilde{\mu}_1^-$ and $\tilde{\mu}_2^+$. The first factor reflects the fact that the spin operator cannot connect spin-singlet states. Moreover, for the range of photon momentum $\vec{k}$ involved in most of the existing NNγ experiments, both $f(\vec{p}\,',\vec{p}-\vec{k}/2)$ and $f(\vec{p}\,',\vec{p}+\vec{k}/2)$ have the same sign. Also, to leading order $\tilde{\mu}_1^- = \tilde{\mu}_2^+$ for identical particles so that the above matrix element will be suppressed to a large extent if a spin-singlet state is present in either the initial or final state as has been pointed out in ref.[60]. In particular, the strong $^1S_0$ state at small NN c.m. energies only contributes significantly to the ppγ cross section for those kinematical conditions where the cancellation in eq. (65) is minimized. In the coplanar geometry this happens at forward and backward photon emission angles and for small proton scattering angles. Therefore, put it the other way around, the sensitivity of the cross section to the $P_{S=0}$ component observed in Fig. (8) is only due to the very strong off-shell interaction in this component at small incident energies. We, therefore, see that in terms of absolute value the cross section is much more sensitive to the variation of the tensor component than to that of the central spin-singlet component. The sensitivity of the ppγ cross section to the tensor force has been also noticed in earlier works [60, 61] from a partial-wave analysis. The analyzing power is mostly sensitive to the tensor part of the interaction, as has been also found in ref.[60] from a partial-wave analysis, followed by the central spin-triplet component. This observable is insensitive to the spin-singlet and also, surprisingly, to the spin-orbit components. The spin correlation coefficients [62] are, in general, sensitive to all parts of the interaction. However, they are mostly sensitive to the tensor component.

Higher-order corrections

In this section we discuss briefly all those terms which are regarded as higher-order corrections within the framework of nucleon and meson degrees-of-freedom. These corrections are:

1) Relativistic spin correction (RSC) which is due to V_{rsc} and V_{rcm} in eq. (54).
2) One-body rescattering (RES) or double-scattering contribution (see Fig. (4)).
3) Coulomb correction which affects the pp T-matrix interaction due to the presence of a Coulomb potential.
4) Off-shell NN e.m. form factor effects.
5) Two-body current contribution beyond the SPA.

Currently, most of the existing new calculations take into account these corrections in one way or another, except the last two items.

The RSC has been taken into account by some authors in the past [18, 40, 41] who have shown that this correction reduces the ppγ cross section. The dominant RSC arises from V_{rsc} in eq. (54). However, the physical basis for this reduction has received limited attention until quite recently. In ref.[42], it has been demonstrated that, for energetic photons, the RSC effect is largely dynamical in origin. Indeed, using the basic structure of the NNγ amplitude, it has been shown [42] that this correction is large when the emitted photon has an energy in the region near the maximum value allowed kinematically, even when the incident energy is relatively low (depending on the kinematics involved, it may affect the cross section by as much as a factor of 2). This is due, primarily, to the $^1S_0$ state contribution which is large and strongly energy dependent compared to other states. By considering this energy dependence it is shown [42] how and under what kinematical conditions this state gives rise to such a strong RSC. For the kinematical conditions of the TRIUMF experiment [63], the RSC is shown to reduce the ppγ cross section by $(20 - 30)\%$ [42] and is the most important correction among various higher-order corrections investigated.

Relativistic corrections can be included to non-relativistic calculation via different methods. The standard Foldy-Wouthuysen transformation, for example, generates operators which order by order in $1/m$ expansion decouple the upper from the lower components. The upper part of the decoupled blocks then defines the effective two-component operator to be used in non-relativistic approaches. In most NNγ calculations, the RSC has been included via this method. Another way of obtaining effective two-component operators is via the direct Pauli reduction, i.e., sandwiching the four-component operators with free positive-energy Dirac spinors. In the formalism presented in the previous chapter, this latter method is used to account for the RSC. It has been shown quite recently by Fearing et al. [64] that the two methods mentioned above are not equivalent to each other. While the Foldy-Wouthuysen reduction includes effectively the influence of the intermediate negative-energy states, the direct Pauli reduction method ignores completely this effect. As a result, the two methods yield results which differ beyond first-order perturbation theory. For the kinematics involved in the currently existing ppγ data at very small proton scattering angles the discrepancy in the predictions based on these two methods is of the order of 7% at most [64] (for the TRIUMF geometry [63], there is practically no difference between the predictions of the two methods). As it has been mentioned in ref.[64], however, there is no solid reason to decide which of the above two methods is correct in the case where the fully relativistic Hamiltonian is not known. This is the case for all potential model calculations currently available. In particular, the NN T-matrices derived from meson-exchange models developed by the Bonn group [5] and used in many NNγ calculations are based on the direct Pauli reduction method. In order to include relativistic corrections consistently into the calculation, one has to treat both the e.m. and strong interactions on the same footing, i.e., effects of intermediate negative-energy states should be taken into account in both parts of the interaction Hamiltonian. Such a consistent calculation is not yet available.

The one-body RES contribution to the NNγ reaction has been considered by a number of authors in the past [39, 61] and is known to enhance both the ppγ and npγ cross sections. Until recently, however, there was no calculation in which both the RES and RSC were included simultaneously. Since the RSC reduces the NNγ cross section while the RES enhances it by a similar amount, these corrections should be treated simultaneously. In ref.[42], it is shown that depending on the kinematics involved a significant difference arises in the role of the RES relative to that in an approach where the RSC is ignored. In particular, for kinematical conditions where the photon energy is close to its endpoint energy, the RES contribution reduces the cross section instead of enhancing it as is the case of the RES contribution without the RSC. For the kinematical conditions of the TRIUMF experiment [63], the RES enhances the ppγ cross section by $\sim 20\%$ [42]. However, since the RSC reduces the cross section by $\sim (20 - 30)\%$, the net effect is still a reduction of the ppγ cross section, especially for small proton scattering angles. For proton scattering angles around $\theta_1 \sim \theta_2 \sim 30°$, the RES contribution cancels almost completely the contribution due to the RSC [42].

The Coulomb correction has been also investigated by a number of authors in the past [65, 66]. In particular, Rich and Heller [66] has treated this correction exactly within the framework of a completely non-relativistic approach. Their results show that the Coulomb correction becomes significant for proton scattering angles $\leq 10°$ and reduces the ppγ cross section. The Coulomb force affects dominantly the $^1S_0$ partial-wave state and at low incident energies. Moreover, the interference with the nuclear force is destructive; the NN potential in the $^1S_0$ state is attractive while the Coulomb potential is repulsive. Therefore, we expect that this correction will be important only for geometries in which photons with energies near the endpoint energy are emitted, since in this situation the ppγ reaction samples more efficiently the spin-singlet states (see discussion of the previous section). In the coplanar geometry this corresponds to small proton scattering angles and forward or backward photon emission angles. For the kinematics of the TRIUMF experiment [63], Jetter et al. [58] who also include the Coulomb correction exactly, have shown that this correction affects the ppγ observables by less than 5%.

The effect of the NN e.m. form factor on the ppγ reaction has been investigated by Nyman [67] in the past. He has obtained the half-off-shell form factors $f_2^{(\pm)}$ using dispersion relations in the mass of the nucleon assuming threshold dominance [44]. Using these form factors he has found that the predicted ppγ cross sections are incompatible with the data, and that the threshold dominance assumption should be abandoned. More recently, Naus and Koch [45] have calculated $f_2^{(\pm)}$ in a simple one-pion loop model. Tiemeijer and Tjon [46] have refined this model by introducing vector dominance. The results of ref.[46] may be parametrized as

$$\begin{aligned}
F_2^{(+)}(W) &\equiv f_2^{(+)}(k=0,m,W)/f_2^{(+)}(k=0,m,m) \\
&= 1 + A(W^2/m^2 - 1) + B(W^2/m^2 - 1)^2 \, ,
\end{aligned} \qquad (66)$$

with $A = -0.13$ and $B = -0.04$ corresponding to that of using the pseudo-scalar coupling for the NNπ vertex and, $A = -0.20$ and $B = 0$ corresponding to that of the pseudo-vector coupling. We note that the form factors $f_2^{(\pm)}$ are analytic functions [43]

and, therefore, they can be expanded as in eq. (66). In ref.[68], the sensitivity of the
ppγ observables to the above half-off-shell e.m. form factor is investigated. There, the
values $A = \pm 0.5$ and $B = 0$ are used, corresponding to variations of the form factor of
the order of $\pm 10\%$ from its on-shell value ($F_2(W = m) = 1$) in the considered kinematic
regime. The effect of such an off-shell form factor is found to be negligible for cross
sections and analyzing powers. For the spin correlation coefficient, however, it changes
the results by as much as ± 0.15. It should be mentioned that the NNγ vertex required
in the rescattering term involves not the half-off-shell but the full-off-shell form factor
for which no model estimates are available. A rough estimate of the sensitivity of the
rescattering term to a fully off-shell form factor $F_2(0, W', W)$ with both W' and W
dependence parametrized similarly to that given in eq. (66), shows that this term is
nearly insensitive to such an off-shell modification of the form factor. Therefore, we
believe that the qualitative features discussed above will not be affected by the inclusion
of the rescattering contribution.

So far, no NNγ calculation include the two-body current beyond the SPA in a
consistent way with the NN potential used. As we have seen, in the SPA limit there is
no two-body current contribution to ppγ processes if one evaluates the corresponding
NNγ amplitude in the NN c.m. system. In the npγ reaction, the dominant (two-body)
exchange current arises from the one-pion exchange which gives rise to a local potential.
The ρ-meson also yields a predominantly local potential. Therefore, the currents due to
these mesons do not contribute to ppγ reactions even beyond the SPA. However, other
mesons such as the σ- and ω-meson which generate the non-locality of the NN potential
(see eq. (1)) do contribute to the two-body current in ppγ reactions beyond the SPA.
Although we would expect such contributions to be relatively small, they might affect
the observables in a significant way in view of the relatively small uncertainties involved
in the currently available potential model calculations within the framework of nucleon
and meson degrees-of-freedom. In particular, the interference of the dominant one-body
magnetic current with the two-body magnetic current, which is present oniy beyond
the SPA, may become considerable for energetic photon productions.

Comparison of NN potential models

Having dicussed some of the general features of the ppγ process and the uncertainties involved in the current theory, we now want to compare the ppγ results based on
different modern realistic interactions in order to investigate their off-shell differences.
The NN T-matrix interactions we will consider in this section are based on the OBEPQ
potential [5], the energy-dependent OBEPT potential [5], and the Yukawa parametrized
version of the PARIS potential [8].

First of all, we mention that as far as the NNγ reaction is concerned these interactions are on-shell equivalent over a wide region of incident energy [68].

Before comparing the off-shell behavior of different NN interactions we observe
that different off-shell behavior of NN interactions do not necessarily lead to different
bremsstrahlung results. This is because the effective two-body current depends on the
particular NN interaction. For example, the two-body current generated by the OBEPT
potential leads (among other diagrams) to a NNγ process in which a photon is emitted
from one of the interacting nucleons while exchanging a meson (recoil diagrams). This
is due to the fact that the OBEPT potential is based on a time ordered perturbation theory and takes into account meson retardation effects. This is not the case, e.
g., of the OBEPQ potential where mesons are exchanged instantaneously. Therefore,
different interactions generate different two-body currents which may compensate at
the end the differences in the off-shell behavior of these interactions. However, since

all presently available calculations of ppγ reactions neglect the two-body current (this current vanishes identically in the SPA for ppγ processes, as discussed in chapter 3), possible off-shell differences should lead to different predictions of NNγ observables in such calculations.

Bearing in mind this fact, we show in Fig. (6) a comparison of the half-off-shell behavior of different NN interactions for fixed incident energies of $T_{lab} = 20$ and $280 MeV$. At the incident energy of $T_{lab} = 20 MeV$ all three interactions go, practically, through the same on-shell point in all channels; at this energy these interactions are practically on-shell equivalent. Also, they exhibit a quite similar off-shell behavior. Appreciable off-shell differences may be seen only beyond $p' = 1 fm^{-1}$, especially between the OBEPT based T-matrix and the other two interactions. In the central spin-triplet ($P_{S=1}$) component these interactions exhibit relatively large off-shell differences beyond $p' = 2.5 fm^{-1}$. We also observe that, in the coplanar geometry at small proton scattering angles of $\theta_1 \sim \theta_2 \sim 10°$, one goes as far off-shell as $p' \sim 2 fm^{-1}$. This means that we may expect the off-shell differences exhibited by these interactions to show up in the ppγ reaction under such a kinematical condition, especially in the spin observables which are dominantly sensitive to the tensor components of the interaction (see Fig. (8)). Indeed, as we shall show later, this is the case.

At an incident energy of $T_{lab} = 280 MeV$, non-negligible off-shell differences can be seen in the central and tensor components. We observe that the relevant off-shell region for NNγ reactions at this incident energy is the left side of the vertical line. In the coplanar geometry at $\theta_1 \sim \theta_2 \sim 10°$, one reaches as far off-shell as $p' \sim 0.5 fm^{-1}$. Therefore, the relatively large off-shell difference observed in the $P_{S=1}$ component between the OBEPQ based interaction and others is not seen by the NNγ reaction. In the central $P_{S=0}$ component, the PARIS T-matrix differs appreciably off-shell from the other two interactions around $p' \sim 0.5 fm^{-1}$; on-shell, the OBEPQ based interaction differs from the others. Anyway, the NNγ reaction will be rather insensitive to the $P_{S=0}$ component if its strength is of the order of that exhibited at these high incident energies. In the tensor component, the OBEPT based interaction is considerably different off-shell from the other two interactions which is the case also at lower incident energies. The off-shell difference occurs around $p' \sim 0.5 fm^{-1}$ which, as we shall show later, can be seen in the spin observables in ppγ reaction.

At this point it is worth noting that, in inverse scattering theory one can show that T-matrices that are based on local and energy-independent NN potentials and that are on-shell equivalent for all energies will also have the same off-shell characteristics [10]. In spite of the fact that the NN interactions considered above are based on non-local and not strictly on-shell equivalent potentials (the OBEPT potential is also energy-dependent) and, therefore, may be quite different off-shell, to some extent that property mentioned above seems to manifest in the result of Fig. (6), in the sense that those interactions do not differ terribly from each other off-shell at least in the region sampled by the NNγ reaction.

A comparison of the off-shell NN interactions for a specific kinematics of the TRIUMF experiment [63], namely, at $T_{lab} = 280 MeV$ and $\theta_1 = 12.4°$ and $\theta_2 = 12.0°$, is shown in Fig. (7). Columns a) and b) show the T-matrix interaction which enter in the amplitude M_a in eq. (61) describing the NN scattering after the photon emission. In the central spin-singlet component, the OBEPQ based interaction differs from the OBEPT and PARIS based interactions by $\sim 100 MeV fm^3$ in the forward and backward photon emission angles. This corresponds roughly to $\sim 20\%$ discrepancy and it is caused essentially by the difference in the $^1S_0$ partial-wave state [69]. In the central spin-triplet and tensor components, the OBEPT based interaction differs from the other two interactions by $\sim 30 MeV fm^3$ and $\sim 15 MeV fm^3$, respectively. These correspond to a

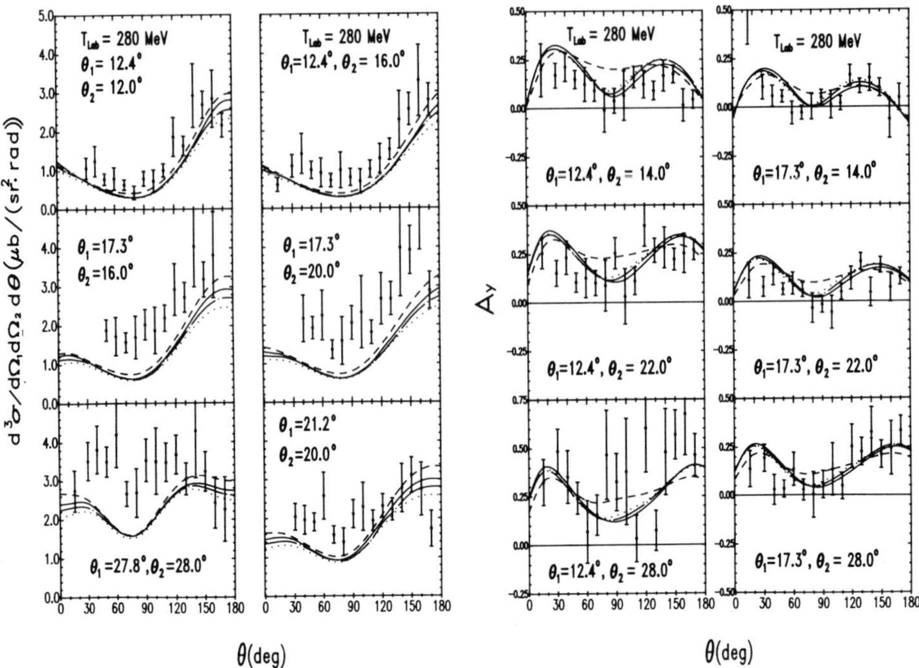

Figure 9. Coplanar-geometry ppγ cross sections and analyzing powers in the laboratory frame as a function of the photon angle θ at an incident energy of $T_{lab} = 280\,MeV$ and for various proton scattering angles, θ_1 and θ_2. The cross-section data [63] contain no arbitrary normalization factor. The analyzing power data [63] have been multiplied by a factor of -1. Labelling of the curves according to Fig. (6). The dotted curves correspond to the results using the Hamada-Johnston potential.

discrepancy of $\sim 25\%$ and $\sim 20\%$, respectively. In the tensor component this difference arises from the off-diagonal $^3PF_2 +^3 FP_2$ partial-wave states while in the $P_{S=1}$ component, it is dominantly caused by the $^3P_0$ state [69]. In the spin-orbit component, we see practically no difference among the three interactions considered. Columns c) and d) show the interactions which enter in the amplitude M_b in eq. (62), where the nucleons interact before the photon emission. In the spin-triplet components, we observe a similar feature to that seen in columns a) and b). In the spin-singlet component, we see practically no difference among these interactions.

In Fig. (9), we compare the different potential model predictions of the ppγ cross sections and analyzing powers together with the data from the TRIUMF group [63] for various proton scattering angles. Here, we include also the prediction using the Hamada-Johnston potential [3]. We see that, for cross sections, no appreciable difference can be seen in the results based on the potentials considered. The ppγ cross sections are not sensitive enough to disentangle the off-shell differences if these are of the order of $\sim 20\%$ (see Fig. (7)). We also note that the all potential models underpredict the data. However, unfortunately, these data contain an ambiguity in the absolute normalization. In fact, originally, they have been published [63] with an arbitrary factor of 2/3 which is removed in the data shown here. In contrast to the cross section, the

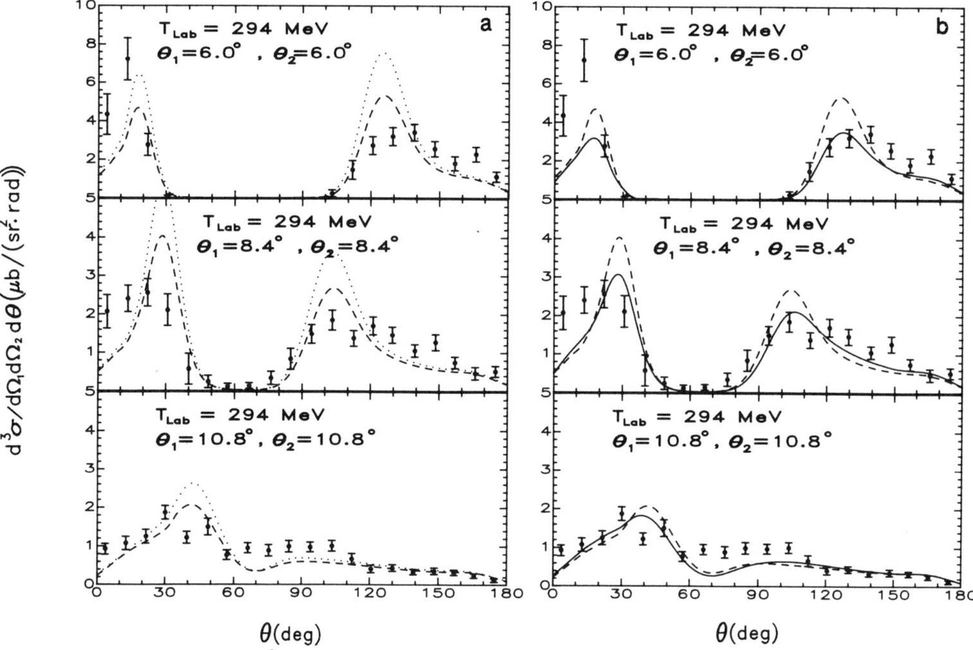

Figure 10. ppγ cross sections in the IUCF non-coplanar geometry [21] at an incident energy of $T_{lab} = 294\,MeV$ and for various proton scattering angles. In (a), the dotted lines correspond to the results based on the OBEPQ potential while the dashed lines to those based on the PARIS potential. In (b), the results with (solid) and without (dashed) Coulomb correction are shown. Here, the PARIS potential is used. The data are from [21].

analyzing power can disentangle these different NN interactions. Moreover, since this observable is dominantly sensitive to the tensor component of the interaction followed by the central spin-triplet component, and insensitive to the spin-singlet part of the interaction (see Fig. (8)), we see that the difference observed between the OBEPT potential model and the other model predictions is due to the difference in the off-shell behavior of the OBEPT based interaction from the other interactions in the tensor and $P_{S=1}$ components as has been shown in Fig. (7). Therefore, the analyzing power is able to discriminate interactions if they differ by more than $\sim 20\%$ in the spin-triplet components (except the spin-orbit), especially in the tensor channel. As has been also shown in refs.[62, 68, 69], the spin correlation coefficients are even more sensitive to off-shell differences in the spin-triplet components of the interaction. It would be interesting to see whether the inclusion of the two-body current beyond the SPA would compensate for the above difference observed between the OBEPT and other potential models.

In refs.[57, 58, 68], the sensitivity of the ppγ observables to off-shell effects of the NN interaction in non-coplanar geometries are investigated. A similar feature to that observed in the coplanar geometry has been found. Therefore, going out of the plane does not seem to improve the sensitivity of the ppγ reaction to off-shell effects of the NN interaction.

Fig. (10a) shows the predictions of the OBEPQ and PARIS potentials for the ppγ cross sections in a non-coplanar geometry of the IUCF experiment [21] at an incident energy of $T_{lab} = 294MeV$ and for very small proton scattering angles, each having a width of $\pm 1.2^\circ$ centered about the angles θ_1 and θ_2 as indicated in the figure. Again, both potentials yield very similar results to each other, although one sees some differences at the peaks. It should be mentioned that in the IUCF experiment the events with photon energies smaller than $20MeV$ and larger than $120MeV$ were not detected. As a consequence, the cross section vanishes at smaller proton scattering angles and photon emission angle, θ, around 80°. We also observe that the incident energy of $T_{lab} = 294MeV$ is a luminosity weighted average between $T_{lab} = 278$ and $325MeV$. This average is not taken into account in the calculated results here. However, the experimental resolution of 1° in the proton scattering angles and the estimated resolution of 10° in the photon angle have been taken into account by folding the calculated results with gaussian distributions with corresponding widths.

Fig. (10)b illustrates the influence of the Coulomb correction at these small proton scattering angles. The effect is larger than what we observe [58] at the proton scattering angles involved in the TRIUMF experiment and improves the agreement with the data except at forward photon emission angles where the data are underpredicted by a factor of ~ 2 even before the inclusion of the Coulomb correction. We mention that, here, the Coulomb correction is treated exactly as in ref.[58] and the result in Fig. (10)b is consistent with that of ref.[58].

CONCLUSIONS

While off-shell effects of the NN interaction can be clearly seen in ppγ reactions [60, ?], most of the modern interactions, although they may be based on quite different NN potential models, differ only slightly in their predictions for observables in these reactions. This feature is a result of the combination of two effects: first, as we have seen in section 4.3, once the NN interactions fit the phase-shifts they do not differ terribly off-shell in the regions sampled by bremsstrahlung, despite the fact that the associated NN potentials are non-local and, in some cases, even energy-dependent. We recall that in inverse scattering theory one can show that T-matrices that are based on local (in each partial-wave state) and energy-independent NN potentials and that are on-shell equivalent for all energies will also be off-shell equivalent. Second, the ppγ reaction is not sensitive enough to the spin-singlet component of the NN interaction where the largest off-shell differences (in absolute magnitude) are present among modern realistic interactions. As we have discussed in section 4.1, this is due to a large suppression of spin-singlet states occuring in the dominant one-body magnetic current contribution. On the other hand, although the ppγ reaction is sensitive to the off-shell behavior of the spin-triplet components, these interactions have a rather similar off-shell behavior in these components in the regions sampled by NNγ processes.

We stress that spin observables in ppγ reactions are sensitive enough to disentangle different NN interactions if they differ off-shell in the tensor and/or central spin-triplet components of the interaction. Indeed, as we have seen in section 4.3 (bearing in mind that the two-body current beyond the SPA is ingored), the OBEPT based interaction, which differs from the other interactions considered in the tensor and $P_{S=1}$ components by $\sim 20\%$, can be clearly distinguished. Therefore, this reaction can be used as a tool for investigating the off-shell behavior of the tensor and central spin-triplet components of the interaction. As we have mentioned before, however, most of the modern NN interactions have a rather similar off-shell behavior in these components in the regions sampled by NNγ processes at energies below the pion production threshold. If one wants to sample off-shell regions where they differ considerably (as shown in Fig. (6)), one has to go up in incident energy. However, these interactions are, strictly speaking, valid only up to pion threshold because they do not account for inelasticities.

Apart from investigating effects of the two-body current beyond the SPA and possibly effects of negative-energy intermediate states (this means a consistent relativistic formulation of both the e.m. and strong Hamiltonian parts) it seems there is little left to do within the framework of existing potential models with only nucleon and meson degrees-of-freedom. Therefore, it appears natural to investigate the influence of subnucleonic degrees-of-freedom in ppγ reactions, namely, effects of Δ-isobar degree-of-freedom. In particular, the inclusion of Δ-isobar is crucial for describing the NN phase-shifts beyond pion threshold. A preliminar calculation [70] using the NN- as well as the NΔ-interaction of terHaar and Malfliet [71] shows that this degree-of-freedom affects the ppγ observables significantly, even at bombarding energies below pion threshold.

Future high-precision experiments are crucial in order to better test our model for the ppγ reaction, in particular, the off-shell behaviour of the NN interaction when subnucleonic degrees-of-freedom are taken into account. This is especially important at energies above pion-threshold, where the potential models are on less firm grounds than at lower energies. In this sense, the ppγ experiments at two-pion threshold scheduled at the COSY-Jülich for the near future are of special interest. At such an energy, we expect the effect of the Δ-isobar on ppγ reactions to be extremely important in view of its role at lower energies. In other words, the ppγ reaction seems to be a promising tool for testing models of subnucleonic degrees-of-freedom.

As one goes to higher energy domains, not only those questions about off-shell effects of the NN interaction but also questions related to the NNγ reaction mechanism itself become more and more important. In particular, contributions of effective two-body currents (including those due to isobar excitations), relativistic effects, and a better knowledge of the structure of the basic NNγ vertex are expected to become more critical than at lower energies.

In this series of lectures we have concentrated our discussions on the ppγ reaction and have not dicussed the npγ reaction. This latter reaction has been also investigated in the past [31, 39, 72] in an effort to extract off-shell information of the NN interaction. The npγ reaction is, however, quite different from the ppγ reaction in many aspects. The most relevant difference is the presence of a strong and dominant (two-body) exchange current in the former reaction [39, 56] for producing high energy photons. While for pp bremsstrahlung the two-body current vanishes in leading order, this is not the case for np bremsstrahlung. Therefore, this reaction is an ideal process for investigating exchange currents in a simple two-nucleon system. Also, more recently, the observation of energetic photons from heavy-ion collisions has motivated an increasing interest in this reaction [27, 56, 73, 74, 75], since this process is the dominant reaction mechanism for producing these photons [25, 26, 76]. Energetic photons has been also observed in proton-nucleus [77] as well as in proton-deuteron [78] collisions. The investigation of energetic products in nuclear reactions is of great importance since these products carry information about nuclear reaction dynamics. Among them, the photon is of particular interest because the weakness of the e.m. interaction makes it a clean probe.

ACKNOWLEDGEMENT

I thank the organizers of this school for the invitation and for their warm hospitality. Most of the material presented here has resulted from a close collaboration with Dr. V. Herrmann and also with Drs. F. de Jong, O. Scholten, and H. Arellano. I am particulary indebted to Dr. V. Herrmann for many constructive remarks on this manuscript.

References

[1] S. Takeuchi, K. Shimizu, and K. Yazaki, Nucl. Phys. A504, 777 (1989).

[2] R. Vin Mau et al., Phys. Rev. Lett. 67 1392 (1991).

[3] T. Hamada and I. D. Johnston, Nucl. Phys. 34, 382 (1962).

[4] R. V. Reid, Ann. Phys. (NY) 50, 411 (1968).

[5] R. Machleidt, K. Holinde, and Ch. Elster, Phys. Rep. 149, 1 (1987).

[6] R. Machleidt, in *Adv. Nucl. Phys.*, eds. J.W. Negele and E. Vogt, (Plenum, New York, 1989), Vol. 19.

[7] R. Vin Mau, in *Mesons in Nuclei*, eds. M. Rho and D. Wilkinson, (North-Holland Publ. Comp., 1979), Vol. 1.

[8] M. Lacombe et al., Phys. Rev. C21, 861 (1980).

[9] M. H. Macfarlane and E. F. Redish, Phys. Rev. C37, 2245 (1988).

[10] Th. Kirst, K. Amos, L. Berge, M. Coz and H.V. von Geramb, Phys. Rev. C40, 912 (1989).

[11] H. Arellano, F. A. Brieva, and W. G. Love, Phys. Rev. C42, 652 (1990).

[12] J. Ashkin and R. E. Marshak, Phys. Rev. 76, 58 (1949).

[13] B. Gottshalk, W. J. Shlaer, and K. H. Wang, Phys. Lett. 16, 294 (1965); Nucl. Phys. 75, 549 (1966).

[14] M. I. Sobel and A. H. Cromer, Phys. Rev. 132, 2698 (1963).

[15] An extensive list of references can be obtained from the following articles: M. L. Harbert, in *Proceedings of the Gull Lake Symposium on the Two-Body Force in Nuclei*, eds. S. M. Austin and G. M. Crawly, (Plenum, New York, 1972); M. J. Moravcsik, Rep. Prog. Phys. 35, 587 (1972); E. M. Nyman, Phys. Rep. 9, 179 (1974); M. K. Liou, in *Proceedings of the International Conference on Few-Body Problems in Nuclear and Particle Physics*, eds. R. J. Slobodrian et al., (Les Presses de l'Universite Laval, Quebec, 1975); J. V. Jovanovich, in *Proceedings of the Second International Conference on Nucleon-Nucleon Interactions*, eds. H. W. Fearing, D. Measday, and S. Strathdee, (AIP, New York, 1977); L. Heller, in *Few-Body Systems and Nuclear Forces II*, Lecture Notes in Physics Vol.87, eds. H. Zingl, M. Haftel, and H. Zankel, (Springer-Verlag, Berlin, 1978).

[16] F. E. Low, Phys. Rev. 110, 974 (1958).

[17] P. Kitching et al., Phys. Rev. Lett. 57, 2363 (1986).

[18] R. L. Workman and H. W. Fearing, Phys. Rev. C34, 780 (1986).

[19] P. Kitching et al., Nucl. Phys. A463, 87c (1987).

[20] H. W. Fearing, Nucl. Phys. A463, 95c (1987).

[21] B. v. Przewoski et al., Phys. Rev. C45, 2001 (1991).

[22] E. Kuhlmann, *Investigation of p+p bremsstrahlung*, Proposal (E3) to COSY at KFA-Jülich (1993).

[23] N. Kalantar-Nayestanaki, "Plans for Bremsstrahlung Experiments at AGOR", invited talk presented at the *Workshop on NN Bremsstrahlung and Polarization Observables*, Groningen-NL, 16-18 November 1992.

[24] A. Johanson, "NN-Bremsstrahlung Experiment with the CELSIUS Ring", invited talk presented at the *Workshop on NN Bremsstrahlung and Polarization Observables*, Groningen-NL, 16-18 November 1992.

[25] H. Niefenecker and J. A. Pinston, Annu. Rev. Nucl. Part. Sci. 40, 113 (1990), and references therein.

[26] V. Metag, Ann. d. Physik 49, (1991).

[27] V. Herrmann, J. Speth, and K. Nakayama, Phys. Rev. C43, 394 (1991).

[28] C. Naudet et al., Phys. Rev. Lett. 62, 2652 (1989).

[29] Gy. Wolf, W. Cassing, and U. Mosel, Nucl. Phys. A552, 549 (1993).

[30] A. J. F. Siegert, Phys. Rev. 52, 787 (1937).

[31] R. H. Thompson and L. Heller, Phys. Rev. C7, 2355 (1973).

[32] L. Heller, in *Proceedings of the Gull Lake Symposium on the Two-Body Force in Nuclei*, eds. S. M. Austin and G. M. Crawly, (Plenum, New York, 1972).

[33] D. O. Riska and G. E. Brown, Phys. Lett. B38, 193 (1972).

[34] K. Ohta, Nucl. Phys. A495, 564 (1989).

[35] K. M. Schmitt and H. Arenhövel, Few-Body Syst. 6, 117 (1989).

[36] The following references contain different methods for constructing exchange current operators: M. Chemtob, in *Mesons in Nuclei*, eds. M. Rho and D. Wilkinson, (North-Holland Publ. Comp., 1979), Vol. 1; H. Arenhövel, Z. Phys. A297, 129 (1980); J. Mathiot, Phys. Rep. 173, 63 (1989); D. O. Riska, Phys. Rep. 181, 207 (1989); J. Adam, Jr., E. Truhlik, and D. Adamova, Nucl. Phys. A492, 556 (1989); S. M. Sadeghi, S. S. M. Wong, and S. Bayegan, Nucl. Phys. A554, 620 (1993).

[37] A. Buchmann, W. Leidemann, and H. Arenhövel, Nucl. Phys. A443, 726 (1985); H. Ito, W. W. Buck, and F. Gross, Phys. Rev. C43, 2483 (1991). In these references exchange currents consistent with some particular phenomenological NN potentials are constructed.

[38] J. Adam, Jr., E. Truhlik, and D. Adamova, Nucl. Phys. A526, 560 (1991). This reference discusses the off-shell NN e.m. form factor in exchange currents.

[39] V. R. Brown and J. Franklin, Phys. Rev. C8, 1706 (1973).

[40] M. K. Liou and M. I. Sobel, Ann. Phys. (NY) 72, 323 (1972).

[41] L. S. Celenza, M. K. Liou, M. I. Sobel, and B. F. Gibson, Phys. Rev. C8, 838 (1973).

[42] V. Herrmann and K. Nakayama, Phys. Rev. C46, 2199 (1992).

[43] A. M. Bincer, Phys. Rev. 118, 855 (1960).

[44] E. M. Nyman, Nucl. Phys. A154, 97 (1970).

[45] H. W. L. Naus and J. H. Koch, Phys. Rev. C36, 245 (1987).

[46] P. C. Tiemeijer and J. A. Tjon, Phys. Rev. C42, 599 (1990).

[47] J. D. Bjorken and S. D. Drell, *Relativistic Quantum Mechanics*, (McGraw-Hill, New York, 1964).

[48] M. L. Goldberger and K. M. Watson, *Collision Theory*, (John Wiley & Sons, New York, 1964).

[49] K. Nakayama, S. Krewald, J. Speth, and G. Love, Nucl. Phys. A431, 419 (1984).

[50] A. K. Kermann, H. McManus, and R. M. Thaler, Ann. Phys. (NY) 8, 551 (1959).

[51] K. Nakayama, S. Krewald and J. Speth, Nucl. Phys. A451, 243 (1986).

[52] E. M. Nyman, Phys. Rev. 170, 1628 (1968).

[53] S. L. Adler and Y. Dothan, Phys. Rev. 151, 1267 (1966).

[54] M. K. Liou and M. I. Sobel, Phys. Rev. C3, 1430 (1971).

[55] M. K. Liou and M. I. Sobel, Phys. Rev. C4, 1507 (1971).

[56] K. Nakayama, Phys. Rev. C39, 1475 (1989).

[57] A. Katsogiannis and K. Amos, Phys. Rev. C47, 1376 (1993).

[58] M. Jetter, H. Freitag and H.V. von Geramb, Phys. Scr. 48, 229 (1993).

[59] V.R. Brown, P.L. Anthony and J. Franklin, Phys. Rev. C44, 2199 (1992).

[60] V. Herrmann and K. Nakayama, Phys. Lett. B251, 6 (1990).

[61] V. R. Brown, Phys. Rev. 177, 1498 (1969).

[62] V. Herrmann and K. Nakayama, Phys. Rev. C44, R1254 (1991).

[63] K. Michaelian et al., Phys. Rev. D41, 2689 (1992).

[64] H. W. Fearing, G. I. Poulis, and S. Scherer, preprint TRI-PP-93-12, submitted to Nucl. Phys. A.

[65] D. Marker and P. Signell, Phys. Rev. 185, 1286 (1969).

[66] L. Heller and M. Rich, Phys. Rev. C10, 479 (1974).

[67] E. M. Nyman, Nucl. Phys. A160, 517 (1971).

[68] V. Herrmann, K. Nakayama, H. Arellano, and O. Scholten, preprint (1993).

[69] V. Herrmann and K. Nakayama, Phys. Rev. C45, 1450 (1992).

[70] F. de Jong, K. Nakayama, V. Herrmann, and O. Scholten, to be published.

[71] B. ter Haar and R. Malfliet, Phys. Rep. 149, 207 (1987).

[72] J. H. McGuire and W. A. Pearce, Nucl. Phys. A162, 573 (1971).

[73] C. Dupon et al., Nucl. Phys. A481, 424 (1988).

[74] M. Schäfer et al., Z. Phys. A339, 391 (1991).

[75] F. Malek et al., Phys. Lett. B266, 255 (1991).

[76] K. Nakayama and G. F. Bertsch, Phys. Rev. C40, 685 (1989).

[77] K. Kwato et al., Phys. Lett. B207, 269 (1988); J. Pinston et al., Phys. Lett. B218, 128 (1989); K. Nakayama and G. F. Bertsch, Phys. Rev. C40, 2520 (1989)

[78] J. Pinston et al., Phys. Lett. B249, 402 (1990); J. Clayton et al., Phys. Rev. C45, 1819 (1992); K. Nakayama, Phys. Rev. C45, 2039 (1992).

CONTRIBUTORS

France:

Prof.dr. A. Capella
I.P.N., Batiment 211
Boite Postale 2
F-91405 Orsay Cedex
Telephone number: 33-169417365
FAX number   : 33-169419551

Germany:

Prof.dr. V. Metag
Physikalisches Institut
Universitat Giessen
Heinrich-Buff-Ring 16
D-6300 Giessen
Telephone number: 49-6417022760
FAX number   : 49-64174390

Dr. B. Metsch
Institut fur Theoretische Physik
Universitat Bonn
Nussallee 14-16
D-5300 Bonn
Telephone number: 49-228732378
FAX number   : 49-228733728

Prof.dr. H. Satz
Facultat fur Physik   and   Theory Division
Universitat Bielefeld   CERN
Bielefeld   CH-1211 Geneva 23
Germany   Switzerland
FAX number   : 41-227823914

Prof.dr. B. Schoch
Physikalisches Institut
Universitat Bonn
Nussallee 12
D-5300 Bonn
Telephone number: 49-228732344
FAX number   : 49-228737869

Dr. K. Werner
Theoretische Physik
Universitat Heidelberg
Philosophenweg 10
D-6900 Heidelberg 1
Telephone number: 49-6221569431
FAX number   : 49-6221475733

The Netherlands:

Dr. Tj. Ketel
Faculteit Natuurkunde en Sterrenkunde
Vrije Universiteit
de Boelelaan 1081
1081 HV Amsterdam
Telephone number: 31-205484135
FAX number   : 31-206461459

Prof.dr. H. Loehner
K.V.I.
Zernikelaan 25
9747 AA Groningen
Telephone number: 31-50633614
FAX number   : 31-50634003

U.S.A.

Dr. V. Burkert
CEBAF
12000 Jefferson Avenue
Newport News, VA 23606
Telephone number: 1-8042497540
FAX number   : 1-8042497363

Prof.dr. T.W. Donnelly
Institute for Theoretical Physics
MIT
Cambridge, Mass. 02139
Telephone number: 1-6172534847

Prof.dr. R.D. McKeown
106-38 Kellog Lab.
CALTECH
Pasadena, CA 91125
Telephone number: 1-8183564316

Prof.dr. K. Nakayama
Dept. of Physics and Astronomy
University of Georgia
Athens, GA 30602

Prof. Nakayama is visiting in Juelich this year. His address there is:
Institut fur Kernphysik, Theorie
Forschungszentrum Juelich
D-52425 Juelich
Germany
Telephone number: 49-2461614560

Prof.dr. J.P. Ralston
Dept. of Physics and Atronomy
University of Kansas
1082 Malott Hall
Lawrence, KS 66045
Telephone number: 1-9138644626
FAX number   : 1-9138645262

INDEX